Experimental Methods in Biophysical Chemistry

Edited by

Claude Nicolau

Department of Molecular Biology,
Craiova University, Romania

A Wiley-Interscience Publication

JOHN WILEY & SONS
LONDON . NEW YORK . SYDNEY . TORONTO

Library of Congress catalog card number 72-5720

ISBN 0 471 63795 5

Printed by J. W. Arrowsmith Ltd., Bristol

328740

Preface

The ever increasing number of books appearing year after year somewhat discourages projects of editing or writing new ones. With this book, it obviously was not the case, and that for several reasons.

Biophysical Chemistry, though a subject read in a few universities, may still be considered a young area of Science 'lying between Physics, Chemistry and Biology'. It has developed into a successful realm of the Life-sciences and very interesting work was and is currently reported along this line. It deals mainly with sizes, shapes, structure and interactions of biomolecules, and these results have been of great significance to a large spectrum of disciplines.

One might wonder how the methods described in this book have been selected, why only these and not others. Most of the methods presented here have yielded already very valuable information, are theoretically well-founded and their use will probably be a lasting one. A few others are of significant potential interest and although they have produced a number of interesting data, their full development is, at least partly, to come. A very few of the methods dealt with in this book are slightly less used now than they were sometime ago. But, in those particular cases, we feel that there is a definite chance for them to be used again, in connection with others and that new developments are to be expected.

All the authors in this book are well-known contributors to the areas which they describe and their personalities emerge out of their work. The book is intended as a source of informations and references for those engaged in biophysicochemical research but as well for graduate students and for people working in hospital- and biological laboratories at large.

The speed of publication of a contributed volume is, unfortunately, dictated by the delivery date of the last manuscript. When this was received all contributors were given the opportunity to up-date their manuscripts. It is a great pleasure to acknowledge the lasting patience displayed by the publisher with the editor during the processing of the material as well as the excellent cooperation, which, notwithstanding the distances, characterized our relations.

Craiova, September 1972

CLAUDE NICOLAU

Foreword

Rather than to attempt to define biophysics here, I should like to evoke its general meaning. There can be no better approach, I feel, than to consider its origin and its historical development. It is no paradox to claim that biophysics was born several centuries before this era, under the impulsion of Leucippus, Democritus and Epicurus, who progressively reduced the phenomena of the Cosmos to atoms. It was Lucretius the Roman, a contemporary of Julius Caesar, who brought the character of a universal vision to this hypothesis in his remarkable poem entitled 'De Natura Rerum' in which he describes how atoms which come together in every possible manner during their movement finally adopt stable configurations which correspond to the observable inanimate world and to the living world. What an admirable way of expressing a unitary theory, which to-day, after many vicissitudes, imposes itself more and more.

It is not however possible to follow step by step the evolution of this great idea, which brought biology into the realm of physics; I should like, however, in transferring the scene to the fertile 19th century, to recall a few examples of particularly productive interactions between the two sciences. It was to Julius Robert von Mayer, doctor of medicine and philosopher, and then to his contemporary H. von Helmholtz, who was originally a physiologist and later became an innovator in physics, that we owe the establishment of the principle of the conservation of energy in its universal meaning. Nor can I leave unmentioned the case of Ludwig Boltzmann, who was the founder of statistical physics, and who in establishing this was inspired by the law of natural selection which had been discovered a little earlier by the great biologist Charles Darwin, to account for the evolution of living things.

The latter, had himself drawn from the theories of the British economist Malthus, who worked out his population theory in the first quarter of the nineteenth century. For his part, Ernst Mach, who was a sceptical empiricist, precursor of Einstein, wrote: 'Physics is not the whole world: biology has also its place there, and is essential in any depicting of the world'. These remarks throw a special light on the openness of mind of a great physicist towards biology. Louis Pasteur, Agrégé ès sciences physiques, had already made royal presents to this science, as we know. More recently, Schrödinger with his book entitled 'What is life?' made a considerable contribution to the bringing together of the two sciences. And finally, to-day, the Institut de la Vie of France, amongst other functions, ensures a truly efficacious liaison between the physicists and biologists.

But if, in an individual form, biophysics has always existed, it is only recently that it has been transformed into a truly independent science, comparable to physical chemistry, geophysics, or astrophysics. How has this come about? To understand this, we must once more return to the period towards the end of the nineteenth century, when on the one hand Kékulé and on the other hand van't Hoff succeeded in giving a geometrical form to chemistry by bestowing a spatial structure on such molecules as benzene and methane. Around this period also, first Newlands and then Lothar Meyer and Mendelejeff in working out the periodical system of chemical elements, which they arranged in function of increasing atomic weight produced the 'arithmetical' aspect of chemistry, with the well-known law of octaves. Later, with the discovery of isotopes by Fajans and Soddy, it was seen that the effects of degeneracy completed the whole picture. To understand and interpret these fine discoveries, which in fact establish the whole strategy of the inorganic world, knowledge had to be extended to a universe as yet unexplored. When first G. N. Lewis and then Rutherford penetrated into the electronic and even the nuclear structure of matter, scientists succeeded in accounting for the fundamental characteristics of the periodical system. It then became clear that there existed an ineluctable law which demanded the exploration of the universe firstly to dimensions of 10^{-8} cm and then to 10^{-13} cm, to make possible the interpretation of macroscopic regularities.

And what about biology? The first attempts to spatialize biomolecules by means of X-ray diffraction occurred from the 1940's onwards. The fine work of J. D. Bernal and Astbury, disciples of Bragg, on the structure of DNA, and that of Linus Pauling on proteins, must especially be cited. It was however in 1953 that the structure of the double helix of DNA was established thanks to the imagination and the combined efforts of Wilkins on the one hand, and of Watson and Crick on the other. This was the point of departure for a relationship between tridimensional structure and biology which was of very fundamental importance, and which in principle can be compared with that established in an earlier period between the spatial structure of methane or benzene and chemical activity. Further, the genetic code, with its concept of codon or triade, which gives an arithmetical form to the strategy of life, recalls the octaves of the periodical system. From this, and by analogy, it is not difficult to predict an aspect of to-morrow's evolution. It is on the electronic level that the profound meaning of the genetic code will have to be interpreted, in particular its stability, its specificity, the curious effect of degeneracy, the separation of the two helixes, order-disorder, etc. Work in this direction is already well under way, in particular that done by B. and A. Pullman, who grasped this ineluctable phenomenon very early and who are every day succeeding in bringing the language of wave mechanics more and more into the realm of biology. But beyond the electronic scale, exactly as in in the case of the periodical system, the nuclear scale takes form. The variations in the isotopic ratios were established about ten years ago for the atoms of carbon 12 and carbon 13, according to the species or class, for plants and invertebrates. Quite recently the author and his associates wished to explore the incidence of nuclear properties in the higher animals, in the case of a specific mechanism. With this aim in view, the CO_2 exhaled during breathing was chosen, taking the case of batracians, dog, man and birds. They were surprised to find that the ratio C^{12}/C^{13} is substantially altered through the sequence, so that between the two extremes the difference in the ratios reaches nearly 1 per cent, with a progressive growth in the concentration of C^{13}. The profound meaning of such a result, which throws into relief the effect of the animal class, is still not quite clear. Within the framework of this research, it has recently been discovered by us—so far only in the case of rats—that under the impact of specific pathological conditions the isotopic ratios may also be significantly altered. This new fact awaits further development in the field of medicine. Yet, despite these successive changes in scale, which open up unsuspected horizons, we retain the impression, as the fine mental picture of Newton has it, that we are still like children who are playing on the beach while before them there lies the expanse of the ocean of truth.

Alongside the progress which has taken place in the field of the structure of biomolecules, electron microscopy has brought a new light to bear on the organization of cells, as, for example, the work on mitochondria, which are energy generators, and on ribosomes, sites of protein synthesis. The aim must be to account for this remarkable cellular heterogeneity through the knowledge of the interactions between constituent molecules. This is doubtless one of to-morrow's great problems, and shows up the gulf which separates isolated molecules from large associations, for basically the perspective must be the understanding of life in its most fundamental characteristics. Indeed, the study of biogenesis, of cryobiology and of the phenomena of reviviscence by emphasizing the importance of organization represent very encouraging tendencies. Following up the fine research carried out in France by Paul Becquerel, scientists have been successful in maintaining seeds, lichens and tardigrades in a latent state of life, submitting these to profound dehydration and maintaining them in a vacuum at temperatures near to absolute zero. Thus treated, the complete arrest of metabolism and the occurrence of apparently appreciable modifications of the cellular structures do not prevent reviviscence. This means that the fundamental factor for life is organization. But at once the problem can be posed as to what is the limit beyond which life fades away. Experiments undertaken in the laboratory of the author on dried lichens, dating from about two hundred years up to quite recent periods, brought to light a decrease of the intensity of paramagnetic signals as age increased. The study of this phenomenon, which appears to be tied up with a progressive disorganization of the structure of the whole entity, is still proceeding.

As for biochemistry which is already well established its aims were, and remain, on the one hand the identification of biomolecules and their morphological analysis, and on the other hand an examination of the processes in living organisms in relation to functions. Its mission is carried further by biophysics within the perspective which has just been stated, to the point where these two tendencies, which express in a profound way all that the most refined resources of scientific thought can produce, constitute closely related and complementary branches of what is known as molecular biology. This, then, is a convergence which is entirely analogous to that undergone by physics and chemistry when these condensed around the year 1880, in the form of physical chemistry. This is perhaps the moment to pose the question as to whether the most fascinating properties of life, such as that of consciousness may be interpreted one day thanks to the carrying further of the line of thought which we have just evoked. This means in fact the posing anew of the problem which was stated so authoritatively by Lucretius and in which was implicit the idea that the final goal of physics was to explain all cosmic phenomena, including life itself. Our own replay must contain more nuance than this. It does not seem to me to be evident that there is no limit to our explanatory possibilities, in particular when we endeavour to reach the most subtle expressions of life. Can it be, then, the consciousness, which is the product of a prodigious cosmic emergence, could be reduced to simple concepts when, at the same time, these arise from consciousness itself? To say the least, these questions are irritating ones, and molecular biology does not seem to me to be ready to reply to them so far.

And yet the development of this science enlarges our vision in every direction. I cannot discuss all the aspects of this here, such as those connected with what I would call the defects in nucleic acids, in a sense which is entirely comparable to that concerned with deficiencies in crystalline structures. I still recall the time when it would have been considered a heresy not to believe that a crystal was a product of the implacable repetition of an elementary unit. And yet it is the deviations from these fine regularities, that is to say the defects in the lattice, which have become the agents responsible for exciting crystalline properties. Biological mutations should be set within this framework, and it is then possible to grasp how far we are from having understood the whole wealth of the possibilities connected with DNA. There is no doubt that we now have methods at our disposal which seem to me to be first class, and which are far from having been entirely explored, by which perturbances in DNA can be produced. I should particularly like to draw attention here to phenomena of photodynamic action in biology, the study of which, using electron spin resonance, has been undertaken by the author and his associates and which allows one to envisage the possibility of directed mutations. This would of course be an exciting success!

There is also the whole field of molecular medicine which is still to be developed, as Linus Pauling has pointed out, in spite of some remarkable first successes such as the discovery of 'Sickle cell anaemia', which is connected with molecular perturbances in haemoglobin. And certain pathological states manifest themselves in a very subtle way at the level of the tissues. Thus, cancerous tissues seem to give rise to paramagnetic signals well before the effects of the tumours can be detected by the usual methods. Recent in vitro syntheses, firstly of RNA and secondly of DNA, both of which are biologically active, permit one to be extremely hopeful, perhaps even allowing one to envisage the possibility of correcting certain insufficiencies in the genetic code.

But completely undeciphered fields remain, such as that of olfaction which is perhaps linked to very subtle molecular interactions, as are, for example, charge transfers. The whole realm of pharmacodynamics remains largely to be understood, where analogous forces probably play a significant part, according to recent research by the author and his associates.

But where life is concerned the dominant note, which must never be lost sight of, seems to depend on the fact that the phenomena which occur cannot be isolated from their context, and from the notion that each emergent structure reacts on others, and vice versa. Life, then, would be global phenomenon, the reduction of which into separated forms would represent only an approximation which is valid up to a certain point, beyond which the phenomena of consciousness are probably situated.

The building up of theoretical biology, the extension of the fundamental concepts of statistical mechanics to the interpretation of biogenesis in the direction shown by I. Prigogine and his school, permits us to hope that life may be looked at in its own perspective.

Moreover, the experimental methods of modern physics, not only those based on X-ray diffraction, but also on electron spin resonance, nuclear magnetic resonance, mass spectroscopy, photochemistry techniques, and lasers, are all not only promising, but they have already achieved important progress.

In this book, the Editor has brought together a comprehensive set of articles by eminent scientists which admirably illustrate many essential aspects of biophysics. One realizes however that if the road is still a very long one, we are nonetheless embarked in a vessel, the direction of which will not cease for a long time to stimulate the ardour and the imagination of this new category of research workers.

University of Liège, Belgium JULES DUCHESNE

Contributing Authors

H. ALTMANN   Institute of Biology, Reactor Centre, Seibersdorf, Austria

G. P. ARSENAULT   Department of Chemistry, Massachusetts Institute of Technology, Cambridge, Massachusetts, U.S.A.

L. G. AUGENSTEIN   Late of Biophysics Department, Michigan State University, East Lansing, Michigan, U.S.A.

J. S. COHEN   Physical Sciences Laboratory, Division of Computer Research and Technology, National Institutes of Health, Department of Health, Education, and Welfare, Bethesda, Maryland, U.S.A.

G. ECKHARDT   Institute of Organic Chemistry, University of Bonn, Federal Republic of Germany

G. L. GRIGORYAN   D. I. Ivanovsky Institute of Virology, The U.S.S.R. Academy, Moscow, U.S.S.R.

S. GUINAND   Laboratoire de Biologie Physico-Chimique, Université Paris-Sud, Orsay, France

W. HABERDITZL   Berlin Humboldt University, G.D.R.

U. HAGEN   Institut für Strahlenbiologie, Kernforschungszentrum, Karlsruhe, Federal Republic of Germany

A. E. KALMANSON   D. I. Ivanovsky Institute of Virology, The U.S.S.R. Academy, Moscow, U.S.S.R.

R. L. MUNIER   Institut Pasteur, Paris, France

E. POSTOW   Department of the Navy, Office of Naval Research, Arlington, Virginia, U.S.A.

B. ROSENBERG   Biophysics Department, Michigan State University, East Lansing, Michigan, U.S.A.

H. W. SHIELDS   Wake Forest University, Winston-Salem, North Carolina, U.S.A.

G. SNATZKE   Institute of Organic Chemistry, University of Bonn, Federal Republic of Germany

J. TONNELAT   Laboratoire de Biologie Physico-Chimique, Université Paris-Sud, Orsay, France

J. O. WILLIAMS   Edward Davies Chemical Laboratories, University College of Wales, Aberystwyth, Wales

R. WURMSER   Institut de Biologie Physico-Chimique, Paris, France

P. ZUMAN   Department of Chemistry, Clarkson College of Technology, Potsdam, New York, U.S.A.

Contents

I
Structures of Biomolecules

CHAPTER 1

Mass spectrometry of biomolecules

G. P. Arsenault

Department of Chemistry,
Massachusetts Institute of Technology,
Cambridge, Massachusetts, U.S.A.

I. INTRODUCTION

Mass spectrometry is one of the oldest and possibly the most versatile physical method of analysis. The first mass spectrometer bearing a strong resemblance to instruments in use today was built by Dempster.[1] The investigation of naturally occurring substances by mass spectrometry was begun in the nineteen fifties by K. Biemann, R. I. Reed, R. Ryhage and E. Stenhagen. Great impetus was given to this field of investigation in the nineteen sixties

by the successful development and commercial production of direct introduction systems (section IIID2), gas chromatograph–mass spectrometer
combinations (section IIID3) and high resolution mass spectrometers
(section IIIE), and by the use of digital computers and the development of
computer techniques to record, reduce and interpret the wealth of data
produced by mass spectrometers (Biemann and Fennessey,[2] Hites and
Biemann[3,4]). Today mass spectrometry is available to most organic chemists
as one of several instrumental methods of analysis and should become a
routine tool in biochemical and biomedical laboratories in the nineteen
seventies. The long time span between the birth of mass spectrometry and its
widespread use by everyone is a reflection of the cost, complexity and
commercial unavailability of the instruments.

Mass spectrometry is the most sensitive instrumental method of analysis.
It gives more information per microgram of sample than any other method,
and satisfactory spectra may be obtained on as little as 1×10^{-12} g of
material.

The sensitivity of the mass spectrometer is so great that sample isolation
and handling techniques often lag behind. A common occurrence in taking
the mass spectrum of a sample isolated in amounts of a few milligrams or less
is to find that it contains as much impurity as desired compound. Fortunately,
the sample does not necessarily have to be pure, and herein lies an advantage
of mass spectrometry over other instrumental methods of analysis, since mass
spectrometry provides several alternatives for dealing with interfering
impurities. Firstly, purification without the necessity of further handling of
the sample prior to analysis may be accomplished by the tandem use of the gas
chromatograph and the mass spectrometer (section IIID3). Secondly,
fractionation of microgram samples frequently occurs when they are
vaporized slowly and directly into the ion source of the mass spectrometer
(section IIID2). This is equivalent to a high vacuum fractional distillation or
sublimation in which the mass spectrometer is used to analyse each fraction;
the number of fractions taken is determined by the scanning speed of the
instrument, which is not a limiting factor since complete spectra may be
obtained in less than a few seconds if necessary. Finally, the high resolution
mass spectrum (section IIIE) of a sample, which consists of the exact masses
and, hence, the elemental compositions of all ionic species in the spectrum,
may permit differentiation between ionic species originating from the
impurities and the compound present in the sample.

The principal limitation of mass spectrometry lies in the vapour pressure
requirement placed on the sample to be analysed. The minimum vapour
pressure must be about 1×10^{-6} torr at the maximum temperature at
which the sample may be heated without undergoing decomposition (section
IIID). The vapour pressure of the sample thus drastically limits the molecular
weight of compounds amenable to mass spectrometric analysis. The vast

majority of biomolecules studied by mass spectrometry have molecular weights below 1000 while none have molecular weights above 2000. The volatility of a sample can quite often be improved significantly by preparation of a simple chemical derivative. This is not sufficient for biopolymers which have to be hydrolysed (sections IVA and B) to obtain oligomeric units sufficiently small to be submitted directly or after chemical derivatization to mass spectrometric analysis. A possible but as yet undeveloped approach to circumvent lack of vapour pressure is the use of pyrolysis–gas chromatography (Levy[5]) in combination with mass spectrometry. Such an approach suffers from the disadvantage inherent to extensive and possibly unspecific degradation of sample before mass spectrometric analysis while having the advantage of removing completely the vapour pressure requirement placed on samples for mass spectrometric analysis.

In a chapter of this length it is not possible to cover the subject matter thoroughly. The choice of material included, in particular the examples selected, represents a bias on the part of the author. The chapter consists of a brief discussion on fundamentals of mass spectrometry followed by two other sections, the first dealing with instrumentation and technique, and the second, with applications.

II. SOME FUNDAMENTALS OF MASS SPECTROMETRY

The function of a mass spectrometer is to ionize (section IIIA) the sample under investigation, separate (section IIIB) the ions formed according to their mass-to-charge ratios (m/e), and measure (section IIIC) the ion current of the separated species. Fundamentals of mass spectrometry have been dealt with at length by Beynon[6] and Biemann,[7] and these references should be consulted if a more advanced treatment than is presented herein is required. The conventions for organic mass spectrometry suggested by Budzikiewicz[8] will be followed throughout this chapter.

Ionization of organic molecules is usually accomplished by electron impact, also called electron bombardment. All spectra shown in this chapter are electron impact mass spectra. Ionization of hypothetical sample molecules ABC with energetic electrons may be represented as follows:

$$ABC + e^- \rightarrow ABC^{+\cdot} + 2e^- \qquad (1)$$

$ABC^{+\cdot}$ is referred to as the molecular ion. The symbol $^{+\cdot}$ indicates that the ion carries a single positive charge and has an unpaired electron. There are millions of molecular ions produced in the course of taking the mass spectrum of a compound. Negative ions and multiply charged positive ions are also formed in the ionization process, but in most instruments and for most organic compounds they represent only a small fraction of the total ionization and will not be referred to further.

The electron energy generally used for ionization is in the range of 50 to 80 eV, is much larger than the energy required for the formation of singly charged positive ions, and results in the formation of molecular ions with excess energy which may fragment in several ways. The following equations illustrate some fragmentation pathways for the molecular ion $ABC^{+\cdot}$ and are not intended to include all possibilities

$$ABC^{+\cdot} \rightarrow AB^+ + C^\cdot \tag{2}$$

$$ABC^{+\cdot} \rightarrow AB^\cdot + C^+ \tag{3}$$

$$AB^+ \rightarrow A^+ + B \tag{4}$$

$$ABC^{+\cdot} \rightarrow AC^{+\cdot} + C \tag{5}$$

Equations (2) to (4) are examples of simple cleavage while equation (5) shows a rearrangement in which a bond, absent in molecule ABC, is formed as a result of fragmentation. The abundance of the molecular ion $ABC^{+\cdot}$ depends on its stability. It is sufficiently intense in more than 80 per cent of all spectra to be easily recognized as such and hence to be useful in determining the molecular weight from the mass spectrum. Equations (2) to (5) may all be possible, but their probability of occurrence differ. Fragmentation has been rationalized over the years in terms of mechanistic carbonium ion chemistry and was the subject of a review by McLafferty.[9]

The mass spectrum consists of the abundance of all ionic species separated in the analyser according to their mass-to-charge ratios. This is represented in the form of a bar graph or table, and is illustrated for methane in Table 1 and Figure 1. The graphic representation is the more common and desirable, and shows increasing m/e from left to right on the abscissa, and relative intensity from 0 at the bottom to 100 at the top on the left ordinate. The vertical lines in the graphic representation are referred to as peaks. The most abundant peak in a spectrum is defined as the 'base peak' since it is assigned an intensity of 100 and the intensity of all other peaks is based on it. The term 'base peak', however, should not be construed to have any other physical significance. Metastables*, doubly charged ions occurring at fractional m/e, and peaks below an arbitrarily set abundance limit of 0·5 to 2 per cent of base peak are usually excluded from the graphic representation of spectra. Whenever accurate intensities or completeness of data is desirable, the tabular representation is preferable. The two modes of presentation of spectra should be reserved for 'show' purposes and need not be used at the working

---

* Metastables, also called metastable peaks, are weak, diffuse, roughly Gaussian-shaped peaks which usually occur at non-integral m/e. Beynon[10] has reviewed the usefulness of metastables in interpreting mass spectra. In particular, their presence proves the existence of plausible fragmentation pathways such as

$$ABC^{+\cdot} \rightarrow AB^+ \rightarrow A^+$$

Table 1. Mass spectrum of methane;[a] tabular representation

m/e	Relative intensity
1	3·36
2	0·21
12	2·80
13	8·09
14	16·1
15	85·9
16	100
17	1·11

[a] *Mass Spectral Data*, American Petroleum Institute Research Project 44, serial number 1

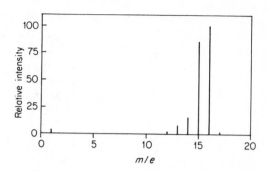

Figure 1. Mass spectrum of methane: graphic representation

level for qualitative organic mass spectrometry because the actual recordings (section IIIC) are sufficient for that purpose.

The mass spectrum of methane (Table 1) shows an intense peak at m/e 16 and a low abundance peak at m/e 17. These two peaks are related in that they are mainly isotope peaks of each other. The nuclidic mass and natural abundance of selected isotopes of elements chosen because of their occurrence in biomolecules and because of their importance in the mass spectrometry of biomolecules are listed in Table 2. A cursory examination shows, for example, that the element carbon occurs in nature as ^{12}C and ^{13}C. Hence, if the minute contribution of ^2H is ignored, naturally occurring methane must be made up of $^{12}C^1H_4$ and $^{13}C^1H_4$ in the ratio of the natural abundance of the carbon isotopes, and the mass spectrum of methane shows $^{12}C^1H_4^{+\cdot}$ at m/e 16 and $^{13}C^1H_4^{+\cdot}$ at m/e 17 in the same ratio. It should be noted that about 1 per cent of the intensity of the peak at m/e 16 is due to $^{13}C^1H_3^+$, the ^{12}C isotope of which is at m/e 15. For the sake of convenience, the lightest naturally occurring isotope of each element, such as ^{12}C, ^1H, ^{14}N and ^{16}O, is usually referred to without mention of its mass number, and, thus $^{12}C^1H_4$ and $^{13}C^1H_4$ becomes CH_4 and $^{13}CH_4$. Molecular weight, as defined by mass spectrometry, is monoisotopic and is based on the lightest naturally occurring isotope of each element. Hence, for methane it is

$$(12\cdot000000 \times 1) + (1\cdot007825 \times 4) = 16\cdot031300 \text{ a.m.u.} \tag{6}$$

A closer examination of the isotopic natural abundance shown in Table 2 reveals that the elements listed may be divided into three categories. The first group includes the three monoisotopic elements, fluorine, phosphorus and

Table 2. Nuclidic mass and natural abundance of selected isotopes
of a few elements[a]

Isotope	Nuclidic mass[b]	Natural abundance[c]
[1]H	1·007825	99·985
[2]H	2·014102	0·015
[12]C	12·000000	98·9
[13]C	13·003354	1·1
[14]N	14·003074	99·6
[15]N	15·000108	0·4
[16]O	15·994915	99·8
[18]O	17·999160	0·2
[19]F	18·998405	100
[28]Si	27·976927	92·2
[29]Si	28·976491	4·7
[30]Si	29·973761	3·1
[31]P	30·973763	100
[32]S	31·972074	95·0
[33]S	32·971460	0·8
[34]S	33·967864	4·2
[35]Cl	34·968854	75·5
[37]Cl	36·965896	24·5
[79]Br	78·918348	50·5
[81]Br	80·916344	49·5
[127]I	126·904352	100

[a] From a compilation by J. Roboz[11] p. 503
[b] Based on $^{12}C = 12·000000$ atomic mass units
[c] Expressed as a percentage

iodine. The presence of any of these three elements in an organic compound
may sometimes be inferred from its mass spectrum because these elements
make no contribution to the ^{13}C isotope peaks. The second group comprises
hydrogen, carbon, nitrogen and oxygen which have a light isotope accounting
for over 99 per cent of the natural abundance but which may occur in such
large number in biomolecules as to give rise to significant peaks due to the
heavier isotope. The most important of these is carbon because of the
abundance of its ^{13}C isotope as well as its presence in considerable numbers
in most organic compounds. An estimate of the relative intensity of any ^{13}C
isotope peak is given by

Rel. int. of peak at M due to ^{12}C × number of C atoms present × 0·011 =
    rel. int. of peak at $M + 1$ due to ^{13}C (7)

The last group is made up of the remaining elements, silicon, sulphur,
chlorine and bromine, which do not occur in large number in organic
molecules but which have abundant isotopes one and/or two atomic mass

units (a.m.u.) heavier than their light isotopes. These elements may be easily recognized in the mass spectrum of compounds in which they occur because they give rise to characteristic clusters of peaks.

The resolution R in qualitative organic mass spectrometry is defined as

$$R = \frac{M}{\Delta M}$$

(8)

where M is the mass of an ionic species, and the meaning of ΔM depends on the definition of resolution being used. In the 'per cent valley' definition commonly in use, M is the mass difference between two equally intense peaks of mass M and $M + \Delta M$ separated by a valley the height of which is expressed as a percentage of the intensity of the peaks. This is illustrated in Figure 2(a)

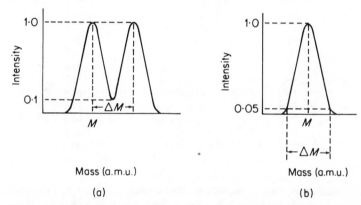

Figure 2. The concept of resolution in mass spectrometry: (a) 10 per cent valley definition; (b) 5 per cent peak width definition

for a 10 per cent valley which is frequently used, although values ranging from 1 to 50 per cent valley are not uncommon. The mass spectrum of methane (Table 1) shows two peaks of about equal height (considered to be equal for our purposes) at m/e 15 and 16, and may be used to further clarify this definition: the resolution necessary to separate these two peaks ($M = 15$; $\Delta M = 1$) with a 10 per cent valley between peaks is 15.

The per cent valley definition of resolution suffers from the requirements of adjacent peaks of equal height, a situation seldom encountered except in mixtures expressly made for this purpose. An alternative method of expressing resolution which does not suffer from this disadvantage is the 'peak width' definition in which ΔM is the width of the peak in a.m.u. at a certain height. Values ranging from 1 to 50 per cent of peak height are used. The 5 per cent peak width definition illustrated in Figure 2(b) gives the same numerical value of resolution as the 10 per cent valley definition, provided the peaks are symmetrical.

III. INSTRUMENTATION AND TECHNIQUE

The individual components which make up a mass spectrometer are shown in the form of a block diagram in Figure 3, and consist of the sample introduction system, the ion source, the analyser, the detector–recorder and the vacuum system, each of which will be discussed separately. The features selected for discussion are those likely to be found in mass spectrometers of recent manufacture which are used for organic chemical applications. The discussion is of necessity rather brief and a more thorough treatment of the subject may be found in standard texts (Beynon,[6] Brunnée and Voshage,[12] Kiser,[13] McDowell[14] and Roboz[11]).

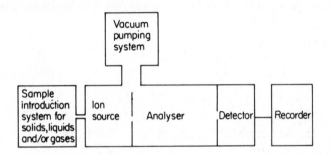

Figure 3. Block diagram of mass spectrometer showing components making up the instrument

Mass spectrometers are operated under high vacuum. The ion source and analyser regions are generally maintained at a pressure below 1×10^{-7} torr. The vacuum pumping system is represented in Figure 3 by a single block for the sake of simplicity but frequently consists of several individual units which are used for different purposes such as to achieve low pressure in the analyser region in the face of somewhat higher pressures in the ion source region. Each unit consists of an oil diffusion pump backed by a mechanical pump. Backstreaming of pump oil into the instrument is generally avoided by the use of a trap cooled with liquid nitrogen.

A. Ion Source

Electron impact is the only means of ionization in general use for organic compounds. Field ionization and chemical ionization are newer means of producing positive ions and are likely to be used increasingly in the future, to complement rather than supplant electron impact ionization. Field ionization and chemical ionization sources are becoming commercially available and will be discussed in some detail along with the electron impact ion source. The radio frequency spark source deserves passing mention here since it may be used to analyse simultaneously for all trace elements in materials of biological origin (Evans and Morrison[15]).

1. Electron Impact Ionization

An electron impact ion source is shown in Figure 4. The filament F is usually maintained at -50 to -80 volts with respect to the ion source block B and is heated by passing through it a current of a few amps. The glowing filament emits electrons which are attracted towards block B because of the potential difference between them. The path followed by the electrons is a tight spiral because of the small magnetic field of a few hundred gauss— magnet poles M and M′ in Figure 4—applied externally to the ion source.

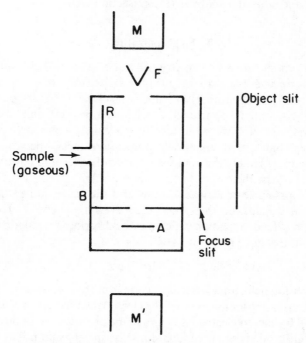

Figure 4. Diagram of an electron impact ion source showing
ion source block B, repeller or repellers R, anode A, filament
F and collimating magnet poles M and M′

The collimated electrons enter the ion source via a slit in B and in the absence of sample travel uninterrupted towards the anode A which is kept at a small positive potential—usually less than 70 volts—with respect to the ion source block B in order to facilitate the collection of electrons and to prevent secondary electron emission. Heaters—not shown in Figure 4—are fastened on the outside of block B to heat the source to temperatures up to 350 °C to prevent condensation of samples of low volatility while the heating elements themselves are not in direct contact with the sample. If sample molecules in the gaseous state collide with electrons within the source, they become ionized (equation (1)). The positive ions formed are expelled from the source by the small positive voltage at which the repeller R is maintained with respect to block B. The positive ions leave the source which is at several thousand volts—the accelerating voltage V_a which is positive—and are accelerated towards the object slit which is at ground potential. The focus slit which precedes the object slit is kept at a potential close to but not exceeding the block potential and serves the purpose of lining up the ion beam with the object slit. As the ions pass through the object slit they are fully accelerated and enter the analyser region (section IIIB).

2. Field Ionization

Field ionization mass spectrometry will not be discussed elsewhere in this chapter. It is thus necessary to include here a discussion of this area of mass spectrometry as a whole rather than limit ourselves to the ion source alone. Field ionization mass spectrometry originated in field ion microscopy. Inghram and Gomer[16] were the first to use a field ion source in a mass spectrometer. Since that time field ionization mass spectrometry has developed rapidly largely through the efforts of Beckey and coworkers who conveniently review their work[17,18,19] periodically.

Positive ions are formed when atoms and molecules are placed within intense electric fields of the order of 3–$50 \times 10^7 \, V \, cm^{-1}$. This is the phenomenon of field ionization which may be described in equation form for an hypothetical sample molecule ABC

$$ABC + \text{Electric field} \ (3\text{–}50 \times 10^7 \, V \, cm^{-1}) \ \rightarrow \ ABC^{+\cdot} + e^- \qquad (9)$$

Equation (9) for field ionization is a counterpart to equation (1) used to describe electron impact ionization. The electric field strength necessary may be produced by curved electrodes of any shape provided the radius of curvature of the electrode is small enough. Initially, sharp metal points (Inghram and Gomer[16]) were used as electrodes but they were later replaced by fine metal wires (Beckey,[20] Robertson, Viney, and Warrington[21]) and sharp edges such as a razor blade (Beckey[20]). The general formula (Beckey and coworkers[19])

$$F_0 = \frac{V_0}{r_0} \times K_i \qquad (10)$$

represents the field strength F_0 in $V \, cm^{-1}$ at the surface of the curved electrode known as the anode or emitter. V_0 is the potential difference between the emitter and the cathode, r_0 is the radius of the curvature of the emitter and K_i is a dimensionless geometry factor which depends on the shape of the emitter. Since K_i varies between 10^{-1} and 10^{-3} and V_0 cannot be greater than 2×10^4 V without causing experimental difficulties, r_0 is in the range of 10^{-4} to 10^{-6} cm. The relative merits of points, wires or edges as electrodes were discussed by Beckey and coworkers[19] who have used all three and favour wires.

A simple field ion source adapted from the one reported by Robertson and Viney[22] is shown in Figure 5, and consists of the emitter E and three parallel plates with narrow slits which are in alignment with each other. The emitter is a razor blade the plane of which is at right angles to the plane of the cathode

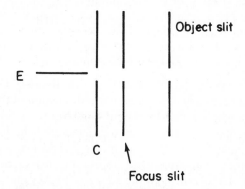

Figure 5. Diagram of a field ion source showing a razor blade as the emitter E and a plate with a slit as the cathode C

C. The sharp edge of the blade is parallel to the long axis of the slit in C, is within about 0·5 mm of the plane of C and is centred with respect to the width of the slit in C. A positive voltage is applied at E and a negative voltage, at C. The object slit is at ground potential and the focus slit is at a positive potential intermediate between the voltage at E and ground potential. The voltage applied at E is the accelerating voltage V_a (section IIIA1) since the object slit is at ground potential, and the potential difference between E and C is V_0 in Equation (10). The ions which pass through the object slit are fully accelerated and enter the analyser region (section IIIB).

The similarity in the designs, in particular of the ion accelerating section, of the electron impact ion source (Figure 4) and the field ion source (Figure 5) should be noted. Indeed, the electron impact ion source of commercially available mass spectrometers has been modified so as to become a combined electron impact-field ion source (Beckey and coworkers,[18] Chait and coworkers,[23] Schulze, Simoneit and Burlingame[24]). Combined sources are now commercially available, and no more than a few minutes are required to change from one mode of ionization to the other.

The field ionization mass spectrum of a compound shows less fragmentation and a more abundant molecular ion than its electron impact counterpart. In some cases too little fragmentation is obtained by field ionization to make this method of ion production useful for anything but determining the molecular weight of biomolecules. However, recent publications (Brown, Pettit, and Robins,[25] Damico, Barron, and Sphon,[26] Krone and Beckey[27]) show that the field ionization mass spectra of some biologically significant organic compounds have sufficiently abundant fragments to permit correlating fragment peaks and structure, an important feature of the application of mass spectrometry to structure determination.

The wide acceptance of field ionization as a complement to electron impact ionization may well depend on technical improvements, some of which are being actively investigated (Beckey,[28] Beckey and coworkers[29]), and on the further development of gas chromatography–field ionization mass spectrometry (Damico and Barron[30]).

3. Chemical Ionization

Chemical ionization as a means of generating positive ions for mass analysis is the newest mode of ionization for organic molecules, and was first reported by Munson and Field.[31] Excellent reviews on this subject have since appeared (Field,[32] Field[33]) and details of the work done by Field and coworkers may be traced back through their latest publication (Field[34]). Chemical ionization evolved from ion–molecule reaction studies by Field and Munson,[35,36] and, for this reason, the discussion of gas phase ion–molecule reactions was delayed until now.

Under the low pressure conditions which usually prevail in an ion source while the mass spectrum of an organic compound is being measured, the collision of ions with neutral molecules is relatively rare. Ion–molecule collision between an hypothetical molecule ABC and its molecular ion may result in the following reaction:

$$ABC^{+\cdot} + ABC \rightarrow [ABC \cdot ABC]^{+\cdot} \rightarrow ABCA^+ + BC^{\cdot} \qquad (11)$$

The mass of the ion–molecule reaction product $ABCA^+$ is greater than the mass of the molecule ABC from which it originated, a fact which must be kept in mind when the molecular weight of a compound is to be established by mass spectrometry. The most common ion–molecule reaction taking place at low pressure in the ion source is the abstraction of a hydrogen radical by the molecular ion, giving rise to a peak one mass unit larger than the molecular ion peak.* Unlike other reactions taking place in the ion source which are

* Not to be confused with the isotope peak of the molecular ion discussed in section II.

unimolecular (for example equations (2) to (5)), ion–molecule reactions are bimolecular (equation (11)) and are recognized easily because their occurrence is proportional to the square of the sample pressure.

Gas phase ion–molecule reaction studies, in contrast to routine determination of the mass spectra of organic compounds, are carried out at relatively high ion source pressure, in the range of 10^{-4} torr to several torr. The low pressure mass spectrum may be altered considerably when the pressure is increased. A case in point is the mass spectrum of methane (Table 1). The primary ions, that is the ions formed as a result of electron impact ionization at low ion source pressure, which account for most of the total ionization, are $CH_4^{+\cdot}$, CH_3^+, $CH_2^{+\cdot}$ and CH^+. As the pressure is increased, the primary ions react with neutral methane molecules to give secondary and higher-order ionic products (Field and Munson[36]), some of which are listed below

$$CH_4^{+\cdot} + CH_4 \rightarrow CH_5^+ + CH_3 \cdot \qquad (12)$$

$$CH_3^+ + CH_4 \rightarrow C_2H_5^+ + H_2 \qquad (13)$$

$$CH_3^+ + 2CH_4 \rightarrow C_3H_7^+ + 2H_2 \qquad (14)$$

$$CH_2^{+\cdot} + 2CH_4 \rightarrow C_3H_5^+ + 2H_2 + H \qquad (15)$$

$$CH_2^{+\cdot} + CH_4 \rightarrow C_2H_4^+ + H_2 \qquad (16)$$

At an ion source pressure of 1 torr of methane (Field[33]) 95 per cent of the total ionization is due to CH_5^+ (48 per cent), $C_2H_5^+$ (41 per cent) and $C_3H_5^+$ (6 per cent), and the mass spectrum thus shows mainly three peaks at m/e 17, 29 and 41.

Chemical ionization is a means of ion production in which molecules of the sample under investigation are ionized by reacting with ions which have been selected as ionizing reactants. This is accomplished in practice by operating the electron impact ion source at a pressure of about 1 torr (Munson and Field[31]). The reactant ions are formed from the reactant gas—a substance which need be a gas only under the ion source operating conditions—by a combination of electron impact ionization and ion–molecule reactions, and the sample for which the mass spectrum is desired is chemically ionized within the same ion source by ion–molecule reactions with the reactant ions. The sample concentration is kept below 0·1 per cent of the reactant gas concentration to prevent either electron impact ionization of the sample or ion–molecule reactions from taking place between sample ions and neutral sample molecules. The chemical ionization mass spectrum of a substance therefore consists of two sets of ions, the ions formed by reacting the substance with the reactant ions, and the reactant ions over and above those used in ionizing the substance. In practice the latter are excluded from the representation (section II) of the chemical ionization mass spectra of the substance.

Any gas may be used as the reactant gas provided it gives a set of reactant ions unreactive towards the reactant gas but reactive towards the sample under investigation. There is thus the possibility of changing the chemical ionization mass spectrum of a substance by changing the reactant gas. Most of the work published thus far has been done with methane at the reactant gas. In that case the reactant ions are mainly CH_5^+, $C_2H_5^+$ and $C_3H_5^+$, all three of which may react as either Bronsted or Lewis acids (Field[33]). The origin of the reactant ions in methane was given in equations (12), (13) and (15).

An electron impact ion source (Figure 4) may be converted (Arsenault[37]) to a combination electron impact–chemical ionization source by gas tightening of the ion source to permit its operation at a pressure of about 1 torr. In addition a power supply to increase the electron energy to several hundred electron volts must be added since the 50–80 eV electrons generally used for electron impact ionization do not penetrate sufficiently into an ion source operated at high pressure to give good sensitivity. Finally, the vacuum pumping system of the instrument should be adequate to handle the gas load imposed by the high pressure technique used for chemical ionization.

Chemical ionization mass spectra in general tend to show less fragmentation and more abundant high mass ions than do electron impact ionization mass spectra. The chemical ionization mass spectra of a relatively small number of organic compounds have been published thus far, some biomolecules being included among them, for example pristane (Gelpi and Oró[38]). The successful development in 1969 (Arsenault,[37] Fales, Milne and Vestal[39]) of combined electron impact–chemical ionization sources having direct sample introduction systems (section IIID2) and the more recent successful combination of gas chromatography and chemical ionization mass spectrometry (Arsenault, Dolhun and Biemann,[40] Schoengold and Munson[41]) are bound to lead to more investigations of biomolecules by chemical ionization mass spectrometry. The results obtained thus far (Arsenault, Althaus and Divekar,[42] Fales, Milne and Vestal[43]) offer great promise for future applications of this technique.

B. Analyser

As the ions pass through the object slit (Figures 4 and 5) they are fully accelerated and enter the analyser region (Figure 3) in which separation of the ions according to mass-to-charge ratios takes place. This is most frequently done in present day instruments used in the investigation of biomolecules by deflecting the ion beam in a magnetic field. The focusing action of a sector magnetic field is illustrated in Figure 6. A divergent beam

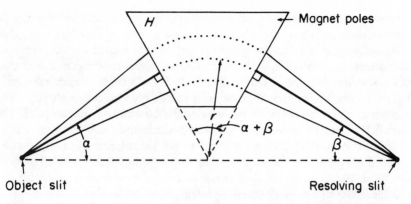

Figure 6. Illustration of the focusing properties of a sector magnetic field: the ions travel from the object slit to the resolving slit in a flight tube which is, in part, inserted in the gap between two magnet poles; the magnetic field H is at right angles to the plane of the paper

of ions, which is homogeneous in mass and energy and which enters and leaves a uniform magnetic field at right angles to the edges of the poles, is focused at the point at which the resolving slit is placed; this is on the extension of a line from the object slit through the centre of curvature of the ion beam. Most mass spectrometers have a symmetrical arrangement in which $\alpha = \beta = 30°$ or $45°$. The mass-to-charge ratio of the ions brought to a focus at the resolving slit is given by the following formula

$$m/e = \frac{H^2 r^2}{2V_a} \tag{17}$$

H is the strength of the magnetic field, r is the radius of curvature of the path of the ions through the magnetic field and V_a is the accelerating voltage mentioned earlier (sections IIIA1 and IIIA2). To obtain the mass spectrum of a substance it is necessary to focus ions of different m/e in succession at the resolving slit behind which the detector is located (section IIIC). This is accomplished by changing either V_a (electric scan) or H (magnetic scan), the latter being more common as well as preferable.

The mass-to-charge ratio separation of ions is not limited to mass spectrometers in which the ions are deflected in a magnetic field. Cycloidal, time-of-flight and quadrupole mass spectrometers, as well as others, use different principles for mass-to-charge separation, are available commercially and may offer advantages over magnetic deflection instruments for some applications.

C. Detector and Recorder

It is necessary to measure and record the intensity of the ion beam which is focused at the resolving slit (Figure 6). This is accomplished with the detector–

recorder component of the mass spectrometer (Figure 3) which must respond rapidly, accurately and reliably to a very wide range of ion abundance, and ultimately must automatically plot a permanent record of the spectrum. The detector is located in the vacuum system of the mass spectrometer after the resolving slit, and consists of a collector electrode, which may be a Faraday cage or the first dynode of a secondary electron multiplier. The analogue signal from the detector is amplified before it is recorded. The recorder is located outside the vacuum system, and may be, in increasing order of complexity and cost, a pen-and-ink recorder, a recording oscillograph, a magnetic tape (analogue or digital), or a digital computer. The most common combination of detector–recorder in instruments used for the routine determination of mass spectra of biomolecules is the electron multiplier and the recording oscillograph. This combination provides sensitivity of detection, speed of response and an ability to handle ion abundance ratios of up to $1:10,000$, which are often encountered in a mass spectrum; it does this by simultaneously recording several spectra at different sensitivities on a single chart of paper, and, provided an ultraviolet lamp and ultraviolet-sensitive paper are used in the recording oscillograph, the combination will also give an immediate visual display of the mass spectrum.

A combination detector–recorder which needs to be mentioned apart from all others is the photographic plate which is an integrating device. It is used today solely in instruments which have a focal plane (section IIIE) rather than a focal point. It permits detecting and recording the complete mass spectrum at once over a wide range of mass, and is widely used in mass spectrometers which use the radio frequency spark source (section IIIA) as well as in high resolution mass spectrometry (section IIIE).

D. Sample Introduction System

The sample under mass spectrometric investigation must have a vapour pressure of no less than about 1×10^{-6} torr under ion source (section IIIA) operating conditions in order to give a satisfactory mass spectrum. Thus, while a sample is being introduced, the pressure in the ion source itself is somewhat higher than it is in the remainder of the instrument. The sample may be a solid, liquid or gas at room temperature and pressure, and is introduced into the mass spectrometer by means of the sample introduction system, a component of the instrument (Figure 3) which may consist of several separate units. The units commonly used for organic chemical applications are the heated inlet system (section IIID1), the direct introduction system (section IIID2) and the gas chromatograph (section IIID3). The choice of unit depends on the problem at hand. The development of direct introduction systems and gas chromatograph–mass spectrometer combinations has given great impetus to the use of mass spectrometry in biochemistry, because both of these means of sample introduction reduced the amount and the vapour pressure of the sample required.

1. Heated Inlet System

A schematic diagram of an heated inlet system, including the means of introducing gases, liquids and solids into the system, is shown in Figure 7. The inlet system may be made of metal or glass, and the maximum temperature at which it is operated is about 150 °C for the all-metal inlet system and about 350 °C for the all-glass inlet system. Temperatures above 150 °C

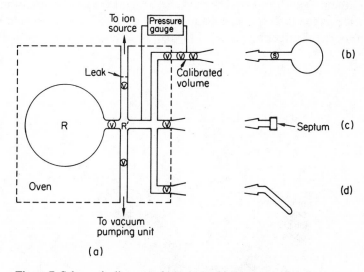

Figure 7. Schematic diagram of (a) a heated inlet system showing means of introducing into the system (b) gases, (c) liquids, and (d) solids. S stands for stopcock; V, for valve; R, for reservoir; and R' for the connecting volume, useful as a small reservoir, between the valves located inside the oven. The various means of sample introduction are connected to the sample ports of the heated inlet system by tapered joints made vacuum tight with teflon rings (not shown). Both (c) and (d) may require separate heaters (not shown) to permit the introduction of samples of low volatility into the inlet system

should be avoided for biomolecules, and the direct introduction system (section IIID2) should be used in most applications for samples of low volatility requiring a heated inlet system temperature greater than 150 °C. The sample is introduced by suitable manipulations of the valves into the reservoir which has a 1–5 litre volume, and, when required, leaked into the ion source through a low conductance molecular leak.

A pressure of 1×10^{-2} torr is required in the heated inlet system in order to give a sample pressure of 1×10^{-5} torr in the ion source. The sample size required to give such a pressure depends on the volume of the reservoir, the molecular weight of the sample and the temperature of the inlet system, and is of the order of 0·1 mg.

2. Direct Introduction System

The direct introduction system consists of a probe such as the one shown in Figure 8 and a vacuum lock. A capillary containing the sample, which is generally a solid, is inserted in the sample heater and evacuated in the vacuum lock to about 1×10^{-5} torr. The valve isolating the vacuum lock from the ion source region is then opened, and the sample pushed into position near the ion source chamber by means of the handle at the end of the probe shaft.

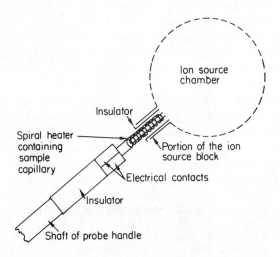

Figure 8. Schematic diagram of the probe of a direct introduction system showing the capillary containing the sample in position at the edge of the ion source chamber

The direct introduction system makes the most efficient use of the sample which is vapourized directly into the ion source chamber by means of the sample heater. A submicrogram sample would be more than sufficient to give a satisfactory mass spectrum for most applications, and mass spectra may be obtained from a sample consumed at the rate of $1 \times 10^{-12}\,\mathrm{g\,sec^{-1}}$. The minimum vapour pressure required of the sample to give a useable mass spectrum must be about 1×10^{-6} torr at the maximum temperature, generally less than 400 °C, at which the sample may be heated without undergoing decomposition.

3. Gas Chromatograph

The gas chromatograph is an ideal instrument to use in combination with a mass spectrometer because it is highly sensitive as well as capable of separating complex mixtures. The broad range of combined use of these two instruments has been reviewed in recent years by McFadden,[44] Leemans and McCloskey,[45] and Watson.[46]

The initial use of gas chromatography in conjunction with mass spectrometry consisted in collecting the material in fractions eluting from a gas chromatographic column in capillary tubes, or in a series of cold traps (Drew and coworkers[47]), and introducing the collected samples into the mass spectrometer after completion of the gas chromatographic separation. Later, direct connections between gas chromatographs and mass spectrometers were made in which a small portion of the effluent from a packed column (Gohlke,[48] Holmes and Morrell[49]) or the total effluent from a capillary column (McFadden[50]) was introduced and analysed by the mass spectrometer while the gas chromatographic separation was in progress. A direct connection of both instruments is also used in the stopped-flow technique in which the material in a fraction is collected, the carrier gas flowing through the column is stopped and the collected sample is analysed by mass spectrometry before resuming the gas chromatographic separation (Beynon, Saunders, and Williams[51]).

The use of gas chromatography and mass spectrometry in tandem (GC–MS) did not reach any degree of importance until the development of pressure reducing-sample enriching devices, known under the generic name of separators, which provide a direct and efficient GC–MS interface. Separators are generally used with packed columns and may also be used with capillary columns, if needed. The effluent from the separator is introduced into the mass spectrometer on the low pressure side of the inlet line between the molecular leak and the ion source (Figure 7), or directly into the ion source chamber as in the case of the direct introduction system (Figure 8). The function of a separator is to remove from the effluent of the gas chromatograph as much carrier gas (usually helium) as possible, while minimizing the loss of sample. Thus, the effluent from the separator contains a larger proportion of sample than its equivalent from the gas chromatograph, even though some loss of sample took place in the separator. Separators may be classified into three groups on the basis of the principle used to carry out their function. Figure 9 shows three separators, each one illustrating a group and each one being available commercially.

Figure 9(a) shows the most commonly used separator, the effusion separator, often called the fritted glass separator, developed by Watson and Biemann.[52] In this separator, the effluent from the gas chromatograph is passed through the hollow core of a cylindrical, ultrafine porous glass tube in which conditions of molecular flow exist. The carrier gas is pumped out through the pores of the ultrafine porous glass tube much more rapidly than the sample. The jet-orifice separator developed for mass spectrometry by Ryhage[53] and Stenhagen[54] is shown in Figure 9(b). In this separator, the gas chromatographic effluent is passed at high speed from a fine aperture through

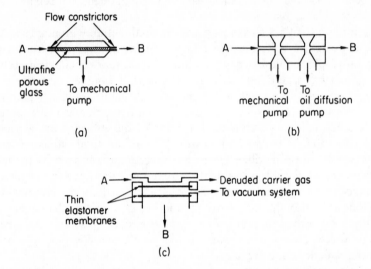

Figure 9. Schematic drawing of separators used in interfacing gas chromatographs with mass spectrometers: (a) effusion type; (b) jet-orifice type; (c) selective permeability barrier type. A: effluent from the gas chromatograph; B: to mass spectrometer

an open space to an orifice, the process being repeated twice. The momentum carries most of the sample forward from the aperture to the orifice while the carrier gas, being lighter, is pumped away. Figure 9(c) shows the selective permeability barrier separator developed by Llewellyn and Littlejohn.[55] In this separator, the sample in the gas chromatographic effluent is adsorbed preferentially over the carrier gas on the high pressure side of a thin elastomer membrane and desorbed on the low pressure side, the process being repeated twice.

Any sample which can be eluted from a gas chromatographic column has sufficient vapour pressure for analysis by mass spectrometry. The overall sensitivity of GC–MS is somewhat less—by one or two orders of magnitude—than the ultimate sensitivity possible with the direct introduction system (section IIID2) because of sample losses in collectors or in separators, or decreased efficiency of ionization due to the presence of some carrier gas in the ion source. In practice, however, a 1 μg sample is sufficient for GC–MS and is a conservative estimate of the sample size required because gas chromatographic peaks do not linger and the mass spectrometer in the GC–MS mode of operation is most often scanned rapidly, at the rate of a spectrum every few seconds.

GC–MS has been used extensively in geochemistry and in flavour and fragrance analysis, and widespread applications may be expected in clinical and biochemical laboratories in the near future. The development of GC–MS spurred the development of data acquisition and data reduction systems ideally suited to minimize human intervention in recording and processing hundreds of mass spectra from a single GC run (Hites and Biemann[3]). The lack of such elaborate systems should not deter anyone, however, from the immediate and extensive use of GC–MS should applications be at hand.

E. High Resolution Mass Spectrometry

The isotopes have nuclidic masses (Table 2) which are not whole numbers or integers. Hence, while the molecular weight of methane is generally stated to be 16, it is more exact to refer to it as 16·031300 a.m.u. (Equation (6)). The net effect of the slight difference of the nuclidic mass of the isotopes from

Table 3. Listing of all possible elemental compositions within ± 0.003 a.m.u. and $C_{0-24}H_{0-50}N_{0-4}O_{0-10}S_{0-1}$ for species of determined mass 292·0947 a.m.u.

Calculated mass[a]	Deviation from determined mass[a]	Elemental composition				
		C	H	N	O	S
292·0960	−0·0013	16	12	4	2	0
292·0974	−0·0027	18	14	1	3	0
292·0933	+0·0014	13	14	3	5	0
292·0947	0·0000	15	16	0	6	0
292·0922	+0·0025	19	16	0	1	1
292·0967	−0·0020	10	18	3	5	1

[a] In a.m.u.

whole numbers is that a number of combinations of elements have the same integral mass, also referred to as nominal mass, but differ in their exact mass. An example to illustrate this point is given in Table 3 which is a listing of all the possible elemental compositions for a species of determined mass*

* The discussion is limited to singly charged species with mass m, collected at m/e. Hence, the mass and mass-to-charge ratio of each species have the same numerical value

292·0947. A table including all possible elemental compositions with nominal mass 292 would be too long to include here; the list given in Table 3 was limited to six possibilities by selecting the mass 292·0947, by placing narrow limits of ± 0.003 a.m.u. on acceptable possibilities and by considering only five different isotopes. The consequence of the difference in the exact mass of different combinations of isotopes is that, provided the mass of a species can be measured with sufficient accuracy, the elemental composition of that species can be established unambiguously.

Table 4 lists a few pairs of elemental combinations differing by less than 5 millimass units. These pairs may be encountered at all masses, and the resolution necessary to separate such doublets, should they exist at all, may be calculated (section II). For example, the pair $C_2H_2O-N_3$ (Table 4)

Table 4. A few pairs of elemental combinations differing by less than 5 m.m.u.[a]

Pairs of elemental combinations[b]		Mass difference (m.m.u.)[a]
CH	^{13}C	4·47
$^{13}CHNO_2-$	C_5	4·08
C_5	N_2O_2	4·02
H_4S	C_3	3·37
C_3N	H_2O_3	2·68
H_5Cl	CN_2	1·83
H_2	^2H	1·55
$^{13}CH_3O_5-$	C_8	1·40
C_2H_2O	N_3	1·34
H_3NP	C_4	0·31

[a] One m.m.u. = 0·001 a.m.u.
[b] The heavier combination of each pair is on the left

is shown in Table 3 as the first two possible elemental compositions, $C_{16}H_{12}NO_2 + (N_3$ or $C_2H_2O)$. The resolution necessary to separate that pair would be $292·0960 \div 0·00134 = 2·18 \times 10^5$. At one half the exact mass under consideration, the same pair would require one half the resolution; at twice the mass, twice the resolution. Such high resolution in mass spectrometry is seldom available and quite unnecessary for almost all applications except the study of deuterium-labelled compounds. Close doublets are rare, in the mass spectra of biomolecules, particularly at the higher masses. What is needed is not high resolving power but the ability to measure a mass accurately enough either to determine unambiguously the elemental composition of an ion or to limit severely the number of possibilities that need to be considered. Thus, for most applications centred around organic molecules, the term high resolution mass spectrometry refers to the ability to determine masses accurately, usually within an error of a few parts per million.

Exact mass measurements are most often made with double focusing mass spectrometers operated at a resolution greater than 10,000, although Beynon[56] showed in his pioneering work that accurate mass measurements of organic ions were possible with a single focusing mass spectrometer of much lower resolving power. The two basic designs of double focusing analysers shown in Figure 10 are commonly used and commercially available. In both

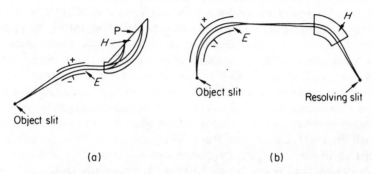

Figure 10. Double focusing analyser designs (a) of Mattauch and Herzog (1934) with focal plane P, and (b) of Johnson and Nier (1953) with a focal point. *E* stands for electrostatic field, and *H*, for magnetic field

designs, an electrostatic field is used to compensate for the energy spread of the ions. The electrostatic field produces velocity focusing of the ions, and is followed by a magnetic deflection analyser which effects direction focusing of the ions. These designs are thus known as 'double focusing' analysers in contrast to the magnetic deflection analyser (section IIIB) which would be 'single focusing'.

The design of Mattauch and Herzog[57] shown in Figure 10(a) is distinctly different from the other because all ions are focused at once in a focal plane where a photographic plate (section IIIC) is located to photograph the complete mass spectrum without the need for scanning. The design of Johnson and Nier[58] shown in Figure 10(b) has a focal point only where a resolving slit is placed and the mass spectrum is obtained by scanning. The Mattauch and Herzog design (Figure 10(a)) may also be used to scan spectra provided a resolving slit is placed at a suitable point in the focal plane.

Occasional, accurate mass measurements using either design of double focusing mass spectrometer have been made for some time by the peak-matching technique described by Quisenberry, Scolman and Nier.[59] This technique, in adroit hands, requires no more than a minute for a single mass measurement but is not generally used to measure more than a few peaks per spectrum because it soon becomes tiring.

Mass measuring of all ions in a mass spectrum was indeed rare until Biemann and his associates used a double focusing mass spectrometer of the Mattauch and Herzog design (Figure 10(a)) and photographic recording to obtain high resolution mass spectra with relative ease; they also developed computer techniques to simplify the task of converting line position on the photographic plate into elemental composition and to present the data in a manner which is more meaningful to the interpreter. Biemann[60] has reviewed the subject of high resolution mass spectrometry recently, and examples illustrating the utility of measuring the exact mass of all ions in a mass spectrum will be given in section IV. Electrical detection, rather than photographic detection, has also been used to obtain complete mass measurements of all ions in mass spectra with both the Johnson and Nier[58] design (McMurray, Greene and Lipsky[61]) and the Mattauch and Herzog[57] design (Burlingame, Smith and Olsen[62]) of double focusing mass spectrometers.

Mass measurements may be made by the peak-matching technique with an error of less than 1 p.p.m. although errors greater than 10 p.p.m. are often reported in the chemical literature. The mass measurement techniques referred to in the preceding paragraph are somewhat less accurate than peak-matching, and errors up to 10 p.p.m. are generally considered to be acceptable.

IV. SELECTED APPLICATIONS

Most mass spectrometric work on organic molecules has been done in the past, and will be done in the near future, with low resolution instruments (herein defined as mass spectrometers with a resolution (section II) of less than a few thousand) having only an electron impact ion source (section IIIA1). Complete, detailed coverage of all applications of mass spectrometry to biomolecules is not possible and is not intended. A much more complete coverage may be found in the books by Biemann,[7] Budzikiewicz, Djerassi, and Williams,[63,64] and Waller[65] which are the best available references to the mass spectra of natural products and biomolecules. The applications selected for discussion in the following three subsections are limited to the use of mass spectrometry in elucidating the structure of biomolecules.

The greatest omission among these applications (sections IVA–C) is possibly the lack of examples of routine uses in which mass spectrometry plays a major role at the initial stage in the solution of a problem without appearing to have done so at a later time because the mass spectrometric evidence has been, by then, largely overshadowed by the total accumulated evidence from all sources. One such example will suffice to illustrate this point. A small amount of the quinonoid pigment (+)-solaniol was isolated from the culture filtrate of *Fusarium solani* D_2 purple. The mass spectrum suggested a molecular weight of 292 for this compound, and the accurate mass of the peak at m/e 292 was measured to be 292·0948 by the peak-matching technique (section IIIE). This corresponds to an elemental composition of $C_{15}H_{16}O_6$ (Table 3), with an error of 0·3 p.p.m. in the measured mass, provided it is assumed that (+)-solaniol contains only the elements C, H, and O. On the assumption that the elemental composition was indeed $C_{15}H_{16}O_6$, structure (1) was proposed for (+)-solaniol and proved by

(1)

synthesis (Arsenault[66]). Thus, in retrospect, the contribution made by mass spectrometry to solving the structure of (+)-solaniol was small, and yet it was of some importance in focusing attention on the path to follow to the ultimate proof of structure.

A. Determination of the Amino Acid Sequence in Oligopeptides

The use of mass spectrometry in determining the amino acid sequence in oligopeptides has an historical development which parallels advances in mass spectrometry, and will be so presented in this subsection.

The first approach to the determination of the amino acid sequence in oligopeptides by mass spectrometry was suggested by Biemann, Gapp and Seibl.[67] These authors reduced oligopeptides (2) to polyamino alcohols (3) with lithium aluminium hydride (Chart 1) to overcome the lack of volatility of peptides at a time when the heated inlet system (section IIID1) was the only available means of sample introduction. Aside from volatility, poly-amino alcohols have the advantage of having simpler mass spectra than the corresponding peptides because the mass spectra of polyamino alcohols (3) are dominated by peaks due to the fragmentation of C–C bonds between adjacent nitrogen atoms, with preferred charge retention on the more highly substituted carbon atom of each fragmented C–C bond. Since R' is known because R is deliberately used as the N-terminal blocking group, for example, if R were CH_3CO, R' would be C_2H_5, the mere recognition of the more abundant fragments labelled $a_1, a_2, a_3, \ldots a_n$ in structure (3) would be sufficient to establish the sequence in the polyamino alcohol starting from the N-terminal, and hence the sequence in the peptide from which it originated.

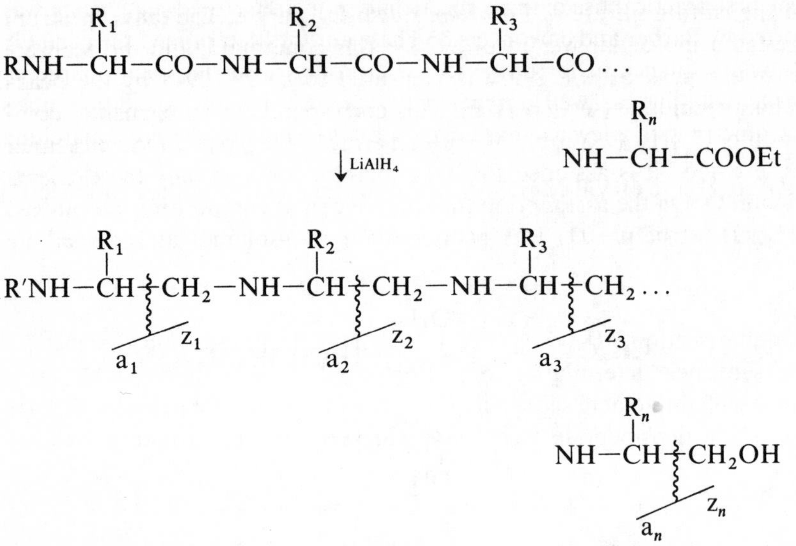

$$\overset{R_1}{\underset{|}{RNH-CH-CO-NH-\underset{|}{\overset{R_2}{CH}}-CO-NH-\underset{|}{\overset{R_3}{CH}}-CO\ldots}}$$

$$\overset{R_n}{\underset{|}{NH-CH-COOEt}} \quad (2)$$

$$\downarrow \text{LiAlH}_4$$

$$\overset{R_1}{\underset{|}{R'NH-CH}}\{CH_2-NH-\overset{R_2}{\underset{|}{CH}}\{CH_2-NH-\overset{R_3}{\underset{|}{CH}}\{CH_2\ldots$$
$$a_1 \quad z_1 \qquad a_2 \quad z_2 \qquad a_3 \quad z_3$$

$$\overset{R_n}{\underset{|}{NH-CH}}\{CH_2OH \quad (3)$$
$$a_n \quad z_n$$

Chart 1

The less abundant fragments labelled $z_1, z_2, z_3, \ldots z_n$ in structure (3) may be used to confirm the established sequence starting from the C-terminal.

Although the idea of amino acid sequencing by mass spectrometry was retained, the specific approach suggested by Biemann, Gapp and Seibl[67] was soon relegated to the background by the development of direct introduction systems (section IIID2). Small peptides (4) with a C-terminal

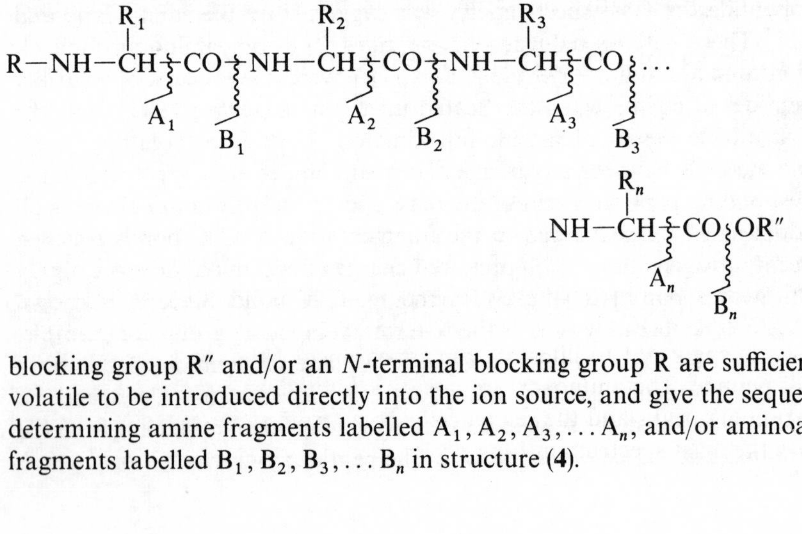

$$R-NH-\overset{R_1}{\underset{|}{CH}}\{CO\}NH-\overset{R_2}{\underset{|}{CH}}\{CO\}NH-\overset{R_3}{\underset{|}{CH}}\{CO\}\ldots$$
$$A_1 \quad B_1 \qquad A_2 \quad B_2 \qquad A_3 \quad B_3$$

$$NH-\overset{R_n}{\underset{|}{CH}}\{CO\}OR'' \quad (4)$$
$$A_n \quad B_n$$

blocking group R'' and/or an N-terminal blocking group R are sufficiently volatile to be introduced directly into the ion source, and give the sequence determining amine fragments labelled $A_1, A_2, A_3, \ldots A_n$, and/or aminoacyl fragments labelled $B_1, B_2, B_3, \ldots B_n$ in structure (4).

The first practical use of mass spectrometry in amino acid sequencing was reported by Barber and coworkers.[68] These authors determined the partially unknown sequence in fortuitine (5), a naturally occurring peptidolipid of unusual volatility as well as chemical structure, by low resolution mass spectrometry with a few accurate mass measurements by the peak-matching

$$CH_3(CH_2)_n\text{-CO-Val-MeLeu-Val-Val-MeLeu-Thr-Thr-Ala-Pro-OMe} \quad (5)$$
$$\underset{Ac}{|} \quad \underset{Ac}{|}$$

$$n = 18, 20$$

technique (section IIIE) to verify the elemental composition assigned to some sequence determining ions. Fortuitine (5) was followed by other examples of amino acid sequencing in cyclic and antibiotic peptides. Among these is the depsipeptide isariin for which the partial structure (6a) was

$$CH_3(CH_2)_8CHCH_2CONHCHCONHCHCONHCHCONHCHCONHCHCO$$

$$\begin{array}{ccccc} | & | & | & | & | \\ H & R_1 & R_2 & R_3 & CH \\ & & & & H_3C \nearrow \searrow CH_3 \end{array}$$

$$R_1, R_2, R_3 = CH_3, CH(CH_3)_2, CH_2CH(CH_3)_2 \qquad (6a)$$

$$R_1 = CH(CH_3)_2; R_2 = CH_2CH(CH_3)_2; R_3 = CH_3 \qquad (6b)$$

known (Vining and Taber[69]) prior to completion of the amino acid sequence by mass spectrometry which established the structure as (6b) (Biemann and coworkers,[70] Wolstenholme and Vining[71]). The mass spectrum of methyl isariate (7), a derivative of isariin used in the sequence determination, is

$$CH_3(CH_2)_8CHOHCH_2CONHCHCONHCHCONHCH$$

$$\begin{array}{ccc} | & | & | \\ H & R_1 & R_2 \end{array}$$

$$CONHCHCONHCHCOOCH_3$$

$$\begin{array}{cc} | & | \\ R_3 & CH \\ & H_3C \nearrow \searrow CH_3 \end{array}$$

$$R_1 = CH(CH_3)_2; R_2 = CH_2CH(CH_3)_2; R_3 = CH_3 \qquad (7)$$

shown in Figure 11 to illustrate the appearance of the mass spectrum of a small peptide. The aminoacyl fragments are the most prominent sequence determining peaks and the sequence deduced from these peaks is indicated above the mass spectrum in Figure 11.

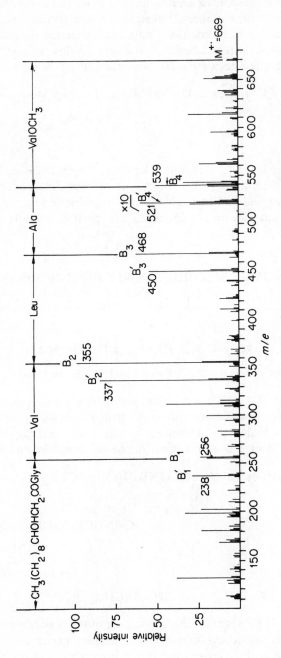

Figure 11. Partial mass spectrum of methyl isariate (7) showing the sequence determining aminoacyl ions labelled B_1, B_2, B_3 and B_4. The peaks labelled B_1', B_2', B_3' and B_4' are due to the facile loss of water in the N-terminal side chain

The development of techniques to measure rapidly the exact mass of all ions in a high resolution mass spectrum (section IIIE) led to a further development in determining the amino acid sequence in oligopeptides, that of computer interpretation of their high resolution mass spectra. This approach takes advantage of the unique elemental composition, and hence the accurate measured mass is also unique, of each sequence-determining fragment and the ease with which a digital computer may perform endlessly the large number of trivial but tedious calculations necessary to consider all possible sequences fitting the available data. Amino acid sequencing by computer interpretation of the high resolution mass spectra of oligopeptides was suggested in 1966 by three different groups (Barber and coworkers;[72] Biemann and coworkers;[70] Senn, Venkataraghavan and McLafferty[73]), and used by Biemann and coworkers[70] to determine the partially unknown sequence in isariin (**6b**) while a more conventional mass spectrometric and degradative approach led to the same sequence (Wolstenholme and Vining[71]).

Most of the work on the amino acid sequence determination by mass spectrometry was done on model peptides modified in some way, except that the amide carbonyl was left intact, to increase their volatility and to provide abundant sequence determining fragments easily detectable among other peaks in their mass spectrum. In that respect, the *N*-methylation of peptides proposed by Das, Gero and Lederer[74] is one of the more successful peptide modifications among those suggested and was used in part to determine the amino acid sequence in the mammalian peptide hormone called feline gastrin (**8**) (Agarwal, Kenner and Sheppard[75]).

$$\ulcorner\text{Glu-Gly-Pro-Try-Leu-(Glu)}_4\text{-(Ala)}_2\text{-Tyr-Gly-Try-Met-Asp-Phe-NH}_2 \quad (\mathbf{8})$$

Amino acid sequencing by mass spectrometry will always be limited to small peptides ranging from dipeptides in unfavourable cases to pentadecapeptides in favourable cases. This was recognized by Biemann and Cone[76] who wrote a computer programme to assemble all the possible sequences of a large peptide from the known sequences of a number of small peptides obtained from the large one in an unspecified manner. Tests of this programme showed that small peptides could be used to determine a unique amino acid sequence for the large peptide provided the small peptides overlapped sufficiently. For example, a test was done using A-chain of beef insulin (**9**) as the assumed unknown sequence and an input of known data to the sequence assembly programme limited to a number of dipeptides and/or tripeptides only. The programme assembled 468 sequences for the A-chain

Gly-Ile-Val-Glu-Gln-Cys-Cys-Ala-Ser-Val-Cys-

-Ser-Leu-Tyr-Gln-Leu-Glu-Asn-Tyr-Cys-Asn (**9**)

of beef insulin when the input data was limited to the twenty possible dipeptides, while fourteen of the nineteen tripeptides were sufficient to give the unique and correct sequence (**9**). The minimum selected input data required to give sequence (**9**) only was determined to be ten dipeptides plus five tripeptides.

The development of GC–MS (section IIID3) to its present state makes it possible, in a short time, to separate a complex mixture of small peptides by gas chromatography and to identify the individual components of the mixture by mass spectrometry. The sequence assembly computer programme mentioned in the previous paragraph is then all that is necessary to complete the sequence of the large peptide from which the complex mixture originated. Biemann and coworkers[77] have turned their attention to the use of GC–MS in their latest work on the determination of amino acid sequences by mass spectrometry. *N*-acetylated small peptides are reduced to polyamino alcohols (**3**); the polyamino alcohols are silylated, separated by gas chromatography and identified by computer-interpretation of their low resolution mass spectra (Biemann and coworkers[77]). It has already been possible using this approach to separate and identify a mixture made up of seventeen dipeptides and tripeptides. The mass spectrum of compound (**10**) obtained by reduction and silylation of compound (**11**) is shown in Figure 12 to illustrate the appearance of the spectrum of silylated polyamino alcohols.

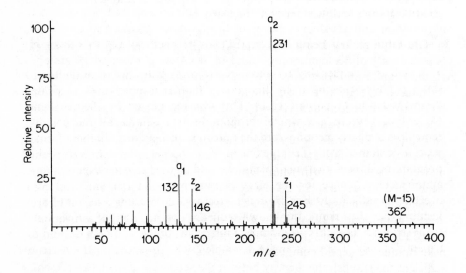

Figure 12. Mass spectrum of the silylated polyamino alcohol (**10**) showing the amino acid sequence determining peaks labelled as in (**3**)

$$CH_3-CH_2-NH-$$

$$-CH-CH_2-NH-CH-CH_2-NH-CH_2-CH_2-O-Si(CH_3)_3 \quad (10)$$

CH$_3$CO-Met-Leu-Gly-OCH$_3$ (11)

The importance as well as the difficulty of determining the amino acid sequence in peptides has led to much work being done in that field by mass spectrometry. Much of this work was omitted or glossed over quickly in this subsection and is more thoroughly reviewed by Biemann.[78]

B. Determination of the Nucleotide Sequence in Oligonucleotides

The application of mass spectrometry to the structure determination of nucleic acid derivatives has been quite limited because these compounds are not sufficiently volatile (section IIID). The mass spectra of purines, pyrimidines, nucleosides, mononucleotides and dinucleotides have been thoroughly reviewed recently by Hignite.[79]

Dinucleotides are the smallest nucleic acid components carrying sequence information and yet do not have sufficient vapour pressure to give mass spectra without first being derivatized. Poly-(trimethylsilyl) derivatives of fifteen dinucleotides commonly found in RNA were reported (Hignite,[80] Hunt, Hignite and Biemann[81]) to have adequate vapour pressure when introduced directly into the ion source of the mass spectrometer (section IIID2) to afford excellent mass spectra. The high mass end of the spectrum of octa-(trimethylsilyl)-adenylyl-(3′, 5′)-guanosine (12) is shown in Figure 13 as an example. The mass spectrum of each poly-(trimethylsilyl)-dinucleotide showed an $M^{+\cdot}$ and an $(M-15)^+$ which are sufficient to establish their molecular weight. Furthermore, their elemental composition may be established by accurate mass measurements (section IIIE). For example, Hignite and Biemann[82] determined the exact mass of the $M^{+\cdot}$ and $(M-15)^+$ ions in compound (12) within an error of 2 p.p.m. Since the four common bases, adenine, cytosine, guanine, and uracil, as well as most of the rare bases have unique elemental compositions, the exact mass of any dinucleotide is sufficient to establish the identity but not the sequence of the bases present.

To obtain the sequence of the bases in dinucleotides, Hunt, Hignite and Biemann[81] suggested using relative intensities of the fragments labelled

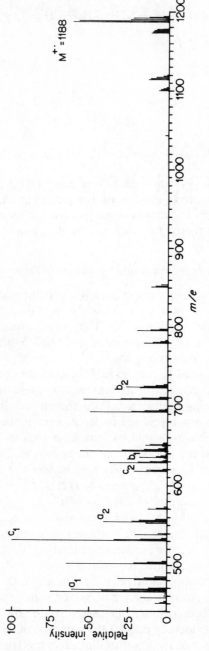

Figure 13. Partial mass spectrum of octa-(trimethylsilyl)-adenylyl-(3′,5′)-guanosine (**12**) (Hignite[80]) showing the nucleotide sequence determining peaks labelled as in (**13**)

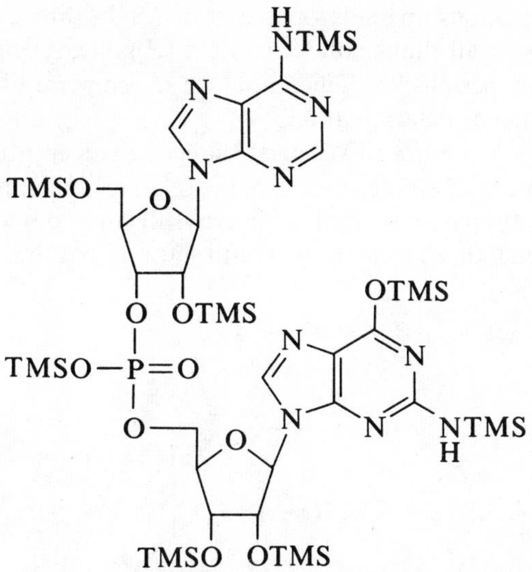

(12)

a_1, a_2, b_1, b_2, c_1 and c_2 in the generalized structure of poly-(trimethylsilyl)-dinucleotides (13) and in Figure 13. Fragments a_1 and a_2 originate from cleavage (section II) of the PO–C-3′ and PO–C-5′ bonds respectively, with charge retention on the carbon atoms. Charge retention on the oxygen atoms accompanied by transfer of two hydrogens to form protonated phosphate ions gives rise to b_1 and b_2. The third set of fragment ions labelled c_1 and c_2 is related to b_1 and b_2 except that a single hydrogen is transferred and is accompanied by the loss of $(CH_3)_3SiOH$ and CH_3, the net effect being shown in (13) as b less the elements $C_4H_{14}OSi$. In general, where applicable, fragments a_1 are less abundant than a_2, b_1 less than b_2, and c_1 more than c_2.

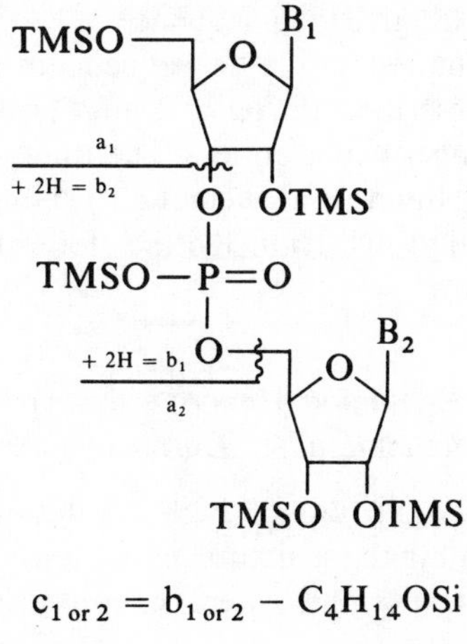

(13)

$$c_{1\,or\,2} = b_{1\,or\,2} - C_4H_{14}OSi$$

This intensity relationship holds for at least two of the three pairs of fragments a's, b's and c's in all dinucleotides studied (Hignite,[80] Hunt, Hignite and Biemann[81]) and permits the differentiation of isomeric dinucleotides and hence their sequence determination.

More recently, Dolhun and Wiebers[83] suggested an approach to dinucleotide sequencing which is independent of the relative abundance of fragments. These authors prepared six poly-(trimethylsilylated)-dinucleotide phenylboronates of generalized structure (14) and showed that the compounds gave

complex mass spectra with, in each case, an $M^{+ \cdot}$ and fragments a_1, a_2. These data alone are sufficient to establish the identity of the bases and, should they be different, to give their sequence.

Dinucleotides have limited utility in sequencing nucleic acids but an extension of the work described above, or a modification thereof, would at least permit the sequencing of trinucleotides. The ability to measure accurately masses up to m/e 1700 (Hignite and Biemann[82]) should help greatly in this matter since poly-(trimethylsilyl)-trinucleotides, for example, have molecular weights in the 1700 region.

C. The Use of High Resolution Mass Spectrometry in Determining the Structure of Antheridiol, a Sex Hormone in Achlya bisexualis

Examples of the extensive use of high resolution mass spectrometry (section IIIE) in determining the structure of biomolecules are still somewhat rare and some of these examples were mentioned by Biemann.[60] The elucidation of the structure of antheridiol was accomplished by making use of all available physical methods, with the emphasis on high resolution mass spectrometry for reasons which will be apparent as this work is described.

Many years of work led in 1965 to the isolation from the aquatic fungus *Achlya bisexualis* of about 1 mg of material which was designated antheridiol (McMorris and Barksdale[84]) after its structure had become apparent. The structure elucidation of antheridiol was an unusual and interesting challenge for three reasons. First, antheridiol was never isolated in any quantity in spite of much effort spent in the attempt. Second, antheridiol had a well known biological function which was originally postulated by Raper in 1939.[85] Antheridiol is secreted by the female mycelium and initiates sexual reproduction in *Achlya bisexualis*. In response to the presence of antheridiol at a concentration level of 2×10^{-8} mg/ml, the male hyphae produce specialized branches with antheridia at their tip and the sexual reproduction cycle is thus underway. Third, antheridiol belonged to an unknown class of compound. Indeed, when work began in 1965 on elucidating the structure of antheridiol no assistance was obtained from the knowledge of prior work on plant sex hormones because very few had been isolated and none had an established chemical structure. Thus, in the absence of any chemical precedent which might have been useful at least in the early stages of the work, antheridiol had to be treated as a totally unknown compound.

The work which led to the proposal of structure (**15**) for antheridiol was carried out by Biemann and the author at the Massachusetts Institute of

(**15**)

Technology working in collaboration with Barksdale and McMorris at The New York Botanical Garden (Arsenault and coworkers[86]). This work was begun by taking a high resolution mass spectrum (section IIIE) of antheridiol to conserve material while securing as much information as possible. It was necessary to establish which element was present in antheridiol before its elemental composition could be determined. This was done by looking at the elemental composition of all ions below the arbitrarily set limit of m/e 80 in the high resolution mass spectrum. Since plausible elemental compositions could be found for all ions below m/e 80 by assuming the presence of only

	CH	CHO	CHO2	CHO3	CHO4	CHO5
150		10/14**	9/10**			
151		10/15**	9/11**			
152	12/ 8**		9/12*******	8/10*		
153	12/ 9***			8/11**		
154	12/10**					
155	12/11***					
156	12/12**					
157	12/13****	11/ 9***				
158	12/14**	11/10**				
159	12/15***	11/11******				
160	12/16**					
161	12/17***	11/13******	10/11**			
163	12/19**	11/15***				
164		11/16*	10/13***			
165	13/ 9**		10/14***			
166	13/10**					
167	13/11**					
168	13/12**					
169	13/13***	12/ 9**				
170	13/14**					
171	13/15***					
172		12/11***	11/11*			
173	13/17***	12/12**	11/13***			
174	13/18***	12/13*****	11/14**			
175	13/19**	12/14*****				
177		12/15****				
178	14/10**	12/17**				
179	14/11**					
180	14/12*					
181	14/13**	13/ 9*				
183	14/15**	13/11**	12/13*			
184	14/16**	13/12*	12/14**			
185	14/17**	13/13*****	12/15***			
186		13/14**	12/16*****			
187	14/19**	13/15******				
189		13/17***				
190		13/18**				
191		13/19**				
192						
193	15/13**					
195	15/15**					
196	15/16*					
197	15/17**	14/13**				
198		14/14**	13/15**			
199	15/19**	14/15***	13/16*			
200		14/16**	13/17*******			
201	15/21*	14/17***				
202		14/18**				
203						
204						
205						
211		15/15**				

m/e	CH	CHO	CHO2	CHO3	CHO4	CHO5
213	16/21*					
215	15/17***					
217	15/19***					
219		14/17**				
225		14/19***				
227	16/17*					
229	16/19***					
231	16/21**	15/17*				
232		15/19*				
239		15/20*				
241	18/25*					
243	17/19***					
244	17/21**	16/20*				
245	17/23**	16/21*****				
251	19/23*					
253	18/21***					
254	18/22**					
255	18/23**	17/23**				
257	18/25*	17/24****				
259						
260						
267	19/23**	18/23**				
268	19/24**	18/24*				
269	19/25*****	18/25**				
271						
272						
273	20/27***					
283						
284		19/24**				
285		19/25*****				
286		19/26***				
287		19/27******				
293	21/25**					
296	21/28*					
297	21/29*					
298	21/30**					
299		20/27*				
311		21/27*****				
314		21/30***				
315		21/31****				
316		21/32*****				
326		22/30******				
329				21/29**		
342				22/30**		
344				22/32********	29/40**	
452						
470						29/42**

Figure 14. Element map of antheridiol (15) abridged by deleting all data below m/e 150

C, H and O in antheridiol and accepting an error limit of ± 0.0005 a.m.u. between the measured and calculated masses of the ions, it was concluded that the compound contained those elements only. The high resolution mass spectral data then showed the elemental composition of antheridiol to be $C_{29}H_{42}O_5$ and is displayed in the form of an element map (Biemann, Bommer, and Desiderio[87]) in Figure 14. The element map of antheridiol was abridged to reduce its length without, in this case, deleting essential data by starting the map at m/e 150.

The hypothetical structure of antheridiol which was arrived at from an examination of the element map (Figure 14) is shown in structure (**16**) with an indication of the origin of some of the fragment ions to be discussed. Note that the CH, CHO and CHO_2 columns are well filled and terminate with

$$-H = C_{22}H_{32}O_3$$

(**16**)

2 Double bonds
2 Oxygen functions

$C_{19}H_{23}$, $C_{21}H_{30}O$ and $C_{22}H_{30}O_2$, and the species $C_{19}H_{25}O$ and $C_{19}H_{27}O_2$ are particularly abundant. This distribution of the fragment ions suggests a steroid having two oxygen functions in the tetracyclic ring system. If the assumption is made that the abundant $C_{19}H_{27}O_2$ ion comes from the simple cleavage of the C-17, C-20 bond in structure (**16**), then the steroid nucleus has to contain two double bonds, since a saturated ring system would have yielded the fragment $C_{19}H_{31}O_2$ instead. Ten carbons, fifteen hydrogens and three oxygens then remain to be accounted for—the difference between $C_{29}H_{42}O_5$ and $C_{19}H_{27}O_2$—and must form a side chain attached at C-17. The scarcity of $C_{20}O$ and $C_{20}O_2$ species, C_{19} and C_{21} species with one or two oxygens being much more abundant, further suggests the presence of a methyl side chain at C-20. The elemental composition $C_{22}H_{32}O_3$ of the most intense peak in the spectrum indicated C-22 as the point of attachment of the third oxygen atom. This tentative assignment was reinforced by the scarcity of CHO_3 ions with less than 22 carbon atoms. $C_{21}H_{29}O_3$ may be explained by the loss of a methyl group—the elements CH_3—from $C_{22}H_{32}O_3$, and $C_8H_{11}O_3$, by simple cleavage of the C-20, C-22 bond with charge retention at C-22. Caution should always be exercised, especially at the early stages of an investigation, when interpreting low abundance, relatively isolated ions such as $C_8H_{11}O_3$ since their presence may well be due to an impurity. Nevertheless, if the origin of $C_8H_{11}O_3$ and $C_{22}H_{32}O_3$ is as indicated in structure (16), then the oxygen atom at C-22 must be present in the form of a hydroxyl group. One further deduction may be made from the element map of antheridiol (Figure 14): the side chain attached at C_{22} must contain the elements $C_7H_9O_2$ which require that there be three double bonds or rings in that side chain.

The work on antheridiol was continued to refine and complete hypothetical structure (16). A total of four derivatives were prepared, an acetate, a hydrogenation product, an ethylene ketal of the hydrogenation product, and a dehydration product. The element map of each of these was used with assistance from the infrared and ultraviolet data on antheridiol and the above derivatives to finally arrive at structure (15) which was in agreement with all available data including the n.m.r. data eventually obtained on antheridiol itself. The position of the angular methyl groups and the stereochemistry shown in structure (15) were based solely on biogenetic analogies.

A mass spectrometric model was then prepared to secure further evidence in favour of structure (15). The C_{22} steroid aldehyde (17) was prepared by McMorris because its structure was expected to be the same as the structure

(17)

of the most abundant ion, $C_{22}H_{32}O_3$ at m/e 344, in the element map of antheridiol (Figure 14). The high abundance as well as the hydrogen content of the $C_{22}H_{32}O_3$ ion suggested that it be formed in a facile process involving the rearrangement of one hydrogen atom away from the C_{22} unit, the double bond functioning as the acceptor as shown in Chart 2. The mass spectra of the C_{22} steroid aldehyde (17) and antheridiol (15) were found to be highly similar, both in the abundance and in the elemental composition of the ions. The low resolution mass spectra of these two compounds, (15) and (17), are shown in Figure 15 to illustrate their similarity. Comparison of the two spectra should be made between m/e 100 and 350 since there should be obvious dissimilarities outside these limits because of the known structural differences between the two compounds. Each group of ions labelled with a

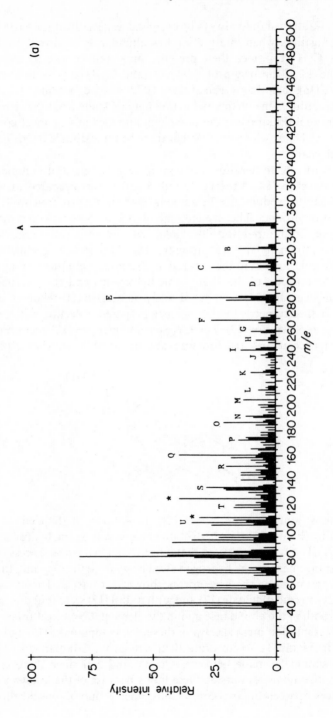

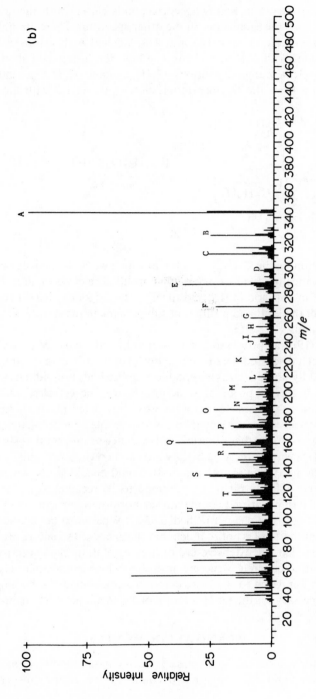

Figure 15. Mass spectra of (a) antheridiol (**15**) and (b) the model steroid aldehyde (**17**). The labelling of peaks with letters and asterisks is explained in the text

Experimental Methods in Biophysical Chemistry

letter in one spectrum in Figure 15 corresponds closely with the group of ions labelled with the same letter in the other spectrum. There are only two exceptions, the peaks at m/e 126 and 111 labelled with asterisks in the spectrum of antheridiol (Figure 15(a)). The high resolution data show these to have the elemental compositions $C_7H_{10}O_2$ and $C_6H_7O_2$, and they probably arise from the fragmentation shown in Chart 2 with, however,

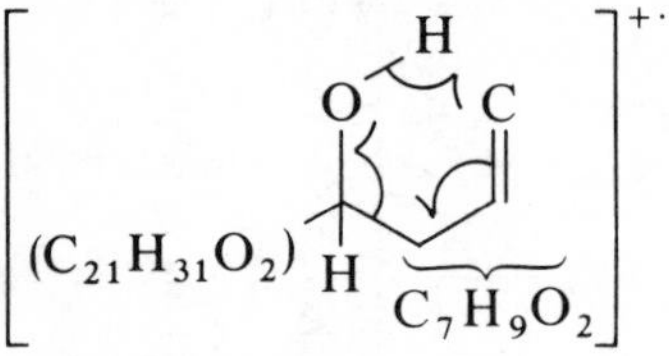

$$\rightarrow [(C_{21}H_{31}O_2)CHO]^{+\cdot} + C_7H_{10}O_2$$
$$m/e\ 344$$

Antheridiol (**15**)

Chart 2

charge retention on the $C_7H_{10}O_2$ fragment which would then give rise to the $C_6H_7O_2$ ion through the loss of a methyl group. The close similarity of the mass spectra of antheridiol (**15**) and the C_{22} steroid aldehyde (**17**) confirms the identity of the kind and position of substituents in the steroid nucleus of both compounds.

Structure (**15**) was proposed for antheridiol on the basis of the evidence which has been discussed herein. While this proposed structure did not rest on evidence fully sanctioned as acceptable by precedents nor did the evidence constitute a rigorous proof of structure, it was nevertheless published (Arsenault and coworkers[86]) because it was anticipated that structure (**15**) could be synthesized more easily than it was possible to accumulate enough antheridiol to secure further experimental evidence to support that structure. Compound (**15**) was synthesized by Edwards and coworkers[88] and it proved to be identical to authentic antheridiol. It should be added that antheridiol is the first and only steroidal sex hormone to be recognized in the plant kingdom and it differs from mammalian sex hormones particularly in that it has a much longer side chain attached at C-17. What must be retained from the work on antheridiol is not so much that the proposed structure proved to be correct but that the extensive use of high resolution mass spectrometry quickly led to a concrete structure proposal which could then easily be verified. This approach to structure determination should not be neglected when it is necessary for the rapid solution of a difficult bio-organic problem.

ACKNOWLEDGMENT

The author is indebted to Professor K. Biemann for his support, his encouragement, and for constructive comments regarding this chapter.

The quality of the manuscript owes much to the expert and devoted assistance of Mrs. Anne G. Arsenault.

REFERENCES

1. A. J. Dempster, *Phys. Rev.*, **11**, 316 (1918)
2. K. Biemann and P. V. Fennessey, *Chimia*, **21**, 226 (1967)
3. R. A. Hites and K. Biemann, In *Advances in Mass Spectrometry*, Vol. 4 (Ed. E. Kendrick), The Institute of Petroleum, London, 1968, pp. 37–54
4. R. A. Hites and K. Biemann, *Anal. Chem.*, **42**, 855 (1970)
5. R. L. Levy, In *Chromatographic Reviews*, Vol. 8 (Ed. M. Lederer), Elsevier Publishing Company, Amsterdam, 1966, pp. 48–89
6. J. H. Beynon, *Mass Spectrometry and its Applications to Organic Chemistry*, Elsevier Publishing Company, Amsterdam, 1960
7. K. Biemann, *Mass Spectrometry–Organic Chemical Applications*, McGraw-Hill, New York, 1962
8. H. Budzikiewicz, *Org. Mass Spectrom.*, **2**, 249 (1969)
9. F. W. McLafferty, *Interpretation of Mass Spectra: an Introduction*, W. A. Benjamin, Inc., New York, 1966
10. J. H. Beynon, In *Advances in Mass Spectrometry*, Vol. 4 (Ed. E. Kendrick), The Institute of Petroleum, London, 1968, pp. 123–138
11. J. Roboz, *Introduction to Mass Spectrometry–Instrumentation and Techniques*, Wiley–Interscience, New York–London, 1968
12. C. Brunnée and H. Voshage, *Massenspektrometrie*, Verlag Karl Thiemig KG, München, 1964
13. R. W. Kiser, *Introduction to Mass Spectrometry and its Applications*, Prentice–Hall, Englewood Cliffs, 1965
14. C. A. McDowell (Ed.), *Mass Spectrometry*, McGraw-Hill, New York, 1963
15. C. A. Evans Jr. and G. H. Morrison, *Anal. Chem.*, **40**, 869 (1968)
16. M. G. Inghram and R. Gomer, *J. Chem. Phys.*, **22**, 1279 (1954)
17. H. D. Beckey, *Advances in Mass Spectrometry*, Vol. 2 (Ed. R. M. Elliott), Pergamon Press, London, 1963, pp. 1–24
18. H. D. Beckey, H. Knöppel, G. Metzinger and P. Schulze, In *Advances in Mass Spectrometry*, Vol. 3 (Ed. W. L. Mead), The Institute of Petroleum, London, 1966, pp. 35–67
19. H. D. Beckey, H. Heising, H. Hey, and H. G. Metzinger, In *Advances in Mass Spectrometry*, Vol. 4 (Ed. E. Kendrick), The Institute of Petroleum, London, 1968, pp. 817–831
20. H. D. Beckey, *Z. Instrumentenkunde*, **71**, 51 (1963)
21. A. J. B. Robertson, B. W. Viney and M. Warrington, *Brit. J. Appl. Phys.*, **14**, 278 (1963)
22. A. J. B. Robertson and B. W. Viney, In *Advances in Mass Spectrometry*, Vol. 3 (Ed. W. L. Mead), The Institute of Petroleum, London, 1966, pp. 23–34
23. E. M. Chait, T. W. Shannon, W. O. Perry, G. E. Van Lear and F. W. McLafferty, *Int. J. Mass Spectrom. Ion Phys.*, **2**, 141 (1969)
24. P. Schulze, B. R. Simoneit and A. L. Burlingame, *Int. J. Mass Spectrom. Ion Phys.*, **2**, 183 (1969)
25. P. Brown, G. R. Pettit and R. K. Robins, *Org. Mass Spectrom.*, **2**, 521 (1969)
26. J. N. Damico, R. P. Barron and J. A. Sphon, *Int. J. Mass Spectrom. Ion Phys.*, **2**, 161 (1969)

27. H. Krone and H. D. Beckey, *Org. Mass Spectrom.*, **2**, 427 (1969)
28. H. D. Beckey, *Int. J. Mass Spectrom. Ion Phys.*, **2**, 500 (1969)
29. H. D. Beckey, E. Hilt, A. Mass, M. D. Migahed and E. Ochterbeck, *Int. J. Mass Spectrom. Ion Phys.*, **3**, 161 (1969)
30. J. N. Damico and R. P. Barron, *Anal. Chem.*, **43**, 17 (1971)
31. M. S. B. Munson and F. H. Field, *J. Am. Chem. Soc.*, **88**, 2621 (1966)
32. F. H. Field, *Accounts of Chemical Research*, **1**, 42 (1968)
33. F. H. Field, In *Advances in Mass Spectrometry*, Vol. 4 (Ed. E. Kendrick), The Institute of Petroleum, London, 1968, pp. 645–665
34. F. H. Field, *J. Am. Chem. Soc.*, **92**, 2672 (1970)
35. F. H. Field, *J. Am. Chem. Soc.*, **83**, 1523 (1961)
36. F. H. Field and M. S. B. Munson, *J. Am. Chem. Soc.*, **87**, 3289 (1965)
37. G. P. Arsenault, In *Seventeenth Annual Conference on Mass Spectrometry and Allied Topics*, Dallas, Texas, May 1969, p. 372
38. E. Gelpi and J. Oró, *Anal. Chem.*, **39**, 389 (1967)
39. H. M. Fales, G. W. A. Milne and M. Vestal, In *Seventeenth Annual Conference on Mass Spectrometry and Allied Topics*, Dallas, Texas, May 1969, p. 275.
40. G. P. Arsenault, J. J. Dolhun and K. Biemann, *Chem. Communications*, 1542 (1970)
41. D. M. Schoengold and B. Munson, *Anal. Chem.*, **42**, 1811 (1970)
42. G. P. Arsenault, J. R. Althaus and P. V. Divekar, *Chem. Communications*, 1414 (1969)
43. H. M. Fales, G. W. A. Milne and M. L. Vestal, *J. Am. Chem. Soc.*, **91**, 3682 (1969)
44. W. H. McFadden, *Separ. Sci.*, **1**, 723 (1966)
45. F. A. J. M. Leemans and J. A. McCloskey, *J. Amer. Oil Chem. Soc.*, **44**, 11 (1967)
46. J. T. Watson, In *Ancillary Techniques of Gas Chromatography* (Eds. L. S. Ettre and W. H. McFadden), Wiley–Interscience, New York–London, pp. 145–225 (1969)
47. C. M. Drew, J. R. McNesby, S. R. Smith and A. S. Gordon, *Anal. Chem.*, **28**, 979 (1956)
48. R. S. Gohlke, *Anal. Chem.*, **31**, 535 (1959)
49. J. C. Holmes and F. A. Morrell, *Appl. Spectroscopy*, **11**, 86 (1957)
50. W. H. McFadden, R. Teranishi, D. R. Black and J. C. Day, *J. Food. Sci.*, **28**, 316 (1963)
51. J. H. Beynon, R. A. Saunders and A. E. Williams, *J. Sci. Instr.*, **36**, 375 (1959)
52. J. T. Watson and K. Biemann, *Anal. Chem.*, **36**, 1135 (1964)
53. R. Ryhage, *Anal. Chem.*, **36**, 759 (1964)
54. E. Stenhagen, *Z. analyt. Chem.*, **205**, 109 (1964)
55. P. M. Llewellyn and D. Littlejohn, Paper presented at the *Pittsburgh Conference on Analytical Chemistry and Applied Spectroscopy* (1966) and quoted extensively by Watson[46]
56. J. H. Beynon, *Nature*, **174**, 735 (1954)
57. J. Mattauch and R. Herzog, *Z. Physik*, **89**, 786 (1934)
58. E. G. Johnson and A. O. Nier, *Phys. Rev.*, **91**, 10 (1953)
59. K. S. Quisenberry, T. T. Scolman and A. O. Nier, *Phys. Rev.*, **102**, 1071 (1956)
60. K. Biemann, In *Topics in Organic Mass Spectrometry* (Ed. A. L. Burlingame), Wiley–Interscience, New York–London, pp. 185–221 (1970)
61. W. J. McMurray, B. N. Greene and S. R. Lipsky, *Anal. Chem.*, **38**, 1194 (1966)
62. A. L. Burlingame, D. H. Smith and R. W. Olsen, *Anal. Chem.*, **40**, 13 (1968)
63. H. Budzikiewicz, C. Djerassi and D. H. Williams, *Structure Elucidation of Natural Products by Mass Spectrometry. Volume I: Alkaloids*, Holden-Day, Inc., San Francisco, 1964

64. H. Budzikiewicz, C. Djerassi and D. H. Williams, *Structure Elucidation of Natural Products by Mass Spectrometry. Volume II: Steroids, Terpenoids, Sugars, and Miscellaneous Classes*, Holden–Day, Inc., San Francisco, 1964
65. G. R. Waller (Ed.), *Biochemical Applications of Mass Spectrometry*, Wiley–Interscience, New York–London, 1972
66. G. P. Arsenault, *Tetrahedron*, **24**, 4745 (1968)
67. K. Biemann, F. Gapp and J. Seibl, *J. Am. Chem. Soc.*, **81**, 2274 (1959)
68. M. Barber, P. Jolles, E. Vilkas and E. Lederer, *Biochem. Biophys. Res. Commun.*, **18**, 469 (1965)
69. L. C. Vining and W. A. Taber, *Can. J. Chem.*, **40**, 1579 (1962)
70. K. Biemann, C. Cone, B. R. Webster and G. P. Arsenault, *J. Am. Chem. Soc.*, **88**, 5598 (1966)
71. W. A. Wolstenholme and L. C. Vining, *Tetrahedron Letters*, 2785 (1966)
72. M. Barber, P. Powers, M. J. Wallington and W. A. Wolstenholme, *Nature*, **212**, 784 (1966)
73. M. Senn, R. Venkataraghavan and F. W. McLafferty, *J. Am. Chem. Soc.*, **88**, 5593 (1966)
74. B. C. Das, S. D. Gero and E. Lederer, *Biochem. Biophys. Res. Commun.*, **29**, 211 (1967)
75. K. L. Agarwal, G. W. Kenner and R. C. Sheppard, *J. Am. Chem. Soc.*, **91**, 3096 (1969)
76. K. Biemann and C. Cone, presented at the *115th National Meeting of the American Chemical Society* held at San Francisco, California, March 31–April 5, 1968, A-001
77. K. Biemann, J. R. Althaus, J. A. Kelley, R. A. Salomone and S. W. Tam, presented at the *Biochemical Conference on Protein Structure and Function* held at Ste. Marguerite, Quebec, March 2–7, 1969
78. K. Biemann, In *Biochemical Applications of Mass Spectrometry* (Ed. G. R. Waller), Wiley–Interscience, New York–London, pp. 405–428 (1972)
79. C. E. Hignite, In *Biochemical Applications of Mass Spectrometry* (Ed. G. R. Waller), Wiley–Interscience, New York–London, pp. 429–447 (1972)
80. C. E. Hignite, Ph.D. Thesis, Massachusetts Institute of Technology, Cambridge, U.S.A., 1969
81. D. F. Hunt, C. E. Hignite and K. Biemann, *Biochem. Biophys. Res. Commun.*, **33**, 378 (1968)
82. C. Hignite and K. Biemann, *Org. Mass Spectrom.*, **2**, 1215 (1969)
83. J. J. Dolhun and J. L. Wiebers, *J. Am. Chem. Soc.*, **91**, 7755 (1969)
84. T. C. McMorris and A. W. Barksdale, *Nature*, **215**, 320 (1967)
85. J. R. Raper, *Am. J. Bot.*, **26**, 639 (1939)
86. G. P. Arsenault, K. Biemann, A. W. Barksdale and T. C. McMorris, *J. Am. Chem. Soc.*, **90**, 5635 (1968)
87. K. Biemann, P. Bommer and D. M. Desiderio, *Tetrahedron Letters*, 1725 (1964)
88. J. A. Edwards, J. S. Mills, J. Sundeen and J. H. Fried, *J. Am. Chem. Soc.*, **91**, 1248 (1969)

CHAPTER 2

Use of activation analysis in molecular biology

H. Altmann

Institute of Biology,
Reactor Centre, Seibersdorf,
Austria

I. INTRODUCTION

Activation analysis is one of the most rapidly developing analytical techniques but, until recently, its application in molecular biology—one of the fastest growing branches of science—was rather restricted. This review will show that lately there has been a growing interest in using activation techniques in this field. It is difficult to determine where the term 'pure' molecular biology should be limited. The origin of life was closely connected with the synthesis of precursors of macromolecules, under different physical conditions, from H_2, NH_3, H_2O and CH_4; often inorganic compounds acted as catalysts. At a later step—the joining of precursors to form macromolecules and subcellular structures—trace metals also played an important role. With the development of more complicated systems it was found that metals combined with proteins had a higher catalytic activity, and now many enzymes contain

49

trace metals as a cofactor and many macromolecules use metals for stabilizing structures. The following chapters will deal mainly with the determination of trace metals in macromolecules, the detection of halogens, ^{18}O and ^{31}P, the use of enriched stable isotopes for labelling molecules and their detection in metabolized form by activation analysis. Most biological molecules do not contain atoms which are easily detectable by activation analysis. Therefore, derivatives of metabolites have been prepared containing atoms suitable for activation. The great difficulty with the use of activation analysis in molecular biology is that in many cases contamination is not totally excluded. The major problems of contamination and loss occur prior to the activation process. It is possible to determine traces of contaminants which could not be detected by any other known method, because activation differs from most other analytical methods in that it is based on the properties of nuclei and not on the behaviour of the outer electrons. The bombarding particles convert some of the atoms present in the material into isotopes of either the same or different elements.

The rate of increase of the literature on activation analysis is very high. Bowen calculated a doubling time of about 3·3 years.[1] In the section on 'Application', a condensed summary in the different fields of molecular biology is given. It must be understood, however, that it was not possible to treat any investigation in detail.

II. THEORY OF ACTIVATION ANALYSIS

If material is irradiated, some atoms of this material will interact with the bombarding particles or rays. This interaction may result in elastic scattering, inelastic scattering, excited states and conversion into different isotopes of the same or different elements, depending on the nature of the reactants. A great number of the isotopes produced are radioactive. Immediately after nuclear interaction a rapid decay occurs. This radiation can also be measured if the daughter-product is not radioactive. Radioactivity from daughter-products is measurable over a much longer time and is a purely random process. The rate at which these isotopes disintegrate follows the following equation:

$$\frac{dN}{dt} = -\lambda N$$

N is the number of atoms of the radioisotope at any time t and λ is a radioactive disintegration constant. The half-life of one radioisotope is a function of λ and is the time in which half of the radioactive nuclei will decay. The emitted radiation in the cases we are concerned with may be a $+\beta$ or $-\beta$ particle, or a γ-ray. The rate of nuclear reaction, R, is given by:

$$R = \Phi\sigma f$$

where Φ is the particle flux, σ is the reaction cross-section and f is the fractional abundance of the nuclide.

When using conventional neutron activation analysis, good results are obtained when the sample is irradiated for a period equal to the half-life of the nuclide which is to be determined. The following formula can be used to calculate the approximate activity induced in any nuclide:

$$A_t = 6 \times 10^{-3} f \sigma \Phi \frac{W}{M} (1 - e^{-0.693t/T}) \cdot e^{-0.693d/T}$$

A_t = activity at time t (dis/sec)
W = sample weight
T = half life of the nuclide
d = time between activation and counting
M = atomic weight
Φ = neutron flux (neutrons per cm^2 and sec)
σ = cross section (barn)
f = per cent abundance of the isotope

But the most usual method is to irradiate the sample simultaneously with known amounts of standard material. Almost all activation analyses performed today employ neutrons as the bombarding particles and probably 95 per cent of all these neutron reactions involve the use of thermal neutrons.[2]

III. IRRADIATION FACILITIES

The nuclear reactor is the most used neutron source for neutron activation analysis. Most research reactors have a maximum flux of about 10^{12} to 10^{13} n.cm^{-2} sec^{-1}. There is normally a spectrum range in the energy of the neutrons between 0.02 eV (thermal n) and 10 MeV (fast n). There are devices available which can sort all reactor neutrons, but normally thermal neutrons are used. Absolute activation analysis can be done only for a few elements whose cross-sections and half-life are well known, but most analyses are carried out by using standards. From all known neutron sources, nuclear reactors produce the highest flux of neutrons. These neutrons are a product of nuclear fission. The power of a reactor depends on how many neutrons are produced per unit time; 3×10^{10} fissions per second produce a watt of energy. A 5 Megawatt reactor has therefore 1.5×10^{17} fissions per second. With the irradiation facilities for neutron activation of such a reactor a neutron flux between 1×10^{12} and 1×10^{14} n.cm^{-2} sec^{-1} can be obtained. With special reactors there exists also the possibility of flux pulses. The pulse-action power level of a 1000 MW peak corresponds to a peak total neutron flux of 3×10^{16} n.cm^{-2} sec^{-1}. For neutron activation analysis also portable neutron sources are available. But all of these produce low fluxes. The sources most frequently used are deuterium–beryllium or antimony–beryllium combinations. Sources based on (αn) reaction instead of (γn), like a Ra–Be or Po–Be source have slightly higher fluxes. ^{252}Californium undergoes spontaneous fission, and a 10 mg source will provide as many as 3×10^{10} neutrons per second.

Accelerators can be used for fast or thermal neutron activation analysis. For thermal neutron activation analysis the target area must be surrounded with a water–paraffin moderator, but interferences from fast neutron reaction must be always considered. Charged particles like protons or deuterons of energies up to 6 MeV may be obtained from Van de Graaff or Cockcroft–Walton accelerators. The Van de Graaff neutron generator normally produces in the (d, t) reaction neutrons of approximately 14 MeV. In order to enable short-lived radioactivities to take advantage of the accessibility of these machines, a pneumatic tube system is normally employed.

IV. HANDLING OF SAMPLES IN NEUTRON ACTIVATION TECHNIQUE

Before activation occurs, handling of samples and chemical steps should be limited to a minimum. Plastic beakers and instruments and ultrapure reagents must be used. Standards are best prepared by weight and irradiated simultaneously with the sample. Canning of samples for activation analysis can be done with polyethylene, polypropylene, silica or aluminium. Following removal from the reactor, non-destructive methods or radiochemical separation methods may be employed before counting the induced radioactivity. Lithium-drifted germanium detectors for high resolution γ-ray spectroscopy reduce the need for wet chemistry in neutron activation analysis.[3] The coupling of neutron activation analysis and gas–liquid chromatography represents a dynamic analytical system, especially useful for short-lived isotopes.[4] Kusaka and Meinke have reviewed rapid radiochemical separation techniques,[5] and automatic group separation systems for the simultaneous determination of a great number of elements in biological material were worked out, to reduce the effort required when many samples have to be processed.[6,7,8] The radiochemistry of most elements is summarized in a series of monographs issued by the United States' National Academy of Science (NAS-NS). Precipitation is still widely used, combined with organic solvent extractions.[9] Green[10] has reviewed the use of liquid ion exchangers, and Eisner and coworkers published the application of semipermeable ion exchange membranes to trace analysis of metal ions by electrochemical and neutron activation techniques.[11] Radiochemical separation steps are reviewed also by Lenihan[12] and Bowen.[13]

V. RADIOACTIVE MEASUREMENTS

For measuring γ radioactivity from sample and standards, the equipment most widely used is a sodium iodide crystal or a Ge(Li) detector, combined with a multichannel analyser. The shape of a γ-spectrum depends, beside of the activated elements, on the crystal or semiconductor, multiplier and electronic equipment. In the energy range from a few electron volts to about 5 MeV there are three basic processes by which γ-photons may interact with matter: the photoelectric effect, Compton absorption and pair production. In practice a γ-spectrum shows the following peaks and shoulders: the photoelectric peak, bremsstrahlung from X-radiation produced in the phosphor, a backscatter peak from photoelectric absorption of quanta scattered through 180°, a Compton edge and escape peaks from pair production. Crouthamel[14] has compiled in his book the γ-spectra from most elements, and are very useful in identifying unknown species. Coincidence and anticoincidence counting combined with computer methods of analysis of γ-spectra have provided a rapid, economical routine method of activation analysis. A system was described which is capable of analysing and assaying samples for eight or more elements simultaneously by entirely instrumental means. The γ-ray spectra of the reactor-activated samples are collected at various times after irradiation with a Ge(Li) detector, and the computer programme processes the time-dependent spectra and identifies the elemental constituents by the energy and half-life of the corresponding peaks in the spectra.[15] In many cases it is not necessary to separate all metals by chemical methods, but separation of ^{24}Na gives much better results. For the removal of radiosodium from biological samples, a simple and rapid radiochemical procedure has been described. The method is based on the precipitation of carrier and active Na as NaCl from a solution of the sample by a n-butanol–HCl mixture. This procedure not only removes active sodium but also reduces the chlorine activity.[16] The current trend of activation analysis goes, however, towards the method of pure instrumental analysis.

VI. APPLICATION

A. Rapid γ-Rays from Radiative Capture of Thermal Neutrons

Rapid radiation technique has received very little attention in molecular biology. However, this method permits the radiation emitted during the decay of excited nuclear states to be counted, thus avoiding the necessity of choosing a nuclear reaction producing a radioactive isotope as a basis for the determination.[17] Rapid γ-spectra are often complex and difficult to evaluate.[18,19] The high cross-section for thermal neutrons (0·025 eV) made it possible to detect elements with low neutron fluxes and minimum destruction of molecules. For example, Cd has a cross-section of 2537 barn, Co 36·3,

W 19·1, Mn 13, V 5, Ni 4·6, Cu 3·8, Cr 3·1 and Fe 2·6.[20] But also isotopes produced after neutron irradiation, which are pure β-emitters, such as phosphorus (0·19 barn) or lead (0·17 barn), have reasonable cross-sections for prompt γ-reactions. The sensitivity of boron determination by the $^{10}B(n\alpha)^7Li$ reaction can be enhanced, using the prompt γ-radiation of ^7Li with 0·478 MeV γ-rays.[21] Iron, an important element in the structure and catalyst in many macromolecules, consists of 4 stable isotopes, ^{54}Fe(5·8 per cent), ^{56}Fe(91·6 per cent), ^{57}Fe(2·2 per cent) and ^{58}Fe(0·33 per cent). Only from ^{54}Fe and ^{58}Fe radioactive isotopes are produced by (nγ) reaction. ^{55}Fe is only a β-emitter and the abundance of ^{58}Fe is very low. But the prompt γ-rays from the reaction $^{56}Fe(n\gamma)^{57}Fe$ and $^{57}Fe(n\gamma)^{58}Fe$ can also be used for the determination of iron. Also γ-photoactivation was tried for the detection of elements because below 4 MeV only γ (γ . γ) reactions occur.[22] Measurement of rapid γ-rays obtained after neutron irradiation with a neutron generator is even more complicated, but nitrogen and carbon in bulk material could be analysed.[23] Prompt γ-rays obtained after irradiation with charged particles is more promising for measuring light elements in molecular biology.[24,25]

B. Stable Isotopes

Protein metabolism studies are normally done by tracing proteins with radioactive isotopes. In some cases, however, especially if the patients are infants or pregnant women, even very small radiation doses should not be applied. For such investigations stable bromine, ^{79}Br or ^{81}Br can be used.[26] Human albumin and γ-globulin labelled in this way appeared to be similar to the turnover of their radiobrominated and their radioiodinated analogues. Other stable isotopes suited for activation and which are used especially in medicine are ^{46}Ca, ^{58}Fe and ^{129}I.[27,28,29] Deuterium incorporated in molecules can be activated by γ-rays and the neutrons obtained by the $^2H(\gamma n)^1H$ reaction can be measured.[30,31]

C. Derivative Activation Analysis

Molecular biology is dealing mainly with molecules containing atoms which are not easy to activate. Organic chemistry described many coupling methods introducing groups with elements well suited for activation. Therefore Steim and Benson described the method of derivative activation chromatography.[32] Halogenation of unsaturated compounds like olefines has been described long ago, but the quantitative determination of very small amounts of Br, I or F in these compounds was difficult. After a chromatographic or electrophoretic separation step, the halogenated molecules can be detected quantitatively by measuring the γ-spectra after neutron irradiation.

SH-groups in very small protein preparations can be measured by coupling SH with *p*-hydroxymercurybenzoate, and activation analysis is a very sensitive method for detection of Hg.[33,34] Mercuriated fatty acids were obtained by treatment of oleic, linoleic and linolenic acid with mercuric acetate in methanol. After paper chromatography and autoradiography, well-defined spots could be detected. For amino acid determination, the *N*-brosylderivatives are produced, the *p*-bromophenacylesters are formed from carboxylic acids and the *p*-bromophenylhydrazones from sugars and keto acids.[32] Besides paper chromatography, thin-layer chromatography was used for the separation of Br-derivatives. A 5 mg sample of silica gel GF had non detectable levels of bromide. For drug assays, bromine-containing derivatives of aspirin and salicylic acid were prepared, but spectrofluorometry or gas chromatography are more sensitive than neutron activation because of background bromine and interfering substances.[35] The estimation of separated serum proteins on polyacrylamide gels after staining with bromphenolblue and activation of the bromine is not quantitative, because this dye is bound differently to different proteins.[36]

A new method of radioactivation analysis is based on the quantitative isotope dilution principle. After the addition of a known amount of radioactive isotopes to the sample, the element in question is determined by measuring the change of specific activity.[37,38] A combination of this method with derivative activation analysis seems to be promising.

D. Nucleic Acids

Activation analysis can be used in nucleic acid research for the determination of phosphorus, trace metal content, ^{18}O determination and the measurement of incorporated halogenated base analogues. Hybridization techniques are used very often for measuring DNA homologues.[39] The complementarity between RNA and DNA can also be detected by the same technique.[40] But for this method, one of the different DNAs or the RNA should be radioactively labelled. In some cases it is difficult to obtain nucleic acids extracted from cells which were grown in media containing radioactivity labelled precursors. DNA, exposed to thermal neutrons was partly degraded, but could be used for the hybridization procedure.[41] The best activation condition used in this experiment was the exposure of dry DNA to 2×10^{13} n.cm^{-2} sec^{-1} for 4 hours. In special cases it is also possible to label *t*-RNA with inactive F-uracil, or DNA with Br-uracil, to do the hybridization procedure according to Gillespie and Spiegelman,[42] and to measure the halogen after neutron activation on the pre-washed nitrocellulose filters. *t*-RNA labelled with F-uracil loses its amino acid specificity only to a very low extent. As the lower limit of detection is dependent on the magnitude of the background activity contributed by the filters, or more generally, by chromatographic paper, it is important to select paper with low content of elements, especially those which should be determined and elements which possess high activation cross-sections.[43,44] A cadmium shield during neutron activation results in an enhancement of the activation of those elements with a significant resonance cross-section structure, compared to those elements without. If the applied dose is higher than 2×10^{16} neutrons per cm^2, the paper loses its shape, Autoradiograms show that the induced activity is not distributed homogeneously over the paper.[45] It is therefore advisable not to take too small paper strips and to avoid contamination by ^{32}P derived from ^{35}Cl by the $^{35}Cl(n\alpha)^{32}P$ reaction, to wash out very thoroughly all chemicals containing chlorine, or better, if possible, to use chlorine- and sodium-free materials.

The phosphorus content of tissues and sections has been determined by activation technique together with autoradiography.[46,47] Nucleic acids in histological preparations are determined by the same method after treatment with specific enzymes.[48] Activation analysis was also used for quantitative determination of RNA and DNA in small samples of protozoa.[49] The extracted DNA and RNA were dried in polyethylene tubes and exposed to a thermal neutron flux of 4×10^{12} n.cm^{-2} sec^{-1} for one hour. 20 days after neutron activation ^{32}P was counted with a lead-shielded low-background Geiger–Mueller counter.

Especially in virus research it appeared necessary to determine base ratios of very small RNA samples. Incorporation of $^{32}PO_4$ is often limited on slow incorporation rates and reduction of non-labelled phosphate in the medium. Activation techniques have been applied to samples containing only about one-tenth of the amount of RNA required for an optical determination of base ratios.[50] After paper electrophoresis, the regions of the paper containing the nucleotide spots were cut out and irradiated at a neutron flux of 2×10^{12} n.cm^{-2} sec^{-1} for 5 days. After 7 days' delay, the activity of ^{32}P was measured. Nucleic acids isolated from various biological sources contain significant amounts of metal ions, some of which are determined by neutron activation analysis.[51,52,53] The nucleic acid samples were irradiated together with standards at an integrated neutron flux of 1×10^{16} or 1×10^{18} n.cm^2 sec^{-1}. Beside the metal ion determination, ^{32}P was counted 14 days after irradiation.

Tobacco-mosaic-virus-RNA determined by this method showed a much higher copper content compared to RNA from tobacco leaves.[54,55] Meristem tissue of plants is normally not infected by viruses and nucleic acids extracted from young leaves showed higher Cu value compared to nucleic acids from old leaves.[56] A good relationship was found between the copper content of RNA and its degradation by RNAase.[57,58] Special attention should be given to the isolation procedures, because very often artefacts can be measured.[59,60] A removal of trace elements from nucleic acids of yeast could be found after ionizing radiation *in vivo*[61] and *in vitro*.[62,63] Incorporation of Br-uracil in DNA was often used for radioenhancing studies. Bromine content easily may be detected by neutron activation.[64,65] This method can also be used for fluorine determination in RNA from F-uracil which can be incorporated in RNA. It is an often used drug in cancer treatment. The use of ^{32}P labelled nucleotides as tracers, especially in membrane penetration studies, has considerable limitations. In some cases, oxygen is a better tracer than phosphorus for the study of biochemical processes inside the cell. Radio-active oxygen isotopes have very short half-life times and only ^{15}O with a half-life of 2 minutes is widely used as a tracer in physiological and medical studies.[66] 29·4 second ^{19}O is formed by neutron irradiation of natural oxygen which emits 0·2 MeV γ-rays. The cross-section of ^{18}O(nγ)^{19}O and the abundance of ^{18}O are both low, therefore activation analysis using ^{19}O is only applicable to the determination of macro-amounts of oxygen.[67] The abundance of the stable isotope ^{18}O is only 0·2 per cent of the natural element, but can be enriched in water up to 80 per cent. Beside the assay by mass spectrometry, activation analysis based on the reaction ^{18}O(pn)^{18}F can be used for measuring the incorporation-rate of oxygen from water into various phosphorus-containing compounds.[68] Stable oxygen isotopes were also used for the study of chemical mechanisms of sonic, acid, alkaline and enzymatic degradation of DNA.[69] Investigations have also been carried out on ^{18}O labelling of DNA during synthesis and stability of the label during replication.[70] The labelling of RNA from *E. coli* grown in glucose-1-^{18}O was reported by Nicholson.[71] In smaller amounts (maximum 100 μg O$_2$) oxygen can be determined by thermal activation, but only by a complicated procedure in a secondary reaction technique. The O$_2$ containing substance must be mixed with lithium and irradiation produces the following reactions:

$$^{6}\text{Li(n, }^{3}\text{H)}^{16}\text{O(}^{3}\text{H, n)}^{18}\text{F} \xrightarrow[112\,m]{\beta+} {}^{18}\text{O}$$

With 14 MeV neutrons, the reaction ^{16}O(np)^{16}N $\xrightarrow[7s]{\beta-}$ occurs, and under special conditions 10 μg ^{16}O or less can be determined.[72]

E. Enzymes and Other Proteins

In many enzymes, metal ions play an important role,[73] participating directly in catalysis through interactions between the metal ion and the substrates[74] or stabilizing protein conformation. Bowen[75] tried to classify metalloenzymes according to the prosthetic group involved: (a) metallo-proteinenzymes which have no prosthetic group and the metal is bound directly to the protein; (b) metalloporphyrin enzymes in which the metal is chelated by a porphyrin prosthetic group and (c) metalloflavin enzymes with a flavin prosthetic group. Different metal enzymes react differently when metal dissociation occurs. For Zn-enzymes, as carboxypeptidase A, the reversible dissociation of metal ions as a function of pH appears to be a simple competition of hydrogen ions for the ligands with no major structural changes in the conformation.[74] On the other hand, carbonic anhydrase undergoes major changes in the secondary and tertiary structure of the pH range where the metal ion dissociates.[76] Alkaline phosphatase binds between 2 and 3 gram atoms of Zn^{II} per mole of neutral pH. Between pH 6 and 4, Zn dissociates from the enzyme. Below pH 4 the molecule undergoes a conformation change, associated with the dissociation of the molecule into subunits.[77] Here the sensitivity of activation allows an analysis of samples of metalloenzymes weighing less than 1 mg for Co, Cu, Mn, Mo or Zn.[78]

In leucine aminopeptidase, crystallized from the eye-lenses, Zn was determined by neutron activation analysis, and 8 to 12 Zn atoms in one molecule of this enzyme could be detected. If manganese is added to the preparation, this element is partly able to replace zinc.[79] Glutamic dehydrogenase is also known as a Zn-enzyme.[75] This enzyme from *Blastocladiella* can be rapidly desensitized to AMP activation when treated with mercury. The allosteric concept of this enzyme was discussed in connection with this desensitization.[80] 0·001 μg Hg can be analysed in biological material with a deviation of ± 12 per cent by neutron activation analysis because of its high cross-section for thermal neutrons.[81] SH-groups in cell wall proteins were detected by this method.[33]

One component of polyphenoloxidase was purified from sweet potato and its Cu content was analysed by neutron activation.[82] The Cu content in another preparation of an *o*-diphenol oxidase, analysed by activation technique, was estimated to be 0·27 per cent.[83] This value is very similar to those of other copper-containing enzymes, such as tyrosinase, ascorbate oxidase and lactase. *o*-Diphenol oxidase activity is inhibited by borate and germanate.[84] Germanate can be detected by activation analysis much better than boron.[85] For the determination of boron several activation methods have been applied, but none of them was very satisfying.[86] Autoradiographic localization of ^{10}B in nuclei or cytoplasm of tumour cells was done by utilizing the neutron capture reaction. The boron α-particles and the recoil lithium nuclei, produced in the instantaneous disintegration of ^{11}B, are detected at the time of the reaction by nuclear track emulsion.[87] Molecular biology data may be confirmed by electron microscopic methods. The fixation time and therefore the penetration of osmium acid in tissue is sometimes very important. In ultramicrotome slices, ^{193}Os with $t_{1/2}$ of 1·3 days can be counted. Also, ^{191}Os can be measured from $^{190}Os(n\gamma)^{191}Os$ by irradiation of the sample for 12 days at a flux of $1·5 \times 10^{14}$ n.cm^{-2} sec^{-1}. For γ-spectroscopy of ^{191}Os the KeV level was used.[88] Caeruloplasmin-Cu was determined by neutron activation in gel-chromatographic fractions.[89] This combination of gel-chromatography and neutron activation can generally be applied to the identification of metal–protein complexes in biological fluids, such as human serum. The same combination is useful also to detect changes in the biosynthesis of different proteins, where non-radioactive Se-methionine acts as tracer.[90]

Quantities of selenium as small as 0·005 μg could be determined measuring ^{77m}Se, but the production of ^{19}O activity with $t_{1/2}$ equal to 29 seconds limits the accuracy and sensitivity of this technique.[91] The protein-bound selenium content of plasma and cell fractions of blood was determined by means of neutron activation analysis, utilizing a corrective calibration procedure based on the oxygen content of the samples.[92]

A possible substitution of Fe with Co in the biosynthesis of haemin, the prosthetic group of haemoglobin, was also studied by neutron activation analysis.[93]

Phosphoryl peptides can be characterized by neutron irradiation of paper electropherograms and detection by autoradiography or use of a chromatogram scanner.[94] In rheumatoid arthritis, gold compounds are used in therapy and could be detected by activation analysis in different serum proteins, separated by electrophoresis on polyacrylamide gels.[95,96] Immediately after gold treatment there is an increase in the biosynthesis of a 19s macroglobulin.[97]

The determination of protein-bound iodine can also be done by activation analysis.[98,99] The combined use of a resin column and activation analysis allows the determination of thyroxin iodine levels in the presence of some organic iodine compounds.[100] Diiodotyrosine, thyroxine and triiodothyronine could be determined in amounts of 10^{-9} gram in fractionated serum.[101]

F. Activation Analysis and Environmental Studies

This short section will deal with the general environmental conditions to which population is exposed and over which the individual has no control. There has been world-wide interest in pesticide residue analysis and in establishing residue tolerance levels for control and ecological research purposes. The potential of activation analysis for the detection and measurement of pesticide residues is being actively explored. Bogner[102] presented an excellent review on this topic. The method of activation analysis is insensitive to chlorine, therefore only in some special cases chlorinated organic pesticides can be determined. If all inorganic chlorine were removed or excluded from a fat sample, the total organic chloride content could be assumed to represent an equivalent amount of chlorinated pesticides.

Bromine-containing pesticide residues result from nematocides and fumigants, and after activation, using the Ge(Li) detector, bromine may be detected at the 5–10 p.p.m. level.[103] In order to study structural changes of pesticides in the cells by activation technique, it becomes necessary, prior to irradiation, to separate the metabolized pesticide from the applied molecules by paper chromatography, paper and gel electrophoresis, or thin-layer chromatography. Zn-, Cu-, Hg- and As-containing fungicides, including bactericides, slimicides and wood preservatives, are good sources for activation technique. The increased air pollution constitutes a great problem in public health. Pb, F, Si, Cu and many other elements are, in higher concentrations, toxic to every cell. Neutron activation analysis was used also for the determination of fluorine in aerosols of fluoro-organic compounds and fluoro-organic chelating agents. F can be detected by $^{19}F(n\gamma)^{20}F$ thermal neutrons or $^{19}F(np)^{19}O(n\alpha)^{16}N$ or $^{19}F(n2n)^{18}F$ by 14 MeV neutron irradiation.[104]

Some years ago great efforts were made to analyse fall-out products, and activation analysis was used for the determination of the dilution factor. Higuchi and coworkers described a simultaneous determination of Sr and Ba by neutron activation analysis with a Ge(Li) detector.[105] ^{129}I is a nuclide which occurs naturally as a result of cosmic ray reactions and spontaneous fission, but has also been found in measurable quantities in debris from nuclear explosions. The amount of ^{129}I in a typical human thyroid gland is about 10^{-11} gram and is below the limit of chemical detection. Neutron activation produces ^{130}I, thus enhancing its specific activity by a factor of almost 10^6, and enables reasonably accurate measurements to be made; this forms a basis for the assessment of long term biological hazards and for studies of the movement of iodine in the biosphere.[106,107] This isotope, which is now also available from a commercial source, was incorporated by iodination into 3,5,3′-L-triiodothyronine and has potential biological and medical usefulness.[108]

G. Drugs

Activation analysis has been employed for detection of drugs in cells, subcellular particles and their binding to macromolecules. This method has been used also for comparison and determination of the source of manufacture of drugs.[109] The results of the determination of trace elements in drugs published, including inorganics, salicylates, steroids, vitamins, antibiotics and chlorinated compounds have shown that they include the following trace elements: Al, Sb, As, Br, Cl, Cu, Ga, Mn, Hg, Sc, Na and Th.[110] For most of the drugs there is no clear connection between their action in the body and their chemical and physical properties. One class of drugs for which such a connection has been established, is known as the chelating agent class. Well-known natural chelates are discussed in the protein section. Commonly used drugs which can act as chelating agents are aspirin, adrenalin, cortisone, terramycin and many others. Chelating antidotes have been found for several metal poisons. A SH-group-containing substance, called BAL, is effective against acute arsenic, mercury and also gold poisoning. All three metal chelates can easily be detected by neutron activation. In Wilson's disease, the copper content of the tissues rises, especially in liver and brain, to intolerable high levels. Penicillamine, a sulphur-containing chelator, is capable of controlling this formerly incurable disease.[111] The trivalent antimony compound tartar emetic is the drug most frequently applied to combat schistosomiasis. The uptake of antimony by schistosomes has also received growing attention. Neutron activation analysis for further studies on the disposition of administered antimony in host and parasite has been emphasized by the World Health Organization.[112]

In medicine, toxicity to manganese and phenothiazine drugs is manifested as dyskinesia. Activation analysis was used for Mn determination in nuclei of rhesus monkeys' brain, influenced by phenothiazine.[113] Cotzias and coworkers have noted that phenothiazines form a semiquinone radical with manganese, and have suggested that manganese may be the cause of neuro-logic changes in workers of manganese mines, similar to Parkinson's disease.[114] Mn might function as a biological generator of free radicals, thus any biological structure which is rich in free radicals might also be rich in Mn. It appears likely that Mn may function in the final autooxidative stage of melanin formation. Analysis of pigmented human hair which is rich in melanin showed concentrations at least twice as high as those found in non-pigmented hair.[115] In many cases the action mechanism of radio-sensitizers is not quite clear. The halogens in 5-Br-uracil or 5-Br-desoxyuridin, 5-Cl-uracil, 5-I-uracil and 5-Br-desoxycytidin can be detected by activation technique when incorporated in nucleic acids, or free in the cytoplasm of the cells.[116]

SH-groups in enzymes can be blocked by mercurials like *p*-hydroxy-mercuribenzoate, or SH-containing radiation-protective molecules can be inactivated. In both cases a radio-enhancing effect was produced.[33] In the prophylaxis of certain pathological symptoms the important role of selenium was determined. The relations between symptoms of vitamin E deficiency and Se content was studied in chicken and Se determined as the 17·5 second ^{77m}Se after neutron irradiation.[117]

1-Ephedrine and norephedrine Ag complex pyridinates were prepared for the synthesis of argentic-8-hydroxyquinolinate. Ag determination was done after irradiation of the preparations in a neutron howitzer containing 3 Ci plutonium-beryllium.[118] Comar[119] investigated vanadium as an important inhibitor for the biosynthesis of cholesterol, by neutron activation analysis.

REFERENCES

1. H. J. M. Bowen, *Chimia*, **21**, 29 (1967)
2. W. S. Lyon, *Guide to Activation Analysis*, Van Nostrand, Princeton, New Jersey, 1964

3. G. L. Schröder, H. W. Kraner, R. D. Evans and T. Brydgea, *Science*, **151**, 815 (1966)
4. S. P. Cram and J. L. Brownlee, *J. Gas. Chrom.*, **6**, 313 (1968); **4**, 353 (1967); **6**, 305 (1968)
5. Y. Kusakoi and W. W. Meinke, in *Rapid Radiochemical Separation*, U.S. Nat. Acad. Sci., Nucl. Sci., Ser. NAS-NS 3104 (1961)
6. K. Samsahl, P. O. Wester and O. Landström, *Anal. Chem.*, **40**, 181 (1968); K. Samsahl, *Analyst*, **93**, 101 (1968)
7. F. Girardi, M. Merlini, J. Pauly and R. Pietra, *Radiochemical Methods of Analysis*, IAEA, Vienna, **2**, 1 (1965)
8. D. Camar and C. Le Poec, in *Radiochemical Methods of Analysis*, IAEA, Vienna, **2**, 15 (1965)
9. G. H. Morrison and H. Freiser, in *Solvent Extraction in Analytical Chemistry*, J. Wiley and Sons Inc., New York, 1962
10. H. Green, *Talanta*, **11**, 1561 (1964)
11. U. Eisner, J. M. Rottschafer, F. J. Berlandi and H. B. Mark, *Anal. Chem.*, **39**, 1466 (1967)
12. J. M. A. Lenihan and S. J. Thomson, in *Modern Trends in Activation Analysis*, Academic Press, London–New York, 1965
13. H. J. M. Bowen and D. Gibbons, in *Radioactive Analysis*, Oxford University Press, Oxford, 1963
14. C. E. Crouthamel, in *Applied Gamma-Ray Spectroscopy*, Pergamon Press, Oxford (1960)
15. O. U. Anders, *Anal. Chem.*, **41**, 428 (1969)
16. M. P. Menon and R. E. Wainerdi, in *Modern Trends in Activation Analysis*, Academic Press, London–New York, 1965, p. 152
17. W. G. Lussie and J. L. Brownlee, in *Modern Trends in Activation Analysis*, Academic Press, London–New York, 1965, p. 194
18. R. C. Greenwood and J. Reed, in *Proc. Intern. Conf. on Modern Trends in Activation Analysis*, Texas A and M Univ., 1961, p. 166
19. R. C. Greenwood, *USAEC Report* A-RF 1193-26 (1963)
20. *Prompt γ-Rays from Radiative Capture of Thermal Neutrons*, I.I.T. Res. Inst. 1193-53 (1965)
21. T. L. Isenhour and G. H. Morrison, *Anal. Chem.*, **38**, 167 (1966)
22. W. S. Leon, *Proc. Symp. in Nuclear Activation Techniques in Life Science*, IAEA, Vienna, 1967
23. W. B. Nelligan and J. Tittman, *Proc. 6th Scintillation Counter Symp.*, IRE Trans. Nucl. Sci., 187 (1958)
24. T. B. Pierce and P. F. Peck, *Proc. SAC Conference, Nottingham W.*, Heffer and Sons Ltd., Cambridge, 1965, p. 159
25. T. B. Pierce, 'Particular aspects of activation analysis with charged particles'. *EURATOM Rep.* EUR-2957, p. 53
26. P. Fireman, D. Borg and D. Gitlin, *Nature*, **210**, 547 (1966)
27. D. C. Borg, *BNL-report*, **10**, 130 (1966)
28. V. P. Guinn, *Isotopes in Experimental Pharmacology*, Ed. L. J. Roth, Univ. of Chicago Press, 1965, Chapter 3
29. D. M. Taylor, *Nuclear Activation Techniques in the Life Sciences*, IAEA, Vienna, 1965, p. 391
30. C. P. Haigh, *Nature*, **172**, 359 (1953)
31. E. Odeblad, *Clin. Chim. Acta*, **1**, 67 (1956)
32. J. M. Steim and A. A. Benson, *Anal. Biochem.*, **9**, 21 (1964)

33. H. Altmann, *Intern. J. Radiation Biol.*, **10**, 294 (1966)
34. A. K. Bruce and W. H. Malchman, *Radiation Res.*, **24**, 473 (1965)
35. W. A. Skinner, M. A. Leaffer and R. M. Parkhurst, *J. Pharm. Sci.*, **57**, 338 (1968)
36. H. Altmann, unpublished work
37. N. Suzuki and K. Kudo, *Anal. Chim. Acta*, **32**, 456 (1965)
38. W. Ruzicka and J. Stary, *Talanta*, **8**, 228 (1961); **9**, 617 (1962)
39. B. J. McCarthy and E. T. Bolton, *Proc. Nat. Acad. Sci. U.S.*, **50**, 156 (1963)
40. B. J. McCarthy, *Bacteriol. Rev.*, **31**, 215 (1967)
41. G. Dennis and H. Searcy, *BBA*, **166**, 360 (1968)
42. D. Gillespie and S. Spiegelman, *J. Mol. Biol.*, **12**, 829 (1965)
43. J. B. Smathers, D. Duffey and S. Lakshmanan, *Anal. Chim. Acta*, **39**, 529 (1967)
44. W. Bock-Werthmann and W. Schulze, *Modern Trends in Activation Analysis*, Academic Press, London–New York, 1965, p. 120
45. P. Patek and H. Sorantin, *Analyt. Chem.*, **39**, 1458 (1967)
46. J. Chanteur and P. Pellerin, *Nature*, **187**, 472 (1960)
47. J. Vincent, S. Hanmont and J. Roels, *J. Cell. Biol.*, **24**, 31 (1965)
48. M. A. Duprez and M. P. P. Grasse, *Acad. Sci. Paris*, **1**, 267 (1968)
49. M. Akaboshi, T. Maeda and A. Waki, *BBA*, **138**, 596 (1967)
50. M. Flikke and E. Steinnes, *Arch. Biochem. Biophys.*, **118**, 82 (1967)
51. G. Stehlik and H. Altmann, *Mh. Chem.*, **94**, 1163 (1963)
52. H. J. M. Bowen, in *Trace Elements in Biochemistry*, Academic Press, London–New York, 1966, p. 84
53. H. E. Haeder, *Diss. Gießen* (1966)
54. H. Altmann, G. Stehlik and K. Kaindl, *3e Coll. Intern. de Biol. de Saclay*, 1963, on *L'Analyse par Radioactivation.*, Presse Univ. de France, 1964, p. 243
55. H. Altmann, *IAEA Techn. Rep. Series*, **66**, 49 (1966)
56. H. Altmann, G. Stehlik and K. Kaindl, *2nd FEBS-meeting*, Vienna, 1965, A23
57. H. Altmann, *Biophysik*, **1**, 329 (1964)
58. G. Stehlik, H. Altmann and K. Kaindl, *SGAE-BL-5* (1964)
59. H. Altmann, in *Phys. Chemie biogener Makromoleküle*, Akad. Verlag, Berlin, 1964, p. 367
60. H. Altmann, K. Kaindl, H. Frischauf and H. Kaindl, *Tagungsbericht Biophysik Wien*, **14–16**, 1964, p. 123
61. H. Altmann, G. Stehlik and K. Kaindl, *Nature*, **199**, 823 (1963)
62. H. Altmann, *Studia Biophysica*, **8**, 69 (1968)
63. K. Kaindl and H. Altmann, *Sonderh. Z. 'Landwirtsch. Forsch.'*, **18**, 101 (1964)
64. L. Kaminski, *Diss. Wien* (1966)
65. H. Altmann, *SGAE-BL-16* (1965)
66. C. T. Dollery and J. B. West, *Nature*, **187**, 1121 (1960)
67. Y. Kamemoto, *Nature*, **203**, 513 (1964)
68. A. Fleckenstein and J. Janke, *3e Coll. Intern. de Biol. de Saclay*, 1963, on *L'Analyse par Radioactivation.* Presse Univ. de France, 1964, p. 267
69. O. C. Richards and P. D. Bayer, *J. Mol. Biol.*, **11**, 327 (1965)
70. O. C. Richards and P. D. Bayer, *J. Mol. Biol.*, **19**, 109 (1966)
71. J. F. Nicholson, L. Ponticorvo and D. Rittenberg, *BBA*, **108**, 519 (1965)
72. L. C. Bate, *Nucleonics*, **21**, 72 (1963)
73. T. Bersin, in *Biochemie der Mineral und Spurenelemente*, Akad. Verlagsges., Frankfurt a.M., 1963
74. B. L. Vallee and J. E. Coleman, *Comp. Biochem. Physiol.*, **12**, 165 (1964)
75. H. J. M. Bowen, *Trace Elements in Biochemistry*, Academic Press, London–New York, 1966, p. 124

76. J. E. Coleman, *Biochemistry*, **4**, 2644 (1965)
77. M. L. Applebury and J. E. Coleman, *J. Biol. Chem.*, **244**, 308 (1969)
78. H. J. M. Bowen, *Nuclear Activation Techniques in the Life Sciences*, IAEA, Vienna, 1967
79. M. Böttger, S. Fittkan, S. Niese and H. Altman, *Acta Biol. Med. Germanica*, **21**, 144 (1968)
80. H. B. Le John and S. Jackson, *Biochem. Biophys. Res. Comm.*, **33**, 613 (1968)
81. H. Altmann, H. Frischauf, D. Adamiker and K. Kaindl, in *Radioaktive Isotope in Klinik und Forschung*, Verl. Urban und Schwarzenberg, München, 1967, p. 390
82. H. Hyodo and I. Uritam, *J. Japan Biochem. Soc.*, **36**, 514 (1964)
83. H. Hyodo and S. Bando, *Agr. Biol. Chem.*, **29**, 763 (1965)
84. U. Weser, *Z. Physiol. Chem.*, **349**, 982 (1968)
85. D. de Soete, R. de Neve and J. Hoste, in *Modern Trends in Activation Analysis*, Academic Press, London–New York, 1965, p. 31
86. W. A. Canatt, H. P. Estey and R. Slott, *Anal. Chim. Acta*, **37**, 545 (1967)
87. L. C. Edwards, *Intern. J. Appl. Radiation Isotopes*, **1**, 184 (1956)
88. S. G. Malmskog and A. Bäcklin, *AE-352 Report*, Sweden (1969)
89. D. J. R. Evans and K. Fritze, *Anal. Chim. Acta*, **44**, 1 (1969)
90. H. Altmann, R. Eberl, W. Pusch, *Radioaktive Isotope in Klinik und Forschung*, Vol. 9, Verl. Urban und Schwarzenberg, Wien, 1970.
91. R. C. Dickson and R. H. Tomilson, *Intern. J. Appl. Radiation Isotopes*, **18**, 153 (1967)
92. R. H. Tomilson and R. C. Dickson, in *Modern Trends in Activation Analysis*, Academic Press, London–New York, 1965, p. 66
93. V. Maxia, S. Meloni, M. A. Rollier and M. T. Valentin, *Intern. J. Appl. Radiation Isotopes*, **18**, 267 (1967)
94. D. C. Show, *Nature*, **215**, 410 (1967)
95. R. Eberl and H. Altmann, *Wien. Klin. Wochschr.*, **81**, 950 (1969)
96. R. Eberl and H. Altmann, *Klinische Chemie und Klinische Biochemie*, **8**, 99 (1970)
97. R. Eberl and H. Altmann, in *Radioaktive Isotope in Klinik und Forschung*, Vol. 9, Verl. Urban und Schwarzenberg, Wien, 1970, p. 126
98. C. Kellershohn, D. Comar and D. Le Poll, *Intern. J. Appl. Radiation Isotopes*, **12**, 87 (1964)
99. H. Frischauf and H. Altmann, in *Radioaktive Isotope in Klinik und Forschung*, Verl. Urban und Schwarzenberg, Wien, 1970, p. 372.
100. M. H. Feldman, J. McNamara, R. C. Reba and W. Welester, *J. Nuclear Med.*, **8**, 122 (1967)
101. H. Stärk and D. Knorr, *Atomkernenergie*, **6**, 408 (1961)
102. R. L. Bogner, in *Radioisotopes in the Detection of Pesticide Residues*, IAEA, Vienna, 1966, p. 78
103. R. E. Wainerdi and M. P. Menon, *Symp. on Nuclear Activation Techniques in the Life Sciences*, IAEA, Vienna, 1967, SM-91/93
104. E. A. M. England, J. B. Hornsky, W. T. Jones and D. R. Terrey, *Anal. Chim. Acta*, **40**, 365 (1968)
105. H. Higuchi, K. Tomura, H. Tahakaski, N. Onuma and H. Hamaguchi, *Anal. Chim. Acta*, **44**, 431 (1969)
106. R. C. Koch, *Nature*, **200**, 641 (1963)
107. B. Keisch, R. G. Koch and A. S. Levine, in *Modern Trends in Activation Analysis*, Academic Press, London–New York, 1963, p. 284
108. J. T. Watson, D. K. Roe and H. A. Selenkow, *Radiation Res.*, **26**, 159 (1965)

109. H. L. Schlesinger, M. J. Pro, C. M. Hoffman and M. Cohen, *J. Off. Agric. Chem.*, **48**, 1139 (1965)
110. M. M. Tuckerman, L. C. Bate and G. W. Leddicotte, *J. Pharm. Sci.*, **53**, 983 (1964)
111. J. Schubert, personal communication
112. M. N. Malokhia and H. Smith, *Bull. World Health Organ.*, **40**, 123 (1969)
113. E. D. Bird, L. G. Grant and W. H. Ellis, *Symp. on Nuclear Activation Techniques in the Life Sciences*, IAEA, Vienna, 1967
114. D. C. Borg and G. C. Cotzias, *Proc. Natl. Acad. Sci. U.S.*, **48**, 617 (1962)
115. G. C. Cotzias, P. S. Papavasilion and S. T. Miller, *Nature*, **201**, 1228 (1964)
116. H. Altmann, H. Frischauf, D. Adamiker and K. Kaindl, *SGAE-BL-17* (1966)
117. K. E. Duftschmid and J. Leibetseder, *SGAE-IBIA-17* (1967)
118. M. L. Borke, P. B. Madan and B. D. Martin, *J. Pharm. Sci.*, **57**, 1788 (1968)
119. D. Comar, in *International Conference on Radioactive Isotopes in Pharmacology, 1967*, Wiley–Interscience, London–New York, 1969, p. 91

CHAPTER 3

Optical rotatory dispersion and circular dichroism of biomolecules

G. Snatzke and G. Eckhardt

Institute of Organic Chemistry,
University of Bonn,
Federal Republic of Germany

I. INTRODUCTION

Though optical rotatory dispersion (ORD) has been known since 1817[1] and circular dichroism (CD) in solutions since 1896,[2] both methods were applied extensively to stereochemical problems only after the revival of ORD in the nineteen fifties by C. Djerassi,[3] who had access to the first commercial spectropolarimeter, produced by Rudolph.[4] Automatic recording of CD curves has been possible since 1960, when Grosjean and Legrand[5,6] developed their ingenious device using the Pockels effect.[7] Today several instruments for measuring ORD and/or CD are commercially available.[8,9]

II. PRINCIPLES AND NOMENCLATURE

According to Fresnel[11] any linearly plane polarized light can be built up from two circularly polarized rays of equal frequency and intensity, but opposite helicity (Figure 1). In a transparent optically active medium these two rays travel with different velocities ('circular birefringence') which leads

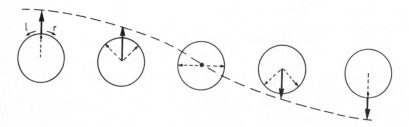

Figure 1. Superposition of right- and left-handed circularly polarized rays of equal frequency, amplitude, and wavelength. View is along the ray at equidistant positions. The sum vector describes a linearly polarized ray

to rotation of the sum vector, i.e., of the plane of polarization. If $\Delta n = n_l - n_r$ is the difference between the indexes of refraction for left and right handed circularly polarized light, the angle of rotation is given by Fresnel's equation[12]

$$\alpha(°) = \frac{1800 \cdot l \cdot \Delta n}{\lambda_{\text{vac}}} \tag{1}$$

when l is the length of the polarimeter tube (in dm), and λ_{vac} the wavelength in vacuum (Figure 2). The specific rotation $[\alpha]_\lambda^T$ and the molecular rotation $[\Phi]$

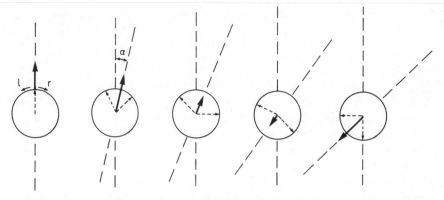

Figure 2. Superposition of right- and left-handed circularly polarized rays in optical active medium : wavelengths are not equal. View is along the ray at equidistant positions. The sum vector is again a linearly polarized ray, but its plane of polarization is rotated by an angle α which increases with the pathlength in medium

for solutions are defined as

$$[\alpha]_\lambda^T = \frac{\alpha}{lc'} \qquad [\Phi]_\lambda^T = \frac{[\alpha]_\lambda^T M}{100} \tag{2}$$

when the concentration c' is measured in g/cm^3 and M is the molecular weight. Subscript λ and superscript T denote that the angle of rotation is dependent on wavelength and temperature.

By going to shorter wavelengths the absolute value of $[\alpha]$ increases, and this behaviour is called 'normalous ORD curve' or 'plain curve' (Figure 3).

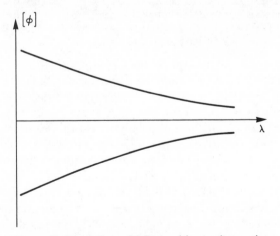

Figure 3. Normalous ORD (positive and negative plain curve)

It can be described by a Drude equation[13] of general form

$$[\alpha] = \sum_i \frac{A_i}{\lambda^2 - \lambda_i^2} \qquad (3)$$

In great distance from absorption bands a one-term Drude equation ($i = 1$) may be used. If, however, the ORD curve is recorded through a band of absorption, an S-shaped curve is superimposed onto the plain curve, giving rise to two extrema, called the 'trough' and the 'peak' (Figure 4). A curve of

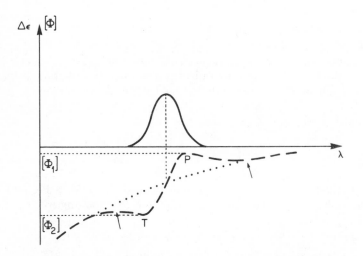

Figure 4. CD and anomalous ORD: ——— CD-curve, whose maximum corresponds to midpoint between P (peak) and T (trough) of ORD-curve (– – – –). Arrows point to pseudoextrema. Background rotation is indicated by . . . .

this type is named 'anomalous', and the reason for it is the fact that within the band the left- and right-handed circularly polarized rays are absorbed to different extents ('circular dichroism'). The CD can be measured directly as $\Delta\varepsilon = \varepsilon_l - \varepsilon_r$, if ε_l and ε_r are the molar decadic absorption coefficients for left- and right-handed circularly polarized light. According to the rule of Bruhat[14] and Natanson[15] we reach the peak first and the trough afterwards by going from long to short wavelengths, if the CD is positive, and the reverse situation follows from a negative CD. The point of inflexion of an anomalous ORD curve corresponds to the (positive or negative) maximum of the CD curve, and the 'amplitude' of the anomalous ORD curve, defined as $a = ([\Phi_1] - [\Phi_2])/100$, is about $40\cdot28 \times \Delta\varepsilon_{max}$ in case of Gaussian shaped CD bands.[16] $[\Phi_1]$ denotes the 'first' extremum, i.e. the extremum at longer wavelength, and $[\Phi_2]$ the 'second' at shorter wavelength. The anomalous ORD and corresponding CD are both known as the 'Cotton effect' (CE).

Linearly polarized light traversing an optically active substance becomes elliptically polarized in the range of absorption bands. This ellipticity is characterized as usual as $\psi = \arctan(b/a)$ if a is the long and b the short axis of the ellipse which is formed by the tip of the light vector. In analogy to specific and molecular rotation the specific $[\psi]$ and molecular $[\theta]$ ellipticities are defined as

$$[\psi] = \frac{\psi}{lc'} \quad \text{and} \quad [\theta] = \frac{[\psi]M}{100} \tag{4}$$

and the relation between $\Delta\varepsilon$ and $[\theta]$ is given by $[\theta] = 3298\Delta\varepsilon$. Arguments in favour of one or the other of these magnitudes have been discussed,[6,9,17] throughout this article $\Delta\varepsilon_{max}$ values will be used only.

With the help of the Kronig–Kramers transform[18,19] (equations 5 and 6) an ORD-curve within the k-th partial band can be computed from the CD-curve and vice versa, if one is known over the complete wavelength range.

$$[\Phi_k(\lambda)] = \frac{2}{\pi} \int_0^\infty [\theta_k(\lambda')] \frac{\lambda'}{\lambda^2 - \lambda'^2} \, d\lambda' \tag{5}$$

$$[\theta_k(\lambda)] = \frac{-2}{\pi\lambda} \int_0^\infty [\Phi_k(\lambda')] \frac{\lambda'^2}{\lambda^2 - \lambda'^2} \, d\lambda' \tag{6}$$

For quantitative correlations and theoretical calculations the rotatory strength (= rotational strength) R_k of the k-th band may be computed from CD-data. It is the area under the wavelength-weighed CD-curve (cf. reference 18).[19,20]

$$R_k = \frac{3hc \cdot 10^3 \ln 10}{32\pi^3 N_L} \int \frac{\Delta\varepsilon}{\lambda} \, d\lambda = 22.9 \times 10^{-40} \int \frac{\Delta\varepsilon}{\lambda} \, d\lambda \tag{7}$$

If μ_e denotes the electrical, μ_m the magnetical transition moment vector during the excitation, R_k is given by

$$R_k = \mu_e \mu_m \cos(\mu_e, \mu_m) \tag{8}$$

CD-curves are usually not corrected for differences in refraction indexes of different solvents, but this should always be done for ORD-values by multiplying with the factor $3/(n^2 + 2)$. These corrected values are called 'rotivities'.[21]

A CE can thus be measured either by recording the angle of rotation or the differential absorption versus wavelengths. Due to their narrower half-widths, their simplicity and the fact, that their signs can always be determined unequivocally even in case of fine structure, CD-curves are easier to interpret than ORD-curves. Whenever absorption bands are accessible with instruments (at present down to about 185 nm) CD will, therefore, be preferred over ORD with few exceptions. A comparison of both methods has been given,[22,23] general reviews are found in the literature.[3,6,10,12,21,23-31]

III. COTTON EFFECT AND STEREOCHEMISTRY OF BIOMOLECULES (EXCLUDING SUGARS, AMINO ACIDS, NUCLEOSIDES AND RELATED COMPOUNDS)

A. General

A CE is characterized by four features, viz. 1. the wavelength of highest $\Delta\varepsilon$ in a CD-curve (λ_{max}), corresponding to the point of inflexion of a partial anomalous ORD-curve, 2. the sign of the CE, 3. its magnitude ($\Delta\varepsilon_{max}$ or amplitude a), and 4. its fine structure. In the case of electrically and magnetically allowed absorption bands λ_{max} of the CD-curve is identical with λ_{max} of the corresponding u.v. curve, and the half band widths of the two are also the same.[18] The partial g'-factor,[24] defined as $g' = \Delta\varepsilon/\varepsilon$ is constant throughout the whole partial band. For electrically forbidden, magnetically allowed bands the maximum of the CD-curve may be slightly redshifted compared to the corresponding absorption curve, and its half band width may be smaller than that of the latter. Thus, in general, no more informations can be obtained from the CD maximum than from the u.v. maximum position, but in several cases small u.v. bands may be buried under more intense ones. These however, may be frequently detected in CD-curves due to sign inversion and/or smaller band width, e.g., the 330 nm band of aliphatic nitro compounds,[32,33] or the 260 nm band of conjugated acids (esters, lactones).[34]

The sign of the CE of many chromophores is now predictable from empirical or even theoretical rules, from the geometry of a compound and is mainly used for the determination of the stereochemistry (conformation or configuration).

The magnitude of a CE can only in special cases be used for the elucidation of the stereochemistry because of lack of quantitative rules. It may be used, however, to differentiate between the two classes of chromophores, which Moscowitz[20] has proposed, viz. the inherently (intrinsically) dissymmetric chromophores and those, which are inherently symmetric, but dissymmetrically perturbed. In the first type the absorbing electron system is chiral in itself, which causes rotatory strengths of the order of 10^{-38}, as e.g. in hexahelicene (1), or in some β, γ-unsaturated oxo compounds (see below), whereas in the second class the chromophore is locally achiral, but incorporated into a chiral environment. Most ketones belong into this second class. The rotatory strength does usually not exceed 10^{-40}. Changes

(1)

of $\Delta\varepsilon_{max}$ by epimerizations, etc., have been used more frequently for stereo-chemical correlations of ketones, but even they are not always predictable with certainty.[35]

Finally, in some cases the position of the partial bands in a CE with fine structure can also be used for the determination of stereochemistry.[36,37]

A similar division of chromophores has been used by one of these authors[33,38] (cf. also Klyne[39]): the molecule is divided into 'spheres', starting with the chromophore. This itself forms the first sphere, the ring, into which it is incorporated, the second one, rings or groups bound directly to this the third, etc. That dissymmetric sphere which is nearest to the

Table 1. Examples of inherently dissymmetric chromophores

Chromophore	λ_{max} (nm)
C=C—C=C (not coplanar)	~250
C=C—C=O, $\pi \rightarrow \pi^*$-band (not coplanar)	~250
C=C—C—C=O, $n \rightarrow \pi^*$-band (special geometry, cf. Table 4)	~300
—S—S—	~200 (260)

chromophore determines the sign and even to a great extent the magnitude of the CE.

Some inherently dissymmetric chromophores important for biomolecules are summarized in Table 1, some inherently symmetric, but dissymmetrically perturbed chromophores in Table 2. Groups, as —OH, —NH_2, which are not absorbing in the accessible wavelengths range of present instruments may be converted into chromophoric or 'cottonogenic' derivatives (see Table 3). Some years ago acids also had to be transformed into such derivatives, but modern instruments are capable of scanning through their $n \rightarrow \pi^*$-absorption band (some of them are included in Table 3, cf. also the compilation by Sjöberg[40]).

Table 2. Some examples of inherently symmetrical, but dissymmetrically perturbed chromophores

Chromophore	λ_{max} (nm)
$C=O$	300
$C=C-C=O$, $n \rightarrow \pi^*$-band	350
$-C(=O)O-$	210
$-C(=O)NR-$	210
$C=C$	200
many aromates	depending on substitution and structure
unsaturated heterocycles	depending on substitution and structure

Table 3. Examples of chromophoric derivatives of groups absorbing at short wavelengths (cf. reference 40)

Group	Chromophoric derivative	λ_{max} (nm)
$-OH$	$-O-C(=S)-SCH_3$	355
	$-O-NO$	320–40
$-NH_2$	$-NH-C(=S)-SCH_3$	330
	$-N=CH-C_6H_4-OH$ (ortho-hydroxyphenyl)	315–400
	$-N$ (phthalimide)	300–330
	$-NH$ (dimedone enone derivative)	280
$\diagdown$NH$\diagup$	$\diagdown$N$-NO$\diagup$	370
	$\diagdown$N$-Cl$\diagup$	260
$-C(=O)-OH$	$-C(=S)-NHCH_3$	325–60
	$-C(=O)-NH-C(=S)-NR_2$	340

B. Oxo Compounds

The oxo group is one of the best investigated chromophores because of its ubiquity, and its weak absorption.

It was, therefore, the first chromophore examined extensively by Djerassi and his collaborators[3] (by ORD), as well as by the Roussel group[6] (by CD). If the first and second spheres are symmetric (e.g. cyclohexanone in chair conformation) it is only the distribution in space of substituents and/or rings around the chromophore which determines the sign and magnitude of the CE. For such compounds the famous octant rule[3,41] was developed on an empirical basis, though theoretical approaches are possible.[42,43] Polar axial substituents in β-position may, however, give 'inverse' contributions to the CE.[35] It must be emphasized that the octant rule in its original

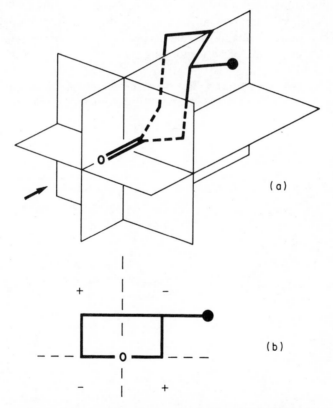

Figure 5. Octant rule. (a) The three nodal planes building up the eight octants. Arrow indicates the direction of projection. (b) Octant projection of back octants. Signs correspond to contributions of substituents in those four octants (for exceptions see text)

form can only be applied to oxo compounds with symmetric first and second spheres.

In the $n \to \pi^*$-transition of the C=O group an electron is promoted from the nonbonding p_y-orbital of oxygen to the antibonding π^*-orbital. The nodal spheres of these orbitals (assumed to be planes) divide the space around the carbonyl group into octants (Figure 5(a)), and the contribution of any substituent to the CE changes its sign by going through one of these nodal planes, groups lying in one of these planes giving to a first approximation no contribution. Usually only the four back (rear) octants are drawn in a so-called 'octant projection' (Figure 5(b)), by looking from the oxygen to the carbon atom of the C=O group. Substituents in the right upper or left lower back octants (with the exception of some β-axial polar groups[35] and, perhaps, fluorine[44]) give according to the octant rule negative, such in the left upper and right lower back octants positive contributions to the CE. For substituents in front octants these signs are inverted, though some theoretists assume a 'quadrant' rule according to which the signs of the contribution of groups in corresponding back and front octants are the same.[42]

A classical example of the application of the octant rule is the comparison of the CE of a 2- and a 3-keto 5α-steroid[41] (Figure 6). In 2-keto compound (2)

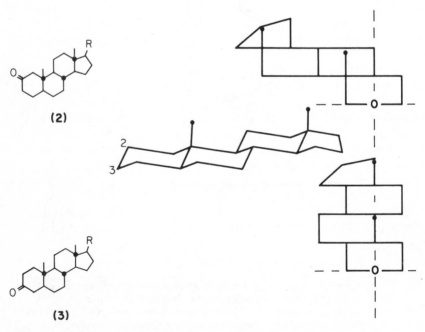

Figure 6. Octant projections of compounds (2) (top) and (3) (bottom)

all C-atoms (the contributions of the H-atoms are usually neglected, cf., however, reference 43) besides those of ring A (which lie either in nodal planes or compensate each other) and C-6 are in a positive octant, and accordingly a relatively high positive amplitude ($a = +121$) in the ORD-curve is obtained. On the contrary, in (3) contributions of the atoms in rings A and C are either zero or compensate each other, furthermore both methyl groups as well as the side chain at C-17 are in a nodal plane, so that only the positive contributions of C-atoms 6, 7, 15 and 16 (the latter two are very distant from the chromophore and have, therefore, only minor influence) remain. The CE is again positive, but the amplitude is approximately only half of that of (2) ($a = +55$).

As the magnitudes of the CE of steroids with keto groups in all possible ring positions are known,[3,6,23,41] these can be used to localize a C=O group in this system, as long as there are no substituents near to the chromophore, which could change its CE. Thus, e.g., 1-keto-5α-steroids give a very weak negative CE ($a = -25$) superimposed on a steep characteristic positive background rotation.[45] ORD gives only apparently more information in this case than CD, because the latter is also highly characteristic, it shows a double humped curve, (negative around 310, positive around 280 nm) whereas nearly all other keto steroids give only one band.[46,47] A better method to determine the position of a keto group in a steroid skeleton was developed by Djerassi, Mitscher and Mitscher[48] by taking advantage of the fact that hemiketal formation in acidified methanol is highly dependent on steric hindrance. The magnitude of the CE is first recorded in methanol solution, then a drop of HCl is added and the CE is rerun. It decreases by about 70 per cent in case of a 3-keto-steroid, by 12 per cent for a 2-ketone, by 10 per cent for an 11- and not for a 12-ketone.

With the aid of the octant rule or by comparison of the data with those of known analogues the configuration of hundreds of steroids and terpenoids have been determined. An example from the diterpene field was given by Yoshikoshi.[49] Dolabradiene (4) was degraded to the ketones (5) and (7). The CE of the first was identical with that of a 17-keto steroid (6), that of the latter was the mirror image of that of a 3-keto-5α-steroid (3). Compound (5) gave thus information about the configuration at C-5 and C-10, (7) about C-8 and C-9. Early ORD work revealed also that some diterpenoids do not have the usual absolute configuration of the skeleton. Thus cafestol (8) was degraded into the ketone (9), which gave a CE almost enantiomeric to that of the 4α-ethyl derivative of (3).[50] Similarly eperuic acid (10) and labdanolic acid (11) gave nor-ketones of type (12) whose CE were nearly mirror images.[51]

For ketones with dissymmetric second[52] or first sphere the octant rule has to be replaced by other rules; the more important ones are shown in Table 4. Thus, e.g., axial Cl, Br or I in α-position to a carbonyl gives a very strong CE whose sign is in accord with a formally applied octant rule.[53] α, β-Epoxy and

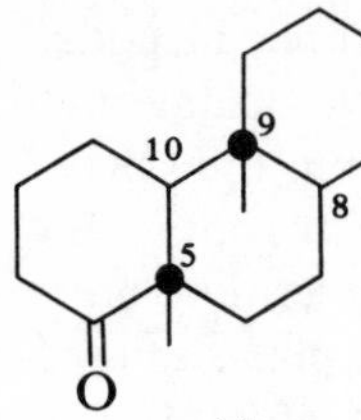

(4)

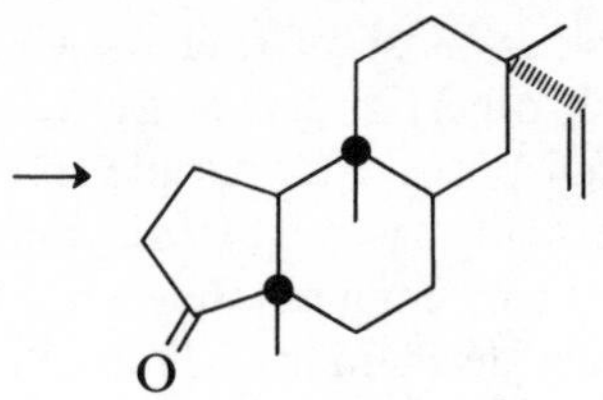

(5)

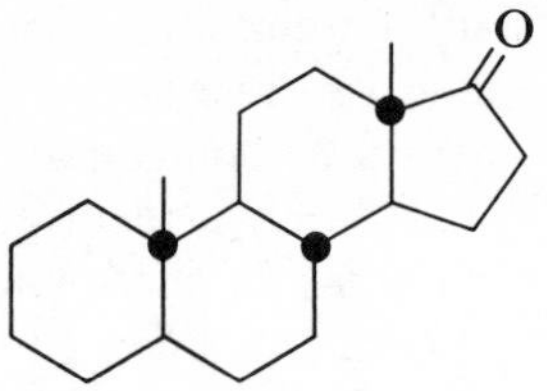

(6)

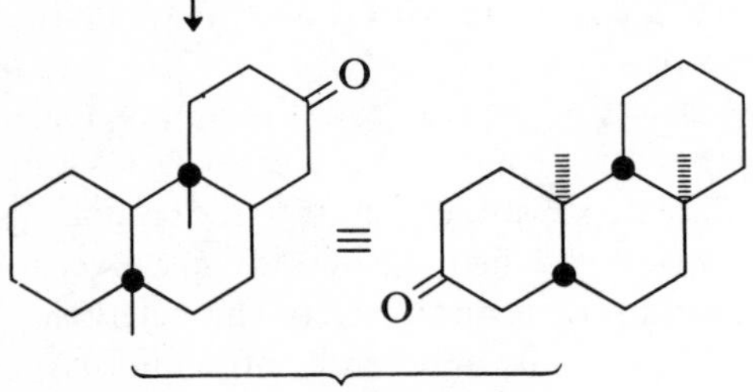

(7)

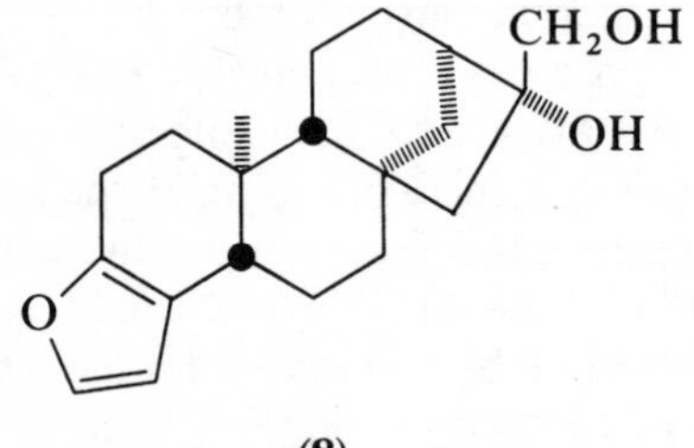

(8)

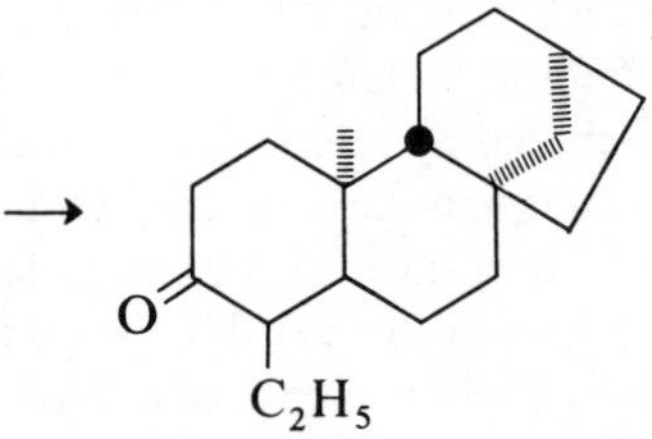

(9)

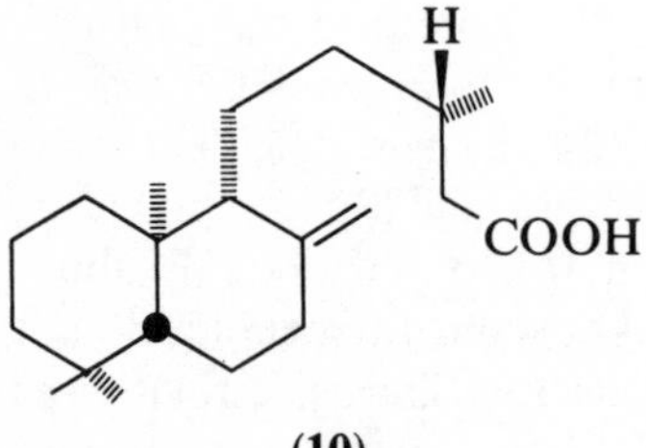

(10)

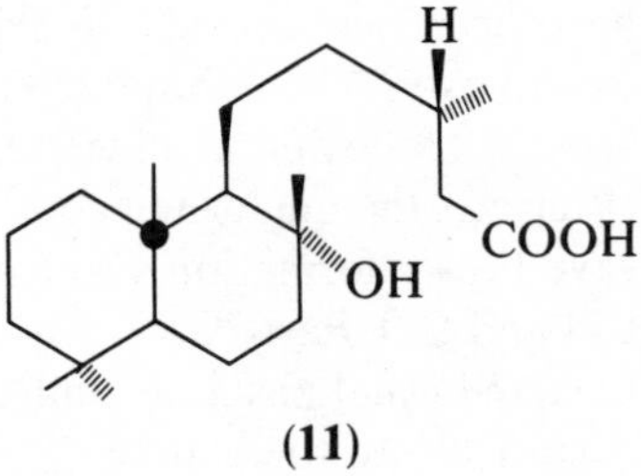

(11)

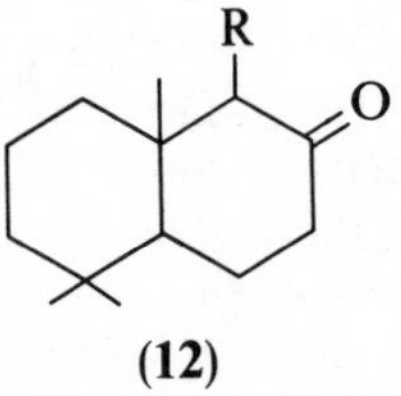

(12)

Table 4. Schematic representation of some rules for ketones leading to a negative CE

General formula		Octant type projection	Reference
$(CH_2)_n$ ⌬–R, O	n arbitrary, C_s-symmetry of ring or open chain	R ◄O►	41
$(CH_2)_n$ ⌬, O	$n = 4, 5, \ldots$ twist conformation, C=O in 'point' of twist	◁O▷	39, 52
$(CH_2)_n$ ⌬, C–C, X O	n arbitrary X = axial Cl,Br,I, (inverse for F)	◄O► X	53
$(CH_2)_n$ ⌬, X O	$n = 2, 3, \ldots$ X = \CR_2 or \O/ $n \rightarrow \pi^*$-band	X ⌐O►	54
$(CH_2)_n$ ⌬, O	$n = 3, 4$ or open chain (inverse for $n = 2$ in most cases) $n \rightarrow \pi^*$-band	⌐O►	38, 55
$(CH_2)_n$ ⌬, H_2C C–C, O	$n \rightarrow \pi^*$-band	π O►	38

Table 4.—cont.

General formula	Octant type projection	Reference
$n \rightarrow \pi^*$-band		38
$n \rightarrow \pi^*$-band	$\sim 100°$	56

α, β-methylene ketones follow (with few exceptions) an 'inverse rule', i.e., if the oxygen or methylene group is in a 'positive octant' (Figure 5) it gives a negative CE, and vice versa. The R-band ($n \rightarrow \pi^*$-transition) of conjugated transoid enones follows a similar 'inverse rule' (with the exception of cyclopentenones).[38,55] The more the enone grouping deviates from coplanarity, the greater becomes $\Delta\varepsilon_{max}$. For example, in the 1-en-3-one (**14**) $\Delta\varepsilon_{max}$ is about -1, and this value drops to -0.81 for (**13**), which lacks the repulsion of the angular methyl group C-19 with the 4β- and 6β-hydrogen atoms. Introduction of a 4β-methyl group (as in (**15**)) increases the deviation from coplanarity and $\Delta\varepsilon_{max}$ becomes -1.8. Presence of an additional 8β-methyl group (**16**) augments this repulsive effect and increases the CD to -3.5.

The K-band ($\pi \rightarrow \pi^*$-transition) of such conjugated enones shows in general a sign of its CE which is inverse to that of the R-band.[57] With their

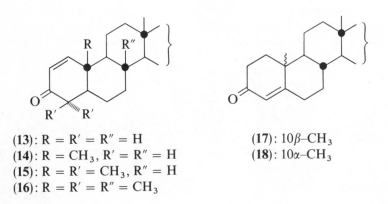

(**13**): R = R′ = R″ = H
(**14**): R = CH₃, R′ = R″ = H
(**15**): R = R′ = CH₃, R″ = H
(**16**): R = R′ = R″ = CH₃

(**17**): 10β–CH₃
(**18**): 10α–CH₃

new dichrograph the Roussel–Uclaf chemists[58] have in addition to this CD band found another one at about 210 nm, which is usually greater than the $\pi \to \pi^*$ band and whose origin is not known with certainty. In the case of testosterone (17) and its 10-epimer (18), these bands can be better used to determine the stereochemistry at C-10, as both R-bands are negative and of nearly equal magnitude though shifted one against the other. The 210 nm CD band of (17) is, however, strongly positive $(+11)$, whereas that of (18) is negative (-12). The sign of the $\pi \to \pi^*$-band CD has formerly been thought to be unequivocally determined by the helicity of the $C{=}C{-}C{=}O$ moiety,[57] recent investigations have shown, however, that a better correlation is possible by taking into account the allylic hydrogen atoms.[58a] The sign of the 210 nm-band CE has been correlated with the stereochemistry of the axial hydrogen atom at the α'-position.[58a]

The CD within the K-band of 'conjugated' cyclopropyl ketones in general is of opposite sign to that of their R-band CD,[54] whereas for α, β-epoxy ketones these two CE have identical signs.[54]

β. γ-unsaturated oxo compounds show an exceptional high CE if the geometry meets the conditions of the figure in Table 4 (or its mirror image). First thought to be due to a charge transfer from the $C{=}C$ to the $C{=}O$ bond,[59] it is believed now that the reason is an overlap of localized orbitals.[56] $\Delta\varepsilon_{max}$ may be as high as $+36$ (santonide (19) and parasantonide (20)), and

(19): α-CH$_3$
(20): β-CH$_3$

such enones are examples of the rare cases when it is allowed to determine conformation (from the high magnitude) and absolute configuration (from the sign) from one measurement.

C. Acids and Their Derivatives (including Lactones)

Saturated lactones have been treated in different ways, emphasizing either the dissymmetry of the second[60–63] or the third (fourth, . . .) sphere,[64,65] leading to 'helicity rules' or 'sector rules'. A recent compilation[63,66] and theoretical reasons[67] indicate that sector rules may—as in the case of ketones —be applied only if the lactone ring is coplanar, otherwise the helicity rule determines the sign of the CE (Figure 7). For example, the CD of the 13β-Me

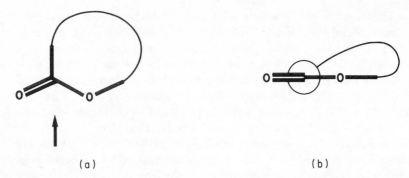

Figure 7. Rule for saturated lactones. Thick bonds are coplanar. Arrow in (a) denotes direction of projection of (b). The figure corresponds to a negative torsion angle (i.e. to a positive CE)

lactone (**21**) is −0·56, that of its 13-epimer (**22**) is +0·57, which corresponds to the rule of Figure 7.

(**21**): 13β-CH$_3$
(**22**): 13α-CH$_3$

Carboxylic derivatives with dissymmetric first or second sphere can be treated like the corresponding ketones.[68] Thus the configuration of jaborosalactone A (**23**) at C-22 was determined by comparison with the CD of parasorbic acid (**24**), which, in turn, follows the rule for transoid enones, if

CH$_2$OH

(**23**)   (**24**)

(as in all other lactone cases) the $-C(=O)-O-C-$ moiety is assumed to be coplanar.[68,69]

D. Chromophoric Derivatives of Groups Absorbing at Short Wavelengths

In Table 3 is given a compilation of some derivatives of alcohols, amines and acids, used for the determination of the stereochemistry of such compounds. Thus, e.g., the configuration of an OH-group at C-20 of a pregnane derivative with $-CH(OH)-CH_3$ side chain can be deduced from the sign of the CE of its xanthate[70] (positive for 20β-(25), negative for 20α-configuration (26)) or nitrite[32,71] (positive for 20α, negative for 20β). The corresponding

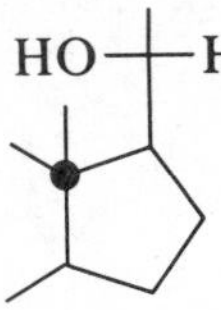

(25) 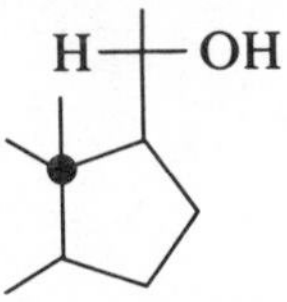   (26)

primary amino derivatives can be differentiated by the CE of their salicylidene derivatives (positive at 320 nm for 20α-, negative for 20β-amines) etc.[72] Acids may e.g. be transferred into their N-methyl-thionamides, whose CE at about 325–360 nm is positive for (S)- and negative for (R)-configuration.[73]

E. Solvent and Temperature Effects

Change of solvent may influence the CE directly (hydrogen bonding, solvent–solute complex formation, dipole–dipole interaction, etc.) or indirectly (change of conformational equilibria, different polarizability of solvent molecules compared to substituents, etc.)[74] In the case of conformational equilibria very often 'double humped' CD curves are observed,[75] though they may also be due to vibronic coupling in absence of equilibria.[76] Rassat[77] has for example investigated bicyclic monoterpene ketones and found drastic changes of the CD by going from polar to unpolar solvents (Figure 8). Kuriyama and coworkers[78] correlated the change of the CD due to a shift of an equilibrium between two conformers of the bromo ketone (28)

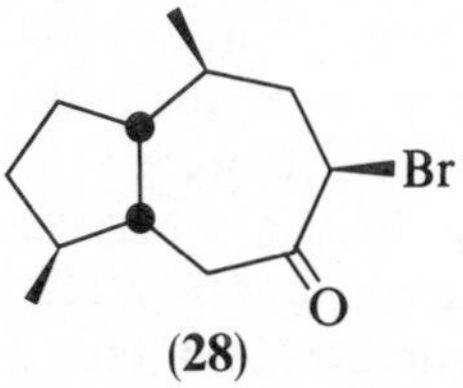

(28)

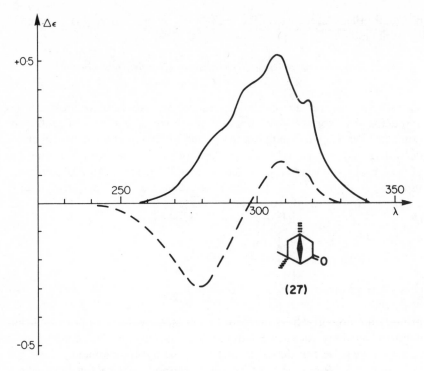

Figure 8. CD of compound (**27**) in cyclohexane (———) and ethanol (– – – –)

with the dielectric constant of the solvent and obtained by this method similar results as have been found from temperature variation studies.[29]

Variation of the temperature can also influence the CE in several ways, as by changing solvent–solute equilibria, by shifting conformational equilibria, by changing the contribution of 'hot bands', by altering the ratio of allowed to forbidden vibronic transitions,[74,79,80] etc. An estimation on theoretical basis leads to the conclusion that the change of the CD between say, $-190°$ and $+20°$ due to all effects besides conformational ones may not be greater than about 15–20 per cent relative.[79] Thus the method of temperature variation was mainly used in studies of conformational equilibria, and under the assumption that only two species are present and that $\Delta S = 0$, rotatory strengths for the individual conformers and their free energy differences ΔG may be obtained[81] from plots of the rotatory strengths *vs.* $[1 + \exp(-\Delta G/RT)]^{-1}$. In a more qualitative way temperature variation studies have also been used to determine the amount of steric hindrance to rotation of groups like e.g. NO_2,[33] which lead to the conclusion that the barrier to free rotation of an equatorial 4α- (**29**) or 6α- (**31**) nitro group (by the peri-standing 6α- or 4α-hydrogen) is of the same magnitude as that of the

(29): 4α-NO$_2$ (31): 6α-NO$_2$
(30): 4β-NO$_2$ (32): 6β-NO$_2$

axial epimers (30) and (32) by methyl C-19 and two axial hydrogens in 1:3 position.

Usually the fine structure of a CD- or ORD-curve increases by lowering the temperature (suppression of hot bands, etc.) and frequently small blue shifts are observed. With keto steroids this method can be used as a diagnostic means, as only some 11-keto-5α-derivatives show an inversion of the sign of the CD at liquid nitrogen temperature.[47,82] With compounds of mobile chains low temperature CD measurements are sometimes the only possibility to detect a CE at all, as at room temperature the contributions of all the conformers present may cancel each other accidentally. Thus, at room temperature practically no CE was measured for tetrahydro-ionone (33),

(33)

whereas at $-179°$ a negative CD-band with fine structure ($\Delta\varepsilon_{max} = -0.23$) was observed, which was in agreement with predictions from the octant rule for the most preferred conformation.[29]

IV. COTTON EFFECT AND STEREOCHEMISTRY OF AMINO ACIDS AND PROTEINS

A. Amino Acids

With earlier instrumentation it was not possible to traverse the absorption band of an amino acid ($n \rightarrow \pi^*$ of carboxylic group) and, therefore, N- or C-derivatives have been investigated. Besides those mentioned in Table 3 there have been used e.g. N-thioacyl[83] and N-thiono carbethoxy derivatives,[84] selenoaryl esters,[85] and 3-phenyl-2-thiohydantoins[84] from α-amino acids.

As the CE of these compounds is very often greatly influenced in wavelength, magnitude, and even sign by changing the solvent, considerable caution must be observed in assigning configurations solely on the basis of the sign of the CE, when the solute–solvent interaction is not sufficiently well studied.[40,83,86] Nowadays the CE of the α-amino acids can be seen directly, and it is positive for all investigated simple acids of the L-series.[87–94] In general this positive CE becomes stronger by going to acidic solutions, which is in accord with the long known Lutz–Jirgensons rule[95] which states that the rotation at the NaD line of an L-α-amino acid is more positive in acidic solution than at pH 7 or in alkaline medium. The CD-maximum of the zwitter ion in water solution is centred around 203 nm and is shifted to about 209 nm in acidic and to 213 nm in alkaline solution.[94] An exception makes L-proline, but this is due to the appearance of a second CD-band of uncertain origin.

Amino acids containing other absorbing groups, as cystine, phenyl alanine, tyrosine, or tryptophan may show additional CE. Whereas optical activity is easily detected in the ^1L_b transition of tyrosine,[92,96–102,102a] phenyl alanine gives only a very weak CE near 260 nm,[92,101,102,102b] which is hardly detectable. According to Moscowitz and coworkers[101] this can be explained by the fact that due to the local C_{2v} symmetry of the benzene ring electrical and magnetical transition moment vectors are perpendicular to each other, so that the rotatory strength becomes zero. In tyrosine mixing with $n \rightarrow \pi^*$ transitions (n-electrons from oxygen atom of OH-group) destroys this orthogonality to some extent, whereas in phenyl alanine only $3d$- or σ-orbitals can mix in, which is a much more ineffective way to produce rotatory strength. Cystine shows a negative CE near 260 nm,[92,103] and in acidic solution there appear two more CD bands, a positive one at 219 and a negative one at 196 nm.[92] The aromatic bands of tryptophan near 270 nm give also rise to a positive CD.[92,102,103]

B. Oligopeptides

The CE of di- and oligopeptides has been investigated by ORD and CD. Legrand and Viennet[104] prepared dipeptides in which either the C- or the N-terminal amino acid was the achiral glycine and found that the CD around 200 nm was positive and approximately five times greater than in the free amino acids with aminoacyl glycines, whereas a sign inversion was observed (at about 195 nm) in case of glycyl amino acids. This band is assigned a $\pi \rightarrow \pi^*$ transition of the amide group[104,105] and the other $n \rightarrow \pi^*$ band is also seen in both cases at about 220 nm (positive besides in L-leucyl glycine). Dipeptides comprising two L-amino acids give CD-bands which may deviate appreciably from the calculated ones; with L-lysyl-L-glutamic acid the 200 nm band has even the opposite sign compared to a 'synthetic curve' obtained by addition of the CD of L-lysyl-glycine and glycyl-L-glutamic acid

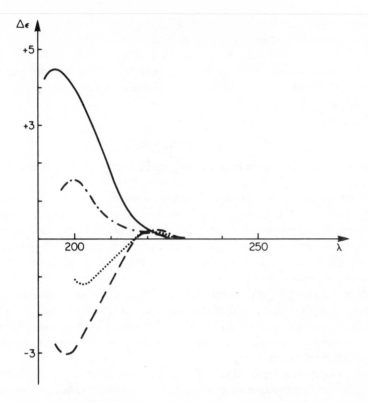

Figure 9. CD of dipeptides. L-Lysyl-glycine (———), glycyl-L-glutamic acid (– – – – – –), L-lysyl-L-glutamic acid, experimental (.....) and calculated (– · — · — · — ·)

(Figure 9). These strong vicinal effects can be attributed to conformational changes.

Beacham and coworkers[106] measured the ORD of all possible diastereo-meric tri- and tetra-peptides of alanine and serine, respectively, and found that the (mostly) plain curves (last measurements at 227 nm) of these peptides can be composed from 'increment curves' for N-terminal (nLL: N-terminal amide between two L-centres; nLD: between L-centre (N-terminal) and D-centre), middle (mL: chirality centre next to CO of amide groups is L-) and C-terminal (cL: carboxylate ion next to L-centre) contributions by arithmetic addition. For example, for LDDL-tetrapeptide in water the sum of contribu-tions would be equal to nLD − mL − mL + cL. This treatment is similar to one used for calculating the rotation at the NaD line of oligopeptides.[107,108] In view of the results of Legrand and Viennet[104] it cannot be expected, however, that such an additivity rule holds in the neighbourhood or within absorption bands. Protection of the NH_2-, COOH- and/or OH-groups in these oligopeptides does not appreciable alter this additivity.*[109]

The ORD- and CD-behaviour within the absorption bands of the alanine tetrapeptide has also been investigated recently by Balasubramanian and Wetlaufer[110] and they found CE curves which were very similar to those of the α-helix.

C. Cyclopeptides

Diketopiperazines have been investigated as very simple models of cyclic peptides by several groups (e.g. references 110, 112–115). Their CE also resemble those of the α-helix and several authors[110,114,115] have pointed out that the three CE 'characteristic' for the α-helix can also be given by other geometric arrangements of interacting amide groups. A striking example is the diketopiperazine of L-proline, which at least in the $n \to \pi^*$-region gives the 'typical' value for the α-helix, though this compound cannot at all adopt the torsion angles of the helix backbone.[115] Schellman[116] has derived a quadrant rule for the prediction of the sign of the CE of the amide group and later also applied this to diketopiperazines.[30] Bláha and Frič[112] used for the same purpose Jones and Eyring's[117] treatment of a chromophore and got agreement with experiment only if they assume a boatlike conformation for the diketopiperazine ring.

In diketopiperazines from two different amino acids the configuration of one seems to be dominant over the other with an order following the polarizabilities (Tyr(Me) > Tyr > Phe > Leu > Lys).[112] Thus, the ORD-curves of cyclo-L-Phe-L-Leu and cyclo-D-Phe-L-Leu are nearly mirror images, whereas the ORD-curves of cyclo-L-Tyr(Me)-L-Leu and cyclo-L-Tyr(Me)--D-Leu are very similar. They give two CE at 215 and 225 nm of same sign, and with *cis*-compounds the band at shorter wavelength is more intense, in the *trans*-isomer less than the 225 nm band.

The CD in the range of the ^1L_b-band of the aromatic chromophore of the Tyr-Tyr-diketopiperazines is several times larger than that of the linear Tyr-Tyr-dipeptide (and twice as high as that of Tyr-Gly-diketopiperazine).[113] These results can be explained by assuming exciton interaction between the aromatic and the amide chromophore due to a fixed conformation of the ring.

Cyclic peptides with a greater number of amino acid units may give rise to two new types of stereoisomerism:[119] two peptides are 'cycloenantiomeric' if by reflexion only the direction of the peptide linkages is reversed without changing the pattern of chiral units (Figure 10). If two cyclopeptides differ only in the direction of the peptide bonds without being mirror images

---

* Other increments have to be used in these cases, of course.

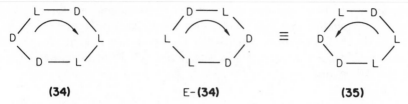

Figure 10. Cycloenantiomerism : (**34**) and (**35**) are cycloenantiomeric to each other

of each other, they are called 'cyclodiastereomeric' (Figure 11). Above 230 nm the ORD-curves of the cycloenantiomeric cyclohexaalanines (**34**) and (**35**) (Figure 10) were practically mirror images.[119]

Cyclohexapeptides containing two Phe residues give observable CE in the ^1L_b-range of the aromatic chromophore only in those cases where both Phe had identical absolute configuration.[112] They give two bands at 210–218 and 225–230 nm, ascribed to $n \rightarrow \pi^*$ transitions, and the sign of the first is

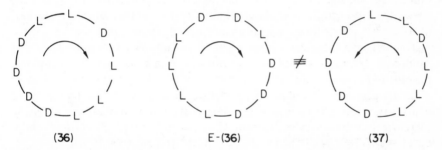

Figure 11. Cyclodiastereoisomerism : (**36**) and (**37**) are cyclodiastereomeric to each other

determined solely by the configuration of the Phe residues, whereas the second one is influenced also by the sequence of the amino acids. It seems possible, therefore, that the 215 nm CD-band can be ascribed to the ^1L_a-transition of the aromatic ring. The $\pi \rightarrow \pi^*$ amide CD-bands (at 190 and 200 nm) are governed solely by the sequence, regardless of whether the side chains are aliphatic or aromatic.

Gramicidin S, a cyclodekapeptide containing two D-Phe residues, gave no detectable CE in the range of the ^1L_b-band of the aromatic chromophore.[120–122] Whereas the reported ORD-data are very similar, the CD spectra differ somewhat in shape. Different conclusions about the preferred conformation of the cyclopolypeptide were, therefore, drawn from these data.

The CD of oxytocin and several analogues at different pH-values have been measured.[123] All these compounds give a positive CE about 250 nm, which must be ascribed to the chiral disulphide bond. Its magnitude is strongly influenced by the presence of the α-amino group of oxytocin and its analogues. Its band position indicates a torsion angle of the —S—S— bridge of about 90°. A CD-band at about 280 nm may in part be due to the disulphide chromophore, in part to the Tyr residue (if present). A positive band at 225 nm with oxytocin and vasopressin could be attributed to the Tyr chromophore.

D. Polypeptides and Proteins

1. Theory

In 1955 Cohen[124] suggested that a change of α-helix content of a polypeptide or protein is echoed by a change of the rotation of plane polarized light, and shortly afterwards Moffitt[125–127] discussed this problem from a theoretical point of view. By assuming exciton coupling[128] he showed that strong CD-bands of opposite signs (termed 'couplet' by Schellman[30]) should originate from each strong electronic transition such as the $\pi^0 \to \pi^-$ transition at 190 nm, and these have parallel $(R_{\parallel})$ and perpendicular $(R_{\perp})$ polarization with respect to the axis of the α-helix. It was pointed out, however, that Moffitt[125,127] neglected some important terms;[129] though it has been estimated later, that these terms are not more than minor 'end-corrections'.[28,130] Tinoco[131–134] showed that for an infinitely long helix they cause the appearance of another couplet located at $v_{\perp} \pm \Gamma/\sqrt{2}$ ($v_{\perp}$ being the frequency of the perpendicularly polarized absorption band, and Γ a 'damping factor', which can be estimated empirically). This refined exciton treatment predicts, therefore, four bands for each strong absorption. For the α-helix they should occur at 185, 189 $(R_{\perp})$, 193 and 195 $(R_{\parallel})$ nm,[134] and furthermore the $n_1 \to \pi^-$-transition of the peptide bond was also predicted to be optically active near 225 nm (Schellman and Oriel[42,116,135]). Present instrumentation does not allow to resolve these five bands completely, especially below 190 nm. Despite the more complex results of the refined theory, the original Moffitt treatment in general describes satisfactorily the CD- and ORD-behaviour of the α-helix, and polarization measurements[135] confirmed this. Reviews about the CD and ORD of polypeptides and proteins are found, e.g., in references 3, 103, 105, 111, 136–142, 142a, 142b.

A simplified version of Moffitt's[125,127] treatment of the α-helix (following reference 141) is given in Figure 12. In a right-handed α-helix (P-helix) 3·6 peptide groups are arranged in one turn. The direction of the electric transition moment vector of the $\pi^0 \to \pi^-$ band is known from polarization spectra of single crystals and lies in the direction of the small arrows of Figure 12. According to exciton theory only two types of cooperative transitions are possible: (a) all transitions in phase (Figure 12(a)), and (b) each out of phase from the next by the angle δ (this means, two 'up' and two 'down', as in Figure 12(b)). The sum of all electric transitions moment vectors

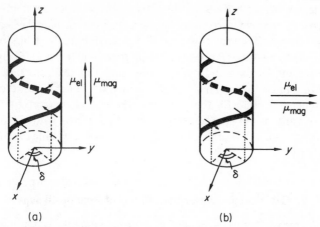

Figure 12. Simplified illustration of parallel (a) and vertical (b) polarized CD-band of a polypeptide. For details see text

for case (a) points in the direction of the positive z-axis, and the magnetic transition moment vector (right-hand rule) is antiparallel to it. According to equation (8) this must lead to a negative CD for this parallel polarized band. In case (b) the electric transition moment vectors point along the $+y$-axis, as does μ_{mag}, and the CD for this perpendicular band is, therefore,

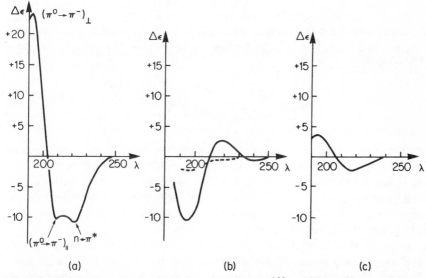

Figure 13. Typical standard CD-curves of α-helix (a),[158a] random coil according to Timasheff[156] (not entirely disordered) (———), and Quadrifoglio and Urry[158b] (— — — —) (b), and antiparallel β-structure[158c] (c)

positive (depending on the beginning of the pairing these vectors may either point in the y- or the x-direction. As the rotational strength of each such transition is only half of that of the parallel polarized band, their sum equals the latter). Exciton theory predicts the perpendicular band to appear at shorter wavelengths than the parallel one, and this leads to a couplet as indicated in Figure 13(a). That the experimental value[143] for the 192 nm band is higher than that for the 204 nm band may come from the fact that the positive wing of the second couplet[131-134] overlaps with the perpendicular band. In Figure 13(a) the $n \rightarrow \pi^*$ CD-band is also visible for the α-helix with P-helicity.

2. ORD outside Absorption Bands of Peptide Groups

As at the time of Moffitt's calculations CD-instruments were not available, he transformed these results with the aid of the Kronig–Kramers transform[18,19] into rotation values. Thus he obtained for the rotivity[21] per amino acid residue (reduced mean residue rotation) a two-term Drude equation (9) by assuming only an equilibrium between the α-helix and the random coil (neglecting the $n \rightarrow \pi^*$-band) outside the absorption bands

$$[\Phi]_{res} \cdot \frac{3}{n^2 + 2} = [m'(\lambda)] = a_0 \frac{\lambda_0^2}{\lambda^2 - \lambda_0^2} + b_0 \frac{\lambda_0^4}{(\lambda^2 - \lambda_0^2)^2} \tag{9}$$

a_0, b_0 and λ_0 are constants; a_0 depends on the solvent, the side chains, etc., b_0 mainly on the helix content, and λ_0 is usually assumed as 212 nm. By a simple transformation (9) can be linearized to (10), the famous 'Moffitt-Yang equation':[126]

$$[m'(\lambda)] \cdot \frac{\lambda^2 - \lambda_0^2}{\lambda_0^2} = a_0 + b_0 \frac{\lambda_0^2}{\lambda^2 - \lambda_0^2} \tag{10}$$

By plotting $[m'(\lambda)] \cdot (\lambda^2 - \lambda_0^2)/\lambda_0^2$ versus $\lambda_0^2/(\lambda^2 - \lambda_0^2)$, a_0 is the intercept and b_0 the slope of the straight line; b_0 is about -630 for 100 per cent P-helix, and 0 for the random coil. From such data the helix content of polypeptides and proteins have been determined quite frequently, though one must be very cautious, as other 'rigid' geometric arrangements may give CD-bands very similar to those of the pure α-helix (cf. p. 95).

Blout[138,144-146] took into consideration the CE of the random coil, and from ORD-measurements inferred, that the α-helix has got 2 CE at 193 and 225 nm (the latter may include the $n \rightarrow \pi^*$ transition), the random coil such mainly at 198 and 225 nm. The corresponding Moffitt equation (11) was then simplified by combining the two 225 nm terms as well as the 193 with the 198 nm term to a 'modified Drude equation' (12), which can be linearized to (13):

$$[m'(\lambda)] = \frac{A(\alpha)_{193} \cdot 193^2}{\lambda^2 - 193^2} + \frac{A(\alpha)_{225} \cdot 225^2}{\lambda^2 - 225^2} + \frac{A(\rho)_{198} \cdot 198^2}{\lambda^2 - 198^2}$$

$$+ \frac{A(\rho)_{225} \cdot 225^2}{\lambda^2 - 225^2} \tag{11}$$

$$[m'(\lambda)] = A(\alpha, \rho)_{193} \cdot \frac{193^2}{\lambda^2 - 193^2} + A(\alpha, \rho)_{225} \cdot \frac{225^2}{\lambda^2 - 225^2} \tag{12}$$

$$[m'(\lambda)] \cdot \frac{\lambda^2 - 193^2}{193^2} = A(\alpha, \rho)_{193} + A(\alpha, \rho)_{225} \cdot \frac{225^2}{193^2}$$

$$+ A(\alpha, \rho)_{225} \frac{225^2 - 193^2}{193^2} \cdot \frac{225^2}{\lambda^2 - 225^2} \tag{13}$$

The Greek letters α and ρ refer to the α-helix and the random coil, respectively, and all A's are constants. $A(\alpha, \rho)$ indicates that these coefficients are depending on the α-helix and the random coil. $A(\alpha, \rho)_{193}$ and $A(\alpha, \rho)_{225}$ can be easily extracted from the straight line corresponding to equation (13), and for aqueous solutions they are linearly related by equation (14)

$$A(\alpha, \rho)_{225} = -0.55\, A(\alpha, \rho)_{193} - 430 \tag{14}$$

The helix content H_λ can be determined from either constant according to (15) as

$$H_{193} = \frac{A(\alpha, \rho)_{193} + 750}{36.5}; \; H_{225} = -\frac{A(\alpha, \rho)_{225} + 60}{19.9}(\%) \tag{15}$$

In organic solvents these equations are altered to (14a) and (15a)

$$A(\alpha, \rho)_{225} = 0.55\, A(\alpha, \rho)_{193} - 280 \tag{14a}$$

$$H_{193} = \frac{A(\alpha, \rho)_{193} + 600}{36.2}; \; H_{225} = -\frac{A(\alpha, \rho)_{225}}{19.0}(\%) \tag{15a}$$

From (15) and (15a) one can derive a new equation (16) which is then applicable in all cases independent of the solvent used:

$$H = \frac{A(\alpha, \rho)_{193} - A(\alpha, \rho)_{225} + 650}{55.8}(\%) \tag{16}$$

The Blout equation may be applied in the range between about 600 and 280 nm, i.e. outside the absorption bands. If the calculated content H_{193} and H_{225} (equation 15 or 15a) are identical within 5 per cent it may be inferred that no other structures than the α-helix and the random coil have to be considered. If they deviate, however, more than this, one of the β-structures or other conformations have to be present. It should be emphasized that mathematically the Moffitt equation (9) and the Blout equation (12) are equivalent, and that, therefore, the helix content should equally well be determinable from a_0 and b_0. From the special choice of constants, however, it follows that the helix content determined from A's is less sensitive to small errors than H determined from a_0 and b_0. Assuming $\lambda_0 = 212$ in the Moffitt equation there holds the relationship

$$a_0 = 0.829 \, A(\alpha, \rho)_{193} + 1.13 \, A(\alpha, \rho)_{225} \tag{17}$$

$$b_0 = -0.142 \, A(\alpha, \rho)_{193} + 0.142 \, A(\alpha, \rho)_{225} \tag{18}$$

From this is apparent, that the b_0 for 100 per cent helix changes with the assumed λ_0, as does the wavelength range where the Moffitt equation is applicable.[136]

The calibration of these constants was done with the aid of poly-Glu, which adopts an α-helix at pH 4·3 and a random coil conformation at pH 7·0. The application of the Blout equation is demonstrated by Table 5, which shows examples of polypeptides and proteins in aqueous solution.[138,144]

Table 5. Determination of helix content of polypeptides and proteins in aqueous solution[138,144]

Substance	pH	$A(\alpha, \rho)_{193}$	$A(\alpha, \rho)_{225}$	H_{193}	H_{225}
poly-L-Glu	4	+2900	−2050	100	100
poly-L-Glu	7	−750	−60	0	0
poly-(L-Glu-L-Lys) (1:1)	3	+1400	−1180	59	57
poly-(γ-morpholinyl-L-Gln)	7	−610	−90	4	2
poly-L-Ser	7	+10	−80	21	1
poly-L-Pro II	7	−2680	−590	−53	27
Paramyosin	5–8	+2780	−1940	97	95
Tropomyosin	5–8	+2410	−1750	87	85
Myosin	5–8	+1535	−1200	63	58
Fibrinogen	5–8	+580	−680	35	31
Fibrinogen (9M urea)	5–8	+100	−500	24	22
β-Lactoglobulin	5–8	+420	−420	32	18
Pepsinogen	5–8	+60	−250	22	10

For poly-L-Ser, poly-L-Pro II, β-Lactoglobulin and Pepsinogen it follows immediately from these data, that other structures besides the α-helix and the random coil are present. As has been mentioned earlier (cf. p. 88) some diketopiperazines or cyclic peptides give CD-curves which are very similar to those of the α-helix, and should then also give parameters in the Moffitt or Blout equation simulating such a structure. Nevertheless for many proteins, especially globulins, ORD-data in solution and X-ray studies in the crystalline state gave comparable results with regard to helix content.

The ORD-curve outside the absorption range of the β-form can also be fitted a Moffitt equation.[147] The a_0 is mostly positive, whereas the b_0 seems to be close to zero. These values are still very uncertain.[148] A recent calculation[149] gave a theoretical value of $b_0 = -30$. For the poly-L-Pro II structure two CE have to be assumed (large negative at 207, positive at 221 nm) to fit the ORD-curve above 300 nm.[146]

3. ORD and CD in the Range of the Amide Absorption

Recent instrument developments have made possible the penetration of the range below 250 nm (down to about 185 nm) for ORD- and CD-measurements. Those CE which have been inferred from ORD measurements above 250 nm from Drude type equations (Moffitt, Blout, etc.) can now be investigated directly and there is agreement in general that CD-measurements are easier to interpret than ORD-data; using the Kronig–Kramers transform[18,19] one can calculate the ORD-curve from CD-data and if there is no good fit with the experimental ORD-curve this indicates other optically active bands below 185 nm. Their position and magnitude cannot, however, be calculated with certainty as this would need an assumption about the number (and rough position) of these transitions.

Several authors have published ORD- and CD-values for the α-helix, random coil, β-structure and other conformations, and Tables 6 and 7

Table 6. ORD-data of various polypeptide conformations

Conformation	Reference	Extrema given as $\lambda([m'])$ [a]
(P)-α-Helix	148	232–3 (t, -15000), 198–199(p, $+68000$), 182–184 (t, neg.).
	150	233 (t, -14000), 198 (p, $+78000$).
Random coil	148	238 (t, w neg.), 228 (p, w neg.), 205 (t, -15000), 189 (p, $+17000$).
	150	242 (t, w neg.), 222 (p, w neg.), 204 (t, -16000), 190 (p, $+12000$).
β-Form (anti-parallel)	148	229–230 (t, -5000), 205 (p, $+24000$), ~ 190 (t, -17000).
	151	230 (t, -6220), 205 (p, $+29200$).
poly-L-Pro I	152	223 (p, $+47500$), 208 (t, -90000).
poly-L-Pro II	146	216 (t, -27000), 196 (p, $+24000$).
	152	216 (t, -41000), 192 (p, $+28000$).
	153	216 (t, -32000), 194 (p, $+19000$).

[a] λ in nm, $[m']$ in deg cm^2 decimole^{-1}; abbr.: p peak, t trough, w weakly, neg. negative

Table 7. CD-data of various polypeptide conformations

Conformation	Reference	Maxima given as $\lambda(\Delta\varepsilon)^a$
(P)-α-Helix	105	$222(-12\cdot1), 206(-11\cdot2), 189(+23\cdot3)$.
	143, 158	$222(-9\cdot0), 208(-8\cdot5), 190(+20\cdot1)$.
	148	$222(-11\cdot5), 209-210(-10\cdot9), 191(+25\cdot8)$.
	154	$221(-12\cdot1), 207(-11\cdot8), 190\cdot5(+21\cdot3)$.
	158a	$222(-10\cdot8), 208(-10\cdot0), 191(+23\cdot2)$.
Random coil	105	$220(+0\cdot8), 202(-14\cdot5)$.
	143, 158	$243(-0\cdot7), 220(+1\cdot0), 198(-8\cdot5)$.
	148	$238(-0\cdot6), 218(+1\cdot4), 199(-10\cdot6)$.
	156	$238(-0\cdot06), 217(+0\cdot9), 196(-10\cdot0)$.
	158b, 191a	$215(-0\cdot50), 198(-2\cdot2)$.
β-Form (anti-parallel)	148	$218(-6\cdot1), 195-197(+14\cdot5)$.
	154	$217(-5\cdot8), 195(+8\cdot5)$.
	155	$218(-7\cdot0)$.
	156, 157	$217(-3\cdot6), 197(+12\cdot8), 190(+11\cdot2)$.
	158c, 191a	$217(-2\cdot20), 195(+3\cdot7)$.
	159	calculated: $218(-), 198(+), 195(+)$.
β-Form (parallel)	159	calculated: $216(-), 181(+)$.
poly-L-Pro I	156	$236(-1\cdot2), 214(+17\cdot6), 200(-8\cdot9)$.
poly-L-Pro II	146	$229(+0\cdot5)$.
	156	$224(+1\cdot1), 205(-16\cdot0)$.

^{a} λ in nm. All values (besides those in reference 159) are measured maxima and not the maxima of resolved CD-bands; they are not corrected for solvent refractivity

summarize some of these results. The differences come from the fact, that different substances have been used for these determinations, but it must be emphasized that below 220 nm instrumental noise is already great and makes these values not so accurate as at longer wavelengths. For the random coil several authors described quite appreciable $\Delta\varepsilon$-values (cf. Table 7); these obviously correspond, however, not to a real 'random' coil but to a still somewhat ordered structure. Poly-L-serine in 8M LiCl seems to be the best model for the completely disordered coil[158b] and indeed only small CD-values have been found for it. Experimental CD-curves of these two different 'random coils'[156,158b] and the antiparallel β-sheet structure[158c] are shown in Figure 13(b) and (c), respectively.

4. Influence of Amino Acid Side Chains

The side chains can influence the chiral–optical properties[160] of a polypeptide either by changing the conformation or by giving rise to new CE, if they contain chromophores. Thus, the bulky nonpolar isobutyl residues stabilize the α-helix of poly-L-Leu in polar solvents due to 'hydrophobic interactions' more than that of poly-L-Ala, whereas in organic solvents the latter is more stable to disruption by TFA, and similar effects have been found in copolymers.[161-163] Also with the help of ORD it has been shown that poly-N^5-[HO(CH$_2$)$_n$]—L—Glu in water forms 65 per cent helix for $n = 4$, 20 per cent for $n = 3$ and no helix for $n = 2$. TFE does not disrupt the helices in all three cases, whereas in formic acid only random coils have been found.[164] Analogously has been explained the difference in helix stability of poly-L-Orn and poly-L-Lys.[165]

Aromatic side chains on one hand will affect the conformation of polypeptides by steric and electronic interactions with the helix backbone,[166] but may also give rise to CE of their own. In the ^1L_b-region they are usually quite small but can become stronger in the short wavelength region so that they may obscure partly or completely the typical CE of the different polypeptide conformations (cf. the review by Goodman and colleagues[140]). Thus although poly-L-Tyr in dimethyl-formamide and in pyridine gives a positive b_0 of the Moffitt equation,[167] it forms a (P)-α-helix in these solvents.[99,100,168] In other solvents poly-L-Tyr may, however, form (M)-helices.[140] Weak Cotton-effects have also been reported within the ^1L_b-band in the ORD and CD spectra of copolymers of L-Phe, and a negative trough at 237 nm suggested a (P)-α-helix,[169-171] but this must be taken with reservation. Poly-L-Trp gives a positive b_0, but is also assumed to form a (P)-α-helix in solution.[172] The ORD- and CD-spectrum in the 260–300 nm range is very complex due to optically active indole transitions, furthermore there are three CE at 226, 210 and 190 nm.[173] No conclusions about presence or sense of an α-helix can be drawn from these data, however, as strong indol bands are also present in this range. By ORD and CD it was demonstrated, that poly-Bzl-L-Asp can adopt left- and right-handed α-helices in different solvents.[140] Introduction of a p-NO$_2$ group into the benzyl residue reverses the sense of the helix, and from the CD-band at 330 nm it was inferred that the nitrobenzyl residues interact pairwise if they are separated by four amino acid units.[174]

The disulphide chromophore gives rise to two CE about 260 and 200 nm in chiral environment,[103,143,175] which cannot always be seen in polypeptides and proteins. In arginine vasotocin and 8-L-ornithine vasopressin the 200 nm effect dominates the ORD properties in this range and has the same sign as in the model compound N,N'-diacetyl-L-cystinebismethylamide, whereas the 260 nm effects have opposite signs in the ORD-spectrum of this and arginine vasotocin.[175] With several disulphide-containing polypeptides, however, (except in one case) only the 260 nm band was observable in CD. It was concluded from these facts,[175] that only the 200 nm band has the properties of an inherently dissymmetric chromophore, whereas the 260 nm CE is mainly governed by the chiral environment of the —S—S— group.

5. Complexing with Dyes, Substrates, etc.

Certain achiral organic dye molecules such as acridine orange form complexes with polypeptides and show CE in the range of their absorption bands, if the polypeptide is in an ordered conformation.[176,177] The signs of these CE reflect the sense of the helix and could, therefore, be used for its determination. It is not necessary (at least in all cases) to assume some sort of 'superhelix' around the polypeptide helix or another such ordered structure built up from many dye molecules, as already one molecule per peptide helix (covalently fixed to its end) gave this effect. Coleman[178] has used an azosulphonamide dye as 'probe' to investigate carbonic anhydrase–sulphonamide complex formation. In actinomycin D (**38**) all absorption bands of the

(38)

chromophore actinocin are optically active by induction from the two identical oligopeptide–lactone rings.[179]

Many authors have investigated porphyrine proteids, as cytochrome c, etc. Thus, for example, Urry[180] used small heme peptides, obtainable by peptic digestion, as model compounds for the natural proteids, and Vinogradov and Zand[181] measured the CD of ferri and ferro forms of cytochrome c from different sources. Especially within the Soret band the CE are strongly influenced by aggregation. Magnetooptical rotation measurements (MORD) of haemoglobin and myoglobin derivatives give additional information about structural changes and type of ligands in these proteids.[182] H^+ and OH^- denaturation caused disappearance of the MORD bands. Correlation with Mössbauer effect data proved the iron ion to be the source for some of these bands.

Whereas FAD and $FADH_2$ show only a negligibly small CE, the holoenzyme D-amino acid oxidase gives well observable CD bands and anomalous ORD in the range of the flavin chromophore.[183] Helix content, determined either from the Moffitt–Yang equation or the CD band in the low u.v. of the holo- and apo-enzyme are, however, very similar, so that it was concluded that the binding site of the protein for the apoenzyme is in the non-helical region. The positions and magnitudes of the optically active absorption bands in the non-heme iron proteins xanthine oxidase and spinach ferredoxin are found to be very similar.[184] These bands must, therefore, have their origin in ligand field transitions of the iron, whereas the molybdenum and the flavin chromophores apparently are bound to a protein site introducing only low asymmetry. From the CD it was inferred that xanthine oxidase contains four identical pairs of iron atoms, while spinach ferredoxin contains only one such pair.[184]

Differences in the ORD- and CD-properties of enzymes in presence and absence of their substrates have been found in several cases, but must not necessarily be ascribed to direct influence of complexing. Thus the change in b_0 of D-glyceraldehyde-phosphate dehydrogenase containing bound NAD^+ by adding glyceraldehyde, with which it forms a rapidly hydrolysable complex, is stable for about 2 hours, and is due to a change of conformation of the protein by reduction of the coenzyme to NADH.[185]

Peroxidase[186] does not change its CE properties in the low u.v. by forming substrate complexes with H_2O_2 or $NaNO_2$. Similarly acetylcholinesterase from the electric eel does not appreciably change its ORD in the low u.v. by complexing with acetyl homocholin chloride as a substrate, whereas other inhibitors give drastic alterations.[187] Substrate binding with luciferase shows, on the other hand, from ORD a decrease of helix structure.[188] The CD of human serum low-density lipoprotein is not altered much by succinylation or by changing the solvent, whereas the lipidfree protein lacks this resistance.[189] From this it was assumed that the lipids stabilize the protein conformation of the native lipoprotein complex.

In cases of very small changes of the chiroptical properties, as during complexing of enzymes with their substrates, the technique of difference spectropolarimetry was used with success[190] in the case of lysozyme and its complex with an inhibitor. A summary of ORD- and CD-studies of enzymes was given by Torchinskiĭ.[191]

6. Determination of Helix Content

As mentioned above, helix content H may be determined either from ORD- or CD-values in the range above or below about 260 nm, and Carver, Shechter and Blout[146] have carefully compared the diverse methods. If there are no aromatic or disulphide chromophores present, H may be obtained equally well from the Moffitt-Yang equation (10) through b_0 or the Blout equation (12). Both are applicable only if no other conformations than the α-helix and the random coil are present. Deviations from the linear equations (14) or (14a), or differences of H found from the pairs of equations (15) or (15a) indicate that other structures play a role; this fact is not so easily seen by applying just the usual Moffitt–Yang treatment.

Direct measurements of the amide Cotton-effects below 250 nm are possible today and give better information about the type of conformation present. Determination of H can be made by using the figures of Tables 6 or 7, one must, however, always keep in mind that unusual rigid polypeptide structures may give rise to CE very similar to those of, e.g. the α-helix (cf. p. 95). Determination of H from the $n \to \pi^*$-band at about 222 nm alone are not very reliable.

Recently Rosenkranz and Scholtan[191a] have published a curve fitting method for the determination of α-helix, random coil, and β-structure content in a polypeptide or protein. Using the CD-curves of poly-L-lysine in water for the α-helix,[158a] of poly-L-serine in 8M LiCl for the random coil,[158b] and of poly-L-lysine in one per cent NDS for the β-structure[158c] as standard curves (cf. Figure 13) they got excellent agreement between results of their method and those from X-ray data (in the crystalline state).

Timasheff and colleagues[156] use a combination of IR and CD (or ORD) spectra; the first give information about the predominant conformation, and from the chiroptical properties quantitative values may be obtained by curve resolution and curve synthesis. Jirgensons[192] uses the response of the CE to treatment of polypeptides and proteins in solution with anionic detergents like decyl and dodecyl sodium sulphates and acids: proteins of high α-helix content become somewhat disordered, those which are present mainly in the random coil acquire some new order, and the β-conformation partially is changed to the α-helix.

Presence of other absorbing groups in many cases obscures these determinations of H because e.g. all aromatic chromophores give strong CE also below 220 nm if in a chiral environment. These will equally affect b_0- or $A(\alpha, \rho)$-values as they will alter the shape, magnitude and even sign of the overall CD-bands in the low u.v. (cf. reference 140). Chiroptical methods alone will not give very reliable results in this case.

During the last few years many hundred papers appeared on the determination of helix content in polypeptides and proteins. Only a few representative examples could have been cited in this chapter.

V. COTTON EFFECT AND STEREOCHEMISTRY OF NUCLEOSIDES, NUCLEOTIDES AND NUCLEIC ACIDS

A. Nucleosides and Nucleotides

The heterocyclic bases present in nucleosides are achiral and acquire optical activity only by being linked to a sugar moiety (dissymmetry of third sphere). Whereas with pyrimidine derivatives most published data are consistent with each other, in the purine series obviously sometimes wrong data (due to their low g-factor) or misinterpretations lead to contradictory results. The situation is complicated by the fact that in the accessible range up to four $\pi \to \pi^*$ and one $n \to \pi^*$ CD-band(s) can be detected, which cannot be resolved satisfactorily in the ORD-curves.

Clark and Tinoco[193] showed that the spectra of pyrimidines and purines can be classified similar to the benzene bands into ^1L_b, ^1L_a and ^1B, and that this is more than pure formalism was shown recently[194] by comparing the CD-spectra of several nucleosides with their thiono analogues: between 185 and 300 nm these give the same CD-bands as the corresponding oxo compounds. In addition to these $\pi \to \pi^*$-bands an $n \to \pi^*$ CD-band has been detected by its characteristic dependence on solvent and pH in several nucleosides[195] (cf. p. 118).

1. Pyrimidine Nucleosides

Miles and colleagues[195] have published ORD, CD and u.v. data of several pyrimidine derivatives, and Figure 14 shows the CD of uridine (**39**) and cytidine (**40**). A positive CD at 267 ($+4\cdot2$), assigned a ^1L_b transition, a negative at 240 ($-2\cdot0$, ^1L_a), another negative at 210 ($-2, 2$) and a positive at 190 nm

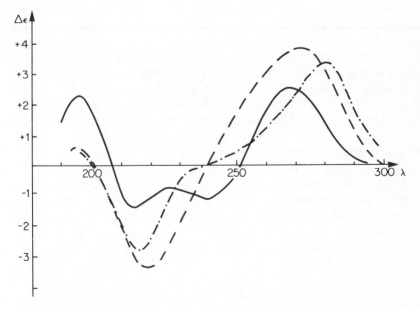

Figure 14. CD of uridine (————), and cytidine at pH 7 (– – – –) and pH 1 (– · – · – · – · –)

$(+4, 0)$ are found for uridine (39) in neutral and acidic (pH 1) medium. The two bands of opposite sign at 210 and 190 nm are assumed to be the couplet due to splitting of the otherwise degenerate ^1B band (of benzene).[30,196] Cytidine

(39)   (40)

(40), which is known to be protonated at N-3 gives on the contrary slightly different CD curves at pH 7 and 1 (Figure 14). The ^1L_b and ^1L_a bands are both positive and merge together, whereas the two branches of the ^1B couplet have the same signs as in uridine. The $n \rightarrow \pi^*$ transition gives rise to a negative CD band at about 230 nm which disappears in acidic solution, and which is not separated from the long wavelength branch of the ^1B couplet.

ORD gives less information in both cases, but was used extensively to correlate the stereochemistry of the glycosides with the sign of the CE. Mostly only a single CE was found around 260–270 nm, which in general will incorporate both the ^1L_b and ^1L_a CD band. Its sign was shown to reflect the stereochemistry at the anomeric C-atom of the furanose ring in both deoxyribosides[195,197–199] and ribosides[198–205]: β-D- and α-L-glycosides give a positive, α-D- and β-L-isomers a negative CE. The configuration of the OH group at C-2' does not change the sign of this CE but influences the magnitude; if it is *cis* with respect to the pyrimidine nucleus the ORD-amplitude is enhanced. The configuration of the OH-group at C-3' does not alter this CE, and nucleotides give the same sign as the corresponding nucleosides. Acetonide formation, acetylation or alkylation does also not change the sign and only slightly the amplitude.

Synthetic pyranosides (of pentoses and hexoses) also follow this rule[201,204] with one apparent exception[204] (α-D-glucopyranosylthymine). CD measurements[206] revealed, however, that (in water) there is indeed a small negative band (-0.14) at 287 nm which was overlooked in the ORD because of the presence of a bigger positive band $(+0.85)$ at 255 nm. The long wavelength branch of the ^1B couplet was found at 216 nm $(+1.62)$. In methanol solution the ^1L_b CD band becomes bigger $(-1.83$ at 275 nm) than the ^1L_a CD $(+1.44$ at 243 nm). This again is an excellent example of the advantage of CD over ORD in the nucleoside field.

The enhancement of the ORD amplitude of the 'single' CE in the 1', 2'-*cis*-nucleosides has been interpreted usually as being due to steric hindrance to rotation around the glycoside bond. Phosphorylation at $O^{5'}$ in α-D-nucleosides decreases the CE, whereas no such influence has been found in the β-D-anomers.[205] As there cannot be a direct steric interaction between the $O^{5'}$-substituent and the base in the α-D-compounds, these changes of the amplitude have been ascribed to alteration of the furanose ring conformation. As CD-measurements clearly showed that two bands of the same or opposite sign (cf. Figure 14) are present between 240 and 270 nm, which usually cannot be separated by ORD-measurements, such statements must be treated with reservation, as long as there are not more CD-data available. Though it is reasonable to assume that both the ^1L_b and the ^1L_a CD band will increase by better fixation of the chromophore, situations may be envisaged where fortuitously the third-sphere-contributions cancel each other in such a rigid conformation.

In order to determine the preferred conformation of the pyrimidine nucleus in solution from CE-data cyclic model compounds have been investigated, whose conformations are known.[200,201] The amplitude* of uridine (39) is +117, of thymidine +95, that of the O^2-2'-cyclouridine (41) +266, of the O^2-3'-cyclouridine (42) +43, and of O^2-5'-cyclothymidine (43) bigger than −299. From these and some similar data an empirical rule was

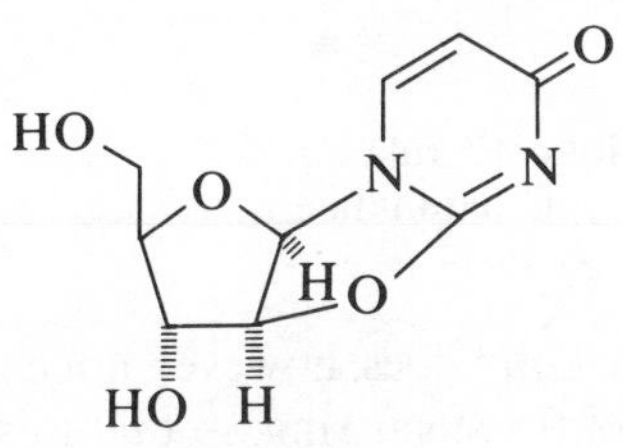

(41)

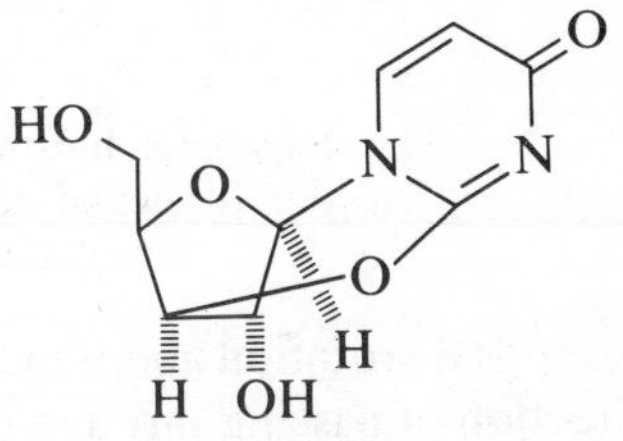

(42)

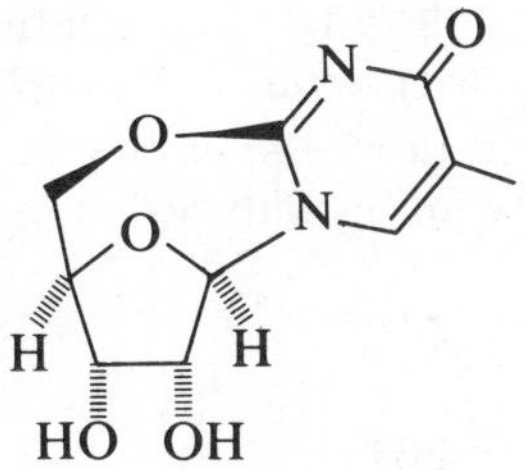

(43a)

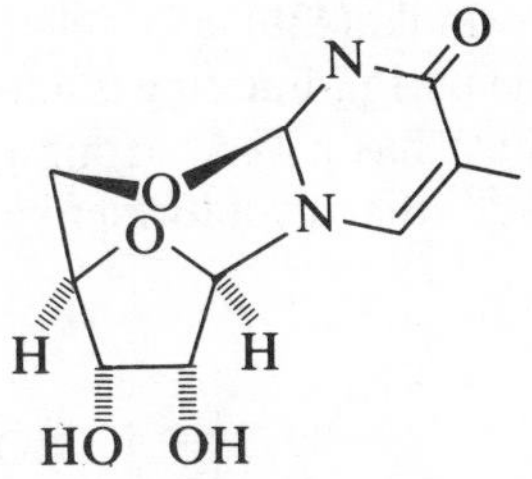

(43b)

* As can clearly be seen from the published[201] ORD-curve two CE are present between 220 and 280, but the amplitude was taken over the combined and not of one of the resolved effects.

derived[201] which is shown in Figure 15. This rule can be applied if the pyrimidine ring has a preferred conformation due to restricted rotation about the glycosidic bond. One has to draw a line from the C^4-carbonyl (C^4-NH_2 in cytosine derivatives) passing through the C^2-carbonyl; if this hits the plane of the furanose ring from above, the CE is positive, provided that it does not pass through $C^{5'}$, in which case it is strongly negative. In any

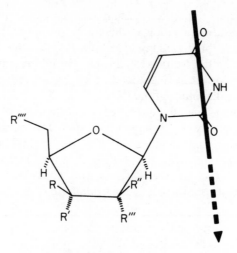

Figure 15. Illustration of Ulbricht's rule for the 'principal ORD-band' of pyrimidine nucleosides. Details see text

β-D-furanoside rotation about the glycosidic bond does, however, not change the direction of passing through the plane of the sugar ring, so that this rule is equivalent with the statement, that a β-D-glycoside gives a positive CE (except for type (**43**)).

Taking the plane of the pyrimidine nucleus of compound (**43**) (O^2 bent inwards, formula (**43b**)) as a reference plane, the angle between this and the plane of the pyrimidine ring is about 110° in (**41**), about 135° in (**42**) and 180° in (**44**), which has a very strong positive CE ($a = +490$). It was, therefore, concluded,[201] that a conformer having O^2 pointing 'outwards' (angle greater

(**44**)

than 90°) gives a positive, one having O^2 pointing 'inwards' (angle smaller than 90°) a negative CE, and this rule (which contradicts the above one in case of O^2-5'-cyclocompounds) was then extrapolated to nucleosides without such an additional ring.[201] The positive CE of β-D-isomers was taken as an indication, that the anti-conformation (as given in Figure 15) is preferred. This agrees with n.m.r. results[207–209]* but seems to be quite fortuitous, because e.g. in cytidine the ^1L_b and the ^1L_a CD-band have the same sign, whereas in uridine they have opposite signs. The ORD-amplitude, as measured hitherto, takes, however only the sum of both and it is not clear which band one can correlate in a simple way with the conformation. Furthermore, it has been pointed out that the second possible conformation of O^2-5'-cyclothymidine (**43a**) taken as reference would lead to other torsion angles about the glycosidic bond. Another complication is the fact that in the common nucleosides O^2 is present as a $C=O$ group, but in the cyclocompounds as an 'enol ether'.

Recent CD-measurements of many uridine and cytidine derivatives[210a] have indeed shown that these generalizations must be taken with care and that now several exceptions to Ulbricht's rule are known. On the basis of our earlier assumptions[38] the CE of nucleosides and cyclonucleosides cannot be compared directly, because in the latter one bond of the chromophoric ring is incorporated into a chiral ring, thus leading do dissymmetry of the second sphere. It is, therefore, not the torsion angle about the glycosidic bond, which determines the sign of the CE in cyclonucleosides, but the chirality of this additional ring.

Figure 16 shows projections (from the pyrimidine ring towards the newly formed ring) for (**41**), (**42**), and the two conformations of (**43**). Due to the *cis*-fusion of two five-membered rings the second sphere of (**41**) may either be symmetric or twisted slightly, though the sense of twist cannot be predicted from models unequivocally. Only the conformation with coplanar second sphere is arbitrarily given in Figure 16. In any case, only a small CE will be expected, and this is true for the ^1L_b-band (-0.3),[210a] which in the first place should be prone to such a correlation. The ^1L_a-band, on the other hand, is very strong, and both are merging together in ORD, thus leading to wrong interpretations.

The projection of (**42**) shows that there are two chiral second spheres, a six-membered and a seven-membered ring, and their chiralities are of opposite signs, their contributions to the CE will, therefore, cancel to a great extent, and this explains the small amplitude in ORD (this argument will even hold if both bands merge together).

In the 'exo'-form (**43a**) of the O^2-5'-cyclo compound there are again two chiral second spheres present, of which only the seven-membered ring should

---

* Orotidine according to recent n.m.r.-measurements prefers the syn-conformation.[209a]

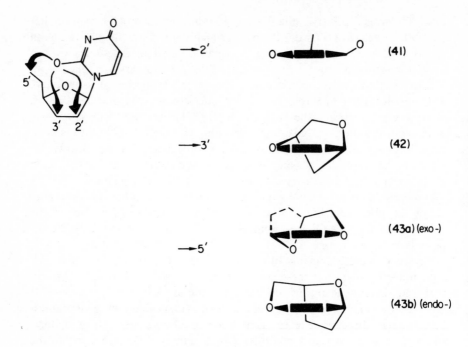

Figure 16. Schematic projections of second spheres in cyclic pyrimidine nucleoside derivatives

have greater influence upon the CE, as the eight-membered is by far less dissymmetric. The CD (of the ^1L_b-band) is thus expected to be greater for (**43**) than for (**41**), which is indeed the case.[210a] That the signs of the CE (ORD-measurements) of (**43**) and (**44**) are opposite is in agreement with general experience, that a change of the oxygen substitution pattern of an aromatic system may have a fundamental influence upon its CD.[210] In the 'endo'-form, which seems less probable from molecular models, the second sphere has nearly C_{2v}-symmetry (besides distribution of oxygen atoms and eight-membered ring).

Miles and coworkers[210a] have given curves for the dependence of the ellipticity of the ^1L_b-band on the torsion angle around the C-1′—N— bond for α- and β-anomers of uridine and cytidine derivatives. The cyclic derivatives mentioned do not follow this rule, which fact reflects the importance of second sphere influences, as discussed here.

2. Purine Nucleosides

Purine nucleosides also give several optically active bands between 190 and 300 nm, which have been investigated by ORD and CD. Figure 17

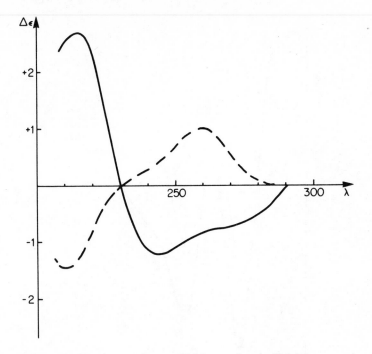

Figure 17. CD of guanosine at pH 7 (————) and pH 0·5 (------)

shows the CD-curves of guanosine (**45**) (two pH values)[211] and Figure 18 those of a cyclo derivative (**46**). According to Miles, Robins and Eyring[195] between 240 and 300 nm there appear the ^1L_b, ^1L_a and an $n \to \pi^*$ band in (**46**); these merge together into a 'principal band' in ORD and its sign is positive for α-D-furanosides and negative for β-D-furanosides.[198,205,212] The g'-factor of purine nucleosides is smaller than for corresponding pyrimidine derivatives, presumably because of a smaller barrier to rotation about

(45)

(46)

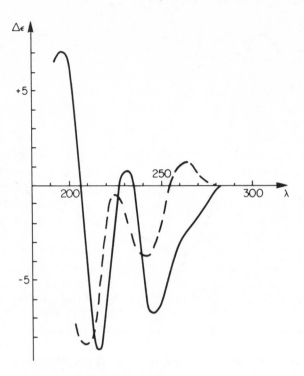

Figure 18. CD of **(46)** at pH 7 (————) and pH 1
(– – – – – –)

the glycosidic bond.[213] Cyclic derivatives have here also been used to determine from the sign of the CE the preferred conformation.[195,207,212,214–217] The same principal reservations against such comparisons as have been outlined already in the above chapter are valid with purine derivatives, too. Furthermore, in some cases for the same compounds completely different ORD curves have been published (e.g. for a cycloadenosine derivative),[212,214] and from identical ORD results contradictory conclusions about preferred torsion angles have been derived (syn-[207,215] against anti-[212]). Contrary to the results in the pyrimidine series cyclic derivatives with a bridge from C-8 to C-2′, C-3′ and C-5′ give always a positive CE increasing in the given order.[217] Replacement of S for O in the bridge results in a bathochromic shift and in enhancement of the amplitude. From n.m.r. data it was concluded that the bridge adopts 'exo' conformation (cf. **(43a)**) in the 8,5′-cyclic derivatives.[217] N.m.r. also favours anticonformation for purine nucleotides.[208]

The influence of solvents and substitutions on the CE and conformation of guanosine derivatives has recently been investigated.[217a]

The amplitude of the CE is smaller in the nucleotides than in the nucleosides,[205] and the adenosine glycosides are exceptions to the rule that the CE is smaller in $1'$. $2'$-*trans*-compounds than in their *cis*-isomers. Acetylation, formation of isopropylidene derivatives, replacement of the $2'$-OH by H or Cl, and substitution in the purine ring does not change the sign of the 'principal CE'.[212] Replacement of $5'$-OH by sulphur leads to inversion of the sign of the CE.[207]

Protonation (assumed to take place at N-7) drastically changes the CE of guanine glycosides[195,198,211] (cf. Figure 17) and this was ascribed[211] to a change of torsion angle about the glycosidic bond. In the CD-spectrum of the cyclic derivative (**46**) a similar inversion is observable only at about 270 nm and is assumed to be due to presence of a strong negative $n \to \pi^*$ band which completely hides the positive ^1L_b band at neutral pH, which disappears after protonation.[195] This explanation may also hold for the sign inversion of guanosine.[211] In concentrated solution guanosine-$5'$ phosphate forms aggregates, which show enhanced optical rotation; from CD and polarization measurements it was deduced, that this aggregation is due to a formation of a (P)-helix and not to layered structures.[218]

3. Dinucleotides

Older ORD-data[219-21] of oligo- and polynucleotides can be described by a one-term Drude equation. CD[222-7] show that there are at least two CE between 230 and 290 nm, and their magnitudes are temperature dependent. Increasing temperature decreases the ellipticities, and in general it is assumed that base stacking already takes place in dimers. The CD at low temperature can be taken as this of the ordered structure, and at higher temperatures the shape and magnitude of the CD-curves approach those of the monomers. For $3'$. $5'$-dimers two types of curves have been observed: (i) 'conservative' and (ii) 'nonconservative' (Figure 19). In case (i) there appears a positive band at longer, a negative one at shorter wavelengths, the crossover point approximately coinciding with the λ_{max} of the u.v. spectrum. The rotatory strength of both is equal. In case (ii) there are 2–3 bands in the same wavelength range which do not compensate to zero. The first behaviour is found with ApA, ApC, CpA, CpU, etc., the latter e.g. with GpA, GpC, CpC, though the u.v. spectra of both groups are very similar. Calculations[227-9] according to the exciton model lead for this $\perp$ polarized band to conservative behaviour, if one assumes nearest neighbour interactions only,[133] and, for example, the CD curve of ApA (Figure 19) must correspond to a (P)-helix. A distance of base planes of 3·4 Å and a number of 10 bases per turn were assumed in these calculations. The reason for the nonconservative type of CD spectra was attributed to far-u.v. contributions[224] and not to such facts as tilting of the bases or difference of bases in the dimers. These interactions

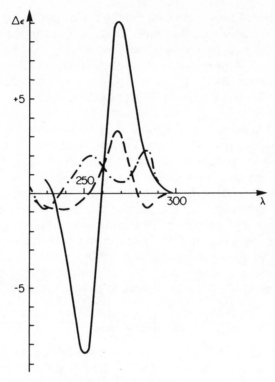

Figure 19. CD of ApA (pH 7·4, ————), GpA
(pH 7, – – – – – –) and GpC (pH 7, – · – · – · – · –)

must be great in G- and C-containing dimers, but small in such containing only A and U.

'Melting curves' (ellipticity of long wavelength band vs temperature) show typical sigmoid shape for all 3′ . 5′-dimers, from which thermodynamical data can be calculated. The standard state enthalpy change is about 6 to 7·5 kcal/mol and the standard state entropy change 20 to 25 e.u./mol and no fundamental differences were found which would allow the division into different categories. Dimers containing G or U are, however, slightly less favoured in the stacking conformation than those of A and C. The CD of the 2′ . 5′-dimers is much smaller than that of the 3′ . 5′-isomers, but similar in shape. In the temperature range between − 20° and + 80° the change of ellipticities is only about 10–40 per cent and linear, which was interpreted[223] as being due to presence of mainly random coil even at − 20°. The stability of the stacked conformation was attributed to a hydrogen bond between 2′-OH and phosphate group (using the arabinose isomers[225]). Recent u.v. measurements[230] showed, however, that typical hypochromism and excimer formation is already possible in model compounds which cannot form such hydrogen bridges, as e.g. in Base-CH_2,$CH_2 \cdot CH_2$-Base. These newer CD-data parallel earlier ORD work (e.g. reference 231).

B. Higher Oligonucleotides and Polynucleotides

Assuming only nearest neighbour interactions, Cantor and Tinoco[232,233] derived equation (19) for the molar rotation per residue

$$[\Phi_{IJK}(\lambda)] = \frac{2[\Phi_{IJ}(\lambda)] + 2[\Phi_{JK}(\lambda)] - [\Phi_{J}(\lambda)]}{3} \tag{19}$$

if $[\Phi_{A}(\lambda)]$, $[\Phi_{AB}(\lambda)]$ and $[\Phi_{ABC}(\lambda)]$ are the molar residual rotations of the monomers, dimers and trimers, respectively. Very good agreement at neutral pH could be obtained for most trinucleotides,[226,231–234] but about the importance of a terminal 3′-phosphate disagreeing assumptions have been published.

Using exciton theory Bush and colleagues[227,229,235,236] calculated from this model of nearest neighbour interactions[133] that the N-mer should lead to N bands (perpendicularly polarized), whose positions are given by the equation

$$v_K = v_0 + \frac{2V}{h} \cos \frac{K\pi}{N+1} \tag{20}$$

where v_0 is the band position in the monomer, and V is the interaction energy. The rotatory strengths of these bands are

$$R_{\pm} = \mp \frac{\pi Z}{2\lambda_0} \mu_{el}^2 \sin \gamma \tag{21}$$

if Z is the vertical distance between the planes of two bases (3·4 Å) and γ is the angle of rotation of successive bases about the helix axis (good results with γ between 30 and 45°). The band at shortest wavelength is always negative, and though correct calculation of the complete CD-curve is complex, crossover points of it can be predicted easier. Figure 20 gives a schematic drawing of calculated band positions, rotatory strengths and crossover points for some A-oligomers and the observed crossover points (in water). The observed shift of the crossover point to shorter wavelengths by increasing N is correctly predicted.

A relative simple theory taking into account all base–base interactions and not only those of nearest neighbours has recently been developed by Johnson and Tinoco[236a] for the CE of polynucleotides. It correctly predicts the conservative CD-spectrum of DNA as well as the nonconservative one of RNA. The latter is due to tilting of the basis with respect to the helix axis.

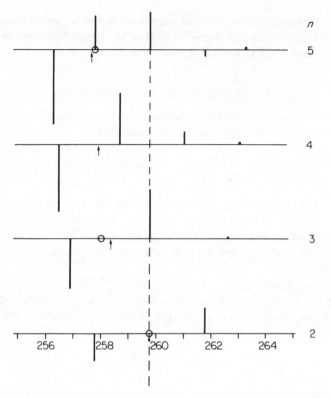

Figure 20. Schematic drawing of positions and rotational strengths of partial bands of small adenylate oligomers (modified from reference 229). Circles denote observed crossing points of the CD-curves, arrows calculated crossover points.
n: Number of monomer units

The contributions to the CE due to interactions with the second and third nearest neighbours together may be stronger than that with the nearest one. This theory leads, however, also to some wrong predictions for a few polymers.

A review on the different theories about the CD of polynucleotides and nucleic acids has recently appeared.[236b]

The rotational strength per mole of residue increases smoothly with the length of the oligomer chain at pH 7·4 for oligo-A. At pH 4·5, however, there is a big jump at approximately $N = 7$ to higher values, furthermore there appears a shoulder in the CD-band at 270–280 nm. poly-N^6-Hydroxy-ethyl-adenylic acid[227] shows the same CD behaviour in neutral and acidic medium, as poly-A at pH 7·4, and analogous conformations should, therefore, be present. As the former cannot build up a double-stranded helix, a single-stranded helix is assumed for both.

In acidic solution poly-A is known, however, to adopt a double-stranded parallel helix,[237] and such a conformation leads to enhancement of the positive partial CD-band at the expense of the negative one. Obviously such a double helix can be formed beginning with $N = 7$ in acid.

The 'melting' of the ordered structure of such oligomers can be observed in a similar way in the CD-curves, as for the dimers. It gives (for poly-A) a sigmoid curve and is independent of chain length, which indicates non-cooperative melting. At 0° there is already a break at each 8th residue of the stacking, and at room temperature only about two-thirds of the bases are stacked. Changes of conformation by change of the solvent can also easily be followed by CD. Results with other polynucleotides[238–42] parallel to those found for poly-A.

C. Nucleic Acids

The CD of nucleic acids was investigated thoroughly by Brahms and his colleagues.[236,241,244] All RNAs gave a similar CE of nonconservative type with a positive band at 265 nm at neutral pH (Figure 21). Two negative CD-bands appear below 250 nm[245,246] and a very faint negative one, which is not seen in synthetic polynucleotides (with the possible exception of poly-I), was found at about 295 nm.[245,246] Tentatively $n \rightarrow \pi^*$ origin was assigned to

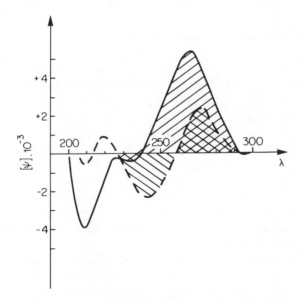

Figure 21. CD of yeast sRNA (————) and calf thymus DNA (– – – – – –). 'Conservative' and 'non conservative' behaviour refers to shaded bands

it.[245] The intensity of the main band at 265 nm decreases gradually with raise of temperature; this band must, therefore, be related to the helix structure. The nonconservative behaviour of RNA was ascribed[244] to the 20° tilting of the bases from the perpendicular plane to the helix axis.[247,248] The low ellipticity of RNA compared to polynucleotides could at least in part be due to compensation by folding and winding of the single strand upon itself.[244] The 295 nm band is more, the 235 nm band much less sensitive to heating than the 265 nm band.[245]

Various DNAs also have similar CD-curves to each other, but they are of conservative type as expected from the theory of Tinoco[131] in the range between 230 and 290 nm:[244–249] a positive maximum is found at 273 and a negative at 243 nm, and their intensity is smaller than that of the main band in the RNA CD-spectra (Figure 21). Still smaller CD bands appear below 230 nm, but their signs depend on the composition of the nucleic acid. Thus DNA from calf thymus (42 mol % G + C)[245] and *Cytophaga johnsonii*[250] have a positive band at 220 and a negative one at 210 nm, whereas DNA from *Streptomyces chrysomallus*[251] and *Micrococcus lysodeikticus*[250,251] (both with 72 mol % G + C) show a relative strong negative band at about 210 nm and a positive at shorter wavelengths. A very faint negative CD band at 310 nm was found in case of calf thymus DNA.[245]

The CD of DNA is comparatively small because it forms a double anti-parallel helix and the CDs of both strands compensate each other to a great extent.[244]

Raising the temperature to about 45° increases the ellipticity of the 273 nm band at the expense of the 243 nm band in a double stranded DNA,[244] whereas at higher temperatures both bands decrease (Figure 22). This was interpreted as being due to an intermediate structure of DNA, which was seen even better in 80 per cent ethanol solution, where the CD is also non-conservative.[244] At 80° this CD-curve resembles that of a mononucleotide and indicates thus complete rupture of the helix at that temperature. The single stranded DNA from $l_\varphi - 7$ phage shows such an intermediate only in presence of salts or organic solvents.[249]

ORD-data[252] of RNA and DNA of different sources are in agreement with these CD curves, and the temperature variation of the 290 nm peak can also be used for following the denaturation by heating. Especially sensitive is the ORD-profile below 250 nm.[252]

Taking into account only nearest neighbour interactions,[232] ORD properties of RNA were calculated from the ORD of mono- and dinucleotides.[253,254] If n_i is the mole fraction (number of times monomer i appears in the polymer divided by the chain length) of monomer i, n_{ij} that of the dimer $i–p–j$, $[\Phi_i]$ the molar rotation of monomer i (or $p–i$, or $i–p$) and $[\Phi_{ij}]$ that of dimer $i–p–j$, the molar rotation of the polymer will be given by

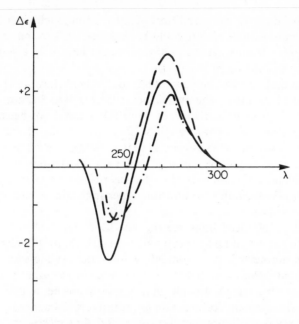

Figure 22. CD of calf thymus DNA in 0·01 M NaCl, 0·01 M
tris buffer, 0·001 M EDTA at 20° (————), 45° (- - - - - - -),
and 80° (- · — · — · — · -)

equation (22)

$$[\Phi] = 2 \sum_{i=1}^{4} \sum_{j=1}^{4} n_{ij}[\Phi_{ij}] - \sum_{i=1}^{4} n_i[\Phi_i] \qquad (22)$$

where the summation has to be done over all four bases A, U, C, and G. As
the sequence is not known, (22) was replaced by (23) which assumes random
distribution along the chain, and this gave reasonable results in absence of
salts at neutral pH.

$$[\Phi] = 2 \sum_{i=1}^{4} \sum_{j=1}^{4} n_i \cdot n_j[\Phi_{ij}] - \sum_{i=1}^{4} n_i[\Phi_i] \qquad (23)$$

In case of a homopolymer (22) or (23) simplifies to

$$[\Phi] = 2[\Phi_{ii}] - [\Phi_i] \qquad (24)$$

Double stranded helix formation was taken into account by a modification
of equation (23) (limited to the case of A—U and G—C pairs).[253] Application
to tobacco mosaic virus RNA gave agreement with the experimental result

(double helix formation by addition of salt). From such calculations the conformation of alanine sRNA could be deduced.[253] The shift of peak and trough by going from single to double stranded helix may also be used to determine helix content.[254]

Viruses contain DNA and a protein coat. Their chiroptical properties are determined mainly by the nucleic acid above 250 to 280 nm, below this mainly by the protein. Three types of ORD-curves have been found,[255] but only in case of an osmotically shocked virus (which leaves intact the protein coat but releases the nucleic acid) the ORD-curve was the sum of the respective curves of the protein and the DNA. Deviation from additivity is due to changes of conformation, of local pH, etc., and these difference ORD-spectra were similar for all three types indicating similar packing of the DNA in viruses.

pH has a pronounced influence on the CD of DNA.[246,249–251,256–259] Between pH 3 and 4 all DNA spectra show drastic changes, being due to the appearance of a new CE at about 260, whose intensity increases with G–C content. It has been attributed to protonation of cytosine in the double helix[251,257] or to change of conformation of guanosine units from anti to syn.[246,250] At low temperature a small negative CD-band at 330 nm was observed,[251] which disappeared already at 20°. For rRNA and sRNA the integrated band intensities follow the cytosine ionization fairly closely; it was concluded that single stranded structures are disrupted in the acid titration range.[258]

In alkaline medium the ORD-bands of DNA and RNA both were slightly diminished and redshifted.[258]

Binding of dyes like acridine orange to DNA gives rise to CE in the range of the dye absorption.[260–263] It is generally assumed that the dye molecules are intercalated between base pairs of the double helix, but there is still some controversy whether optical activity can be induced by the helix onto one single unit or whether interaction of two neighboured dye molecules (in chiral disposal to each other) is necessary.

The low optical activity of actinomycin D (38) (cf. p. 100) is greatly enhanced by binding to DNA in native or denatured state.[264,265] Both lactone rings of the antibiotic are required for complexing, and in this case it was proven that augmentation of optical activity is not based on exciton interaction between two molecules of the dye. Binding to a nucleic acid induces also CD-bands in the absorption regions of (47). Around 360 nm this band is positive for complexes with RNA, but negative for such with DNA.[265a]

Complexing with compounds without absorption bands in the range of investigation may also lead to a change of the chiroptical properties of nucleic acids. Thus from such measurements it could be concluded, that 3.20-diamino steroids form two different types of complexes with DNA.[266]

NO₂ structure

$$NO_2$$

$$NH$$
$$(CH_2)_nN^+(CH_3)_2(CH_2)_3N^+(CH_3)_3.2Br^-$$

(47)

$$n = 2 \text{ or } 3$$

D. Magneto-CD of Nucleosides

The MCD of several nucleosides and their achiral purine and pyrimidine bases has been measured recently by Djerassi and colleagues[267] (Figure 23). Purine nucleosides give very strong MCD in the range of the ^1L_b and ^1L_a band of opposite signs, which are, therefore, much more easily resolved than in the 'natural' CD spectrum. Pyrimidine nucleosides on the other hand give weaker MCD bands which have no advantage over CD spectra. This

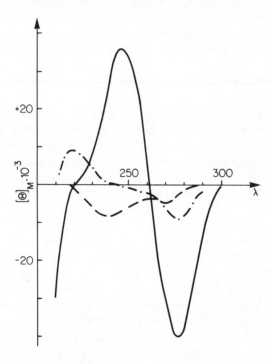

Figure 23. MCD of guanosine (———), and cytidine at pH 7 (– – – –) and pH 2 (–·–·–·–·–)

characteristic feature can be used to differentiate purine and pyrimidine nucleosides in very small quantities. The $n \rightarrow \pi^*$ origin of the 230 nm band of cytidine (**40**) (cf. Figure 14) was also clearly demonstrated by this technique (Figure 23).

VI. CHIROPTICAL PROPERTIES AND STEREOCHEMISTRY OF SUGARS AND THEIR DERIVATIVES

Usually sugars do not contain a chromophore (in the restricted sense), but nevertheless polarimetry has ever since its introduction into organic chemistry played an important role in sugar chemistry. Most measurements have been restricted, however, to the NaD line, and numerous types of 'saccharometers' have been used in industry for the quantitative determination of glucose.

Reducing sugars show 'mutarotation' in solution, i.e. the rotation changes after dissolution slowly until it reaches an end value. This fact is due to the equilibration at the anomeric C-atom, and the final rotation is, therefore, strongly dependent on the solvent, concentration and temperature.

Van't Hoff[268] suggested, that in a sugar the contribution of each individual asymmetric C-atom to the specific rotation is independent of the configuration at the other chiral centres (principle of optical superposition), and this has been extended by Hudson to his isorotation rule:[269] changes of the structure of the remainder of a sugar derivative do scarcely influence the contribution of atom C-1 (the anomeric C-atom) to the rotation. On the other hand, changes at C-1 do nearly not influence the contribution of the remainder of a sugar molecule to the rotation. Application to steroid glycosides lead e.g. to Klyne's rule[270] that D-sugars are bound β-glycosidic and L-sugars α-glycosidic. An explanation of the isorotation rule has been given by the theory of pairwise interaction of Kauzmann and coworkers,[271] and Whiffen[272] and Brewster[273] published rules how to estimate the molecular rotation at the NaD line of sugars of different configuration and in different conformations.

The ORD-curves of sugars are plain, D-galactose and D-talose develop, however, some time after dissolving a pseudoextremum at 208 nm, which is most probably due to the superposition of steep positive and negative partial rotations.[274] By subtraction of the ORD-curves of two sugars differing only in configuration at a single chiral centre, partial curves for the latter have been obtained, and a rule has been put forward for the correlation of the stereochemistry with the chiroptical properties.[275] The spectral range has been divided into two regions, 'A' between 600 and 250 nm, and 'B' around 200 nm. The sign of the contributions of axial groups (OH or CH_2OH) in the C1 conformation are given in Figure 24. According to reference 275, equatorial groups show smaller contributions, which are in general opposite

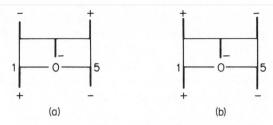

Figure 24. Contribution of axial OR (or CH$_2$OR) groups to ORD curve of pyranose in C1 conformation. (a) wavelength range 'A', (b) wavelength range 'B'. Details see text

to those of the axial ones. The contribution of an equatorial substituent at C-5 depends, however, on the configuration and conformation around C-4.

Derivatives of sugars containing the free aldehyde (ketone) group give, of course, a CE around 300 nm,[276,277] and the same is the case for such sugars like erythropentulose, which does not form a hemiacetal in solution.[278] The fact that fructose 1.6-diphosphate develops besides a CE at shorter wavelengths one at about 300 nm after addition of Mg^{2+}- or Mn^{2+}-ions was taken as an indication that chelated forms like (**48**) (and enediols) are

$$\begin{array}{l} H_2C-O-PO_2^{\ominus} \\ \qquad\qquad\diagdown \\ \qquad\qquad\quad O \\ \qquad\qquad\diagup \\ C=O\cdots H \\ HO-C-H \\ H-C-OH \\ H-C-O-H \\ \qquad\qquad\cdots \\ \qquad\qquad O^{\ominus} \\ \qquad\qquad\diagup \\ H_2C-O-PO_2H \end{array}$$

(**48**)

present in solution.[279] The π-system of glycals[280] and other unsaturated sugar derivatives[281] gives also rise to new CE.

Sugar lactones have also been investigated very intensely. At the beginning of this century Hudson[282] presented his 'lactone rule', which stated originally that γ-lactones of aldonic acids are strongly dextrorotatory at the NaD line if the OH engaged in the lactone ring is on the right in a Fischer-type projection. This rule has later on been extended also to δ-lactones and to lactones in other series,[283] where one usually takes the difference of the molar rotations for the lactone and the corresponding hydroxy acid or the glycol, or even the difference of the full ORD-curves.[284] The sign of the CE within the carboxyl absorption is, however, strictly determined by the configuration at C-2, and it is positive for the (2S)-series.[285] Thus e.g. D-ribono lactone (49) has a specific rotation at the NaD line of $+43°$, whereas its 2.3-acetonide (50) is laevorotatory ($[\alpha]_D = -80°$). The CE in the u.v. is, however, negative for both compounds.[286]

(49)

(50)

Beecham[287] has interpreted the sign of these CE by taking into account the noncoplanarity of the lactone ring and using the rule of Wolf[60] or Legrand.[61] δ-Lactones can be treated in a similar way.[287] Uronic acid derivatives, which contain the carboxyl chromophore, show also a CE in the accessible spectral range; from difference curves contributions of individual C-atoms could be determined[288] similar as with simple sugars.[275]

Bound to an achiral alcohol through a glycosidic bond sugars induce optical activity in the absorption bands of the former. Many examples for this have been discussed above in the section on nucleosides (see p. 100), and even simple phenyl glycosides show CE in the 260–274 nm region. The sign of this multiple effect is positive for α-glycosides.[289] A xanthate group, bound either to the glycosidic or another OH, shows also CE within its absorption bands,[290] as does the *p*-phenylazophenyl-group (extremum not reached completely).[291] Sugar nitrate esters have been investigated by Japanese workers[292] and they found that the CE around 270 nm is positive, if the $-ONO_2$ group is fixed to a C-atom of L-configuration in an α-D-glycoside. In the β-isomer the sign is inversed. 6-Nitrates of α-D-glycosides show a positive CE, the influence of the substituents upon this effect decreases in the series C-2 > C-3 > C-4 > C-5. The nitrates of some dianhydruosugars gave an additional CD-band at about 230 nm,[293] and with some steroid nitrates we found even three such optically active bands.[294]

Complexes of sugars with molybdate at pH 5 to 6 show Cotton effects in the range between 350–220 nm, which are due to metal transitions.[295,296] Such molybdate complexes are not only formed if three vicinal hydroxyl groups are present in the *cis–cis* arrangement (so as to allow for the conformation axial-equatorial-axial; type A), but also if only two axial OH groups are present (type B). The free glycosidic OH is necessary e.g. in glucose, mannose and xylose. Complexes of type A give three to four bands, and the one at 264 nm is negative, if the OH groups at C-1 and C-3 are axial in the 1C-conformation (β-D-mannose, β-D-lyxose). If this conformation can be adopted, however, only in the C1-conformation, the same band is positive (α-D-ribose, β-L-rhamnose). Complexes of type B give only two CD-bands around 275 and 240 nm, and their ellipticity is smaller than that of the main bands of type A. The molybdate complexes of sugar alcohols can also be used for the determination of the stereochemistry of these compounds.[297]

Amino-deoxy sugars give also only plain curves in the ORD-spectrum.[298] Their *N*-acetates show, however, a CE in the short wavelength range,[288,299] which is of similar $n \rightarrow \pi^*$-origin as in peptides. A conformation analysis of bloodgroup substances of this type has been tried[299] with the help of Schellman and Oriel's[42] quadrant rule for amides. *N*-Salicyliden aminosugars[300] and such with a *C*-phenylazo grouping[301,302] give CE in a better accessible spectral range.

1-Deoxy-1-nitro sugar alcohols, important intermediates in the homologization of sugars, show two CD-bands at 310 and 280 nm of the same sign, which is positive, if the configuration is (S) at C-2, regardless of whether there is present an OH- or an acetamino-group at this atom. The configuration at the other chiral centres does not influence the sign of this CE.[303]

Azido sugars have been investigated by Paulsen.[304] The $n \rightarrow \pi^*$-band around 290 nm can be treated according to a rule developed for this chromophore[305] and the results are in good agreement with the usual conformational analysis. The same author[306] has also made use of the 300 nm band of azomethines to determine the equilibrium between the piperidinose (**51**) and the piperideine (**52**) in solution.

(**51**)   (**52**)

In order to determine systematically the absolute configuration at different C-atoms of a sugar some of the chiral centres are destroyed by incorporation into a chromophore, which in most cases is a heterocycle. This method has already been used long time ago, when only $[\alpha]_D$-values have been measured, and a very good review is given by Chilton and Krahn.[307] The sign of the rotation of such a derivative can be related to the absolute configuration at the chiral centre nearest to the chromophore, and *inter alii* has been used for benzimidazoles,[308] furanes,[309] osotriazols,[310] and pyrroles.[311] CD- or ORD-measurements permit a much safer correlation, and the following rules have been put forward:[307] benzimidazol derivatives (53) have a positive CE around 245 nm, if the sugar has (S)-configuration at C-2. Similarly a positive CE about 315 nm is found for quinoxalines (54), if the original sugar has (3S)-configuration. Configuration at C-4 may be inferred from the CD of flavazol derivatives (55), and that at C-5 from anhydroosazones (56).

(53)   (54)   (55)

(56)

The plain curves of oligosaccharides have been treated[312] in the same way as those of monosaccharides.[275] They can approximately be built up from partial curves of their constituent monomeric residues irrespective of the nature of the substituent at C-1, as long as the stereochemistry (α or β) is taken correctly into account.

The CE of chondroitinsulphates A, B, C, D, and E has been measured; besides C all gave a trough near 220 nm in the ORD.[281,313] The CE of several unsaturated disaccharides obtained by digestion with chondroitinase ABC has already been mentioned above.[281] Some oligo- and polysaccharides of immunobiological interest have been recently investigated.[299,314] Their CE is due to the N-acetylamino group present, and some generalizations have been found. Thus any substitution on an acetylglucosamine (AGA) residue enhances strongly its CE, especially if the AGA unit is β-linked. Particularly powerful in this respect is a substitution at C-3 (of a disaccharide). Substitution by an α-L-fucosyl residue leads to a large intensification of the amide $n \rightarrow \pi^*$-band.

Heparin displays also a CE between 185–220 nm, and this is influenced by pH changes or the addition of biogenic amines (change of conformation).[315] Heparin complexes with methylene blue in acidic solution, and the absorption bands of the dye become optically active,[316] which fact is reminiscent of the ordered complexes between proteins or nucleic acids and similar dye molecules (pp. 98, 116). Such induced rotational strengths have also been found e.g. with the inclusion compounds of γ-cyclodextrin and congo red[317] and the iodine–starch complex.[318] In the latter case the rotational strength increases strongly with time, though the absorption remains constant.

REFERENCES

1. J. B. Biot, *Mém. Acad. Roy. Sci. Inst. France*, **2**, [2], 41 (1817)
2. A. M. Cotton, *Ann. Chim. Phys.*, **8** [7], 347 (1896)
3. C. Djerassi, *Optical Rotatory Dispersion*, McGraw-Hill, New York, 1960
4. H. Rudolph, *J. Opt. Soc. Am.*, **45**, 50 (1955)
5. M. Grosjean and M. Legrand, *Compt. Rend.*, **251**, 2150 (1960)
6. L. Velluz, M. Legrand, and M. Grosjean, *Optical Circular Dichroism*, Verlag Chemie, Weinheim, 1965
7. Cf. B. H. Billings, *J. Opt. Soc. Am.*, **39**, 797 (1949)
8. G. Snatzke, *Z. Instrumentenk.*, **75**, 111 (1967)
9. F. Woldbye in reference 10, p. 85
10. *Optical Rotatory Dispersion and Circular Dichroism in Organic Chemistry* (Ed. G. Snatzke), Heyden, London, 1967
11. A. Fresnel, Memoirs N° XXVIII (1822), N° XXX (1823). A review of Fresnel's works is given in reference 12
12. T. M. Lowry, *Optical Rotatory Power* (*Republication*), Dover Publications, New York, 1964
13. P. Drude, *Lehrbuch der Optik*, Hirzel, Leipzig, 1900
14. G. Bruhat, *Ann. Phys.*, **3** [9], 232, 417 (1915); **13** [9], 25 (1920)
15. L. Natanson, *Bull. Acad. Sci.* (*Krakau*), **1908**, 764; *J. Phys.*, **8** [4], 321 (1909)
16. W. Kuhn and E. Braun, *Z. Physik. Chem.*, **B 8**, 281 (1930)
17. C. Djerassi and E. Bunnenberg, *Proc. Chem. Soc.*, **1963**, 299
18. W. Moffitt and A. Moscowitz, *J. Chem. Phys.*, **30**, 648 (1959)
19. C. A. Emeis, L. J. Oosterhoff and G. de Vries, *Proc. Roy. Soc.* (*London*), **A 297**, 54 (1967)
20. A. Moscowitz, *Tetrahedron*, **13**, 48 (1961)
21. Cf. W. J. Kauzmann, J. E. Walter and H. Eyring, *Chem. Rev.*, **26**, 339 (1940)
22. P. Crabbé, *Tetrahedron*, **20**, 1221 (1964); p. 126 in reference 10
23. P. Crabbé, *Optical Rotatory Dispersion and Circular Dichroism in Organic Chemistry*, Holden-Day, San Francisco, 1965; enlarged French edition, Gauthier-Villars, Paris, 1968
24. W. Kuhn, Theorie and Grundgesetze der optischen Aktivität, in *Stereochemie* (Ed. K. Freudenberg), Deuticke, Leipzig, 1933

25. S. Mitchell, *The Cotton Effect*, Bell, London, 1933
26. J. P. Mathieu, *Les Théories Moléculaires du Pouvoir Rotatoire Naturel*, Gauthier-Villars, Paris, 1946
27. W. Klyne, Optical Rotatory Dispersion and the Study of Organic Structures, in *Advan. Org. Chem.* (Ed. R. A. Raphael, E. C. Taylor and H. Wynberg), Interscience, New York, 1960, Vol. I, p. 239
28. S. F. Mason, *Quart. Rev.*, **17**, 20 (1963)
29. G. Snatzke, *Angew. Chem.* **80**, 15 (1968); *Intern. Ed. Engl.*, **7**, 14 (1968)
30. J. A. Schellman, *Accounts Chem. Res.*, **1**, 144 (1968)
31. P. Crabbé, Recent Applications of Optical Rotatory Dispersion and Optical Circular Dichroism in Organic Chemistry, in *Topics in Stereochemistry* (Ed. N. L. Allinger and E. L. Eliel), Interscience, New York, 1967, Vol. I, p. 93
32. C. Djerassi, H. Wolf and E. Bunnenberg, *J. Am. Chem. Soc.*, **85**, 2835 (1963)
33. G. Snatzke, D. Becher and J. R. Bull, *Tetrahedron*, **20**, 2443 (1964); J. R. Bull, J. P. Jennings, W. Klyne, G. D. Meakins, P. M. Scopes and G. Snatzke, *J. Chem. Soc.*, **1965**, 3152; G. Snatzke, *J. Chem. Soc.*, **1965**, 5002
34. U. Weiss and H. Ziffer, *J. Org. Chem.*, **28**, 1248 (1963)
35. Cf. G. Snatzke and G. Eckhardt, *Tetrahedron*, **24**, 4543 (1968)
36. Cf. A. I. Scott, G. A. Sim, G. Ferguson, D. W. Young and F. McCapra, *J. Am. Chem. Soc.*, **84**, 3197 (1962)
37. G. Snatzke in reference 10, p. 208.
38. G. Snatzke, *Tetrahedron*, **21**, 413, 421, 439 (1965)
39. W. Klyne, *Tetrahedron*, **13**, 29 (1961)
40. B. Sjöberg in reference 10, p. 173
41. W. Moffitt, R. B. Woodward, A. Moscowitz, W. Klyne and C. Djerassi, *J. Am. Chem. Soc.*, **83**, 4013 (1961)
42. A. Moscowitz, *Advan. Chem. Phys.*, **4**, 67 (1962); O. E. Weigang, Jr. and E. G. Höhn, *J. Am. Chem. Soc.*, **88**, 3673 (1966); E. G. Höhn and O. E. Weigang, Jr., *J. Chem. Phys.*, **48**, 1127 (1968); H. P. Gervais, *Thesis*, Paris (1966); J. A. Schellman and P. Oriel, *J. Chem. Phys.*, **37**, 2114 (1962); J. A. Schellman, *J. Chem. Phys.*, **44**, 55 (1966); S. F. Mason, *Mol. Phys.*, **5**, 343 (1962); G. Wagnière, *J. Am. Chem. Soc.*, **88**, 3937 (1966)
43. Y.-H. Pao and D. P. Santry, *J. Am. Chem. Soc.*, **88**, 4157 (1966)
44. C. Djerassi, J. Osiecki, R. Riniker and B. Riniker, *J. Am. Chem. Soc.*, **80**, 1216 (1958)
45. C. Djerassi, W. Closson and A. E. Lippman, *J. Am. Chem. Soc.*, **78**, 3163 (1956)
46. S. Bory, M. Fétizon and P. Laszlo, *Bull. Soc. Chim. France*, **1963**, 2310
47. K. M. Wellman, R. Records, E. Bunnenberg and C. Djerassi, *J. Am. Chem. Soc.*, **86**, 492 (1964)
48. C. Djerassi, L. A. Mitscher and B. J. Mitscher, *J. Am. Chem. Soc.*, **81**, 947 (1959)
49. A. Yoshikoshi, *Nippon Kagaku Zasshi*, **85**, 390 (1964)
50. C. Djerassi, R. Riniker and B. Riniker, *J. Am. Chem. Soc.*, **78**, 6362 (1956); C. Djerassi, M. Cais and L. A. Mitscher, *J. Am. Chem. Soc.*, **81**, 2386 (1959)
51. C. Djerassi and D. Marshall, *Tetrahedron*, **1**, 238 (1957)
52. C. D. Djerassi and W. Klyne, *Proc. Natl. Acad. Sci. U.S.*, **48**, 1093 (1962)
53. C. Djerassi and W. Klyne, *J. Am. Chem. Soc.*, **79**, 1506 (1957)
54. C. Djerassi, W. Klyne, T. Norin, G. Ohloff and E. Klein, *Tetrahedron*, **21**, 1631 (1965); M. Legrand, R. Viennet and J. Caumartin, *Compt. Rend.*, **253**, 2378 (1961); K. Schaffner and G. Snatzke, *Helv. Chim. Acta.*, **48**, 347 (1965); K. Kuriyama, H. Tada, Y. K. Sawa, S. Itô and I. Itoh, *Tetrahedron Letters*, **1968**, 2539

55. W. B. Whalley, *Chem. Ind. (London)*, **1962**, 1024
56. A. Moscowitz, K. Mislow, M. A. W. Glass and C. Djerassi, *J. Am. Chem. Soc.*, **84**, 1945 (1962); A. Moscowitz, A. E. Hansen, L. S. Forster and K. Rosenheck, *Biopolymers*, Symposia No. 1, 75 (1964)
57. C. Djerassi, R. Records, E. Bunnenberg, K. Mislow and A. Moscowitz, *J. Am. Chem. Soc.*, **84**, 870 (1962)
58. L. Velluz, M. Legrand and R. Viennet, *Compt. Rend.*, **261**, 1687 (1965)
58a. A. W. Burgstahler and R. C. Barkhurst, *J. Am. Chem. Soc.*, **92**, 7601 (1970)
59. R. C. Cookson and S. MacKenzie, *Proc. Chem. Soc.*, **1961**, 423; R. C. Cookson and J. Hudec, *J. Chem. Soc.*, **1962**, 429
60. H. Wolf, *Tetrahedron Letters*, **1965**, 1075; **1966**, 5151
61. M. Legrand and R. Boucourt, *Bull. Soc. Chim. France*, **1967**, 2241
62. K. Kuriyama and T. Komeno in reference 10, p. 366
63. A. F. Beecham, *Tetrahedron Letters*, **1968**, 3591
64. J. P. Jennings, W. Klyne and P. M. Scopes, *Proc. Chem. Soc.*, **1964**, 412; J. P. Jennings, W. Klyne and P. M. Scopes, *J. Chem. Soc.*, **1965**, 7211, 7229
65. G. Snatzke, H. Ripperger, C. Horstmann and K. Schreiber, *Tetrahedron*, **22**, 3103 (1966)
66. A. F. Beecham, private communication
67. for an extensive discussion cf. G. Snatzke, *Fortschr. Chem. Forsch.*, in press
68. G. Snatzke, H. Schwang and P. Welzel, in *Some Newer Physical Methods in Structural Chemistry* (Eds. R. Bonnett and J. G. Davis), United Trade Press, London, 1967, p. 159
69. R. Tschesche, H. Schwang, H.-W. Fehlhaber and G. Snatzke, *Tetrahedron*, **22**, 1129 (1966)
70. C. Djerassi, H. Wolf and E. Bunnenberg, *J. Am. Chem. Soc.*, **84**, 4552 (1962)
71. C. Djerassi, I. T. Harrison, O. Zagneetko and A. L. Nussbaum, *J. Org. Chem.*, **27**, 1173 (1962)
72. D. Bertin and M. Legrand, *Compt. Rend.*, **256**, 960 (1963)
73. J. V. Burakevich and C. Djerassi, *J. Am. Chem. Soc.*, **87**, 51 (1965)
74. A. Moscowitz, K. M. Wellman and C. Djerassi, *Proc. Natl. Acad. Sci. U.S.*, **50**, 799 (1963)
75. K. M. Wellman, P. H. A. Laur, W. S. Briggs, A. Moscowitz and C. Djerassi, *J. Am. Chem. Soc.*, **87**, 66 (1965)
76. O. E. Weigang, Jr., *J. Chem. Phys.*, **43**, 3609 (1965)
77. A. Rassat in reference 10, p. 314
78. K. Kuriyama, T. Iwata, M. Moriyama, M. Ishikawa, H. Minato and K. Takeda, *J. Chem. Soc. (C)*, **1967**, 420
79. O. E. Weigang, Jr., private communication
80. G. Snatzke in reference 10, p. 335
81. A. Moscowitz, K. M. Wellman and C. Djerassi, *J. Am. Chem. Soc.*, **85**, 3515 (1963)
82. K. M. Wellman, E. Bunnenberg and C. Djerassi, *J. Am. Chem. Soc.*, **85**, 1870 (1963); G. Snatzke and D. Becher, *Tetrahedron*, **20**, 1921 (1964)
83. B. Sjöberg, B. Karlén and R. Dahlbom, *Acta Chem. Scand.*, **16**, 1071 (1962); S. Yamada, K. Ishikawa and K. Achiwa, *Chem. Pharm. Bull. (Tokyo)*, **13**, 1266 (1965); E. Bach, A. Kjaer, R. Dahlbom, T. Walle, B. Sjöberg, E. Bunnenberg, C. Djerassi and R. Records, *Acta Chem. Scand.*, **20**, 2781 (1966); G. C. Barrett, *J. Chem. Soc. (C)*, **1966**, 1771, **1967**, 1
84. C. Djerassi, K. Undheim, R. C. Sheppard, W. G. Terry and B. Sjöberg, *Acta Chem. Scand.*, **15**, 903 (1961)

85. K. Bláha, I. Frič and H.-D. Jakubke, *Coll. Czech. Chem. Commun.*, **32**, 558 (1967)
86. S. Yamada, K. Ishikawa and K. Achiwa, *Chem. Pharm. Bull. (Tokyo)*, **13**, 892 (1965)
87. J. P. Jennings and W. Klyne, *Biochem. J.*, **1963**, 12 P
88. I. P. Dirkx and F. L. J. Sixma, *Rec. Trav. Chim.*, **83**, 522 (1964)
89. W. Gaffield, *Chem. Ind. (London)*, **1964**, 1460
90. J. P. Jennings, W. Klyne and P. M. Scopes, *J. Chem. Soc.*, **1965**, 294
91. A. Fredga, J. P. Jennings, W. Klyne, P. M. Scopes, B. Sjöberg and S. Sjöberg, *J. Chem. Soc.*, **1965**, 3928
92. M. Legrand and R. Viennet, *Bull. Soc. Chim. France*, **1965**, 679, **1966**, 2798
93. J. Cymerman Craig and S. K. Roy, *Tetrahedron*, **21**, 391 (1965)
94. L. I. Katzin and E. Gulyas, *J. Am. Chem. Soc.*, **90**, 247 (1968)
95. O. Lutz and B. Jirgensons, *Ber. dtsch. chem. Ges.*, **63**, 448 (1930); **64**, 1221 (1931)
96. M. Billardon, *Compt. Rend.*, **251**, 535, 1759 (1960)
97. E. Iizuka and J. T. Yang, *Biochemistry*, **3**, 1519 (1964)
98. T. M. Hooker and C. Tanford, *J. Am. Chem. Soc.*, **86**, 4989 (1964)
99. G. D. Fasman, E. Bodenheimer and C. Lindblow, *Biochemistry*, **3**, 1665 (1964)
100. S. Beychok and G. D. Fasman, *Biochemistry*, **3**, 1675 (1964)
101. A. Moscowitz, A. Rosenberg and A. E. Hansen, *J. Am. Chem. Soc.*, **87**, 1813 (1965)
102. A. Rosenberg, *J. Biol. Chem.*, **241**, 5119 (1966)
102a. T. M. Hooker, Jr. and J. A. Schellman, *Biopolymers*, **9**, 1319 (1970)
102b. J. Horwitz, E. H. Strickland and C. Billups, *J. Am. Chem. Soc.* **91**, 184 (1969)
103. S. Beychok, *Science*, **154**, 1288 (1966)
104. M. Legrand and R. Viennet, *Compt. Rend.*, **262**, 943 (1966)
105. G. Holzwarth and P. Doty, *J. Am. Chem. Soc.*, **87**, 218 (1965)
106. J. Beacham, V. T. Ivanov, P. M. Scopes and D. R. Sparrow, *J. Chem. Soc.*, **1966**, 1449
107. H. Sachs and E. Brand, *J. Am. Chem. Soc.*, **76**, 1811 (1954)
108. M. Goodman, F. Boardman and I. Listowsky, *J. Am. Chem. Soc.*, **85**, 2491 (1963)
109. P. M. Scopes, D. R. Sparrow, J. Beacham and V. T. Ivanov, *J. Chem. Soc. (C)*, **1967**, 221
110. D. Balasubramanian and D. B. Wetlaufer in reference 111, p. 147; *J. Am. Chem. Soc.*, **88**, 3449 (1966)
111. *Conformation of Biopolymers* (Ed. G. N. Ramachandran), Academic Press, London, 1967
112. K. Bláha and I. Frič, *Proc. Pept. Symp.*, **1968**, 40
113. H. Edelhoch and R. E. Lippoldt, *J. Biol. Chem.*, **243**, 4799 (1967)
114. P. M. Bayley, E. B. Nielsen and J. A. Schellman, in press
115. P. M. Bayley and J. A. Schellman, *Lecture A3, 5th Intern. Symp. Chem. Natl. Prod. (IUPAC) London*, 8.-13.7.1968; J. A. Schellman and E. B. Nielsen in reference 110, p. 109
116. B. J. Litman and J. A. Schellman, *J. Phys. Chem.*, **69**, 978 (1965)
117. L. L. Jones and H. Eyring, *Tetrahedron*, **13**, 235 (1961)
118. V. Prelog and H. Gerlach, *Helv. Chim. Acta*, **47**, 2288 (1964)
119. H. Gerlach, J. A. Owtschinnikow and V. Prelog, *Helv. Chim. Acta*, **47**, 2294 (1964)
120. M. A. Ruttenberg, T. P. King and L. C. Craig, *J. Am. Chem. Soc.*, **87**, 4196 (1965)
121. D. Balasubramanian, *J. Am. Chem. Soc.*, **89**, 5445 (1967)
122. F. Quadrifoglio and D. W. Urry, *Biochem. Biophys. Res. Commun.*, **29**, 785 (1967)
123. S. Beychok and E. Breslow, *J. Biol. Chem.*, **243**, 151 (1968)
124. C. Cohen, *Nature*, **175**, 129 (1955)
125. W. Moffitt, *J. Chem. Phys.*, **25**, 467 (1956)

126. W. Moffitt and J. T. Yang, *Proc. Natl. Acad. Sci. U.S.*, **42**, 596 (1956)
127. W. Moffitt, *Proc. Natl. Acad. Sci. U.S.*, **42**, 736 (1956)
128. J. G. Kirkwood, *J. Chem. Phys.*, **5**, 479 (1937)
129. W. Moffitt, D. D. Fitts and J. G. Kirkwood, *Proc. Natl. Acad. Sci. U.S.*, **43**, 723 (1957)
130. S. F. Mason, *Nature*, **199**, 139 (1963)
131. I. Tinoco, Jr., *Advan. Chem. Phy.*, **4**, 113 (1961); *Radiation Res.*, **20**, 133 (1963); *J. Am. Chem. Soc.*, **86**, 297 (1964)
132. I. Tinoco, Jr., R. W. Woody and D. F. Bradley, *J. Chem. Phys.*, **38**, 1317 (1963)
133. D. F. Bradley, I. Tinoco, Jr., and R. W. Woody, *Biopolymers*, **1**, 239 (1963)
134. Cf. R. W. Woody, Dissertation, University of California, Berkeley, Calif., 1962
135. W. B. Gratzer, G. M. Holzwarth and P. Doty, *Proc. Natl. Acad. Sci. U.S.*, **47**, 1785 (1961)
136. P. Arnes and P. Doty, *Advan. Protein Chem.*, **16**, 401 (1961)
137. G. D. Fasman, *Methods Enzymol.*, **6**, 928 (1963)
138. E. R. Blout in reference 10, p. 224
139. S. Beychok, in *Poly-α-Amino Acids—Protein Models for Conformation Studies* (Ed. G. D. Fasman), Marcel Dekker, New York, 1967
140. M. Goodman, G. W. Davis and E. Benedetti, *Accounts Chem. Res.*, **1**, 275 (1968)
141. A. D. McLachlan, *Proc. Roy. Soc. (London)*, **A 297**, 131 (1967)
142. W. B. Gratzer, *Proc. Roy. Soc. (London)*, **A 297**, 163 (1967)
142a. B. Jirgensons, *Optical Rotatory Dispersion of Proteins and Other Macromolecules*, Springer, Berlin–Heidelberg–New York, 1969
142b. M. Goodman, A. S. Verdini, N. S. Choi and Y. Masuda, Polypeptide Stereochemistry, in *Topics in Stereochemistry* (Ed. N. L. Allinger and E. L. Eliel), Interscience, New York, 1970, Vol. V, p. 69
143. M. Legrand and R. Viennet, *Compt. Rend.*, **259**, 4277 (1964)
144. E. Shechter and E. R. Blout, *Proc. Natl. Acad. Sci. U.S.*, **51**, 695, 794 (1964)
145. E. Shechter, J. P. Carver and E. R. Blout, *Proc. Natl. Acad. Sci. U.S.*, **51**, 1029 (1964)
146. J. P. Carver, E. Shechter and E. R. Blout, *J. Am. Chem. Soc.*, **88**, 2550, 2562 (1966)
147. K. Imahori, *Biochim. Biophys. Acta*, **37**, 336 (1960)
148. J. T. Yang in reference 111, p. 157
149. M. V. Vol'kenshtein and V. A. Zubkov, *Biopolymers*, **5**, 465 (1967)
150. E. R. Blout, I. Schmier and N. S. Simmons, *J. Am. Chem. Soc.*, **84**, 3193 (1962)
151. B. Davidson, N. Tooney and G. D. Fasman, *Biochem. Biophys. Res. Commun.*, **23**, 156 (1966)
152. F. A. Bovey and F. P. Hood, *J. Am. Chem. Soc.*, **88**, 2326 (1966)
153. E. R. Blout, J. P. Carver and J. Gross, *J. Am. Chem. Soc.*, **85**, 644 (1963)
154. R. Townend, T. F. Kumosinski, S. N. Timasheff, G. D. Fasman and B. Davidson, *Biochem. Biophys. Res. Commun.*, **23**, 163 (1966)
155. P. K. Sarkar and P. Doty, *Proc. Natl. Acad. Sci. U.S.*, **55**, 981 (1966)
156. S. N. Timasheff, H. Susi, R. Townend, L. Stevens, M. J. Gorbunoff and T. F. Kumosinski in reference 111, p. 173
157. E. Iizuka and J. T. Yang, *Proc. Natl. Acad. Sci. U.S.*, **55**, 1175 (1966)
158. L. Velluz and M. Legrand, *Angew. Chem.*, **77**, 842 (1965)
158a. N. Greenfield and G. D. Fasman, *Biochemistry*, **8**, 4108 (1969)
158b. F. Quadrifoglio and D. W. Urry, *J. Am. Chem. Soc.*, **90**, 2760 (1968)
158c. L. K. Li and A. Spector, *J. Am. Chem. Soc.*, **91** 220 (1969)
159. E. S. Pysh, *Proc. Natl. Acad. Sci. U.S.*, **56**, 825 (1966); *J. Mol. Biol.*, **23**, 587 (1967)
160. U. Weiss, *Experientia*, **24**, 1088 (1968)

161. W. B. Gratzer and P. Doty, *J. Am. Chem. Soc.*, **85**, 1193 (1963)
162. G. D. Fasman, C. Lindblow and E. Bodenheimer, *Biochemistry*, **3**, 155 (1964)
163. H. E. Auer and P. Doty, *Biochemistry*, **5**, 1716 (1966)
164. N. Lotan, A. Yaron and A. Berger, *Biopolymers*, **4**, 365 (1966)
165. S. R. Chaudhuri and J. T. Yang, *Biochemistry*, **7**, 1379 (1968)
166. E. R. Blout, in *Polyamino Acids, Polypeptides and Proteins* (Ed. M. A. Stahman), University of Wisconsin Press, Madison, Wis., 1962, p. 275
167. A. Elliot, W. E. Hanby and B. R. Malcolm, *Nature*, **180**, 1340 (1957)
168. Y.-H. Pao, R. Longworth and R. C. Kornegay, *Biopolymers*, **3**, 516 (1965)
169. T. Ooi, R. A. Scott, G. Vanderkooi and H. A. Scheraga, *J. Chem. Phys.*, **46**, 4410 (1967)
170. J. Applequist and T. G. Mahr, *J. Am. Chem. Soc.*, **88**, 5419 (1966)
171. A. J. Sage and G. D. Fasman, *Biochemistry*, **5**, 286 (1966)
172. G. D. Fasman, M. Landsberg and M. Buchwald, *Can. J. Chem.*, **43**, 1588 (1965)
173. A. Cosani, E. Peggion, M. Terbojevich and A. Portolan, *Chem. Commun.*, **1967**, 930; A. Cosani, E. Peggion, A. S. Verdini and M. Terbojevich, *Biopolymers*, in press (cited from reference 140).
174. M. Goodman, C. M. Deber and A. M. Felix, *J. Am. Chem. Soc.*, **84**, 3770 (1962); M. Goodman, A. M. Felix, C. M. Deber, A. R. Brause and G. Schwartz, *Biopolymers*, **1**, 371 (1963); M. Goodman, A. M. Felix, C. M. Deber and A. R. Brause, *Biopolymers Symp.*, **1**, 409 (1964); D. F. Bradley, M. Goodman, A. M. Felix and R. Records, *Biopolymers*, **4**, 607 (1966)
175. D. L. Coleman and E. R. Blout in reference 111, p. 123
176. E. R. Blout and L. Stryer, *Proc. Natl. Acad. Sci. U.S.*, **45**, 159 (1959); L. Stryer and E. R. Blout, *J. Am. Chem. Soc.*, **83**, 1411 (1961); E. R. Blout, *Biopolymers Symp.*, **1**, 397 (1964)
177. I. A. Bolotina and M. V. Vol'kenshtein, *Molekul. Biofiz., Akad. Nauk SSSR Inst. Biol. Fiz., Sb. Statei*, **1965**, 27, 36
178. J. E. Coleman, *J. Am. Chem. Soc.*, **89**, 6757 (1967)
179. H. Ziffer, K. Yamaoka and A. B. Mauger, *Biochemistry*, **7**, 996 (1968)
180. D. W. Urry, *J. Am. Chem. Soc.*, **89**, 4190 (1967); D. W. Urry and J. W. Pettegrew, *J. Am. Chem. Soc.*, **89**, 5276 (1967)
181. S. Vinogradov and R. Zand, *Arch. Biochem. Biophys.*, **125**, 902 (1968)
182. M. V. Vol'kenshtein, Y. A. Sharonov and A. K. Shemelin, *Mol. Biol.*, **1**, 467 (1967); B. A. Atanasov, M. V. Vol'kenshtein, Y. A. Sharonov and A. K. Shemelin, *Mol. Biol.*, **1**, 477 (1967)
183. K. Aki, T. Takagi, T. Isemura and T. Yamano, *Biochim. Biophys. Acta*, **122**, 193 (1966)
184. K. Garbett, R. D. Gillard, P. F. Knowles and J. E. Stangroom, *Nature*, **215**, 824 (1967)
185. I. A. Bolotina, D. S. Markovich, M. V. Vol'kenshtein and P. Zavodsky, *Biochim. Biophys. Acta*, **132**, 260 (1967)
186. E. H. Strickland, *Biochim. Biophys. Acta.* **151**, 70 (1968)
187. R. J. Kitz and L. T. Kremzner, *Mol. Pharmacol.*, **4**, 104 (1968)
188. M. De Luca and M. Marsh, *Arch. Biochem. Biophys.*, **121**, 233 (1967)
189. A. Scanu and R. Hirz, *Nature*, **218**, 200 (1968)
190. B. J. Adkins and J. T. Yang, *Biochemistry*, **7**, 266 (1968)
191. Yu. M. Torchinskiĭ, *Usp. Biol. Khim.*, **8**, 61 (1967)
191a. H. Rosenkranz and W. Scholtan, *Z. physiol Chem.*, **352**, 896 (1971)
192. B. Jirgensons, *J. Biol. Chem.*, **240**, 1064 (1965); **241**, 4855 (1966)
193. L. B. Clark and I. Tinoco, Jr., *J. Am. Chem. Soc.*, **87**, 11 (1965)

194. K. H. Scheit, *IUPAC Lecture H 17, 5th Intern. Chem. Symp. Nat. Prod.*, London, 8–13 July 1968 and private communication
195. D. W. Miles, R. K. Robins and H. Eyring, *Proc. Natl. Acad. Sci. U.S.*, **57**, 1138 (1967)
196. D. J. Caldwell and H. Eyring, *Ann. Rev. Phys. Chem.*, **15**, 281 (1964)
197. J. T. Yang and T. Samejima, *J. Am. Chem. Soc.*, **85**, 4039 (1963)
198. J. T. Yang, T. Samejima and P. K. Sarkar, *Biopolymers*, **4**, 623 (1966)
199. T. L. V. Ulbricht, J. P. Jennings, P. M. Scopes and W. Klyne, *Tetrahedron Letters*, **1964**, 695
200. T. L. V. Ulbricht, T. R. Emerson and R. J. Swan, *Biochem. Biophys. Res. Commun.*, **19**, 643 (1965)
201. T. R. Emerson, R. J. Swan and T. L. V. Ulbricht, *Biochemistry*, **6**, 843 (1967)
202. G. D. Fasman, C. Lindblow and L. Grossman, *Biochemistry*, **3**, 1015 (1964)
203. M. R. Lamborg, P. C. Zamecnik, T.-K. Li, J. Kägi and B. L. Vallee, *Biochemistry*, **4**, 63 (1965)
204. I. Frič, J. Smekal and J. Farkaš, *Tetrahedron Letters*, **1966**, 75
205. T. Nishimura, B. Shimizu and I. Iwai, *Biochim. Biophys. Acta*, **157**, 221 (1968)
206. I. Frič, K. Bláha and G. Schmidt, private communication
207. W. A. Klee and S. H. Mudd, *Biochemistry*, **6**, 988 (1967)
208. M. P. Schweizer, H. P. Broom, P. O. P. Ts'o and D. P. Hollis, *J. Am. Chem. Soc.*, **90**, 1042 (1968); F. E. Hruska and S. S. Danyluk, *J. Am. Chem. Soc.*, **90**, 3266 (1968)
209. S. S. Danyluk and F. E. Hruska, *Biochemistry*, **7**, 1038 (1968)
209a. F. E. Hruska, *J. Am. Chem. Soc.*, **93**, 1795 (1971)
210. F. Šantavý, J. Hrbek and G. Snatzke, unpublished results.
210a. D. W. Miles, M. J. Robins, R. K. Robins, M. W. Winkley and H. Eyring, *J. Am. Chem. Soc.*, **91**, 824, 831 (1969)
211. W. Guschlbauer and Y. Courtois, *FEBS Letters*, **1**, 183 (1968)
212. T. R. Emerson, R. J. Swan and T. L. V. Ulbricht, *Biochem. Biophys. Res. Commun.*, **22**, 505 (1966)
213. J. Donohue and K. Trueblood, *J. Mol. Biol.*, **2**, 363 (1960)
214. A. Hampton and A. W. Nichol, *J. Org. Chem.*, **32**, 1688 (1967)
215. M. Ikehara, M. Kaneko, K. Muneyama and H. Tanaka, *Tetrahedron Letters*, **1967**, 3977
216. M. Ikehara, M. Kaneko and Y. Nakahara, *Tetrahedron Letters*, **1968**, 4707
217. M. Ikehara, M. Kaneko and M. Sagai, *Chem. Pharm. Bull.*, **16**, 1151 (1968)
217a. D. W. Miles, L. B. Townsend, M. J. Robins, R. K. Robins, W. H. Inskeep and H. Eyring, *J. Am. Chem. Soc.*, **93**, 1600 (1971)
218. R. B. Homer and S. F. Mason, *Chem. Commun.*, **1966**, 332
219. J. R. Fesco, *Tetrahedron*, **13**, 185 (1961)
220. B. H. Levedahl and T. W. James, *Biochim. Biophys. Acta*, **26**, 89 (1957)
221. P. O. P. Ts'o, G. K. Helmkamp and C. Sander, *Biochim. Biophys. Acta*, **55**, 584 (1962)
222. C. A. Bush and I. Tinoco, Jr., *J. Mol. Biol.*, **23**, 601 (1967)
223. J. Brahms, J. Maurizot and A. M. Michelson, *J. Mol. Biol.*, **25**, 481 (1967)
224. C. A. Bush and J. Brahms, *J. Chem. Phys.*, **46**, 79 (1967)
225. J. C. Maurizot, W. J. Wechter, J. Brahms and C. Sadron, *Nature*, **219**, 377 (1968)
226. G. B. Zavil'gel'skiĭ and L. Li, *Mol. Biol.*, **1**, 323 (1967)
227. K. E. Van Holde, J. Brahms and A. M. Michelson, *J. Mol. Biol.*, **12**, 726 (1965)
228. M. M. Warshaw, C. A. Bush and I. Tinoco, Jr., *Biochem. Biophys. Res. Commun.*, **18**, 633 (1965)
229. J. Brahms, A. M. Michelson and K. E. Van Holde, *J. Mol. Biol.*, **15**, 467 (1966)

230. N. J. Leonard and T. D. Browne, *Lecture A 16, 5th Intern. Symp. Chem. Natl. Prod. (IUPAC) London* 8.–13.7.1968
231. M. M. Warshaw and I. Tinoco, Jr., *J. Mol. Biol.*, **13**, 54 (1965); **20**, 29 (1966)
232. C. R. Cantor and I. Tinoco, Jr., *J. Mol. Biol.*, **13**, 65 (1965)
233. C. R. Cantor and I. Tinoco, Jr., *Biopolymers*, **5**, 821 (1967)
234. Y. Inoue, S. Aoyagi and K. Nakanishi, *J. Am. Chem. Soc.*, **89**, 5701 (1967)
235. J. Brahms and K. E. Van Holde, Ordered Fluids and Liquid Crystals, in *Advances in Chemistry Series, Am. Chem. Soc.*, **63**, 253 (1967)
236. J. Brahms, *Proc. Roy. Soc. (London)*, A **297**, 150 (1967)
236a. W. C. Johnson, Jr. and I. Tinoco, Jr., *Biopolymers*, **7**, 727 (1969)
236b. J. Brahms and S. Brahms, in *Fine Structure of Proteins and Nucleic Acids* (Ed. G. D. Fasman and S. N. Timasheff), M. Dekker, New York, 1970, p. 191
237. A. Rich, D. R. Davies, F. H. C. Crick and J. D. Watson, *J. Mol. Biol.*, **3**, 71 (1961)
238. H. Hashizume and K. Imahori, *J. Biochem. Tokyo*, **61**, 738 (1967)
239. J. Brahms and C. Sadron, *Nature*, **212**, 1309 (1966)
240. P. K. Sarkar and J. T. Yang, *J. Biol. Chem.*, **240**, 2088 (1965)
241. J. Brahms, *J. Mol. Biol.*, **11**, 785 (1965)
242. J. Brahms, *J. Am. Chem. Soc.*, **85**, 3298 (1963)
243. J. Brahms, J. Maurizot and A. M. Michelson, *J. Mol. Biol.*, **25**, 465 (1967)
244. J. Brahms and W. F. H. M. Mommaerts, *J. Mol. Biol.*, **10**, 73 (1964)
245. P. K. Sarkar, B. Wells and J. T. Yang, *J. Mol. Biol.*, **25**, 563 (1967)
246. W. Guschlbauer, Y. Courtois, C. Bové and J. M. Bové, *Molec. Gen. Genetics*, **103**, 150 (1968)
247. R. Langridge and P. J. Gomatos, *Science*, **141**, 694 (1963)
248. M. Spencer, W. Fuller, M. H. F. Wilkins and G. L. Brown, *Nature*, **194**, 1014 (1962)
249. B. N. Il'yashenko and G. B. Zavil'gel'skiĭ, *Biofizika*, **12**, 586 (1967)
250. Y. Courtois, P. Fromageot and W. Guschlbauer, *Eur. J. Biochem.*, **6**, 493 (1968)
251. G. Luck, Ch. Zimmer and G. Snatzke, *Biochim. Biophys. Acta*, **169**, 548 (1968)
252. T. Samejima and J. T. Yang, *Biochemistry*, **3**, 613 (1964); T. Samejima and J. T. Yang, *J. Biol. Chem.*, **240**, 2094 (1965)
253. C. R. Cantor, S. R. Jaskunas and I. Tinoco, Jr., *J. Mol. Biol.*, **20**, 39 (1966)
254. C. A. Bush and H. A. Scheraga, *Biochemistry*, **6**, 3036 (1967)
255. M. F. Maestre and I. Tinoco, Jr., *J. Mol. Biol.*, **23**, 323 (1967)
256. G. Luck and Ch. Zimmer, *Studia Biophysica*, **2**, 163, (1967); G. Luck and Ch. Zimmer, *Biochim. Biophys. Acta*, **169**, 466 (1968)
257. Ch. Zimmer, G. Luck, H. Venner and J. Frič, *Biopolymers*, **6**, 563 (1968)
258. W. B. Gratzer, *Biochim. Biophys. Acta*, **123**, 431 (1966)
259. W. Guschlbauer, *Proc. Natl. Acad. Sci. U.S.*, **57**, 1441 (1967)
260. D. M. Neville and D. F. Bradley, *Biochim. Biophys. Acta*, **50**, 397 (1961)
261. A. Blake and A. R. Peacocke, *Nature*, **206**, 1009 (1965).
262. B. J. Gardner and F. S. Mason, *Biopolymers*, **5**, 79 (1967)
263. K. Yamaoka, *Biochim. Biophys. Acta*, **169**, 553 (1968)
264. K. Yamaoka and H. Ziffer, *Biochemistry*, **7**, 1001 (1968)
265. Y. Courtois, W. Guschlbauer and P. Fromageot, *Eur. J. Biochem.*, **6**, 106 (1968)
265a. E. J. Gabbay, *J. Am. Chem. Soc.*, **90**, 6574 (1968)
266. H. R. Mahler, G. Green, R. Goutarel and Q. Khuong-Huu, *Biochemistry*, **7**, 1568 (1968)
267. W. Voelter, R. Records, E. Bunnenberg and C. Djerassi, *J. Am. Chem. Soc.*, **90**, 6163 (1968)
268. J. H. van't Hoff, *The Arrangement of Atoms in Space*, London 1898, p. 160

269. C. S. Hudson, *J. Am. Chem. Soc.*, **31**, 66 (1909)
270. W. Klyne, *Biochem. J.*, **47**, xli (1950)
271. W. Kauzmann, F. B. Clough and I. Tobias, *Tetrahedron*, **13**, 57 (1961)
272. D. H. Whiffen, *Chem. Ind. (London)*, **1956**, 964
273. J. H. Brewster, *J. Am. Chem. Soc.*, **81**, 5475 (1959)
274. N. Pace, C. Tanford and E. A. Davidson, *J. Am. Chem. Soc.*, **86**, 3160 (1964)
275. I. Listowsky, G. Avigad and S. Englard, *J. Am. Chem. Soc.*, **87**, 1765 (1965)
276. H. Hudson, M. L. Wolfrom and T. M. Lowry, *J. Chem. Soc.*, **1933**, 1179
277. W. C. G. Baldwin, M. L. Wolfrom and T. M. Lowry, *J. Chem. Soc.*, **1935**, 696
278. T. Sticzay, C. Peciar, K. Babor, M. Fedoroňko and K. Linek, *Carbohyd. Res.*, **6**, 418 (1968)
279. R. W. McGilvery, *Biochemistry*, **4**, 1924 (1965)
280. R. J. Ferrier and G. H. Sankey, *J. Chem. Soc.*, *C*, **1966**, 2339
281. S. Suzuki, H. Saito, T. Yamagata, K. Anno, N. Seno, Y. Kawai and T. Furuhashi, *J. Biol. Chem.*, **243**, 1543 (1968)
282. C. S. Hudson, *J. Am. Chem. Soc.*, **32**, 338 (1910)
283. W. Klyne, *Chem. Ind. (London)*, **1954**, 1198
284. W. Klyne, P. M. Scopes and A. Williams, *J. Chem. Soc.*, **1965**, 7237
285. T. Okuda, S. Harigaya and A. Kiyomoto, *Chem. Pharm. Bull. (Tokyo)*, **12**, 504 (1964)
286. R. J. Abraham, L. D. Hall, L. Hough, K. A. McLauchlan and H. J. Miller, *J. Chem. Soc.*, **1963**, 748
287. A. F. Beecham, *Tetrahedron Lett.*, **1968**, 2355
288. I. Listowsky, G. Avigad and S. Englard, *Carbohyd. Res.*, **8**, 205 (1968)
289. T. Sticzay, C. Peciar und S. Bauer, *Tetrahedron Lett.*, **1968**, 2407
290. cf. e.g. B. S. Sjöberg, D. J. Cram, L. Wolf and C. Djerassi, *Acta Chem. Scand.*, **16**, 1079 (1962); C. Djerassi, H. Wolf and E. Bunnenberg, *J. Am. Chem. Soc.*, **84**, 4552 (1962); Y. Tsuzuki, K. Tanaka and K. Tanabe, *Bull. Chem. Soc. Jap.*, **35**, 1614 (1962); Y. Tsuzuki, K. Tanabe, M. Akagi and S. Tejima, *Bull. Chem. Soc. Jap.*, **37**, 162 (1964); Y. Tsuzuki, K. Tanaka, K. Tanabe, M. Akagi and S. Tejima, *Bull. Chem. Soc. Jap.*, **37**, 730 (1964); T. Maki, N. Nakamura, S. Tejima and M. Akagi, *Chem. Pharm. Bull. (Tokyo)*, **13**, 764 (1965)
291. W. A. Bonner, *J. Am. Chem. Soc.*, **71**, 3384 (1949)
292. Y. Tsuzuki, K. Tanabe and K. Okamoto, *Bull. Chem. Soc. Jap.*, **39**, 761 (1966); Y. Tsuzuki, K. Tanabe, K. Okamoto and N. Yamada, *Bull. Chem. Soc. Jap.*, **39**, 1391 (1966); Y. Tsuzuki, K. Tanabe, K. Okamoto and S. Suzuki, *Bull. Chem. Soc. Jap.*, **39**, 2269 (1966)
293. L. D. Hayward and S. Claesson, *Chem. Commun.*, **1967**, 302
294. G. Snatzke, H. Laurent and R. Wiechert, *Tetrahedron*, **25**, 761 (1969)
295. L. Velluz and M. Legrand, *Compt. rend.*, **263**, 1429 (1966)
296. W. Voelter, E. Bayer, R. Records, E. Bunnenberg and C. Djerassi, *Justus Liebigs Ann. Chem.*, **718**, 238 (1968)
297. W. Voelter, E. Bayer, R. Records, E. Bunnenberg and C. Djerassi, *Chem. Ber.*, **102**, 1005 (1969)
298. K. Brendel, P. H. Gross and H. K. Zimmerman, Jr., *Justus Liebigs Ann. Chem.*, **691**, 192 (1966)
299. S. Beychok and E. A. Kabat, *Biochemistry*, **4**, 2565 (1965)
300. M. L. Wolfrom and D. L. Minor, *J. Org. Chem.*, **30**, 841 (1965)
301. M. L. Wolfrom, A. Thompson and D. R. Lineback, *J. Org. Chem.*, **27**, 2563 (1962)
302. E. O. Bishop, G. J. F. Chittenden, R. D. Guthrie, A. F. Johnson and J. F. McCarthy, *Chem. Commun.*, **1965**, 93

303. C. Satoh, A. Kiyomoto and T. Okuda, *Chem. Pharm. Bull.* (*Tokyo*), **12**, 518 (1964); *Carbohyd. Res.*, **3**, 248 (1966); C. Satoh and A. Kiyomoto, *Chem. Pharm. Bull.* (*Tokyo*), **12**, 615 (1964); *Carbohyd. Res.*, **3**, 248 (1966)
304. H. Paulsen, *Chem. Ber.*, **101**, 1571 (1968)
305. C. Djerassi, A. Moscowitz, K. Ponsold and G. Steiner, *J. Am. Chem. Soc.*, **89**, 347 (1967)
306. H. Paulsen, F. Leupold and K. Todt, *Justus Liebigs Ann. Chem.*, **692**, 200 (1966)
307. W. S. Chilton and R. C. Krahn, *J. Am. Chem. Soc.*, **89**, 4129 (1967)
308. N. K. Richtmyer and C. S. Hudson, *J. Am. Chem. Soc.*, **64**, 1612 (1942)
309. F. Garcia-Gonzales, *Advan. Carbohyd.*, **11**, 97 (1956)
310. H. El Khadem, *J. Org. Chem.*, **28**, 2478 (1963); J. A. Mills, *Aust. J. Chem.*, **17**, 277 (1964)
311. F. Garcia-Gonzales and A. Gomez, *Advan. Carbohyd.*, **20**, 313 (1965)
312. S. Englard, G. Avigad and I. Listowsky, *Carbohyd. Res.*, **2**, 380 (1966)
313. E. A. Davidson, *Biochim. Biophys. Acta*, **101**, 121 (1965)
314. E. A. Kabat, K. O. Lloyd and S. Beychok, *Biochemistry*, **8**, 747 (1969)
315. A. L. Stone, *Nature*, **216**, 551 (1967)
316. A. L. Stone and H. Moss, *Biochim. Biophys. Acta*, **136**, 56 (1967)
317. F. Cramer and H. Hettler, *Naturwissenschaften*, **54**, 625 (1967)
318. R. C. Schulz, R. Wolf and H. Mayerhöfer, *Kolloid-Z. Z. Polym.*, **227**, 65 (1968)

II
Shape, Size and Weight of Biopolymers

CHAPTER 4

Light-scattering

J. Tonnelat and S. Guinand

Laboratoire de Biologie Physico-Chimique,
Université Paris-Sud,
Orsay, France

When a beam of light passes through a medium considered as perfectly transparent, a decrease of transmitted energy is observed and a part of light is scattered, i.e. reemitted in all the directions. This phenomenon of scattering is analogous to an absorption, the extinction coefficient of which is called the turbidity, τ, $I = I_0 \, e^{-\tau l}$, with I_0, intensity of incident light, I, intensity of transmitted light and l, length pathway in cm. The scattering is due to the vibrations of the electrons of the atoms induced by the incident electromagnetic wave. After summation of the different atom vibrations, a global effect is observed, the intensity of which depends on the volume and geometry of the scattering molecule and therefore on its molecular weight.

The application of the light-scattering to the molecular weight measurements of macromolecules in solution goes back about 30 years. In 1935, Putzeys and Brosteaux[1] obtained the molecular weight of different proteins by a semi-empiric formula derived from Rayleigh's theory. But the development brought about P. Debye,[2] in 1944, has made a generalization of the method possible, principally in the field of biological macromolecules.

I. THEORETICAL BASIS

A. Small Particles

1. Molecular Basis

A classical theory was outlined by Lord Rayleigh in the case of gases, i.e. when all particles can be regarded as entirely independent of each other.

If the particles are *isotropic and small compared with the wavelength* of the incident radiation (less than about $\frac{1}{20}\lambda$) each particle may be assimilated to a single oscillating dipole emitting a spherical wave with the same frequency as the incident light and an amplitude which is the sum of the amplitudes of waves emitted by each atom.

It has been demonstrated that, for an *incident light radiation polarized in a plane perpendicular to the plane of observation*, the light scattered is

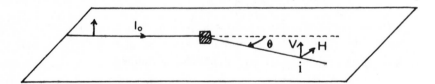

Figure 1. Incident and scattered beam in an horizontal plane

vertically polarized (Figure 1) and the intensity for a molecule at a distance r from the observer is given by Rayleigh's equation:

$$i = I_0 \frac{16\pi^4}{r^2 \lambda_0^4} \alpha^2 \tag{1}$$

where α is the molecular polarizability which characterizes the ability for the electrons to undergo vibrations; α is proportional to $(\varepsilon - \varepsilon_0)/\varepsilon_0$, ε and ε_0 are the dielectric constants of the solute and the surrounding medium. The intensity of the scattered light is proportional to the square of the volume of the particle. It is independent from the angle θ between the incident beam and the direction of observation (Figure 2(a)).

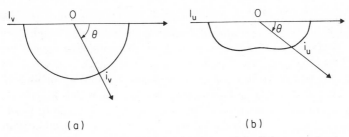

Figure 2. Angular distribution, in one plane, of the light scattering by small isotropic particles for incident light ($L < \lambda/20$): (a) vertically polarized; (b) unpolarized

With incident light polarized in the plane of observation, the scattered intensity varies as $\cos^2 \theta$. For unpolarized light, the scattered intensity is the sum of the scattered intensity produced by two incident radiations polarized at right angles

$$i = i_{90°}(1 + \cos^2 \theta)$$

$i_{90°}$ being the scattered intensity for $\theta = 90°$ (Figure 2(b)). For the sake of simplicity, we will only consider the case where the incident light is *polarized perpendicularly* to the plane of observation. For a given apparatus, the distance r is a constant. It is convenient to introduce the reduced intensity per unit volume or Rayleigh ratio

$$R_\theta = \frac{i_\theta}{I_0} r^2$$

For unpolarized light, the Rayleigh ratio may be related to the turbidity τ by integrating over the surface of a sphere of a radius r

$$\tau = \frac{16\pi}{3} R_\theta (1 + \cos^2 \theta)^{-1}$$

The depolarization of the scattered light is not the same as that of the incident wave except for the vertically polarized incident light. The degree of depolarization is defined by the factor $\rho = H/V$, where H is the intensity of the horizontal component and V that of the vertical component of the scattered wave.

2. Fluctuation Theory

A thermodynamic theory has been established by Smolukowski and by Einstein: the scattered light is admitted as due to fluctuations in density for the liquids, or in concentration for the solutions. These fluctuations, independent of each other, continuously change in time, but the changes are so slow, relatively to the frequency of incident light, that the inhomogeneity of density are regarded, at each time, as frozen and statistically equivalent at every time.

If the scattering medium is a perfectly homogeneous solid, the molecules are regularly and rigidly disposed. They can always be paired in such a way that the scattered waves in the direction of observation are out of phase except for $\theta = 0$. For each pair there is destructive interference and it is the reason why a perfect crystal, at absolute zero, would not scatter light.

In an 'homogeneous' liquid, the number of molecules per unit volume, within a small volume element, does not stay constant; this is due to Brownian motion, and thus the local density fluctuates around a mean value.

In solution, similar fluctuations exist in the local concentration of the solute. As a first approximation, it may be assumed that these fluctuations are independent of the solvent density fluctuations. In the case where there are only two components (the solvent and the solute), it may be assumed that the scattered light energy is the sum of the energies radiated separately by the solvent and the solute. It follows that the intensity scattered by the solute is equal to the difference between the scattered intensities of the solution and the solvent:

$$i \, (\text{solute}) = i \, (\text{solution}) - i \, (\text{solvent})$$

In the rest of the discussion only the 'scattering intensity excess' due to the solute will be considered.

In order to establish the relationship between the light scattered by the solute and its molecular weight, a thermodynamic treatment can be used, the details of which may be found in numerous publications;[2,3,4,5,5a] only the fundamentals will be indicated here. The mean square of the concentration fluctuations is expressed as a function of the partial molar free enthalpy of the solvent. Its value varies with the molar concentration of the solute. A relationship is obtained between the scattering power and the chemical potential of the solvent μ_1:

$$R = \frac{4\pi^2 n_0^2 (\partial n / \partial c)^2 c}{N_A \lambda_0^4 (\partial \mu_1 / \partial c) \cdot (-1 / V_1 k T)} \tag{2}$$

where the wavelength of the incident light in a vacuum is λ_0, the partial molar volume of the solvent V_1, the Boltzman constant k, the absolute temperature T, the solvent refractive index n_0, the refractive index increment $\partial n / \partial c$* and the concentration of solute c in grams per cubic centimetre, N_A is Avogadro's number.

---

* The index, at constant temperature and pressure, depends on concentration only; $\partial n / \partial c$ will be replaced by dn/dc in the text following.

The chemical potential is expressed in terms of successive powers of the molar concentration (virial equation) which is replaced by weight concentration so that the molecular weight of solute M is introduced. Usually only the first two terms of the series are used:

$$K\frac{c}{R_\theta} = \frac{1}{M} + 2Bc \tag{3}$$

for vertically polarized light. For unpolarized light

$$K\frac{c}{R_\theta}(1 + \cos^2 \theta) = \frac{1}{M} + 2Bc$$

$$K = \frac{2\pi^2}{N_A\lambda_0^4} \cdot n_0^2\left(\frac{dn}{dc}\right)^2 \tag{3}$$

The *second virial coefficient B* is identical to the one used in the calculation of osmotic pressure. When no other interaction is present, the coefficient B is proportional to the excluded volume, owing to the impossibility of introducing another solute molecule into a portion of the solute volume when a molecule is present in that same portion. The effect of the excluded volume is equivalent to a short-distance repulsion.

For a compact spherical molecule with a radius r, the excluded volume is equal to the volume of a sphere with a radius $2r$, so that:

$$B = \frac{4\bar{v}}{M}$$

where $\bar{v}$ is the partial specific volume of the solute.

This value is very small and can often be neglected. This is not the case for rigid rods of length and diameter d, the excluded volume of which is

$$u = \frac{2LM\bar{v}}{N_A d}$$

and then

$$B = \frac{L\bar{v}}{Md}$$

When the molecules are charged, as it is generally the case for biological macromolecules, they exert electrostatic interactions between themselves.

Moreover, biological macromolecules are usually in solution with an electrolyte. Other interactions exist with different ions. All of these are accounted for in the expression of the coefficient B given by Scatchard

$$B = \frac{1000\bar{v}_1}{M_2^2}\left[\frac{Z^2}{4m_3} + \frac{\beta_{22}}{2} - \frac{\beta_{23}m_3}{4 + 2\beta_{33}m_3}\right] \tag{4}$$

where the β's are derivatives of activity coefficients $\ln \gamma$; the subscript 1 refers to the pure solvent, the subscript 2 to the macromolecule with molecular weight M_2 and charge Z and the subscript 3 refers to the third component of the solvent. m_3 is the molality of the various components in the solution and $\bar{v}_1$ the partial specific volume of the solvent.

When the solute preferentially binds one of the solvent components, the relation (4) may no longer be applied. The complete theory for a three components[6,7,8] system leads to the relation[9]

$$K \left(\frac{c}{R} \right)_{c=0} = \frac{1}{M_2} \frac{1}{(1 + D)^2} \tag{5}$$

$$1 + D = 1 + \frac{(\partial n/\partial m_3)_{m_2}}{(\partial n/\partial m_2)_{m_3}} \left(\frac{m_3}{m_2} \right)_{\mu_3}$$

where μ_3 is the chemical potential of component 3 and m_3, its molality which is kept identical with that of the reference solvent.

If the solution has been first dialysed against the solvent, the chemical potential of component 3 is the same in the solution and in the solvent (only its weight concentration is different); the solution then obeys to the same law as a two-components system. In order to obtain a correct value of the molecular weight by extrapolation to zero concentration, the refractive index increment and the scattered intensity are measured against the dialysate.

However, when the solute has been dissolved in the solvent without dialysing, an 'apparent molecular weight' is obtained which allows us to know the value of preferential solvation.[9,10]

B. Larger Particles

If one dimension of the particles is *greater than one-tenth of the wavelength* of the incident radiation, each one of them can no longer be assimilated to a single dipole. Thus, the waves emitted by two elements of a molecule which are not in phase interfere. The result is a decrease in the scattered intensity.

For $\theta = 0$, the path length of the two beams coming from each element of the molecule are equal and no interference occurs (Figure 3). The scattered intensity in this direction is the same as if the molecule could be assimilated to a single dipole. But when θ increases, i decreases and for each angle θ, we have

$$i_\theta = i_0 P(\theta) \tag{6}$$

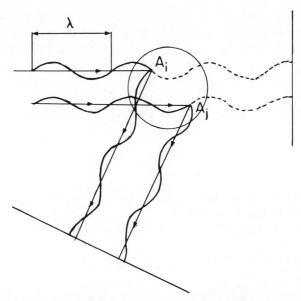

Figure 3. Representation of the phase differences between rays scattered at two points of a macromolecule which is not small compared to the wavelength of light ($> \lambda/10$)

$P(\theta)$ being a factor smaller than unity and depending both on molecular shape and size (Figure 4).

A general relationship between $P(\theta)$ and the molecular shape can be derived. In order to obtain an expression of $P(\theta)$, we consider two elements

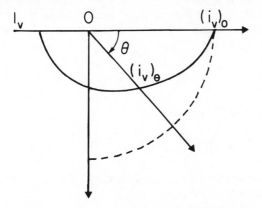

Figure 4. Angular distribution of the scattered intensity of the vertically polarized component $(i_v)_\theta$ for the molecules with one dimension such as $L > \lambda/10$

A_i and A_j of a large scattering particle, each one of which is considered as an autonomous resonator. The intensity in the direction of observation is the sum of the intensity of the light scattered by each pair A_i, A_j:

$$P(\theta) = \frac{1}{n^2} \sum_i \sum_j \frac{\sin h r_{ij}}{h r_{ij}}$$

where r_{ij} is the distance between two of the n scattering elements i and j, with

$$h = \frac{4\pi}{\lambda} \sin \frac{\theta}{2}$$

λ being the wavelength of the incident light in the solution ($\lambda = \lambda_0/n_0$ if λ_0 is the wavelength in vacuum).

Sin $\theta/2$ may be expanded in terms of a power series to give:

$$\lim P(\theta)_{\theta \to 0} = 1 - \frac{h^2}{3!} \sum_i \sum_j r_{ij}^2 \qquad (7)$$

The double summation may be expressed as a function of the radius of gyration of the particle. As a matter of fact all the molecules may be considered as an assembly of elements of mass m_i situated at a distance r_i from the centre of gravity so that $\sum m_i r_i = 0$. The radius of gyration is defined as:

$$R_G^2 = \frac{\sum m_i r_i^2}{\sum m_i}$$

If the elements A_i are identical, $m_i = m_j$ whatever the values of i and j. Instead of considering the distance r_i of an element to the centre of gravity, we can consider the distance r_{ij} between two elements and some geometric considerations show that

$$R_G^2 = \frac{1}{2n^2} \sum_{i=1}^{n} \sum_{j=1}^{n} r_{ij}^2$$

The relation (7) may be written

$$\lim[P(\theta)]_{\theta=0} = 1 - \frac{16\pi^2}{3\lambda^2} R_G^2 \sin^2\left(\frac{\theta}{2}\right) \qquad (8)$$

A simple expression of the factor $P(\theta)$ may be given for the molecules of well defined geometric shapes.

For a sphere of radius $R = L/2$

$$P(\theta) = \left[\frac{3}{h^3 R^3} (\sin hR - hR \cos hR) \right]^2 \qquad (9)$$

For a rod of length L

$$P(\theta) = \frac{2}{hL} \int_0^{hL} \frac{\sin u}{u} \, du - \frac{\sin^2(hL/2)}{h^2 L^2/4} \tag{10}$$

For a gaussian chain where L^2 is the mean square end-to-end distance

$$P(\theta) = \frac{2}{u^2}[u - 1 + \exp(-u)] \tag{11}$$

setting $u = h^2 L^2/6$.

The expression of the factor $P(\theta)$ has been derived for other models[11] and the values have been tabulated in the case of spheres, rods and random coils.[4]

Except in the direction of the incident radiation, the scattered intensity in all other directions is smaller than if the whole molecule was assimilable to a single resonator. This is due to the discordance between the waves emitted by two elementary resonators of the molecule. The scattered energy is proportionally smaller than for a small molecule. This weakening which results from the waves out of phase depends on the wavelength so that the scattered intensity is no longer proportional to λ^{-4} but to λ^{-n}, with $2 < n < 4$.

C. Anisotropic Particles

When the properties of the particle are not the same in all directions, the particle is *anisotropic*, it may no longer be assimilated to a linear oscillator. The depolarization factor depends on the molecular anisotropy.

The thermodynamic theory does not take into account of anisotropy. A correlation factor (Cabannes factor) should be introduced which, at $\theta = 90°$ and for a small particle, is equal to

$$\frac{6 + 6\rho_u}{6 - 7\rho_u}$$

when the incident radiation is unpolarized, ρ_u being the depolarization factor. If incident light is vertically polarized, ρ_v being the depolarization factor, the correction term is

$$\frac{3(1 + \rho_v)}{3 - 4\rho_v}$$

For globular proteins this factor at $\theta = 90°$ has a value close to 0·96; for the polymers, 0·92. If the molecules are of a size close to the wavelength, it is necessary to extrapolate to angle $\theta = 0$.

When an isotropic molecule has some symmetric elements so that its polarizability may be characterized by the polarizability α and β, in two rectangular directions, the molecular anisotropy may be defined by the relation

$$\delta = \frac{\alpha - \beta}{\alpha + 2\beta}$$

It is shown that

$$(\rho_u)_{90°} = \frac{6\delta^2}{5 + 7\delta^2} \qquad (\rho)_v = \frac{3\delta^2}{5 + 4\delta^2}$$

If the particles are large and anisotropic, it must be taken into account. The correction factor is small in most cases. Nevertheless, it is necessary to modify the equation giving M, in the case of a solution of rods.[12]

D. Mixture of Molecules

When the solution contains a *mixture of molecules* with different molecular weight M_i, of weight concentration c_i, a global intensity is measured from which a *weight-average molecular weight* is determined:

$$\overline{M}_w = \frac{\sum M_i c_i}{\sum M_i}$$

If a polydispersity in the molecular dimensions exists, the average value of R_G^2 is given by

$$\overline{R}_G^2 = \frac{\sum c_i M_i (R_i^2)}{\sum c_i M_i}$$

In the case of 'gaussian chain', R_i^2 is a linear function of M_i and the radius of gyration obtained is that of a sample with a z-average molecular weight

$$\overline{M}_z = \frac{\sum c_i M_i^2}{\sum c_i M_i}$$

In the case of rods, the expansion of $P(\theta)$ (equation (10)) can be expressed by the first two terms of a power series when the values of hL are not too large[13]

$$P(\theta)^{-1} = 1 + \frac{1}{9}\left(\frac{h\overline{L}_z}{2}\right)^2$$

and

$$K\left(\frac{c}{R_\theta}\right)_{c=0} = \frac{1}{M_w}\left[1 + \frac{1}{9}\left(\frac{h\overline{L}_z}{2}\right)^2\right]$$

where $\bar{L}_z$ is a z-average of the lengths. If hL is large ($hL > 2$), $P(\theta)$ can be replaced by its asymptotic expression[14]

$$P(\theta)^{-1} = \frac{2}{\pi^2} + \frac{hL}{\pi}$$

and

$$K\left(\frac{c}{R}\right)_{c \to 0} = \frac{2}{\pi^2 M_n} + \frac{h}{\pi} \cdot \frac{1}{M/L}$$

where $\bar{M}_n$ is a number-average molecular weight. The curve must be plotted *vs.* $\sin(\theta/2)$ instead $\sin^2(\theta/2)$; the slope is M/L, which is a mass per unit length of the particles, i.e. a parameter independent of any polydispersity of lengths.

E. *Quasi-Elastic Light-Scattering*[15,16,17]

In the applications above, it has been admitted that the frequency was the same for the scattered light and incident light.

But, because of Brownian motion of molecules, a spectral broadening of Rayleigh line arises. Debye has established a relationship between the spectral distribution of energy and the diffusion coefficient of solute molecules. The width of spectrum is very narrow; for the macromolecules, the diffusion coefficient of which is from 10^{-8} to 10^{-6} cm^2/sec; we expect that the half width of the spectra will be in the range of frequency from 100 Hz to 10 kHz. Such measurements are not possible except by using a laser as source for excitation of scattered light.

The approximation which has been made earlier by admitting the fluctuations as frozen is justified since, only the global scattered intensity is involved in the determination of molecular parameters.

II. UTILIZATION OF THE METHOD

A. *Quantities to be Measured*

1. Scattered Intensity

In order to measure the scattered intensity, there are several types of commercially available apparatuses (Phoenix, Fica) which are based on very similar principles.

A light source sends into the medium a beam of radiations as parallel as possible, so that the observation angle θ is clearly defined and the beam is rendered monochromatic by a filter. By using a LASER as source of excitation these conditions are satisfied. The scattered intensity is measured with a photomultiplier.

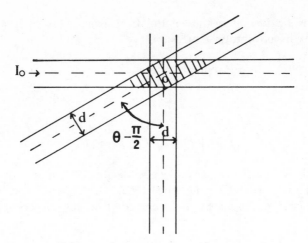

Figure 5. Angular dependence of scattering volume of the
solution

The scattering volume is defined by a series of conveniently disposed
diaphragms; v varies with the angle of observation (Figure 5).

$$v_\theta = v_{90°} \sin \theta$$

Moreover, because the refractive index changes by passing from the air
to the solutions, the scattering volume is not the same for liquids of different
refractive index. Thus a correction factor must be introduced when an
aqueous solution is compared with an organic solvent.

The use of a vertically polarized incident light is preferable to unpolarized
light because the scattered intensity is independent of the angle θ for the
small isotropic particles.

(a) Direct measurements of the reduced intensity of the solution R_θ
(generally less than 10^{-6}) are in practice extremely difficult. Then the
experimental value of the scattering intensity, i_θ, is compared to that of a
standard, the Rayleigh ratio of which is known. The standard most often
employed is the benzene, the Rayleigh ratio of which is $R_b = 48 \times 10^{-6}$ at
4,358 Å and 16×10^{-6} at 5,460 Å for $\theta = 90°$. $R_b = i_b/I_0 \times r^2$, where i_b is
scattering intensity of benzene with unpolarized incident light.

By taking into account that the refractive index is different for the benzene
(n_b) and the solution studied (n_0), a correction factor $(n_b/n_0)^2$ has to be
introduced:

$$R_\theta = i_\theta \left(\frac{n_0}{n_b}\right)^2 \frac{R_b}{i_b}$$

(b) Instead of determining the absolute value of molecular parameters, a more reliable and much easier method consists in using as intensity standard a substance of known molecular weight, with a shape similar to that of the molecule investigated. The depolarization factor of this standard has nearly the same value as the sample so that a suitable choice of the standard allows us to eliminate the Cabannes factor.

2. Refractive Index Increment

The refractive increment of a solution is the increase of refractive index relative to the solvent. As this term is squared, the accuracy of its determination greatly influences the results. Interferential refractometers are usually employed to measure it.

Theoretically the increment varies with the concentration. Practically it remains constant at low concentrations of the solute (less than 2 per cent). Moreover, the refractive index depends on:

(a) the wavelength according to the empirical relation of Perlmann and Longsworth,[18]

(b) the charge Z

$$\left(\frac{dn}{dc}\right)_{z=z} = \left(\frac{dn}{dc}\right)_{z=0} (1 + aZ)$$

where a is an experimental coefficient,

(c) the ionic strength of the solvent

$$\left(\frac{dn}{dc}\right)_{n'} = \left(\frac{dn}{dc}\right)_{n} - \bar{v}(n' - n)$$

n and n' being the refractive index of the solvent for the two ionic strengths μ and μ' and $\bar{v}$ the partial specific volume of the solute. This formula is only valid if $\bar{v}$ is the same at both ionic strengths.

As the refractive index increment depends essentially on the atomic composition, it might be expected that all the proteins of analogous composition in the same conditions would have similar values for their refractive increment.

For example, for the β-lactoglobulin, the value of dn/dc is 0·182 at 5460 Å and 0·189 at 4358 Å with a precision of 0·4 per cent.

However data in the literature[19] show wide variations from one protein to another in the same conditions and also for different preparations of the same protein.

When there is a 'preferential binding' of the ions or one component of the solvent (as in the case of binary solvent), it has been pointed out that the refractive index measurements must be carried out with the dialysate as a reference, after thorough dialysis.

B. *Interpretation of Data*

If one considers the general equation which relates the scattered intensity to the different molecular parameters:

$$K \frac{c}{R_\theta} P(\theta) = \frac{1}{M} + 2Bc \qquad (12)$$

where K is a function of n_0^2, $(\mathrm{d}n/\mathrm{d}c)^2$ and λ_0^4, from the measurements of R_θ at different concentrations c and under different angles, then from the equation the following information can be obtained:

(a) the molecular weight M of the solute, if it is the only component besides the solvent,

(b) the size and the shape of the molecule (and in a more general way, its radius of gyration) from the determination of $P(\theta)$, if the particle dimensions are greater than one-tenth of the wavelength,

(c) the interactions between the molecules of the solute and the solvent from the values of B,

(d) the constant of equilibrium between several species of the solute molecules of molecular weight, if these molecules associate with each other. This can be done by following the variation of the 'average molecular weight' $\overline{M}_w$ under different conditions.

It is worth noting that the information obtained falls into two classes, namely that results which are related to the molecular configuration by extrapolation to infinite dilution and those that are related to the interactions between components as a result of operating at concentrations different from zero.

To determine the molecular parameters, it is seen from equation (12) that if Kc/R is plotted against c, the intercept with the ordinate axis yields the molecular weight value, the slope being a measure of the second virial coefficient B.

Since $R_{\mathrm{solute}} = R_{\mathrm{solution}} - R_{\mathrm{solvent}}$, the accuracy of this method is limited, at low concentrations, by the value of the intensity of the light scattered by the solvent. For aqueous solvents, this value is hardly reproducible. For solutes of low molecular weight, giving low scattering intensity, it is important to have the same value of the scattered intensity of the solvent for different concentrations of the solute. These conditions can be achieved by adding to the solvent small volumes of the concentrated solution of solute (cleaned of all dust beforehand).

(a) When the *particles are small* compared to the wavelength the angular distribution is clearly symmetrical about 90°: $P(\theta) \simeq 1$. Therefore, it is not possible to obtain information on the size of the particles if the main dimension L is below 250–300 Å.

The measurements can be carried out at any angles θ, but preferably at one which is larger than 90° so that the influence of extraneous macromolecules be less important.[20]

When R is plotted against c, a straight line is obtained for low concentration values, the slope of which is proportional to the molecular weight

$$M = K\left(\frac{R_\theta}{c}\right)_{c\to0}$$

If a molecular weight standard M_e, with both the same refractive increment and depolarization factor as the sample, is used

$$M = M_e\frac{(R_\theta/c)}{(R_\theta/c)_e}$$

(b) If the *molecules are larger*, the scattered intensity varies with angle θ. In order to determine molecular parameters, there are two ways to proceed; these are outlined in the next two sections.

1. Extrapolation Method of Zimm

The molecular weight is obtained from the equation (12)

$$\frac{1}{M} = K\lim\left(\frac{c}{R_\theta}\right)^{c=0}_{\theta=0}$$

At each value of θ, the experimental data are extrapolated to zero concentration and these values are plotted against $\sin^2(\theta/2)$. Then extrapolation to zero angle yields the reciprocal of the molecular weight. The double extrapolation may be directly done on the same graph (Figure 6). Kc/R is plotted against $\sin^2(\theta/2) + kc$ where k is an arbitrary constant selected so as to provide convenient spacing of the data (Zimm plot[21]). Extrapolating to $\theta = 0°$, we obtain a plot of Kc/R_0 against c (curve 1, Figure 6) which gives the intercept $1/M$, and the slope $2B$. Extrapolating to $c = 0$, we obtain a plot against $\sin^2(\theta/2)$ (curve 2, Figure 6), the initial slope of which gives the radius of gyration by combining the equations (8) and (12).

$$K\left(\frac{c}{R}\right)_{c=0} = \frac{1}{M}\left(1 + \frac{16\pi^2}{3\lambda^2}R_G^2\sin^2\left(\frac{\theta}{2}\right)\right) \tag{13}$$

and

$$R_G^2 = \frac{3\lambda^2}{16\pi^2}\frac{\text{Initial slope}}{\text{Intercept}}$$

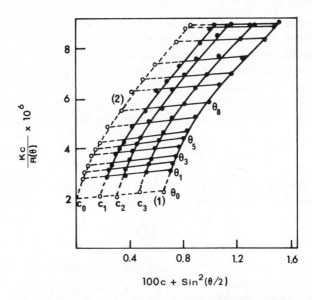

Figure 6. Zimm plot of light scattering data for calf skin tropocollagen (CSC) in citrate (pH 3·5, ionic strength 0·15 M) concentration c in g/ml. The θ angles are covering the range from 27° to 135° (from reference 26, copyright Academic Press 1964)

Using this radius of gyration it is possible to determine the molecular dimensions, if a model is assumed. For a sphere of a radius r

$$R_G^2 = \tfrac{3}{5} r^2 \tag{14}$$

For an ellipsoid of revolution of half-axis a and transverse half-axis b

$$R_G^2 = \tfrac{1}{5}(a^2 + 2b^2) \tag{15}$$

For a cylindrical rod of length L and a diameter $2r$

$$R_G^2 = \frac{L^2}{12} + \frac{r^2}{L} \tag{16}$$

For a flexible gaussian chain

$$R_G^2 = \frac{L^2}{6} \tag{17}$$

L^2 being the mean square average end-to-end.

For a so-called 'worm-like' chain,[22] which is intermediate between the rod and the gaussian chain and characterized by length L when it is stretched and 'persistence length' in setting $x = L/q$ the mean square average end-to-end distance of the chain L^2 is

$$\bar{L}^2 = 2q^2[x - 1 + \exp(-x)]$$

and

$$R_G^2 = q^2\left[x - 1 + \frac{2}{x^2}\{1 - \exp(-x)\}\right]$$

It has been shown that if there is different molecular species in the solution, the measurements of the radius of gyration give a complex average value.

But with some highly scattering polymers, even in small amounts, they will become prominent especially if the scattering intensity of the solutions is weak and at low angles; then a rapid change in slope at low angles is observed. The extrapolation is uncertain, therefore in this case the values of the molecular weight and the radius of gyration are difficult to obtain.

Moreover any upwards curvature suggests a dissociation of the molecule into subunits.

Whereas the initial part of the scattered intensity curve as a function of $\sin^2(\theta/2)$ (Figure 6) allows us to determine the molecular weight and the radius of gyration, by examination of the asymptote of the curve Kc/R (for θ close to π) it is possible to distinguish between different shape models for the molecules of the solute, provided that they are very large. For a thin rod, R_θ varies as $\sin(\theta/2)^{23}$ (Figure 7) for flexible gaussian chain as $\sin^2(\theta/2)$ and for spheres as $\sin^4(\theta/2)$.

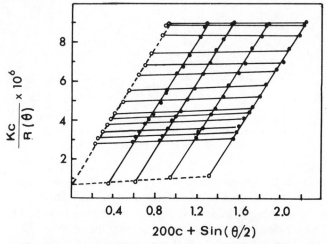

Figure 7. Data of Figure 6 plotted to illustrate conformity with asymptotic scattering behaviour for long rods (from reference 26, copyright Academic Press 1964)

In the case of a rigid rod or zigzag chain, if Kc/R is plotted against $\sin(\theta/2)$ (cf. p. 145) the slope of the asymptote yields the mass per unit length M/L which is a parameter independent of any polydispersity (Figure 7). The latter relation is valid only for molecules of sufficient length: for $\lambda_0 = 5460$ Å and $\theta = 150°$, $h < 2\cdot9 \times 10^{-3}$, i.e. $L > 1400$ Å.

2. Method of Dissymmetry

The experimental results may be interpreted in another way, which allows us to determine at the same time the molecular weight and the size.

The dissymmetry coefficient, defined as the ratio $i_\theta/i_{\pi-\theta}$ depends on shape and size. The values of this dissymmetry have been calculated for different geometrical models and different values of the dimension, from the theoretical relation. These have been plotted on the graphs for $\theta = 45°$ (Figure 8).

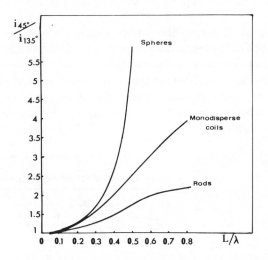

Figure 8. Dissymmetry coefficient $i_\theta/i_{\pi-\theta}$, for $\theta = 45°$ as a function of the ratio L/λ, L being the main dimension of the macromolecule

The value of $P(\theta)$, corresponding to experimental value of dissymmetry and the value of the ratio of the dimension to wavelength L/λ can be obtained from graphs of the function which have been calculated.[4]

For the determination of molecular weight, the relation 12 is applied with the value of $P(\theta)$ obtained above and with $\theta = 90°$. This method gives more reliable results than the extrapolation, since the influence of extraneous highly scattering impurities is eliminated.

III. EXPERIMENTAL RESULTS RELATIVE TO SOME
BIOLOGICAL MACROMOLECULES

A. Proteins

Most of the globular proteins are approximate to spheres (or ellipsoids of revolution with a small eccentricity) with a diameter from 30 Å to 60 Å. In this case, it has been shown that the angular distribution of scattering intensity is symmetrical about 90°. No information can be obtained on the geometric conformation, only the molecular weight can be determined. It is worth noting that the spherical particles of radius of 100 Å and hence a molecular weight of 3×10^6, also have a symmetrical angular distribution giving a symmetrical scattering diagram.

Even the serum macroglobulins (IgM and α_{2M} with a molecular weight close to 10^6), which consist in more or less flexible chains staying on the inside of a spherical envelope, also have a symmetrical scattering diagram.

However proteins such as fibrinogen (axial ratio from 6 to 7) present a diagram, symmetrical if the wavelength of the incident light is 4358 Å, and asymmetrical when a light radiation of 3650 Å is used. The value of the dissymmetry allows us to calculate the length of the particle if it is assumed to be a rod-like: the length is 450 Å.[24]

The so-called fibrilar proteins are highly asymmetrical with axial ratio from 25 to 180 and are generally approximate to rigid rods or elongated ellipsoids.[25] Their polydispersity make the determination of molecular weight or size difficult. On the other hand, this determination is also limited by the uncertainty of the extrapolated values, for example, in the case of particles greater than 3000 Å and with measurements in the angular range 25°–140°. Only some of these proteins need be mentioned, for example tropocollagen,[26] tropomyosin B[27] and paramyosin.[28]

B. Polyamino Acids

In order to examine the change of polypeptide conformation, optical rotatory dispersion and circular dichroism measurements are currently used. But light scattering has also been applied to study some amino acid polymers in dilute solution, for example polybenzyl-L glutamic acid (PBLG). In chloroform-methyl formamide, this polyamino acid adopts a helical structure. Doty and coworkers[29] have calculated the pitch of the helix as a function of the radius of gyration

$$H = \frac{m_0}{M}(12R_G^2)^{1/2}$$

m_0 being the molecular weight of the unit monomer, M that of the polymer. The experimental results have been shown that R_G is independent of M;

this leads to the conclusion that the molecule exists in the form of a cylindrical rigid rod when the molecular weight is below 300,000 and would be a slightly flexible chain above this value. The calculated value of the pitch is 1.5 Å, i.e. that of an α-helix.

More recently, Luzzati and coworkers[30] have collected the data obtained by a number of authors for PBLG in various solvents where H, the pitch of helix is plotted against M. The results are such that M, and thus L, increase at a lower rate than H. Therefore it seems to be a slightly flexible rod. To obtain the true value of the pitch, it is necessary to extrapolate to zero molecular weight. The value of H would be, in this case, 2 Å, whereas the pitch of an α-helix is 1.5 Å.

These results have been subjected to criticism by Benoit[31] and discussed from the examination of asymptotic behaviour. Taking into account the anisotropy and setting $\delta = 0.1$, an α-helix gives the best agreement with the experimental data.

C. Deoxyribonucleic Acids (DNA)

Many studies have been devoted to deoxyribonucleic acids in solution above all by Sadron and his collaborators, using hydrodynamic and optical methods, especially light scattering.[14]

The curve Kc/R against $\sin^2(\theta/2)$ gives a great uncertainty for the extrapolated value to zero angle, if the measurements are not carried out at very low angles: this comes from the contraction of the scale of the abscissa in the region of small values of θ. Moreover the samples are often polydisperse.

Some authors had proposed to plot Kc/R against $\sin(\theta/2)$ and to investigate the asymptotic values when $\theta \to \pi$. By using this interpretation Sadron[32] has proposed a molecular model: the molecule would be a zigzag chain which consists of N rod-like elements, the length of each being larger than $\frac{1}{20}$ of λ and having various possible relative orientations with a random distribution.

The slope of the asymptote gives the value of M/L (Figure 9). Moreover to obtain an accurate value of the molecular weight, the anisotropy δ must be taken into account, as for the polypeptides; in first approximation M/L_{app} is obtained.

Recent investigations on the high molecular weight DNA have shown it was interesting to work at low angles, below 30°, in order to obtain a correct molecular weight, by extrapolating the plot against $\sin^2(\theta/2)$ to zero angle.

So that the results demonstrate a halving of molecular weight by denaturation, acid,[34] alcalin and thermal,[35] whereas, in earlier experiments, it has

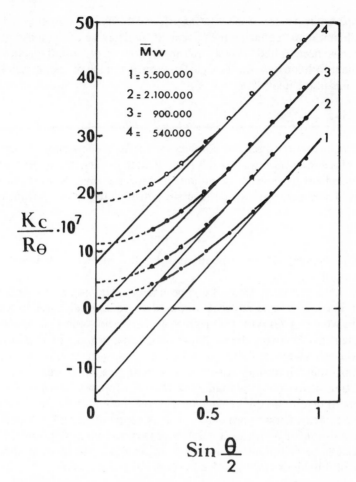

Figure 9. Scattered intensity by enzymatically degraded DNA molecules $K c/R_\theta$ versus $\sin(\theta/2)$
1. Native sample: zigzag chains with N rods
4. Final sample: rod elementary
(from reference 32)

been found that the value of the molecular weight was the same for native and denatured DNA molecule.

D. DNA–Proflavin Complex

When a dye, such as proflavin, is bound to a DNA molecule, an apparent increase of the length of molecule can be observed by various methods (sedimentation and viscosity, light-scattering by X-rays at small angles);

this apparent increase is due to the insertion of the dye between the nucleotide bases. From light-scattering measurements, a decrease of the mass per unit length has been found with increasing amounts of bound proflavin, in agreement with the hypothesis of partial insertion in the planes perpendicular to the axis of the helix.[36]

E. Associating Systems

One of the most important applications of light-scattering concerns the study of self-associating systems. As a matter of fact, many proteins are made of identical subunits (monomers) associated by non-covalent inter-actions. These proteins can be reversibly dissociated under certain conditions.

$$nP_1 \rightleftharpoons P_n, \qquad n = 2, 3, 4, \ldots$$

where P_1 represents the subunit and P_n, the molecule undergoing the dissociation reaction.

What is determined from the experimental data is not the molecular weight of a well defined molecule but an average molecular weight $\overline{M}_w$ of a mixture which varies with the degree of dissociation and is concentration dependent. By following the variation of $\overline{M}_w$, the proportion of dissociated and non-dissociated molecules can be determined, at each concentration; from this, the equilibrium constant may be derived, and this leads to the calculation of the thermodynamic parameters. In some cases it would be possible to determine the arrangement of the monomers in the polymer.

The effect of different parameters such as temperature, pH, ionic strength or the fixation of a ligand, on the thermodynamic equilibrium of these systems can be investigated and the results can give information about the nature of the interactions involved between the monomers.

$$K_d = \frac{2c(2M_1 - \overline{M}_w)^2}{M_1^2(\overline{M}_w - M_1)} \tag{19}$$

where M_1 is the molecular weight of the monomer, c the total concentration in protein and $\overline{M}_w$ the molecular weight of the mixture monomer–dimer at that concentration in g/1; K_d is expressed in moles/l.

The plots of $1/\overline{M}_w$ against c are shown (Figure 10) for a series of monomer–dimer equilibria which have various values of K_d.

But what is obtained from the measurements of scattered intensity at the concentration c is an 'apparent' value of $\overline{M}_w$

$$\left(\frac{M_1}{\overline{M}_w}\right)_{app} = \frac{M_1}{\overline{M}_w} + 2BM_1c$$

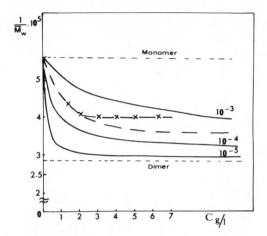

Figure 10. Plots of $1/\overline{M}_w$ versus C for a self associating system monomer–dimer (molecular weight of monomer $= 18,000$); ——— calculated curves for various values of K_d; $\times - \times - \times - \times$ from the experimental data on β-lactoglobulin at pH 8·8 ($\mu = 0·1$) $K_d = 2·4 \times 10^{-4}$ M (data from reference 40); $-----$ calculated curve for the above value of K_d

The values of $\overline{M}_w$ and $(\overline{M}_w)_{app}$ are not identical except if the term $2Bc$ is negligible, i.e. the solutions are ideal. The case of non-ideal solutions has been presented in detail by Adams *et al.*[37] It is worth noting that when the value of K is about the same as that of $2BM_1$ the effect of dissociation on the value of $\overline{M}_w$ is cancelled by the contribution of the second virial coefficient. Then an independence of the concentration is observed, as it is the case of β-lactoglobulin dissociation (Figure 10).

The relation (19) can be generalized in the case where two forms exist in the solution, the monomer and the polymer with one equilibrium constant K. When intermediate forms are in equilibrium, there are $n - 1$ constants. The mode of calculation has been given by Steiner in this general case.[38]

Only a few examples will be given in order to illustrate the advantages to study polymerizing systems by light-scattering:

(a) It is known for a great number of polymeric enzymes, the ligand binding involves a dissociation linked to transconformation. The *β-lacto-globulin* can be considered as a 'structural model' for the polymeric enzymes. As a matter of fact, β-lactoglobulin consists of two identical subunits ($M = 18,000$). Light-scattering investigations have shown that the molecules undergo a reversible dissociation as a function of pH, i.e. by the binding of protons (Figure 11).

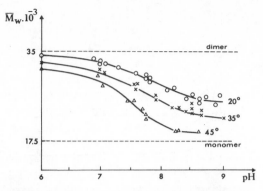

Figure 11. Variations of $\overline{M}_w$ of β-lactoglobulin B,
as function of pH, at different temperatures (ionic
strength 0·1) (Reproduced from reference 40)

Below pH 3·5, the dissociation is due to increasing non-specific electrostatic repulsion between the two subunits.[39] The determination of
equilibrium constant value has permitted the estimation of the magnitude
of free energy of attraction. The magnitude of this free energy could be
accounted for by the presence of hydrophobic bonds in the association.

Above pH 5·5, a dissociation still exists,[40] but it is linked to a change
of conformation and the ionization of two anomalous carboxylic groups.[41]
The calculation of thermodynamic constants has been deduced from light-
scattering data and it has been shown that the difference between the values
of enthalpy at pH 7 and pH 9 is equal to the value of the enthalpy involved
in the change of conformation.

(b) The *genetic species A* of the *same protein* undergoes a reversible
aggregation between pH 3·5 and 5·2, at low temperature, below 10 °C.

From light-scattering measurements it has been found that the association
is limited to tetramers.[42] The equilibrium constant determination has
permitted one to establish that dimers and trimers are never present in
significant amounts. Equilibrium exists only between the monomer (36,000)
and the tetramer (144,000).

(c) The *glutamate dehydrogenase* (GDH) from mammalian liver is
polymerized in the native state; the molecule is formed of enzymatically
active units, each unit ($M = 320,000$) consisting in 6 elements (55,000)
associated in a cylinder. Evidence of these elementary subunits has been
shown by light-scattering in a dissociating solvent.[43]

It has been shown by Eisenberg and Tomkins that the polymerization
is indefinite; $\overline{M}_w$ is concentration dependent and the length of the polymer
increases in a similar manner as the molecular weight when the concentration
is increased.[44] By applying Steiner's mode of calculation, it has been found

that there are several equilibrium constants with similar values.[46] As a result, one can deduce that the polymerization is linear.

From a comparative study between polarimetric and light-scattering measurements, it appears that the 'measured' quantities did not vary similarly with the concentration. There is a good correlation between the results of both methods only with this hypothesis of an end-to-end association of the molecule, i.e. a linear polymerization.[46]

(d) The association state of the *allosteric enzymes* can be modified by the binding of ligand such as substrate or effectors.

In the case of GDH from liver, the dissociation effect produced by GTP (guanosine triphosphate) has been shown by light-scattering.[44,45]

Among the other allosteric enzymes, we will mention Buc and Buc's study of phosphorylase b from rabbit muscle.[47] The information concerning the quaternary structure of the enzyme molecule and the interactions between the sites of binding of ligands obtained by this method is a striking test to its availability.

F. Kinetics of Association

Because any change in scattering properties can be registered instantaneously without either destroying or losing solute, the kinetics of any process involving the changes of molecular weight can easily be followed by the classic method if the process is slow enough. In the case of trypsin, which undergoes maximum self-association at pH 4·8, the kinetics have been investigated by light-scattering. The order of the kinetics is concentration dependent.[48] It has been observed that the rate of variation of $\overline{M}_w$, i.e. of the polymerization, increases with the concentration and becomes independent of the concentration at the high concentrations. Therefore there must be a preliminary step before the association; this must be an isomerization and is shown by the following scheme

$$X \rightarrow Y$$

$$nY \rightleftharpoons Y^n$$

If the kinetics of dissociation or association reactions are fast, the study can be made with the *stopped flow method*, using the spectrophotometer to measure the changes in turbidity, i.e. absorbance of the solutions. The concentration dependent dissociation of bovine-liver glutamate dehydrogenase has been investigated by this method.[49]

The temperature jump relaxation method is also used by utilizing scattered light intensity as detector of dissociation phenomenons.[50] In the studies of kinetics of dissociation of lobster hemocyanin, it has been found three relaxation times, the first, during the heating period (15–35 μsec) is not specific of the dissociation. The others are in the range 100–145 μsec and 4·7–7·5 msec. These results are consistent with those obtained by ultra-centrifugation.

G. Interactions Solute–Solvent

The interactions between macromolecular solute and solvent can generally be estimated by the measurement of coefficient B. In a two-component solvent the macromolecular solute can show a preferential attraction for one of them. The interactions can be measured from the light-scattering data if the two-solvent components have non-identical refractive indices. But these measurements result only in preferential interactions of one component relative to the other. Absolute interactions between the macromolecular solute and one component of solvent can be calculated.[52]

Inoue and Timasheff[9] have shown by this method that β-lactoglobulin A, in mixtures of water with chloroethanol, has a preferential interaction with alcohol which changes to hydration when the alcohol concentration is increasing. This interaction is correlated with a change of the molecular structure as observed by optical rotatory dispersion and circular dichroism. The binding of a non-polar component to the protein is parallel to its unfolding.

Another interesting application of the method is the study of the helix–coil transition of Poly-L-Benzyl-Glutamic acid (PBLGA),[51] in a binary solvent. Since the preferential interaction of one of the solvent components is different in the two forms, helix and coil, it is observed that interaction, measured by light-scattering, varies exactly in the same way as the α-helix contents with increase in concentration of this component.

IV. CONCLUSIONS

This brief review on light-scattering has attempted to deal with the manifold application of the technique to the study of macromolecules of biological origin. The most frequent use of the method is the determination of molecular weight. The method can also give, although only for macro-molecules with optical asymmetry, a measure of the size and an indication of the shape.

The value obtained for the molecular weight, although an absolute one, is a weight-average; therefore high molecular weight impurities, even in small amounts, are liable to distort the results. For small particles this drawback is eliminated by measuring scattered intensity at angles above 90°. But for asymmetric particles, the extrapolated values at zero angle lead to erroneous results because of these impurities; the same effect is observed on the radius of gyration. Therefore, it is essential to check that the solution is free of extraneous impurities. On the other hand, the results obtained for polydisperse systems can be interpreted only when the proportion of the components (in equilibrium or not) are known by independent measurements, for example by analytical ultracentrifugation.

In spite of the restrictions mentioned above, 'light-scattering' is one of the best methods for determination of molecular weights in the range from 10^4 to 2 or 3×10^6.

However, it should be pointed out that the use of this method is not limited to the determination of molecular parameters in a 'static' state.

More interesting information can and has been obtained about 'dynamic' systems. The method is used as a comparative one and permits us to study the variation of molecular parameters with changing temperature, pH, ionic strength and especially with ligand binding such as coenzyme or substrate, on the enzyme which undergoes reversible association–dissociation reactions.

The kinetics of the processes can be satisfactorily studied even when changes of scattered intensity are fast. Recently the temperature jump method has been employed to study the rapid reversible interaction, using light-scattering to detect the process of interactions of macromolecules.

There has been noted a recent application of light-scattering thanks to the use of laser which permits us to obtain the relaxation times correlated with the shape and size of the molecule; this application is still uncommon.

REFERENCES

1. P. Putzeys and J. Brosteaux, *Trans. Faraday Soc.*, **31**, 1314 (1935)
2. P. Debye, *J. Appl. Phys.*, **15**, 338 (1944)
3. B. H. Zimm, R. S. Stein and P. Doty, *Polymer Bulletin*, **1**, 90 (1945)
4. K. A. Stacey, *Light-scattering in Physical Chemistry*, Butterworth Scientific Publications, London (1956)
5. C. Tanford, *Physical Chemistry of Macromolecules*, J. Wiley, New York–London (1961)
5a. J. Tonnelat, *Biophysique, Tome II*, Masson et Cie, 1972
6. G. Scatchard, *J. Am. Chem. Soc.*, **68**, 2315 (1946)
7. J. G. Kirkwood and R. J. Goldberg, *J. Chem. Phys.*, **18**, 54 (1950)
8. W. H. Stockmayer, *J. Chem. Phys.*, **18**, 58 (1951)
9. H. Inoue and S. N. Timasheff, *J. Am. Soc.*, **90**, 1890 (1968)
10. C. Strazielle and H. Benoit, *J. Chim. Phys.*, **58**, 675 (1961)
11. E. P. Geiduschek and A. Holtzer, *Adv. Biol. Med. Phys.*, **6**, 431 (1958)
12. P. Horn and H. Benoit, *C.R. Acad. Sci., Paris*, **251**, 222 (1960)
13. M. E. Reichman, *Canad. J. Chem.*, **37**, 489 (1959)
14. C. Sadron, *The Nucleic Acids* (Ed. E. Chargaff and J. N. Davidson), Academic Press, New York–London (1960)
15. P. Debye, *Phys. Rev. Letters*, **14**, 783 (1965)
16. S. Dubin, J. H. Lunaceck and G. B. Benedek, *Proc. Natn. Acad. Sci., U.S.A.*, **57**, 1164 (1967)
17. S. Candau, *Ann. Phys.*, **4**, 21 (1969)

18. G. E. Perlman and L. G. Longworth, *J. Am. Chem. Soc.*, **69**, 1193 (1948)
19. M. Halwer, G. C. Nutting and B. A. Brice, *J. Am. Chem. Soc.*, **73**, 2786 (1951)
20. S. Guinand and P. Joliot, *J. Chim. Phys.*, **54** 239 (1957)
21. B. H. Zimm, *J. Chem. Phys.*, **16**, 1093 (1948)
22. O. Kratky and G. Porod, *Rec. Trav. Chim.*, **68**, 1106 (1949)
23. A. Guinier and Fournet, *Small Angle X-rays Scattering*, J. Wiley, New York–London
24. F. Capet-Antonini and S. Guinand, *Biochim. Biophys. Acta*, **200**, 486 (1970)
25. M. D. Stern, *Biochem.*, **52**, 2556 (1966)
26. R. V. Rice, E. F. Casassa, R. E. Kerwinand and M. D. Moser, *Arch. Biochem. Biophys.*, **105**, 409 (1964)
27. A. Holtzer, R. Clark and S. Lowey, *Biochem.*, **4**, 240 (1960)
28. S. Lowey, J. Kucera and A. Holtzer, *J. Mol. Biol.*, **7**, 234 (1963)
29. P. Doty, J. H. Bradbury and A. Holtzer, *J. Am. Chem. Soc.*, **78**, 947 (1956)
30. V. Luzzati, M. Cesari, G. Spach, F. Masson and J. M. Vincent, *J. Mol. Biol.*, **3**, 566 (1961)
31. H. Benoit, L. Freund and G. Spach, in *Poly-α-amino acids* (Ed. Fasman), M. Dekker, New York, Vol. 1, p. 108 (1967)
32. C. Sadron, *J. Chim. Phys.*, **58**, 877 (1961)
33. J. A. Harpst, A. U. Krasna and B. H. Zimm, *Biopolymers*, **6**, 985 (1968)
34. A. I. Krasna, J. R. Dawson and J. A. Harpst, *Biopolymers*, **9**, 1017 (1970)
35. A. I. Krasna, *Biopolymers*, **9**, 1029 (1970)
36. Y. Mauss, J. Chambon, M. Daune and H. Benoit, *J. Mol. Biol.*, **27**, 579 (1967)
37. E. T. Adams, Jr. and D. L. Filmer, *Biochem.*, **5**, 2971 (1966)
38. R. F. Steiner, *Arch. Biochem. Biophys.*, **39**, 33 (1952)
39. R. Townend, L. Weinberger and S. N. Timasheff, *J. Am. Chem. Soc.*, **82**, 175 (1960)
40. C. Georges, S. Guinand and J. Tonnelat, *Biochim. Biophys. Acta*, **59**, 737 (1962)
41. D. Pantaloni, *Ph.D. Thesis*, University of Paris (1965)
42. R. Townend and S. N. Timasheff, *J. Am. Chem. Soc.*, **82**, 3168 (1960)
43. E. Appella and G. M. Tomkins, *J. Mol. Biol.*, **18**, 77 (1966)
44. H. Eisenberg and G. M. Tomkins, *J. Mol. Biol.*, **51**, 37 (1968)
45. P. Dessen and D. Pantaloni, *Symposium 'Structure des Proteines et Site Actif des Enzymes'*, Paris, 1971
46. P. Dessen and D. Pantaloni, *European J. Biochem.*, **8**, 292 (1969)
47. M. H. Buc and H. Buc, FEBS Proceeding of the 4th Meeting, Oslo (1967) Universitetsforlaget, Oslo and Academic Press, London–New York
48. A. d'Albis, *Biochim. Biophys. Acta*, **200**, 40 (1970)
49. H. F. Fisher and J. R. Bard, *Biochim. Biophys. Acta*, **188**, 168 (1969)
50. M. S. Tai and G. Kegeles, *Arch. Biochem. Biophys.*, **142**, 258 (1971)
51. E. Marchal and C. Strazielle, *C.R. Acad. Sci.*, *Paris*, **267** (C) 7, 135 (1968)
52. H. Inoue and S. N. Timasheff, *Biopolymers*, **11**, 737 (1972)

CHAPTER 5

Ultracentrifugation

U. Hagen

Institut für Strahlenbiologie,
Kernforschungszentrum,
Karlsruhe, Federal Republic of Germany

I. INTRODUCTION

Ultracentrifugation methods permit the analysis of the movement of a molecule in a gravitational field, which is of utmost importance when investigating the properties of biological macromolecules. The molecule to be analysed may be observed by optical methods during centrifugation, as is the case with an analytical ultracentrifuge. Certain conclusions about the behaviour of molecules during centrifugation may also be drawn when using a preparative ultracentrifuge by determining their position in the tube at the end of the centrifuge run.

This chapter will describe the applications of both the analytical and the preparative ultracentrifuge in molecular biology. An exhaustive description of the various methods will not be possible, but their principles and primarily their applications in biological research will be discussed. Literature references listed at the end of this paper contain details of these methods. A general description of the ultracentrifugation methods now in use may be found in several publications.[1-17] Recently, more advanced studies in theory and application of ultracentrifugal techniques have been published.[18] In the following chapters some references are given to these detailed and extensive studies which may be used by readers more interested in the field. Naturally, the present contribution is intended to give a first introduction to the methods and techniques used in centrifugation. More sophisticated considerations are beyond the scope of this book and also beyond the space allocated.

In most institutes ultracentrifugation serves as an auxiliary science for the determination of specific data. In order to obtain satisfactory and elegant solutions, one must be critical in selecting the proper method from the various methods of ultracentrifugation available. To a certain extent the training courses offered by the manufacturers of the centrifuges can be of help, but these will not serve as a substitute for actual experience.

II. OPTICAL METHODS USED IN THE ANALYTICAL ULTRACENTRIFUGE

To allow observation of macromolecules in solution as they move in a gravitational field, the samples are centrifuged in cells having windows that lie parallel to the plane of rotation of the rotor head. As the rotor turns, the image of the cell is projected by a light source and an optical system. The various optical systems permit observation of the movement of the particles. Ultracentrifuges were first developed by Svedberg in 1923 as a means of generating the high gravitational fields necessary for the sedimentation of large molecules. Later, as high-speed electric motor drives replaced the turbine drives it became apparent that the speed of rotation and the temperature have to be precisely controlled if the sedimentation measurements were to be accurate. For example, if either the temperature varies by $0.4\,°C$ or the speed by 0.5 per cent, the error in the measurement of a sedimentation coefficient will be in the order of 1 per cent.[19]

We need not study in detail here the rotor drive, temperature regulation and vacuum systems of an analytical ultracentrifuge, but it is worthwhile considering the various optical systems, since these have to be selected in accordance with the problem to be studied.

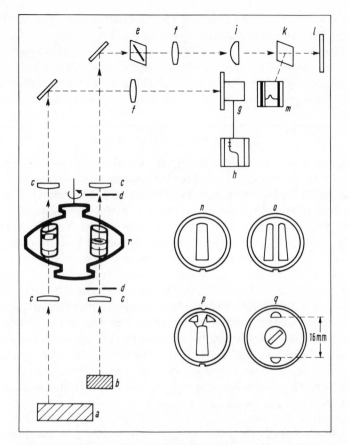

Figure 1. Optical path in the analytical ultracentrifuge (scheme, from drawings of Beckman Instruments):

(a) Monochromator or u.v. light source
(b) Light source for schlieren and interference optical systems
(c) Collimating and condensing lenses
(d) Slits for optical interference system
(e) Phase-plate (schlieren analyser)
(f) Camera lenses
(g) Photographic plate or scanning system
(h) Chart recorder
(i) Cylindrical lens
(k) Swing-out viewing mirror for the schlieren pattern
(l) Photographic plate for schlieren pattern
(m) Schlieren pattern
(n) Single sector centrepiece, upper view
(o) Double sector centrepiece, upper view
(p) Synthetic boundary cell with reservoir, upper view
(q) Counterbalance with reference holes, upper view
(r) Rotor with cell and counterbalance (left side)

Figure 1 illustrates the four different optical systems that are at our disposal: The schlieren and the interference systems, both of which use a single optical path with the same light source, the absorption system, where the light absorption by the sample solution in the cell can be measured either by photographic methods or by the photoelectric scanning system. A monochromator is used to select the desired wavelength of the light.

A. Optical Absorption Method

The concentration of the solution at the various points of the cell is determined by absorption of a light of the appropriate wavelength. This can be accomplished either by measuring the degree of blackening of a photographic film or by the pen deflection of the recorder of the scanning system. Prior to centrifugation, the concentration of the solute is evenly distributed throughout the cell; after exposure to the centrifugal forces for some time, a part of the particles will move to the bottom respectively of the cell periphery and there will be a zone of solvent without particles at the meniscus, i.e. that part of the cell nearest the centre of rotation. The transition line between solvent and solution is called the boundary, its position with respect to the meniscus is the distance each particle has migrated in the cell since the beginning of centrifugation. The region of uniform concentration between the boundary and the bottom of the cell is called the plateau zone.

Figure 2 is a schematic representation of the concentration distribution in the cell and the curves obtained with the film or the photoelectric scanner. The cell should have radial sector walls to ensure a smooth migration of each particle to the bottom of the cell, undisturbed by other particles. The degree of blackening of the film, measured by a special densitometer gives the concentration. When using the film technique, adequate exposure time and development are essential. For good results, the exposure time has to be selected in such a way that even at the reference holes the film does not reach the maximum degree of blackening. It should also be checked to ensure the concentration of the solution is not too high; some light must still be transmitted by the plateau zone. Also, the cell should be evenly illuminated. Under such conditions the position of the boundary can be determined precisely, even though the concentration measurement of the solution is expressed in relative units.

The optical absorption method is greatly improved by the use of a scanning system. A special double sector cell is used (see Figure 4), one sector filled with the solution, the other with the solvent. A photomultiplier compares the light transmission from both sectors of the cell as the rotor is rotating and records the difference. At the same time the photomultiplier covered except for a slit opening, moves slowly along the image of the cell in the direction of its length. Figure 2 gives an example of a scan recording. Before each scan the deflection of the recorder is calibrated with a step switch.

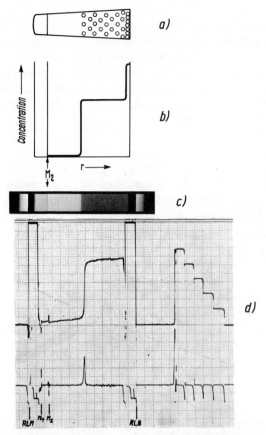

Figure 2. Concentration distribution in the cell of the analytical centrifuge and its evaluation by means of the optical absorption method. Sedimentation of the molecules to the right:

(a) Single sector cell with half-sedimented particles
(b) Concentration as a function of the distance r to the rotor axis
(c) Camera picture
(d) Scan recording. Centrifugation in the double sector cell. At right the calibrating steps, each of 0·2 absorption units. Below the first derivative of the concentration distribution.

T1-DNA was used, 40 μg/ml in 0·165 M NaCl, sedimentation at 15,000 r.p.m., scanning after 200 min. The camera picture with a single sector cell does not match exactly the scan recording of the double sector cell

M_1: Meniscus in the solvent sector
M_2: Meniscus in the solution sector
RLM: Reference line at the meniscus
RLB: Reference line at cell bottom

Several standard-sized pulses are generated corresponding precisely to an absorptivity of 0·2. Thus, it is possible to indicate the absolute concentration of the solution in each region of the cell. A unique feature of the scanning system is that the behaviour of the particles during the run can be observed directly, without having to wait for the development of the film. On the other hand the scanning system requires more care in adjusting the light path, a higher order of cleanliness of the optical surfaces and greater purity of the solutions.

Very small concentrations of macromolecules can be determined with the optical absorption method. When using cells of 12 mm light path, an absorption coefficient of 0·2 is sufficient for measurement, i.e. approx. 150 μg/ml or proteins or 10 μg/ml of nucleic acid. The scanning system ranges from 0·2 to 2·0 absorption units, whereas the film only covers the region 0·1 to 0·6. If the wavelength of light selected does not correspond to that of maximum absorption, even greater concentrations may be analysed, a situation that can be encountered when studying the sedimentation coefficient as a function of concentration. When molecules with different wavelength maximas are present, they can be analysed separately by scanning at the different wavelengths. The various fields of application of the scanning system have been dealt with in a series of papers by Schachman and his colleagues[20–25] (cf. section VIA of this article).

B. Schlieren Optical System

With the schlieren system it is not the concentration c of the solution that is measured, but the first derivative of concentration as a function of the distance r from the axis of rotation, dc/dr. A description of the principles of the schlieren optic is already found (1893) in a paper by Wiener[26] who used it for determination of diffusion coefficients. Figure 3 shows a diagram of this change of concentration with the corresponding schlieren pattern. The light enters the cell through a slit system. If the cell contents are homogeneous, the light beams will stay parallel; but if there is a concentration gradient in some region, the light will be refracted differently, resulting in a distortion of the cell image.

Originally, this distortion was measured with a scale. In the Philpot–Svenson optical system the change of refraction index is transformed into a line by the inclined slit of the schlieren diaphragm (phase-plate). If there is a change of concentration in the cell, this line separates from the base-line and the shape of the concentration gradient, dc/dr, becomes immediately apparent. It should be kept in mind, however, that this curve is the change of the refractive index dn/dr which is proportional to the change of concentration dc/dr. As a consequence of this, the deviation of this line depends

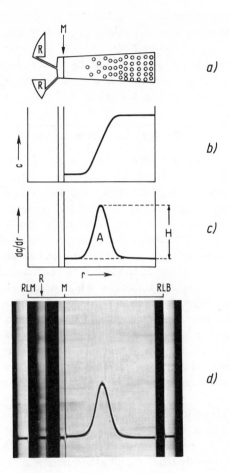

Figure 3. Concentration distribution in a synthetic boundary cell and its evaluation with the schlieren system:
(a) Cell with particles, R: reservoir, connected to the sector by a capillary
(b) Concentration as a function of the distance r to the rotor axis
(c) Concentration gradient as a function of the distance r to the rotor axis
(d) Schlieren pattern (Overlayering of serumalbumin 0·8 per cent in 0·14 M NaCl as solvent)
RLM, RLB and M: see Figure 2; A and H: see text

on the angle of the phase-plate. To obtain the distribution of the concentration c in the cell, the concentration gradient dc/dr resulting from the schlieren image must be integrated. Details about the optical alignment of the schlieren system may be found in various monographs.[3,1,27] For reliable information using a schlieren pattern an exact alignment of the optical system and regular checks of this alignment are essential.[28,29]

C. Rayleigh Interference System

The Rayleigh interference system may be installed in the same optical path used for the schlieren system. The Rayleigh method enables a measurement of the difference in the refractive indices of the sample solution and solvent as they are being centrifuged separately in the double-sector interference cell. Since the refractive index is directly proportional to concentration, the concentration of the solute in the cell can be ascertained immediately. The Rayleigh system operates with double slits between collimating and condensing lenses. The image of the cell shows horizontal bands or fringes in regions of constant concentration, i.e. where $dc/dr = 0$. If the concentration increases, the fringes shift due to refraction. The magnitude of this shift is proportional to the change in concentration. Since only part of the interference pattern is shown on the photographic plate (Figure 4), the various fringes may not be seen in entirety, but the number of the fringes crossing the horizontal are counted. This number corresponds exactly to the increase in concentration from the meniscus over the boundary and into the plateau zone.

Compared to the schlieren system, more calculation time is necessary with the interference system, but on the other hand, there are some advantages with this system. The concentration measurement is much more accurate and may be read directly from the photographic plate (integration is thus eliminated). Changes in concentration of protein as small as 0·05 per cent can still be measured accurately. The interference system and its special applications in molecular biology have been described by various authors.[3,30–33] A combination of the schlieren and the Rayleigh fringe interference systems is described by Chervenka:[34] after exposure of the schlieren pattern, the interference pattern is exposed onto the same photographic plate.

III. THEORY OF MIGRATION IN A CENTRIFUGAL FIELD

The rate of sedimentation of a particle in a centrifugal field dr/dt is determined by the centrifugal force, $F = -mr\omega^2$ applied to a particle of mass m, by the diffusion, and by the frictional coefficient f_s. The angular

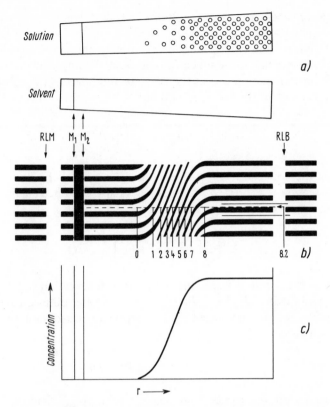

Figure 4. Concentration distribution in the double sector cell and its evaluation with the optical interference system:

(a) Double sector cell. At the top solution with particles, below solvent alone

(b) Scheme of an interference pattern. The number of fringes crossing the horizontal is approximately 8·2 (from drawings of Beckman Instruments)

(c) Concentration as a function of the distance r to the rotor axis

velocity, ω, is expressed in radians per second = (rpm × $2\pi/60$). If dr/dt is considered with respect to the unit centrifugal field $r\omega^2 = 1$, then the sedimentation coefficient s of the particle is

$$s \equiv \frac{dr/dt}{r\omega^2} \tag{1}$$

The units of s are seconds; the basic unit, for convenience, is taken as 10^{-13} sec, which is termed one Svedberg (S). The relationship between s and m will be discussed later (Section VII). Some authors designate the sedimentation coefficient with capital S, when its value is expressed in Svedberg units (S-value).

To determine s, the change of concentration in the radial centrifuge cell in respect to time has to be taken into account. The flow of the solute (J) flowing from one section (r_1) of the cell into the other (r_2) is equal to the change of the rest of the solute with time:

$$(rJ)_{r_1} - (rJ)_{r_2} = \frac{\partial}{\partial t} \int_{r_1}^{r_2} rc \, dr \tag{2}$$

where c is the concentration of the solute. The unit of the flow J across any cross section is expressed in moles per square centimetre per sec.

For the distance r we obtain:

$$\left(\frac{\partial c}{\partial t}\right)_r = -\frac{1}{r}\left[\frac{\partial(rJ)}{\partial r}\right]_t \tag{3}$$

The flow of the solute J is produced by the sedimentation ($J_{sed} = s\omega^2 rc$) and by the diffusion ($J_{diff} = -D(\partial c/\partial r)$), D being the diffusion coefficient of the solute. When introduced into equation (3) we obtain the Lamm differential equation:[35]

$$\frac{\partial c}{\partial t} = -\frac{\partial}{r\partial r}[s\omega^2 r^2 c - Dr(\partial c/\partial r)] \tag{4}$$

To find a solution to s and D, both the concentration gradient $\partial c/\partial r$ with a constant time, and the concentration change in time $\partial c/\partial t$ with r being constant have to be determined. At present, these measurements may not be obtained directly from the centrifuge; for calculations of s or D three limit cases are considered here:

(a) Only the sedimentation is determined, the diffusion is not taken into account. If both the molecular weight and the rotor speed are high enough, this method is acceptable. This method is indicated sometimes by the expression 'Velocity sedimentation'.

(b) In the case of a low molecular weight material in a rotor run at low speed, an equilibrium is reached after an appropriate time between sedimentation and back diffusion in the cell. The concentration gradient $\partial c/\partial r$ in this case is not time dependent and may be used directly for evaluation of the ratio s/D.

(c) At any time a state of equilibrium also exists at the bottom and at the meniscus of the cell, since nothing can diffuse into or sediment out of these points.

The limit cases (b) and (c) will be discussed in sections VIII and IX. If we take limit case (a), not taking diffusion into account, we may determine the sedimentation coefficient. We set $D = 0$ and then integrate equation (4) under the limit conditions ($t > 0$ and r in the plateau zone at time zero). At time t we thus obtain the concentration:

$$c_t = c_0 \exp(-2\omega^2 st) \tag{5}$$

where c_0 is the initial concentration of the solution. For the distance r of a given concentration, say midpoint of the boundary between solution and solvent at time t, we obtain

$$r_t = r_0 \exp(\omega^2 st) \tag{6}$$

r_0 being the distance at time zero (initial position). Determining the distance of the boundary at different times (r_1 at t_1, r_2 at t_2), we may use equation (6) for the calculation of the sedimentation coefficient in the following way:

$$(s_{app})_c = \frac{\ln r_2 - \ln r_1}{60\omega^2(t_2 - t_1)} = \frac{2 \cdot 303}{60\omega^2} \cdot \frac{\log r_2 - \log r_1}{(t_2 - t_1)} \tag{7a}$$

where t is expressed in minutes.

Convenient results may be obtained with this procedure if the distance of the boundary from the axis of rotation is measured at various times and $\log r$ is plotted against time (Figure 5). The apparent sedimentation coefficient may be obtained easily from the slope of the line. Also, irregularities in sedimentation may be spotted at once. Some authors prefer to use regression analysis and calculate the slope with the best possible approximation; in most cases though, a graphic determination is precise enough. Apart from this 'integral way' of calculating the sedimentation coefficient, a differential determination may be effected following the equation:

$$(s_{app})_c = \frac{r_2 - r_1}{60\omega^2(t_2 - t_1)(r_1 + r_2)/2} \tag{7b}$$

The sedimentation coefficient thus obtained applies only to the specific experimental conditions; it is relatively easy though to recalculate for the conditions in pure water at 20 °C.

$$(s_{20,w})_c = (s_{app})_c \cdot \frac{\eta_{t,w}}{\eta_{20,w}} \cdot \frac{\eta_{t,sv}}{\eta_{t,w}} \cdot \frac{(1 - \bar{v}_{20,w} \cdot \rho_{20,w})}{(1 - \bar{v}_{t,sv} \cdot \rho_{t,sv})} \tag{8}$$

where η is the viscosity, $\bar{v}$ the partial specific volume (see section VII) and ρ the density. The subscripts 20 or t refer to 20 °C or the temperature used respectively, sv or w for solvent or water respectively. If centrifugation takes place at 20 °C, only the viscosity and the density of the solvent (e.g. of the buffer) need be taken into account. These may be found in the appropriate tables[6] or may be easily determined.[2,36] With compressive fluids, as for instance organic solvents, both the density and the viscosity of the solvent

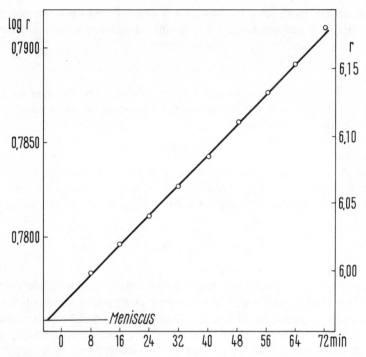

Figure 5. Calculation of the sedimentation coefficient by integration (equations (7a) and (8))
Thymus DNA, 20 μg/ml in 0·2 M NaCl + 0·025 M phosphate buffer, pH 7·3, Temp. 20·0 °C, 17,980 r.p.m.

$$(\eta/\eta_0)_{solv} = 1·025, \; \rho/\rho_0 = 1·0121, \; \bar{v} = 0·556$$

Correction factor = 1·038

$$s_{20,w,c} = \frac{2·303 \, \Delta \log r/\Delta t}{60\omega^2} \times 1·038$$

$$= \frac{2·303 \times 0·00020 \times 1·038}{60(2\pi)^2[17,980/60]^2} \, \sec = 22·4 \times 10^{-13} \, \sec$$

$$s_{20,w,c} = 22·4 \, S$$

change with a change of angular velocity. The applicable corrections are indicated by Schachman *et al.*[3,37] and by Cantow *et al.*[38] Aqueous solutions have a low compressivity and do not require such corrections. For analysing interacting systems, however, care should be taken to avoid pressure effects (cf. section XI).

The sedimentation coefficient thus obtained $(s_{20,w})_c$ is valid only for such molecular concentration used in the particular experiment. Therefore, s_c has to be adjusted to the conditions of an ideal solution which is done by extrapolating to a zero concentration. Thus s^0 is obtained. The concentration dependence of sedimentation obeys the relation:

$$\frac{1}{s_c} = \frac{1}{s^0}(1 + k_s c) \qquad (9)$$

The constant k_s is obtained by sedimentation runs at various concentrations and by plotting $1/s_c$ against concentration (see also Figure 7). If the sedimentation is highly dependent on concentration, then the dilution of the solution in the cell sector must be taken into account as well, as can be calculated[39] from the combination of equations (5) and (6):

$$c_t = c_0(r_0/r_t)^2 \qquad (10)$$

In this case, when expressing $\log r$ against time (see Figure 5), the result is a rising curve. The sedimentation coefficients may be obtained from the difference between the various points of measurement extrapolated to the initial concentration. According to Alberty[40] the concentration dependence of the sedimentation may be determined in this way with a single run. General solutions to the concentration dependence of s were deduced by Cox[41,42,43] from Lamm's equation (4).

IV. DIFFUSION

In order to determine the molecular weight of a macromolecule from the sedimentation coefficient, its shape, state of hydration and flexibility should be known, or at least data such as diffusion or viscosity (see sections V and VII) that will give information about these unknown values. Therefore, when describing the size and the shape of a macromolecule, diffusion values are just as important as sedimentation analysis. The diffusion coefficient may be determined from Lamm's equation (4). If we centrifuge very slowly, there is no sedimentation, and we obtain for the change of concentration with respect to time at the boundary

$$\frac{\partial c}{\partial t} = D\frac{\partial^2 c}{\partial r^2} \qquad (11)$$

The boundary between solvent and solution can be obtained by overlayering the solution with solvent before centrifuging, or by forming a synthetic boundary layer in a special diffusion cell, called 'synthetic boundary cell' (cf. Figure 3).[2,44] D may be determined from formula (11) by measuring the concentration gradient $\partial c/\partial r$ over a certain length of time with a schlieren optical system on the position of the synthetic boundary layer r_0 (see Figure 3):

$$D = \frac{c_0^2}{4\pi t(\partial c/\partial r)^2} \qquad (12a)$$

where c_0 is the initial concentration of the overlayered solution. In the schlieren optical system, c_0 is expressed by the area of the schlieren peak above the basis line (A) and ($\partial c/\partial r$) by its height (H) (see Figure 3). From this data, D can be computed by the relation

$$D = \left(\frac{A}{H}\right)^2 \frac{1}{4\pi t F^2} \tag{12b}$$

whereby F is the magnification factor of the schlieren diagram to the original size in the centrifuge cell. Other methods of calculation are described by Elias.[6] For precise measurements the Rayleigh interference optical system is preferable.

When measuring diffusion in very dilute solutions with the optical absorption system, it must be kept in mind, that with the small concentration difference used only small differences in density between solution and solvent are to be expected. Therefore it may be difficult to obtain a sharp boundary when overlayering the solvent. Stable boundaries may be obtained when the salt concentration of the solution is higher than that of the solvent (e.g. 0.1 M Na^+ versus 0.05 M Na^+).[44] This is possible, however, only when using the optical absorption system. Strassburger and Reinert[45] proceeded in a similar way when measuring the diffusion of high molecular weight DNA. The DNA was dissolved in D_2O and overlayered with H_2O. Diffusion is much less concentration dependent than sedimentation. Nevertheless, it is recommended to perform diffusion measurements over a large range of concentration and to extrapolate the obtained values to zero concentration.[6,44] A new way to determine diffusion coefficients has been found by Lang and Coates.[46] The value of D_0 for DNA molecules was calculated from kinetic studies of the absorption of the molecules onto a cytochrome surface film; the absorbed DNA molecules were then counted with an electron microscope.

V. VISCOSIMETRY

The importance of viscosimetry for the analysis of macromolecular substances was stressed by the work of Staudinger and coworkers.[47,48] Macromolecules in solution lead to a noticeable increase of viscosity which is due to their size, for example as globular protein or a coiled chain-molecule. The viscosity is the factor of proportionality between the flow-rate of a solution and the shearing force to which it is subjected. The unit of viscosity is the poise, 1 poise $= 1$ g cm^{-1} sec^{-1}. At 20 °C, water has a viscosity of 0.01002 poise. As a result of the forces of attraction between the macromolecules, the viscosity of a solution will increase with the number of macromolecules in solution and with the friction coefficient. It appears from the above that the viscosity gives only indications concerning the friction between the molecules and not the molecular weight itself. The relationship between viscosity and molecular shape is explained more in detail in the text-books on Macromolecular Chemistry.[8,9]

For viscosity measurements the first step is to determine the specific viscosity (η_{sp}) of the solution, i.e. the difference between the viscosity of the solution (η) and that of the solvent (η_0)

$$\eta_{sp} \equiv \frac{\eta - \eta_0}{\eta_0} \tag{13}$$

This specific viscosity depends on the concentration c, therefore the 'reduced specific viscosity' η_{sp}/c must be determined. Since the dissolved molecules interfere with each other during the measuring process, the η_{sp}/c value still depends on the concentration. Therefore, to determine the viscosity in an ideal solution, the values obtained at various concentrations have to be extrapolated to zero:

$$[\eta] = \lim_{c \to 0} \eta_{sp}/c \tag{14}$$

This limit value of the reduced specific viscosity may be used to characterize a macromolecule; it is called Staudinger index or 'intrinsic viscosity'. Its dimension is expressed in units of $cm^3 \, g^{-1}$. Various authors[8,49] have also expressed the concentration in $g/100 \, cm^3 = g/dl$, and thus $[\eta]$ in dl/g which can lead to confusion. For that reason, the unit of viscosity $[\eta]$ will always be specified in the equations hereafter.

Viscosity measurements normally do not present special difficulties. It involves only a relative comparison between viscosity of the solvent and viscosity of the solution, thus errors due to the measuring instruments are largely eliminated. Temperature, however, must be kept absolutely constant. Reliable low viscosity values may be obtained with the capillary viscosimeters following OSTWALD or UBBELOHDE.[2,36] The time t is measured during which a given quantity of solution flows through a capillary. This value is compared with the time taken by the solvent (t_0) and we obtain η_{sp}, since t is proportional to η.

It must be noted, though, that the values of $[\eta]$ obtained by using capillary viscosimeters may only be used for the calculation of molecular constants if the configuration of the molecule is not affected when flowing through the capillary. In the case of globular proteins or short linear polymers, no problems are encountered. Long chain molecules, however, such as those found in most DNA preparations, are subject to deformation and to an alignment in the direction of the flow; this leads to a decrease of the measured viscosity. Staudinger[50] designated this effect 'structural viscosity'. In order to avoid changes in the configuration of long chain molecules, the shearing force has to be kept as small as possible during the measuring process. The rotating viscosimeter of HATSCHEK–COUETTE (see Kern and Mehren)[36] allows a noticeable reduction of the shearing force during measurement, from $5 \, dyn \, cm^{-2}$ to $0.1 \, dyn \, cm^{-2}$. DNA preparations up to a molecular weight of approximately 10^7 may thus be analysed. Very low shearing forces (0.001 to $0.004 \, dyn \, cm^{-2}$) are generated in the rotary viscosimeter of ZIMM and CROTHERS.[51] The inner cylinder, under the influence of a rapidly rotating external magnetic field, moves at a very low speed (1–2 rpm), so that the viscosity of very large molecules up to 10^8 daltons can be determined reliably.[51,52,53] The influence of the shearing force on the viscosity may also be eliminated by making the measurements with proper capillary viscosimeters under varying pressures, and extrapolating the values thus obtained to zero pressure.[54,55] But this method is rather tedious and the extrapolation is often unreliable.

$[\eta]$ may be used together with the sedimentation coefficient to determine the molecular weight (see section VII). When analysing similar molecules of varying chain length, e.g. nucleic acid preparations, the molecular weight may be determined from the viscosity alone, if the constants of equation (15) are known:

$$[\eta] = K_\eta \cdot M^{a_\eta} \tag{15a}$$

(see Figure 7). For polymers forming a random coil, $a_\eta = 0.5$; for rod-like particles, $a_\eta = 2$ (Elias[6]). Analysing native DNA molecules, K_η and a_η is constant only for a small range of the molecular weight, as long molecules behave like random coils and small ones like rods. More details about the determination of the molecular weight of DNA are given in section VII.

VI. SEDIMENTATION ANALYSES IN MOLECULAR BIOLOGY

A. Sedimentation of Proteins and Nucleic Acids

For the centrifugation of biological macromolecules, the optical system used must be chosen in accordance with the problems to be solved. The schlieren optical system can be used to obtain a rapid indication of the sedimentation coefficient of a protein and of the homogeneity of the dissolved particles. Whilst with the optical absorption system it is difficult to obtain a clear separation of two close boundaries, the schlieren method will give two clearly separated peaks. But the schlieren method will not allow accurate evaluation of the quantity of each migrating substance. For this determination the absorption or the interference method is preferred.[14] With the interference method very small differences in sedimentation coefficients of two proteins may be determined when centrifuging the solutions in the two compartments of the double sector cell. If the migration speeds of both proteins are the same, then the interference image will show horizontal fringes only, even over the boundary. If there is but a small difference in speed, then the fringes will be curved.

The photoelectric scanning system is useful also to study the binding of small molecules to macromolecules, particularly when the maximum of the absorption of the small ones differs from that of the macromolecules. The binding of nicotinamide adenine dinucleotide (NAD) to an enzyme may be studied for instance scanning the mixture at 280 mμ as well as at 260 mμ. The bounded NAD detected by its absorption at 260 mμ will move in the cell together with the protein whereas the free NAD remains at the meniscus. Another example is given in Figure 6 (Gerhart and Schachman). Undissociated Aspartate Transcarbamylase (ATCase) sediments with about

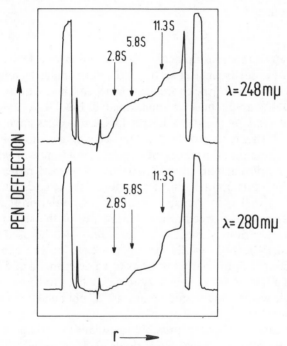

Figure 6. Binding of paramercuribenzoate (PMB) to Aspartate Transcarbamylase (ATCase)
Scanning of a mixture of ATCase with PMB in 0·04 M potassium phosphate buffer (pH 7·0) at a wavelength of 280 and 248 mμ. Centrifugation at 60,000 r.p.m.
For more details see text. (Redrawn from experiments published by J. C. Gerhart and H. K. Schachman, *Biochemistry*, **7**, 538 (1968). Copyright 1968 by the American Chemical Society. Reprinted by permission of the copyright owner)

11·3 S. After adding paramercuribenzoate (PMC) to the enzyme, the enzyme dissociates into subunits with a S-value of 5·8. The concentration distribution of a mixture of ATCase and PMB in the cell was measured at 248 and 280 mμ. At 280 mμ the light absorption is due almost entirely to the protein, whereas at 248 mμ the absorption by protein was about one-half its 280 mμ value and PMB absorbed strongly, whether alone or as its mercaptide complex. Comparing the pen deflection in the scans at different wavelengths, it is apparent in Figure 6 that the PMC is bounded only to the subunit of the enzymes, not to the undissociated enzyme.

When measuring the sedimentation of nucleic acids, special attention must be paid to the concentration dependence of the sedimentation coefficient. The concentration dependence is an almost linear function of the viscosity of the DNA molecule, so that equation (9) can be written in the following manner[56,57]

$$s^0 = s_c \cdot (1 + k'_s[\eta]c) \tag{16}$$

Since there is a relationship between the viscosity of a DNA sample and its sedimentation coefficient (see section VII), when analysing under standard conditions, say 0·2M Na$^+$, the concentration dependence of the sedimentation of a given DNA sample may be approximated. For DNA molecules up to 20×10^6 daltons, k'_s is about 0·8. Larger DNA molecules have a k'_s value of 1·0–1·1.[56,58,59] This is due to the fact that the configuration of these molecules with increasing size comes nearer to a random coil while smaller molecules are approaching rod-like particles. Spherical molecules have a k'_s value of 1·6—this applies both to random coils (e.g. denatured, single-coiled DNA) as well as to compact spheres (e.g. globular proteins).[56,57]

Figure 7 shows the concentration dependence of the sedimentation for native DNA of various sizes. These curves may also be used to obtain a quick general indication on the molecular weight of a native DNA. The position of point $1/S_c$ in relation to c has to be determined and $1/S$ is extrapolated with a line parallel to the neighbouring lines to zero concentration. The molecular weight corresponding to the $1/S^0$ obtained (see section VII) is read from the left abscissa of Figure 7.

For some time it has been possible to isolate the complete genome of viruses, mainly bacteriophages, as a single DNA molecule. These homogeneous DNA preparations form sharp boundaries during sedimentation, since such large molecules diffuse but very little. A vertical line is obtained both with the film densitometer and with the scanner.[58,59,60] If part of the DNA molecule is broken, the pieces will migrate more slowly and a tail will appear behind the uniformly migrating front. The quantity of ruptured molecules may be quickly determined from the height of both fractions. When centrifuging these large molecules, the angular velocity of the rotor should not be too high (up to approx. 15,000 rpm max.). Otherwise quickly

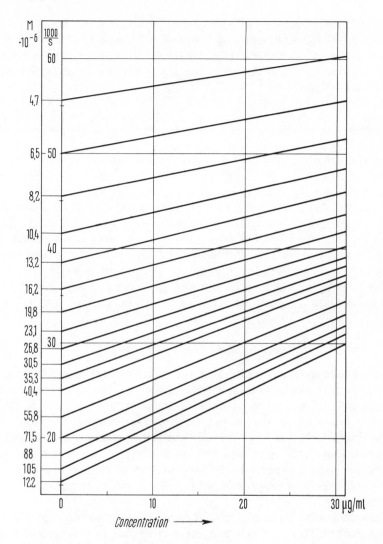

Figure 7. Concentration dependence of sedimentation coefficient of native DNA in $0·1 - 1·0$ M Na^+:

$$S^0 = S_c(1 + k'_s[\eta]c).$$

The relation S^0 to $[\eta]$ was taken from Figure 7.
k'_s for molecules up to 20×10^6 daltons $= 0·8$, then increasing to $1·0$ at 120×10^6 daltons.
$[\eta]$ in $cm^3\,g^{-1}$, c in $g\,cm^{-3}$.
The molecular weight corresponding to the S^0 value is shown on the left hand ordinate. The values for the sedimentation coefficient s^0 are given in this figure in Svedberg Units (S-value)

sedimenting aggregations will be formed and the sharp boundary will disappear.[58,61] Rosenbloom and Schumaker[62] made use of this fact to determine the collision rate of DNA molecules and to calculate the average radius of gyration (r) of the molecular coil. They obtained the following empirical relation:

$$r_{DNA} = [717\,M/200]^{1/2}/\sqrt{6} \times 10^{-8}\,cm \qquad (17)$$

Single-chain denatured DNA is usually centrifuged in high alkaline media. Schumaker and Marano[63] pointed out that buffers with a pH above 8·6 may liberate H_2 from the aluminium of the centrifuge cell which leads to convective disturbances of the solution and interferes with the sedimentation. In such cases cells of synthetic materials should be used (Kel F, Al-filled epon). On the other hand, our own experience has shown that such cells are not recommended for the centrifugation of high molecular weight native DNA at low angular velocity. Quite often, interactions between DNA and cell wall will occur which result in stopping or even in a back migration of the boundary. In this particular case, aluminium centrepieces are preferred.

B. Analysis of the Shape of the Boundary

During sedimentation of biological macromolecules there will be in most cases a gradual transition from solvent to solution, but no uniform, sharp boundary. Such 'broad' boundaries may be due to various causes, so that it becomes necessary to analyse the shape of the boundary carefully. The shape will be influenced by diffusion and heterogeneity of the molecule's size; both have a broadening effect on the boundary, whereas the self-sharpening effect as well as the Johnson–Ogston effect[4] produce a narrowing of the boundary as a result of the concentration dependence of sedimentation.

First of all it should be ascertained whether the broadening of the boundary is due to diffusion or to heterogeneous particle size. If there are reasons to assume that within a given boundary all particles have the same s-value and the same diffusion constant (as would be the case for a uniformly migrating protein fraction), then the diffusion constant may be calculated from the width of the boundary.[3] This is not possible, though, if the molecules are heterogeneous in size.

In order to estimate the heterogeneity, the influence of diffusion on the width of the boundary must be eliminated. The theory of boundary spreading in velocity ultracentrifugation of poly-disperse solutes was recently described in detail by Fujita.[65] A theoretical approach to this problem was given by Weiss and Dishon.[66] Since the broadening of the boundary by diffusion is proportional to $\sqrt{t}$, while the separation of the molecules by differing sedimentation coefficients is proportional to t, the contribution of diffusion to the shape of the boundary can be neglected if the sedimentation distribution curves are extrapolated to infinite time.[67] One must first calculate the integral sedimentation distribution curves, as they are obtained at various times after the start of centrifugation, by stepwise evaluation of the boundary, and then extrapolate for various mass fractions of the sedimentation distribution $(g_i)_s$ the s-values against $1/t = 0$. For large DNA molecules it will be found in most cases that after a short time, say 20–40 min., the distribution becomes independent of the time of measurement, and therefore extrapolation to infinite time is no longer necessary.[67,68] When calculating $(g_i)_s$, the dilution of the solution due to the radial shape of the cell should be taken into account according to equation (10).

The shape of the boundary, as obtained after eliminating diffusion, will not give the true concentration distribution of the particles of varying sedimentation velocity—the boundary is, in fact, too sharp. This is mainly due to the concentration dependence of sedimentation. Particles on the solvent side of the migrating boundary are much less concentrated than those on the solution side. Therefore they migrate faster than the latter. For this reason, the boundary peak is sharper than that of an ideal solution. This is termed 'self-sharpening' effect if the particles stay behind the migrating boundary because of diffusion, and 'Johnston–Ogston effect' if the heterogeneity of the molecules causes the broadening of the boundary.

The influence of the 'self-sharpening' effect may be calculated for homogeneous samples.[70] The influence of the Johnston–Ogston effect may be calculated for a paucimolecular mixture.[64] In case of polymolecular mixtures, the analysis of both effects requires the use of a computer.[41,42] Both effects may be eliminated though, if the sedimentation distribution curves obtained for different concentrations are extrapolated to zero concentration. Then one can proceed according to equation (9) in the calculation of the s_i^0 for each mass fraction g_i. If the concentration dependence of a polymer is known in accordance with equation (16), a single distribution curve is enough to determine distribution at zero concentration.[59,71,72,73] The optical absorption method can be recommended for the determination of such distribution curves, since very low concentrations may be measured. The schlieren method may also be used.

In general, very careful analyses are needed to prove the homogeneity of a macromolecular substance. The schlieren diagram must show not only a uniform symmetrical boundary, but the boundary should include the entire quantity of the substances in solution. Furthermore, the concentration dependence of the sedimentation should also be determined, if the width of the boundary is to be evaluated with certainty. This applies both to the schlieren as well as to the absorption techniques.

C. Band Centrifugation

If only limited quantity of sample is available, the sedimentation coefficient may still be determined with sufficient reliability by means of band centrifugation. This method requires very small quantities. For example only 10 μl of a sample with an adsorption of 0·5 are needed (0·25 μg nucleic acid). This method was developed originally for use in the preparative ultracentrifuge. The sample is laid on top of a sucrose solution, and the position of the bands determined after centrifugation (see section XI). Such an overlayering procedure may also be used in an analytical centrifuge.

Special band-forming centrepieces are used which, like the synthetic boundary cells for diffusion measurements, have a small sample well next to the sector shaped cell. This sample well is connected by a capillary to the inside of the radial cell. With increasing centrifuge speed, the sample (approximately 10 μl) moves into the meniscus of the solvent and migrates then as a band through the cell. In passing through the capillary, even very large DNA molecules resist the shearing force and remain intact. Optical absorption methods are used to observe the migration; the position of the maximum concentration in the band is determined and from this the sedimentation coefficient is calculated.[74] For successful overlayering it is essential that the solvent for the sample has a lower density than the overlayering liquid in the sector shaped cell. In general, a difference in density of 0·02–0·04 g cm^{-3} will be enough.[75,76,77] During centrifugation, a gradient is formed by the molecules of the solvent and this acts as a sedimentation stabilizer to prevent a broadening of the band by diffusion. At first a 'diffusion gradient' between the solvent in the sample well and solvent in the radial cell will appear, but this gradient disappears later in the run. This gradient is necessary for stabilization of the band until a 'field gradient' has been formed by the centrifugal field. This field gradient is produced by simultaneous sedimentation and diffusion of the solvent molecules (see section VIII).

Band centrifugation is also very useful in observing the separation of molecules with varying sedimentation coefficients. The absolute mass of the fraction may also be evaluated directly from the area of the bands appearing on the densitometer.[78] The concentration dependence of sedimentation leads to similar phenomena in band centrifugation as it does in boundary centrifugation. The self-sharpening effect produces a narrowing of the migrating band, especially on the back, so that asymmetrical bands are shown in the densitometer.[79]

Because of the high ionic content of the solvent, the sedimentation values obtained with band centrifugation have to be corrected for viscosity and also for the effective partial specific volume. This volume in particular varies with the hydration of the molecules, depending on the ionic concentration of the solvent.[80]

VII. DETERMINATION OF THE MOLECULAR WEIGHT FROM s^0, D AND $[\eta]$

A. General Remarks

The sedimentation coefficient s of a particle is determined by its mass m and its frictional coefficient f_s. Referring to equation (1) we obtain

$$m = s^0 f_s \tag{18}$$

For a particle in a solvent having a density of ρ_1, it must be taken into account that the centrifugal force does not affect the mass m itself, but only the portion m^* which is heavier than the volume of solvent displaced by the molecule:

$$m^* = m - V\rho_1 = m(1 - \bar{v}\rho_1) \tag{19}$$

where V is the volume of the macromolecule and $\bar{v}$ the partial specific volume ($= V/m$) and ρ_1 the density of the solvent. $(1 - \bar{v}\rho_1)$ is the buoyant term. If the densities of the particle and of the solvent are the same, then $(1 - \bar{v}\rho_1) = 0$ and under these conditions no sedimentation will take place during centrifugation. For molecules weighing less than the solvent, $(1 - \bar{v}\rho_1)$ becomes negative: in a centrifugal field the particles migrate towards the meniscus. The molecular weight of a particle M is equal to $m \cdot N$, where N is Avogadro's number, and appears from equations (18) and (19)

$$M = \frac{s^0 f_s N}{(1 - \bar{v}\rho_1)} \tag{20}$$

Thus, in order to obtain the molecular weight from a sedimentation analysis, both the partial specific volume $\bar{v}$ and the friction coefficient f_s must be known. The determination of $\bar{v}$ will be dealt with later (section VIIC). A direct calculation of the frictional coefficient is possible if we are dealing with unsolvated spherical molecules. Their frictional coefficient f_0 is obtained from the Stokes–Einstein equation[1,6]

$$f_0 = 6\pi\eta_0 \left(\frac{3M\bar{v}}{4\pi N} \right)^{1/3} \tag{21}$$

Unsolvated spherical molecules are not encountered frequently; mostly we will have to deal with irregularly shaped molecules such as rod-like particles, prolate ellipsoids or solvated coils. The calculation of the frictional coefficients of such particles is far more complicated. They may be eliminated from equation (20) if other hydrodynamic properties are introduced which already contain the frictional coefficient. Thus various methods may be used to calculate the molecular weight:

   (i) from the sedimentation coefficient s^0 and the diffusion coefficient D.

   (ii) from s^0 and viscosity $[\eta]$. In this case one should have some knowledge about the shape of the molecule.

(iii) from the sedimentation coefficient and the concentration dependence of s_c. As was shown above (equation (16)) the concentration dependence of s_c ultimately depends on viscosity. Therefore, only the first two methods mentioned above shall need further explanation.

The frictional coefficient f_D of diffusion is given by equation

$$D_0 = \frac{RT}{f_D N} \tag{22}$$

where R is the gas constant and T the absolute temperature. If f_s is set equal to f_D, the well-known 'Svedberg equation' is obtained:

$$M_{s,D} = \frac{s^0 RT}{D_0(1 - \bar{v}\rho_1)} \tag{23}$$

This formula is used mainly for the determination of molecular weights of proteins. As demonstrated below (section VIII) it is also possible to determine the quotient s^0/D_0 by equilibrium sedimentation and thus calculate the molecular weight.

B. Determination of Molecular Weight from s^0 and $[\eta]$

For chain molecules, such as synthetic polymers or nucleic acids, the determination of the molecular weight from s^0 and $[\eta]$ has given good results. It is based on considerations developed mainly by Flory and colleagues.[8,81,82] According to them, there is a relation between the frictional coefficient f of a molecule, the molecular weight and the viscosity $[\eta]$ (in dl/g):

$$f/\eta_0 = P\Phi^{-1/3}(M[\eta])^{1/3} \tag{24}$$

where, as in section V, η_0 is the viscosity of the solvent in poise units and P and Φ are geometrical factors which depend mainly on the shape of the molecule and only to a slight extent upon its size. If we allow $(P^{-1} \cdot \Phi^{1/3}) = \beta$, we obtain from Equations (20) and (24)

$$M_{s,\eta} = \left[\frac{s^0[\eta]^{1/3}\eta_0 N}{\beta(1 - \bar{v}\rho_1)}\right]^{3/2} \tag{25}$$

By means of this equation, the molecular weight of a polymer can be calculated reliably. But it presupposes that β is known, i.e. there have to be concrete assumptions as to the shape of the molecule on which β depends.

Mandelkern et al.[81] were able to demonstrate for flexible chain molecules of varying length that P and Φ are general constants; thus, the same value for β may always be used for a random coil. In a detailed examination Eigner and Doty[49,56] demonstrated that for native DNA in a molecular weight range from 0·3 to 150×10^6 daltons, β will be $2·5 \times 10^6$, where $[\eta]$ has to be expressed in dl/g. For denatured DNA, β was found to be $2·3 \times 10^6$.

If however, rod-like particles or prolate ellipsoids are considered, β will depend on the axial ratio a/b of the particle. According to Scheraga and Mandelkern[82] β then increases from $2 \cdot 12 \times 10^6$ with $a/b = 1$, to $3 \cdot 60 \times 10^6$ with $a/b = 300$. By comparison measurements of s^0, $[\eta]$ and of the molecular weight, for example by light scattering β can be determined (equation (25)) and thus the shape of a protein molecule may be found. Evidence on the axial ratio of a particle can also be obtained by relating the values for sedimentation, diffusion and viscosity. This data furnishes evidence as to the friction of the particle. In order to obtain a further parameter for the shape of the molecule the frictional coefficients obtained from equations (20) or (22) are related to the friction which, following equation (21), a spherical molecule of the same size would have (f/f_0). Corresponding equations were described by Svedberg and Pedersen,[1] Elias,[6] and Stern.[83] Sund and Weber[84,85] have demonstrated with several enzymes that all these different procedures give values for the axial ratio which are in reasonable agreement.

Recently, Reinert and coworkers[86,87] made an extensive study on DNA-molecules of various size in the range of $1 \cdot 0$ to 100×10^6 daltons, determining s^0, $[\eta]$ and D_0. The parameter β can be calculated from these data, using equation (23) and (25). It was shown, that β increases slightly with decreasing molecular weight from $2 \cdot 5 \times 10^6$ at 10^8 daltons to $2 \cdot 8 \times 10^6$ at 10^6 daltons.

Following equation (23), the molecular weight is essentially determined by the product $s^0 [\eta]^{1/3}$. This means that a change in the shape of a molecule influencing the sedimentation coefficient will also affect the viscosity. For example, the shape of the DNA molecule changes with the solvent used. High ionic concentration means densely coiled molecules, consequently the s^0 value is high and viscosity low; with low ionic concentration the molecule expands, causing an increase in viscosity and a decrease of the s-values. However, the product $s^0 [\eta]^{1/3}$ stays fairly constant for a given DNA preparation.[49,57]

If s^0 and $[\eta]$ of molecules of different chain lengths are measured under standard conditions, a relation between $[\eta]$ and s^0 will appear that fits the formula $[\eta] = a(s^0)^b$, where a and b are variable constants which depend on chain length. Figure 8(a) shows such a relation for native DNA in $0 \cdot 2\text{M}$ NaCl. Knowing such a relation, the corresponding molecular weight for each point of the curve may be calculated following equation (25), and the values obtained for $M_{s,\eta}$ related to s^0 and to $[\eta]$ (Figure 8(b)). The equations

$$s^0 = K_s M^{a_s} \tag{26a}$$

$$[\eta] \text{ (in dl/g)} = K_\eta M^{a_\eta} \tag{15a}$$

are thus obtained. In this manner, the molecular weight may be obtained from s^0 alone, as well as from $[\eta]$ alone, so long as the same solvent is used.[89]

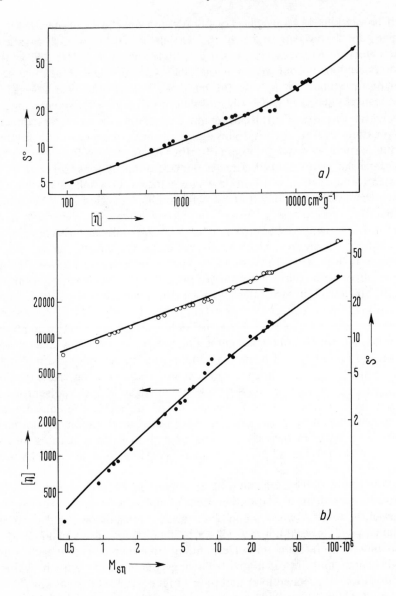

Figure 8. Relationship
(a) between sedimentation coefficient S^0 and viscosity $[\eta]$
(b) between molecular weight M and S^0 and between M and $[\eta]$
for native DNA in $0.1 - 1.0$ M Na^+
●, ○ experimental values according to Weinert and Hagen[88] and Coquerelle and coworkers,[59]
solid line according to Eigner and Doty's equation[49]

The relations of equations (15) and (26) do not always follow a straight line in the double logarithmic scale. Consequently it is necessary to specify the constants K_s, K_η, a_s, a_η for definite ranges of molecular weights only. Corresponding indications for native and denatured DNA are given by various authors.[49,68,87,90] It was suggested by Crothers and Zimm[52] to use instead of equation (26a) the formula:

$$s^0 - q_s = K_s \cdot M^{a_s} \tag{26b}$$

Using a suitable value for the constant q_s, the upward bended curve in Figure 8(b) becomes linear. In this way, the molecular weight of a DNA sample can be easily calculated from the s^0-value. In a corresponding way, the relation between $[\eta]$ and M becomes linear in a logarithmic plot by the equation

$$[\eta] + q_\eta = K_\eta \cdot M^{a_\eta} \tag{15b}$$

using a suitable value for q_η. Reinert and coworkers[86,87] made a detailed description of the mathematical background for these formulae (15b) and (26b); also the numerical values for the various constants are given.

C. Determination of the Partial Specific Volume $\bar{v}$

The accuracy of a molecular weight determination, however, depends essentially on the determination of the partial specific volume $\bar{v}$. $(1 - \bar{v}\rho_1)$ is determined by the density of the solution ρ_3 at a known concentration of the solute c_2:

$$\frac{\rho_3 - \rho_1}{c_2} = \frac{\Delta\rho}{c_2} = (1 - \bar{v}\rho_1) \tag{27}$$

It is advisable to determine the density at various concentrations. However, for these measurements large quantities of substance are needed and such quantities are seldom available for biochemical examinations. A much smaller quantity of sample is needed when using the microbalance accessory that has been specifically designed for buoyancy measurements.[91] A new way to determine the density of small volumes of a solution (0·6 ml) is described by Kratky and coworkers.[92] In their device, a defined volume of the solution takes part in the undamped oscillation of a mechanical oscillator. The oscillation frequency observed in the solution depends on its density which can be computed from the measured data with an accuracy of 10^{-6} g cm^{-3}.

No additional quantities are needed for the determination of $\bar{v}$, if the molecular weight is determined by sedimentation equilibrium in solvents of different concentrations.[93] However, it must be understood that the solvents used will have no influence on the hydration of the macromolecule. Water–deuterium oxide mixtures have proved advantageous in this respect. Gagen[94] elaborated a similar system for measuring the sedimentation in solvents of different density, in order to determine by extrapolation the density in which no sedimentation occurs, i.e. when $(1 - \bar{v}\rho_1)$ = zero. Tanford[9] assembled the values of $\bar{v}$ for various proteins (Tables 22-1 and 23-2); these are between 0·72 and 0·75. The partial specific volume of DNA is indicated[95] as 0·556 at 20°C.

VIII. MEASUREMENTS AT SEDIMENTATION EQUILIBRIUM

The movement of a particle in a centrifugal field is determined, following Lamm's differential equation (4), by sedimentation and back diffusion. After an appropriate centrifugation period, equilibrium is established in the closed cell, that is to say the concentration gradient dc/dr remains stable, dc/dt equals zero.

Figure 9(b) shows a diagram of the concentration distribution. The flow of the solute J equals zero:

$$0 = s\omega^2 rc - D(\partial c/\partial r) \tag{28}$$

By introducing equation (28) into the Svedberg equation (23) we obtain

$$M = \frac{dc/dr}{rc} \cdot \frac{RT}{\omega^2(1 - \bar{v}\rho_1)} \tag{29}$$

The molecular weight can thus be obtained directly from the concentration gradient dc/dr at a distance r and from concentration c in that same area. The steeper the concentration gradient, the more sedimentation will overbalance diffusion and the higher the value of the molecular weight will be. By integrating equation (29) between the distances r_a and r_b, we obtain

$$M = \frac{2RT}{(1 - \bar{v}\rho_1)\omega^2(r_b^2 - r_a^2)} \cdot \frac{c_b - c_a}{c_0} \tag{30}$$

where r_a and r_b represent any distance within the cell, usually r_a is taken as the meniscus and r_b as the bottom. c_a and c_b are the concentrations corresponding to the distance r_a and r_b.

The method of evaluation of these parameters depends on the optical system used. With the schlieren system, the concentration gradient dc/dr is obtained directly. Therefore, equation (29) may be used. Details of this calculation are described by van Holde and Baldwin,[96] Yphantis,[97] and Elias.[6] With the optical absorption or the interference methods, however, the concentration of the solution is obtained and equation (30) is used in the calculations. But this is only possible with a homogeneous substance. In case of heterogeneity, the integration of equation (29) shown below is more appropriate:

$$M = \frac{2RT}{(1 - \bar{v}\rho_1)\omega^2} \cdot \frac{\Delta \ln c}{\Delta r^2} \tag{31}$$

In this method, a series of measurements are made and $\ln c_i$ is plotted against r_i^2. For homogeneous substances the result will be a straight line with the molecular weight represented by the slope. Heterogeneous substances produce a deviation from linearity. The z-average (see equation (42)) of the molecular weight may be calculated from the slope of the curve at the bottom and at the meniscus.[98]

The experimental conditions for sedimentation equilibrium have to be chosen with great care.[15,23,31,99] First of all, the substance to be analysed has to be of sufficient purity, any low molecular weight impurity will alter the results quite severely. In the various measurements it is always the average molecular weight of all absorbing molecules that is obtained. The speed should be adjusted so as to obtain a steep concentration gradient, and in turn to ensure an accurate determination. Since the slope of the gradient is inversely proportional to the molecular size, a lower speed will be needed for large molecules. For example, for molecules of 100,000 daltons, a speed of about 7000 rpm is advisable but for molecules of 1000 daltons, a speed of 70,000 rpm is necessary. Particularly suitable for the analysis of very large molecules is the high speed method, also called the 'meniscus depletion' method, as introduced by Yphantis.[98] A rotation speed is chosen that is high enough to ensure that the concentration near the meniscus is almost zero.

The time t (in seconds) necessary for equilibrium to be achieved is independent of speed; it depends on the diffusion constant of the substance as well as on the height of the solution column in the cell $(b - a)$. Van Holde[99] claims that equilibrium is reached approximately after $t = 0.7 (b - a)^2/D$ sec. For instance, with a substance of molecular weight of 100,000, $D = 0.5 \times 10^{-6}$ and 3 mm column height, about 35 hours are needed. Cells for this purpose have been developed with several channels of very short column height.[97] Furthermore, the time of the run may be shortened if the centrifuge is run initially at rotor speed higher than the ultimate speed, and then slowed down to the desired speed.[100]

Thus it is evident that with long running times the need for constant speed and constant temperature is of primary importance. In order to avoid convection currents, it is advisable to use a rather heavy rotor (titanium), and to switch off the heating assembly (RTIC unit) and throttle the cooling system as soon as the top rotor speed has been reached. A very constant temperature of approximately 16 °C will thus be obtained. The gradients are more stable if the ionic concentration of the solvent is not too low. The formation of a salt gradient does not interfere with the concentration gradient of the dissolved macromolecules.

Equations (29)–(31) show that the accuracy of the molecular weight determination will depend upon the accuracy with which the initial concentration c_0 is determined. This is best done with a separate synthetic boundary run to obtain an exact relation between concentration c and the values as measured with the different optical systems (in the schlieren optical system: area of the boundary; interference system: number of fringes; and in the optical absorption method: deflection of the recorder).

If relatively high concentrations are necessary as is the case with the schlieren system, then the molecular weights determined by equilibrium runs are extrapolated to the initial zero concentration. On the contrary, when either the interference or the optical absorption systems are used, the concentration dependence of molecular weights determined by equilibrium runs may be neglected.[31,96] The influence of the charge of the molecule on the apparent molecular weight obtained was investigated by Casassa and Eisenberg.[101]

In the last years, by the improvement of the technique of ultracentrifugation, the sedimentation equilibrium is widely used in various fields. For instance, the molecular weight of coliphages was determined by the meniscus depletion method very accurately.[102] This method is also suitable for the estimation of the molecular weight in mixtures of two proteins.[103] The knowledge about the interaction of molecules in solution, widely studied by various authors, is based mostly on equilibrium studies (cf. section XIB). Improvements of the computational methods applied to equilibrium experiments will facilitate the evaluation of the data from the centrifuge.[104,105] Readers more interested in the theory of the thermodynamic treatment of equilibrium may study the article by Casassa.[106]

IX. TRANSIENT STATE DURING APPROACH TO SEDIMENTATION EQUILIBRIUM

Whilst the equilibrium between diffusion and sedimentation takes some time to become established over the whole of the cell column, it exists during the entire run of the experiment at the bottom (r_b) and at the meniscus (r_a) of the cell, since nothing can sediment at r_a from above or diffuse at r_b from below: $J(r_a) = J(r_b) = 0$. Together with the considerations based on the flow of the solute (equation (4)), these statements give the conditions of the equilibrium (Archibald state):[107]

$$\left(\frac{1}{rc}\cdot\frac{dc}{dr}\right)_{r=r_a} = \omega^2\left(\frac{s^0}{D}\right)_{r=r_a} \tag{32}$$

The same is true for r_b. The concentration distribution for this transient state is shown by Figure 9(a). The molecular weight can be calculated by means of equation (29), once dc/dr and the concentration at the bottom and at the meniscus are determined. The schlieren optical system enables the concentration c_a or c_b to be determined from the area F and dc/dr from the height y.[108] Extrapolation for determination of y at the meniscus and at the bottom, often rather difficult, is facilitated by the use of a computer.[28,109,110]

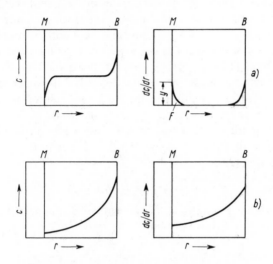

Figure 9.
(a) Concentration (c) and concentration gradient (dc/dr) as a function of the distance r to the rotor axis in the transient state during approach to sedimentation equilibrium
(b) Concentration (c) and concentration gradient (dc/dr) in the cell at equilibrium between sedimentation and diffusion

Centrifugation of heterogeneous substances induced a separation of molecules according to their size. The small particles will concentrate at the meniscus and the large ones at the bottom. Thus, measurements taken at the meniscus will indicate a lower molecular weight than those taken at the bottom. Since there is no separation of the particles at the beginning of the experiment, the real average value of molecular weight may be obtained by extrapolating the various values obtained for M_{app} at different times to the beginning of the experiment. Further modifications of the Archibald procedure to allow greater precision in the determination of the different molecular weights of a paucimolecular mixture were developed by Trautman.[111,112] According to Creeth and Pain,[13] the transient state method is accurate and reliable only at a reasonably high concentration (0·7 per cent). Its advantage, however, is the very short duration of the experiment.

X. ISOPYCNIC CENTRIFUGATION (ISODENSITY EQUILIBRIUM SEDIMENTATION)

If a macromolecule is centrifuged in a solution of very high density, for example a caesium chloride solution, a density gradient will be generated by sedimentation and diffusion of the salt. Under convenient conditions, the macromolecule, after achieving equilibrium, will concentrate at that point where the density of the solution corresponds to its own density, i.e. where the bouyant term $(1 - \bar{v}\rho_1)$ is zero. If the density of the solution at the different points of the cell is known, the density of the macromolecule may be determined from its position in the cell. This method is very accurate, differences in density of $0.001 \, \text{g cm}^{-3}$ can be detected.

The density gradient $d\rho/dr$ is derived from the consideration that in sedimentation equilibrium the flow of solute is zero: $J = dm/dt = 0$, and following equation (28)

$$as^0\omega^2r = D \, da/dr \tag{33}$$

where a represents the activity of the solute. Taking $da/dr = (da/d\rho) \times (d\rho/dr)$ and equation (23) we obtain

$$\frac{d\rho}{dr} = \frac{\omega^2 r}{\beta(\rho)} \tag{34}$$

where

$$\beta = \frac{d \ln a}{d\rho} \cdot \frac{RT}{M_1(1 - \bar{v}\rho_1)}$$

(M_1: molecular weight of the salt)

Equation (34) allows calculation of the density ρ_i at any distance r_i if $\beta(\rho)$ and the density ρ_0 at a given point of the cell, for example r_0, are known. Integrating (34) we obtain

$$\rho_i = \rho_0 + \omega^2/2\beta(r_0^2 - r_i^2) \tag{35}$$

For r_0 the distance r_e within the cell may be conveniently used. This is the point where, after equilibrium has been established, the same concentration c_0 and the same density ρ_0 as those of the initial solution are found[113] (isoconcentration point):

$$r_e = \sqrt{\frac{r_a^2 + r_b^2}{2}} \tag{36a}$$

This relationship is true for the sector shaped cells used in the analytical centrifuge. For the cylindrical tubes of the preparative centrifuge (see section XII), the equation is modified to:

$$r_e = \sqrt{\frac{r_a^2 + r_a r_b + r_b^2}{3}} \tag{36b}$$

Instead of calculating r_e a macromolecular substance of known density (ρ_{ma}) may be added as a marker. Its position indicated by this marker r_{ma} will have the density ρ_{ma}.

For resolution of equation (35) values for $\beta(\rho)$ are required. Ifft *et al.*[113] have calculated $\beta(\rho)$ for various caesium salts and tabulated them with respect to the corresponding density. It must be noted that β also depends on the force exerted on the solution during centrifugation, since this pressure generates a new, additional density gradient. Therefore, $1/\beta_0(\rho)$, while valid for zero pressure, has to be multiplied by a corrective factor β_0/β which is dependent upon the pressure developed in the solution. An accurate derivation of these factors is given by Hearst *et al.*[114] as well as by Dirkx[115] who indicates numerous examples. Thus for caesium chloride of density $\rho = 1.700$, $[1/\beta_0(\rho)] = 8.400 \times 10^{-10}$ and $\beta_0/\beta = 1.090$.

CsCl is most commonly used as a material for generating a density gradient, but other substances easily soluble may also be used, such as LiBr, Cs_2SO_4 or sucrose. The density of the solutions may be determined with a pycnometer, but this method is not usually accurate enough and furthermore requires a substantial amount of sample. Determination of the refractive index n^D is preferred. For CsCl at 25°C we obtain:[113]

$$\rho_{25} = 10.8601\, n_{25}^D - 13.4974$$

With this method, the density of even a small volume may be determined and adjusted if necessary. Density may also be calculated by weighing the salt very accurately.[115,116,10] The time needed to establish equilibrium is determined by the characteristics of the salt. For CsCl approximately 24 hours at a speed of 40,000 to 50,000 rpm is sufficient (see section VIII). After this time has elapsed, the migration of the macromolecule to the position of its density has already ceased.

In density gradient centrifugation it is possible to determine not only the density, but also the molecular weight of a macromolecule. Diffusion works against centrifugal force in localizing the macromolecule at a point corresponding to its density, and a broadening of the band occurs. The method, therefore, is limited to larger macromolecules banding within the cell length. As the cell length does not exceed 1.3 cm, the smallest macromolecules that may be banded in a caesium chloride gradient to give a complete distribution must have a molecular weight between 10,000 and 50,000 (Dirkx[115]). Meselson *et al.*[117] show the distribution of the macromolecules in the band follows a Gaussian curve where the width of the distribution σ depends on the quotient s^0/D. As the latter again depends on molecular weight (equation (23)), the molecular weight M_s of the solvated molecules may be determined from the standard deviation σ:

$$M_s = \frac{RT}{\sigma^2\omega^2 r(d\rho/dr)_r} \tag{37}$$

where r and ρ correspond to the position and to the density of the Gaussian maximum of the band. σ is determined by the width of the band at the position where concentration is 0·607 times the maximum, and $d\rho/dr$ is determined by equation (34).

The molecular weight obtained from equation (37) corresponds to that of the hydrated molecule (M_s); in the highly concentrated salt solution a substantial amount of water has bound to it. Baldwin[119] and Hearst and Vinograd[120] have estimated the extent of hydration Γ' of DNA-molecules, i.e. the quantity of solvent in g which was bound per g of macromolecule

$$\Gamma' = \frac{1 - \rho_0\bar{v}_3}{\rho_0\bar{v}_1 - 1} \tag{38}$$

where $\bar{v}_1$ and $\bar{v}_3$ are the partial specific volumes of the dissolved salt and of the macromolecule respectively. The molecular weight of the anhydrous macromolecule is

$$M_{anh} = M_s(1 - \Gamma') \tag{39}$$

This correction factor is quite considerable. For example, for lysozyme[115] $\Gamma' = 0·201$ and for DNA in $CsCl$[120] $\Gamma' = 0·28$.

If the analysed sample is not macromolecularly homogeneous, a deviation from the Gaussian distribution will be observed. Meselson *et al.*[117] as well as Sueoka[118] have shown how to calculate average values M_w and M_n (see also equations (40)–(42)). Although potentially powerful and promising, the method determining molecular weight of nucleic acids from their equilibrium distribution in density gradients requires a careful interpretation of the data. A new theoretical approach about the interaction of the DNA molecules with the salt molecules was made by several authors.[121,122,123] It is possible with these considerations to obtain reliable results for the molecular weight of DNA, comparable with the data from velocity sedimentation.

XI. ANALYSIS OF MULTICOMPONENT MACROMOLECULAR SYSTEMS

A. Paucimolecular and Polymolecular Mixtures

It is only with highly purified enzymes or with DNA preparations from viruses that all macromolecules in solution are of the same size. In most cases there will be a mixture of various sizes. The molecular weights will either be scattered around an average value (polymolecular mixture) or will consist of several components of different molecular weights (pauci-molecular mixture). We obtain from the ultracentrifuge not only an average value of the molecular weight but also indications as to the distribution of these weights. To characterize a distribution of molecular weights the following average values may be used (consult also the text books on macro-molecular chemistry):[8,9]

Number average
molecular weight
$$M_n = \frac{\sum n_i M_i}{\sum n_i} = \frac{\sum c_i}{\sum c_i/M_i} \tag{40}$$

Weight average
molecular weight
$$M_w = \frac{\sum n_i M_i^2}{\sum n_i M_i} = \frac{\sum c_i M_i}{\sum c_i} \tag{41}$$

z-average
molecular weight
$$M_z = \frac{\sum n_i M_i^3}{\sum n_i M_i^2} = \frac{\sum c_i M_i^2}{\sum c_i M_i} \tag{42}$$

where n_i is the number of molecules of molecular weight M_i and c_i the concentration of the fraction.

The molecular weight values obtained through sedimentation, diffusion or viscosity (M_{sD} or $M_{s\eta}$; equations (23), (25) and (29)) are usually slightly below M_w, provided the molecular weight distribution is not too asymmetric. A general treatment of the relation between M_{sD}, $M_{s\eta}$, M_w and M_n for various forms of molecular weight distributions was given by Gibbons.[124] To a great extent, this relation depends on the configuration of the macro-molecule expressed by the value of the exponent a_s in equation (26).

Whilst a paucimolecular mixture might be recognized with help of the transport method and possibly even analysed, following the method of Trautman and Crampton,[112] the analysis of a polymolecular mixture requires careful attention. A method of obtaining an integral distribution of the sedimentation coefficients s^0 was explained in section VI If the constants K_s and a_s of equation (26) are known for a given substance, the sedimentation distribution may be converted into molecular weight distribution. By adding the separate fractions according to equations (40)–(42) we can obtain the different average values of molecular weight. A more detailed analysis of the transformation of the sedimentation distribution into a molecular weight distribution is given by Reinert and coworkers.[86,87,125]

As the molecular weight of a macromolecule can be determined by equilibrium sedimentation (see Section VIII), several attempts have been made to calculate molecular weight distributions also from equilibrium data. From the concentrations or concentration gradients at a number of positions in the cell at one constant rotor speed M_w and M_z can be calculated.[126] Using the meniscus depletion method, where the concentration of the macromolecules at the meniscus is zero, also M_n is obtained.[127] According to Fujita,[128] the average molecular weights M_n, M_w, and M_z can be computed from equilibrium data relating to only one position in the cell, but to a number of different rotor speeds. A combination of these methods is described by Scholte;[129] with this procedure it is possible to analyse the whole distribution of a paucimolecular or polymolecular mixture with fair accuracy.

B. Associating Systems

A series of substances (P) follow association reactions of the type

$$nP_1 \rightleftharpoons P_n (n = 2, 3, \ldots) \tag{43}$$

or

$$nP_1 \rightleftharpoons qP_2 + mP_3 + \ldots \tag{44}$$

The association may either be limited to a certain number of molecules (discrete self-associating system) or it may continue indefinitely (indefinitive self-associating system). We shall first consider a discrete system, for example the association of a monomer of M_1 molecular weight to form dimers and trimers. If the molecular weight of the complex is determined at increasing concentration say by equilibrium runs, the values obtained will increase initially and then decrease again at high concentration. This is due to non-ideality of the solution, that is the molecules interfere with each other in their mobility. The molecular weight $M_{w,app}$ as obtained with the equilibrium runs may be defined by equation (45):

$$\frac{M_1}{M_{w,app}} = \frac{M_1}{M_{w,(c)}} + BM_1 c \tag{45}$$

where B is the virial coefficient, c the concentration of the solution, $M_{w(c)}$ is the molecular weight of the complex, as would be measured at concentration c in an ideal solution, where B would equal zero but the association of the macromolecules would take place.

From the reaction constant K of equation (44) we obtain the concentration of a given association product c_n:

$$c_n = K_n c_1^n \tag{46}$$

where $n = 2, 3$ and c_1 the concentration of the monomer. The total concentration c, in this case, may be written as follows:

$$c = c_1 + K_2 c_1^2 + K_3 c_1^3 \qquad (47)$$

From this we obtain

$$M_{w(c)} = M_1(c_1 + 2K_2 c_1^2 + 3K_3 c_1^3)/c \qquad (48)$$

Analysis of association systems is directed toward determining the constants of equations (45) and (48), i.e. the reaction constant K_n, c_1 and also B, from the $M_{w,app}$ values obtained at different concentrations. An unequivocal result is possible, however, only by making certain assumptions on the type of association reaction, which in turn have to be verified experimentally.

The first detailed investigations about associating systems were described by Adams and coworkers.[130–134] Since then further studies about associating systems have been published by many authors. Only a short survey on the various procedures of analysis and the various applications can be given, however, in this chapter. A multinomial theory for the calculation of the equilibrium constants in ideal and nonideal systems has been developed by Derechin.[135–137] This theory is applicable to association reactions of any degree of polymerization and is valid irrespective of whether the apparent molecular weight $M_{w,app}$ or the number average molecular weight M_n is used in the calculations. According to Derechin care should be taken analysing with this method biological systems, as secondary aggregations may interfere with the reversible association- and dissociation-reactions.[136] The association of chemically reacting systems with two components A and B of the type $nA + mB \rightleftharpoons A_n B_m$ is discussed by Adams.[138] Interacting systems containing a macromolecule (P) and small molecules (HA) of the type $mP + nHA \rightleftharpoons P_m(HA)_n$ were analysed by Goad and Cann.[139] In this case, the diffusion of the small molecules has to be considered carefully.

Associating systems have been studied extensively with equilibrium methods by Roark and Yphantis.[140] There are two general approaches for such analysis. In the first the totality of observation of the concentration distribution is used which offers greater precision for the parameters of the systems, but direct information is generally lacking about the various species of association products and about their heterogeneity. The other approach examines only limited regions of the concentration distribution in the cell, for example by calculation of local average molecular weights M_n, M_w and M_z in a given position. In this way more information about the nature and details of the system is obtained, but equilibrium constants can be determined with less accuracy. The interpretation of these various point average molecular weights for the type of a self-associating system is described in theory in ideal and nonideal systems.[140] The self-association of

β-lactoglobulin A was studied by these methods. It was shown,[140] that this protein behaves heterogeneously with respect to association properties. When isolating various β-lactoglobulin species by chromatography, mono-mer–dimer association could be observed separately, whereas other species associate to higher polymers. The method of Roark and Yphantis[140] may be recommended therefore to analyse complex biological associating systems.

Association and dissociation reactions should be considered when analysing the quaternary structure of proteins by means of denaturation during which the protein dissociates into subunits. The binding of the coenzyme to the enzyme is another example for association reactions. A detailed description of such analyses on proteins has been published by Sund and coworkers.[143,144] Difficulties are encountered if electrolytes in the solvent interact with the macromolecules. An exhaustive discussion of this problem is found in Casassa[101,106] and Adams.[131] If the self-association continues without limit, a continuous increase in molecular weight with increasing concentration may be observed, as long as the virial coefficient is not too high. Unlimited self-associations of this kind are described for purines and pyrimidines.[141,142] Careful measurements are required in order to distinguish discrete self-associations from indefinitive ones.[134] Associating systems are sensitive to the hydrostatic pressure in the cell and the equilibrium constant therefore depends upon rotor speed as well as upon radial position in the cell. The measurement should be performed at low speed or the pressure effect should be extrapolated to zero pressure or investigated explicitly by the superposition of various levels of aqueous insoluble oil layers on the aqueous column.[145–147]

XII. SEDIMENTATION IN THE PREPARATIVE ULTRACENTRIFUGE

Whilst in the analytical ultracentrifuge the migration of a particle may be observed only when it is characterized by specific absorption or refraction, in the preparative centrifuge further characteristics of the separated macro-molecules may be determined for example enzymatic activity, radioactivity or other properties. Thus, in molecular biology, apart from the analytical centrifuge, more often than not the preparative ultracentrifuge is used for additional characterizations of hydrodynamic properties of the sample.

The tubes of the preparative centrifuge give rise to several factors which considerably interfere with the migration of the particles, so that the deter-mination of the sedimentation coefficient is only possible under certain circumstances. Let us first consider the movement of particles in suspension either in the swinging bucket or in the fixed angle rotor of a preparative centrifuge. During centrifugation the heavy particles displace the lighter molecules of the solvent, so that these migrate upwards. In the fixed angle rotor the convection current of the light particles is favoured by the fact that the heavy particles move first towards the outer wall of the tube and then fall in a joint current to the bottom. In this way centrifugation in a fixed angle rotor is more efficient in the separation effect on particles than in swinging bucket rotors.[16,148,149]

The convection currents in a preparative centrifuge tube do not allow the formation of a boundary, as would be required for the determination of the sedimentation. These currents can be avoided by using shallow density gradients such as those formed by small rapidly diffusing particles, as for instance sucrose. There is a heavier liquid at the bottom of the tube than at the meniscus, and the centrifugal force makes back-flow impossible. This stabilized moving boundary centrifugation is particularly propitious for analysis if the solution to be analysed is overlayered onto an existing density gradient. The particles then migrate through the tube as a band. This zonal or band centrifugation (see also section VIC) allows the determination of the sedimentation coefficient that is sufficiently precise for many experiments. Also, particles of different sizes are well separated. The swinging bucket rotor is generally used for this kind of centrifugation problem.

If the selected solvent has the same density as the particle to be sedimented, we have the so-called isopycnic gradient centrifugation (section X). Here the particles may be mixed in the salt solution. A density gradient will be generated by the centrifugal field. Isopycnic centrifugation may also be performed in a fixed angle rotor.[150]

The experimental technique of band centrifugation in the swinging bucket rotor has been described exhaustively by several authors.[10,151–153] In this preformed density gradient method, sucrose is most often used. It is introduced into the tube in decreasing concentration (say from 20 to 4 per cent). After overlayering with the particles to be analysed, centrifugation is allowed to take place until the particles migrate to about the middle of the tube. Their position may be determined by absorption or refraction methods. A better method is to perforate the bottom of the tube and collect the various fractions drop by drop, as they emerge according to their position in the tube. This enables the determination of the distance of a given fraction from the meniscus. From the time t taken to migrate the distance d (in cm) from the meniscus, the sedimentation coefficient can be calculated according to the approximation formula of Burgi and Hershey:[154]

$$s_{20,w}^0 = \frac{6 \cdot 45 \times 10^{10} \times d}{\omega^2 t} \qquad (51)$$

Furthermore, it is possible to calculate the molecular weight (M_1) of a sample if it is sedimented along with another sample of known molecular weight (M_2), and if both the migration distances d_1 and d_2 are determined.

We have then

$$\frac{d_2}{d_1} = \left(\frac{M_2}{M_1}\right)^k \qquad (52)$$

For native DNA Burgi and Hershey[154] obtained a k value of 0.352 ± 0.005.

Comparing equation (52) with equation (26a) it is evident, that the distance d is proportional to the sedimentation coefficient s of a particle. This is correct, however, only under certain conditions of sucrose sedimentation. Whereas the greater distance from rotation centre induces a faster migration of the particle, the increasing density of the solution towards the bottom of the tube retards the sedimentation. Van der Schans and coworkers[155] have described this phenomena in detail and the corrections necessary to calculate reliable molecular weights are given. Another hindrance to sedimentation is due to the particles hitting the walls during their radial migration. The convection current thus generated is not quite eliminated by the density gradient.

An unimpeded particle migration comparable to that obtained in the analytical centrifuge is possible in the zonal centrifuge developed by Anderson and coworkers.[156–158] Here the solution is centrifuged in a rotating tubular rotor and the particles are quite free to migrate radially in outward direction. It may also be noticed that in the rotating rotor a density gradient is achieved much more rapidly than in the tubes of the swinging rotor. Zonal centrifugation is especially suitable for the separation of larger quantities of solution. The radial band fractions obtained by centrifugation are removed in a step-wise manner by suction. Here again, the position of the bands in the rotor may be deduced from the sequence of the fractions; sedimentation coefficients can be calculated with adequate precision.[159–161]

Zonal centrifugation can be widely used to separate particles from various tissues or cells, for instance nuclei, mitochondria, ribosomes or even plasma membranes.[162] An application of particular interest is the separation of viruses which is promising for high purity and yield of isolation.[163–165]

REFERENCES

1. T. Svedberg and K. O. Pedersen, *The Ultracentrifuge*, Oxford University Press, Oxford, 1940
2. H. K. Schachman, in *Methods in Enzymology*, Vol. IV (Ed. S. P. Colowick and N. O. Kaplan), Academic Press, New York–London, 1957, pp. 32–103
3. H. K. Schachman, *Ultracentrifugation in Biochemistry*, Academic Press, New York–London, 1959
4. R. L. Baldwin and K. E. Van Holde, *Fortschr. Hochpolymer. Forsch.*, **1**, 451 (1960)
5. K. Jahnke und M. Scholtan, *Die Bluteiweisskörper in der Ultrazentrifuge*, Thieme Verlag, Stuttgart, 1960
6. H. G. Elias, *Ultrazentrifugenmethoden*, Beckman Instruments, München, 1961

7. H. Fujita, *Mathematical Theory of Sedimentation Analysis*, Academic Press, New York–London, 1962
8. P. J. Flory, *Principles of Polymer Chemistry*, Cornell University Press, Ithaca, New York, 1953
9. C. Tanford, *Physical Chemistry of Macromolecules*, Wiley–Interscience, New York–London, 1961
10. J. Vinograd, in *Methods in Enzymology*, Vol. VI (Ed. S. P. Colowick and N. O. Kaplan), Academic Press, New York–London, 1963, pp. 854–870
11. N. G. Anderson (Ed.), *The development of zonal centrifuges and ancillary systems for tissue fractionation and analysis*, National Cancer Institute Monograph 21, Bethesda, Md., 1966
12. M. K. Brakke, in *Methods in Virology*, Vol. II (Ed. K. Maramorosch and H. Koprowski), Academic Press, New York–London, 1967, pp. 93–118
13. J. M. Creeth and R. H. Pain, *Progr. Biophys. Mol. Biol.*, **17**, 217 (1967)
14. R. Markham, in *Methods in Virology*, Vol. II (Ed. K. Maramorosch and H. Koprowski), Academic Press, New York–London, 1967, pp. 3–40
15. H. M. Mazzone, in *Methods in Virology*, Vol. II (Ed. K. Maramorosch and H. Koprowski), Academic Press, New York–London, 1967, pp. 41–91
16. G. Kegeles, in *Physical Techniques in Biological Research*, Vol. 2A (Ed. D. H. Moore), Academic Press, New York, 1968, p. 67
17. C. H. Chervenka, *A Manual of Methods for the analytical ultracentrifuge*, Beckman Instruments Inc. Spinco Division, Palo Alto, Calif., 1969
18. D. A. Yphantis (Ed.), *Ann. N.Y. Acad. Sci.*, **164**, 1–306 (1969)
19. H. G. Elias, *Chemie Ingenieur Technik*, **36**, 504 (1964)
20. S. Hanlon, K. Lamers, G. Lauterbach, R. Johnson and H. K. Schachman, *Arch. Biochem. Biophys.*, **99**, 157 (1962)
21. H. K. Schachman, L. Gropper, S. Hanlon and F. Putney, *Arch. Biochem. Biophys.*, **99**, 175 (1962)
22. K. Lamers, F. Putney, J. Z. Steinberg and H. K. Schachman, *Arch. Biochem. Biophys.*, **103**, 379 (1963)
23. H. K. Schachman and S. J. Edelstein, *Biochemistry*, **5**, 2681 (1966)
24. I. Z. Steinberg and H. K. Schachman, *Biochemistry*, **5**, 3728 (1966)
25. J. C. Gerhart and H. K. Schachman, *Biochemistry*, **4**, 1054 (1965)
26. O. Wiener, *Ann. Physik Chem.*, *Neue Folge*, **49**, 104 (1893)
27. Beckman Model E Analytical Ultracentrifuge Instruction Manual (Ed. M. Zeller), Beckman Instruments, Spinco Division, Palo Alto, Calif., U.S.A., 1964
28. F. E. LaBar, *Biochemistry*, **5**, 2368 (1966)
29. L. Gropper, *Anal. Biochem.*, **7**, 401 (1964)
30. H. K. Schachman, *Biochemistry*, **2**, 887 (1963)
31. E. G. Richards, D. C. Teller and H. K. Schachman, *Biochemistry*, **7**, 1054 (1968)
32. F. E. LaBar and R. L. Baldwin, *J. Phys. Chem.*, **66**, 1952 (1962)
33. P. A. Charlwood and M. V. Mussett, *J. Phys. Chem.*, **70**, 3075 (1966)
34. C. H. Chervenka, *Anal. Chem.*, **38**, 356 (1966)
35. O. Lamm, *Arkiv Mat. Astronom. Fysik, Ser. B.*, **21**, No. 2 (1929)
36. W. Kern und W. Mehren, in *Hoppe–Seyler–Thierfelder, Handbuch der physiologisch- und pathologisch-chemischen Analyse, Bd. II* (Ed. K. Lang und E. Lehnartz), Springer Verlag, Berlin–Göttingen–Heidelberg, 1955, pp. 12–23
37. P. Y. Cheng and H. K. Schachman, *J. Am. Chem. Soc.*, **77**, 1498 (1955)
38. M. J. R. Cantow, R. S. Porter and J. F. Johnson, *Makromol. Chem.*, **87**, 248 (1965)
39. R. H. Golder, *J. Am. Chem. Soc.*, **75**, 1739 (1953)
40. R. A. Alberty, *J. Am. Chem. Soc.*, **76**, 3733 (1954)

41. D. J. Cox, *Arch. Biochem. Biophys.*, **112**, 249 (1965)
42. D. J. Cox, *Arch. Biochem. Biophys.*, **112**, 259 (1965)
43. D. J. Cox, *Arch. Biochem. Biophys.*, **119**, 230 (1967)
44. R. Markham, in *Methods in Virology*, Vol. II (Ed. K. Maramorosch and H. Koprowski), Academic Press, New York–London, 1967, pp. 275–302
45. J. Strassburger and K. E. Reinert, *Biopolymers*, **10**, 263 (1971)
46. D. Lang and P. Coates, *J. Mol. Biol.*, **36**, 137 (1968)
47. H. Staudinger and W. Heuer, *Ber.*, **63**, 222 (1930)
48. H. Staudinger and R. Nodzu, *Ber.*, **64**, 721 (1931)
49. J. Eigner and P. Doty, *J. Mol. Biol.*, **12**, 549 (1965)
50. H. Staudinger, *Organische Kolloidchemie*, Viebig, Braunschweig, 1950
51. B. H. Zimm and D. M. Crothers, *Proc. Nat. Acad. Sci. U.S.*, **48**, 905 (1962)
52. D. M. Crothers and B. H. Zimm, *J. Mol. Biol.*, **12**, 525 (1965)
53. J. B. Hays and B. H. Zimm, *J. Mol. Biol.*, **48**, 297 (1970)
54. J. A. V. Butler and A. B. Robins, in *Flow Properties of Blood and Other Biological Systems* (Ed. A. L. Copley and G. Stamsky), Pergamon Press, Oxford, 1966, pp. 337–350
55. J. B. T. Aten, *Proefschrift*, University of Leiden, 1965
56. J. Eigner, C. Schildkraut and P. Doty, *Biochim. Biophys. Acta*, **55**, 13 (1962)
57. J. M. Creeth and C. G. Knight, *Biochim. Biophys. Acta*, **102**, 549 (1965)
58. J. B. T. Aten and J. A. Cohen, *J. Mol. Biol.*, **12**, 537 (1965)
59. Th. Coquerelle, L. Bohne, U. Hagen and J. Merkwitz, *Z. Naturforsch.*, **24b**, 885 (1969)
60. D. Freifelder, *Proc. Nat. Acad. Sci. U.S.*, **54**, 128 (1965)
61. J. E. Hearst and J. Vinograd, *Arch. Biochem. Biophys.*, **92**, 206 (1961)
62. J. Rosenbloom and V. Schumaker, *Biochemistry*, **6**, 276 (1967)
63. V. N. Schumaker and B. Marano, *Arch. Biochem. Biophys.*, **94**, 532 (1961)
64. J. P. Johnston and A. G. Ogston, *Trans. Faraday Soc.*, **42**, 789 (1946)
65. H. Fujita, *Biopolymers*, **7**, 59 (1969)
66. G. H. Weiss and M. Dishon, *Biopolymers*, **9**, 865 (1970)
67. R. L. Baldwin and J. W. Williams, *J. Am. Chem. Soc.*, **72**, 4325 (1950)
68. V. N. Schumaker and H. K. Schachman, *Biochim. Biophys. Acta.* **23**, 628 (1957)
69. U. Hagen, *Biochim. Biophys. Acta*, **134**, 45 (1967)
70. H. Fujita, *J. Chem. Phys.*, **24**, 1084 (1956)
71. J. W. Williams and W. H. Saunders, *J. Phys. Chem.*, **58**, 854 (1954)
72. R. L. Baldwin, *J. Am. Chem. Soc.*, **76**, 402 (1954)
73. H. J. Cantow, *Makromol. Chem.*, **30**, 169 (1959)
74. J. Vinograd, R. Radloff and R. Bruner, *Biopolymers*, **3**, 481 (1965)
75. J. Vinograd and R. Bruner, *Biopolymers*, **4**, 131 (1966)
76. J. Vinograd and R. Bruner, *Biopolymers*, **4**, 157 (1966)
77. J. Vinograd and R. Bruner, *Fractions 1966, No. 1*, Beckman Instruments, Palo Alto, Calif.
78. W. Bauer and J. Vinograd, *J. Mol. Biol.*, **33**, 141 (1968)
79. J. Vinograd, R. Bruner, R. Kent and J. Weigle, *Proc. Nat. Acad. Sci. U.S.*, **49**, 902 (1963)
80. R. Bruner and J. Vinograd, *Biochim. Biophys. Acta*, **108**, 18 (1965)
81. L. Mandelkern, W. R. Krigbaum, H. A. Scheraga and P. J. Flory, *J. Chem. Phys.*, **20**, 1392 (1952)
82. H. A. Scheraga and L. Mandelkern, *J. Am. Chem. Soc.*, **75**, 179 (1953)
83. M. D. Stern, *Biochemistry*, **5**, 2558 (1966)

84. H. Sund, in *Mechanismen enzymatischer Reaktionen, 14. Coll. Ges. physiol. Chemie*, Springer Verlag, Berlin–Göttingen–Heidelberg, 1964, p. 318
85. H. Sund and K. Weber, *Biochem. Z.*, **337**, 24 (1963)
86. K. E. Reinert, *Biopolymers*, **11**, 275 (1971)
87. K. E. Reinert, J. Strassburger and H. Triebel, *Biopolymers*, **11**, 285 (1971)
88. H. Weinert und U. Hagen, *Strahlentherapie*, **136**, 204 (1968)
89. A. H. Rosenberg and F. W. Studier, *Biopolymers*, **7**, 765 (1969)
90. F. W. Studier, *J. Mol. Biol.*, **11**, 373 (1965)
91. K. F. Elgert and K. Cammann, *Z. Anal. Chem.*, **226**, 193 (1967)
92. O. Kratky, H. Leopold and H. Stabinger, *Zschr. angew. Physik*, **27**, 273 (1969)
93. S. J. Edelstein and H. K. Schachman, *J. Biol. Chem.*, **242**, 306 (1967)
94. W. L. Gagen, *Biochemistry*, **5**, 2553 (1966)
95. H. G. Gray and J. E. Hearst, *J. Mol. Biol.*, **35**, 119 (1968)
96. K. E. Van Holde and R. L. Baldwin, *J. Phys. Chem.*, **62**, 734 (1958)
97. D. A. Yphantis, *Ann. N.Y. Acad. Sci.*, **88**, 586 (1960)
98. D. A. Yphantis, *Biochemistry*, **3**, 297 (1964)
99. K. E. Van Holde, *Fractions, 1967, No. 1*, Beckman Instruments, Palo Alto, Calif.
100. E. G. Richards and H. K. Schachman, *J. Phys. Chem.*, **63**, 1578 (1959)
101. E. F. Casassa and H. Eisenberg, *Adv. Protein Chem.*, **19**, 287 (1964)
102. E. C. Bancroft and D. Freifelder, *J. Mol. Biol.*, **54**, 537 (1970)
103. P. D. Jeffrey and M. J. Pont, *Biochemistry*, **8**, 4597 (1969)
104. D. C. Teller, T. A. Horbett, E. G. Richards and H. K. Schachman, *Ann. N.Y. Acad. Sci.*, **164**, 66 (1969)
105. S. P. Spragg and R. F. Goodman, *Ann. N.Y. Acad. Sci.*, **164**, 294 (1969)
106. E. F. Casassa, *Ann. N.Y. Acad. Sci.*, **164**, 13 (1969)
107. W. J. Archibald, *J. Phys. Colloid Chem.*, **51**, 1204 (1947)
108. H. G. Elias, *Angew. Chem.*, **73**, 209 (1961)
109. F. E. LaBar, *Biochemistry*, **5**, 2362 (1966)
110. V. H. Paetkau, *Biochemistry*, **6**, 2767 (1967)
111. R. Trautman, *J. Phys. Chem.*, **60**, 1211 (1956)
112. R. Trautman and C. F. Crampton, *J. Am. Chem. Soc.*, **81**, 4036 (1959)
113. J. B. Ifft, D. H. Voet and J. Vinograd, *J. Phys. Chem.*, **65**, 1138 (1961)
114. J. E. Hearst, J. B. Ifft and J. Vinograd, *Proc. Nat. Acad. Sci. U.S.*, **47**, 1015 (1961)
115. S. J. Dirkx, *Analytical Density Gradient Ultracentrifugation*, Beckman Instruments, München, 1964
116. R. A. Robinson and R. H. Stokes, *Electrolyte Solutions*, Academic Press, New York–London, 1955
117. M. Meselson, F. W. Stahl and J. Vinograd, *Proc. Nat. Acad. Sci. U.S.*, **43**, 581 (1957)
118. N. Sueoka, *Proc. Nat. Acad. Sci. U.S.*, **45**, 1480 (1959)
119. R. L. Baldwin, *Proc. Nat. Acad. Sci. U.S.*, **45**, 939 (1959)
120. J. E. Hearst and J. Vinograd, *Proc. Nat. Acad. Sci. U.S.*, **47**, 1005 (1961)
121. E. Daniel, *Biopolymers*, **7**, 359 (1969)
122. W. Bauer and J. Vinograd, *Ann. N.Y. Acad. Sci.*, **164**, 192 (1969)
123. H. Eisenberg, *Ann. N.Y. Acad. Sci.*, **164**, 25 (1969)
124. R. A. Gibbons, *Biopolymers*, **10**, 411 (1971)
125. K. E. Reinert, *Kolllid Z. and Z. Polymere*, **228**, 68 (1968)
126. W. D. Lansing and E. O. Kraemer, *J. Am. Chem. Soc.*, **57**, 1369 (1935)
127. H. Lütje, *Makromol. Chem.*, **72**, 210 (1964)
128. H. Fujita, *J. Chem. Phys.*, **32**, 1739 (1960)
129. T. G. Scholte, *Ann. N.Y. Acad. Sci.*, **164**, 156 (1969)

130. E. T. Adams, Jr. and J. W. Williams, *J. Am. Chem. Soc.*, **86**, 3454 (1964)
131. E. T. Adams, Jr., *Biochemistry*, **4**, 1646 (1965)
132. E. T. Adams, Jr. and D. L. Filmer, *Biochemistry*, **5**, 2971 (1966)
133. E. T. Adams, Jr., *Biochemistry*, **6**, 1865 (1967)
134. E. T. Adams, Jr., *Fractions 1967, No. 3*, Beckman Instruments, Palo Alto, Calif.
135. M. Derechin, *Biochemistry*, **7**, 3253 (1968)
136. M. Derechin, *Biochemistry*, **8**, 921 (1969)
137. M. Derechin, *Biochemistry*, **8**, 927 (1969)
138. E. T. Adams, Jr., *Ann. N.Y. Acad. Sci.*, **164**, 226 (1969)
139. W. B. Goad and J. R. Cann, *Ann. N.Y. Acad. Sci.*, **164**, 172 (1969)
140. D. E. Roark and D. A. Yphantis, *Ann. N.Y. Acad. Sci.*, **164**, 245 (1969)
141. K. E. Van Holde and G. P. Rossetti, *Biochemistry*, **6**, 2189 (1967)
142. K. E. Van Holde, G. P. Rossetti and R. D. Dyson, *Ann. N.Y. Acad. Sci.*, **164**, 279 (1969)
143. H. Sund, in *New Techniques in Amino Acid, Peptide and Protein Analysis* (Ed. A. Niederwieser and G. Pataki), Ann Arbor Science Publishers, Ann Arbor, Michigan, U.S.A., in the press
144. H. Sund, K. Weber and E. Mölbert, *European J. Biochem.*, **1**, 400 (1967)
145. G. Kegeles, S. Kaplan and J. L. Bethune, *Proc. Nat. Acad. Sci. U.S.*, **58**, 45 (1967)
146. G. Kegeles, S. Kaplan and L. Rhodes, *Ann. N.Y. Acad. Sci.*, **164**, 183 (1969)
147. R. Josephs and W. F. Harrington, *Proc. Nat. Acad. Sci. U.S.*, **58**, 1587 (1967)
148. E. G. Pickels, *J. Gen. Physiol.*, **26**, 341 (1943)
149. Beckman Instruments, *An Introduction to Density Gradient Centrifugation*, Beckman Instruments Inc. Spinco Division, Palo Alto, Calif., 1960
150. W. G. Flamm, H. E. Bond and H. E. Burr, *Biochim. Biophys. Acta*, **129**, 310 (1966)
151. M. K. Brakke, *Virology*, **6**, 96 (1958)
152. V. N. Schumaker, *Adv. Biol. Med. Phys.*, **11**, 246 (1967)
153. C. de Duve, J. Berthet and H. Beaufay, *Progr. Biophys.*, **9**, 325 (1959)
154. E. Burgi and A. D. Hershey, *Biophys. J.*, **3**, 309 (1963)
155. G. P. Van der Schans, J. B. Aten and J. Blok, *Analyt. Biochem.* **32**, 14 (1969)
156. N. G. Anderson and C. L. Burger, *Science*, **136**, 646 (1962)
157. N. G. Anderson, *Fractions 1965, No. 1*, Beckman Instruments, Palo Alto, Calif.
158. N. G. Anderson and G. B. Cline, in *Methods in Virology*, Vol. II (Ed. K. Maramorosch and H. Koprowski), Academic Press, New York–London, 1967, pp. 137–178
159. B. S. Bishop, in *The Development of Zonal Centrifuges and Ancillary Systems for Tissue Fractionation and Analysis* (Ed. N. G. Anderson), National Cancer Institute Monograph 21, Bethesda, Md., 1966, pp. 175–188
160. H. W. Hsu, *Biophysical J.*, **8**, 973 (1968)
161. H. B. Halsall and V. N. Schumaker, *Anal. Biochem.*, **30**, 368 (1969)
162. R. A. Weaver and W. Boyle, *Biochim. Biophys. Acta*, **173**, 377 (1969)
163. E. P. Larkin and R. M. Dutcher, *Appl. Microbiol.*, **20**, 64 (1970)
164. H. E. Bond and W. T. Hall, *J. Nat. Cancer Inst.*, **43**, 1073 (1969)
165. J. A. Elliott and W. L. Ryan, *Appl. Microbiol.*, **20**, 667 (1970)

ACKNOWLEDGMENTS

Thanks are due to Professor A. Müller-Broich and to Professor H. Sund for their stimulating discussions and criticisms. The author is indebted to Mrs. Ch. Heinold for carefully preparing the manuscript.

III
Separation of Biomolecules

CHAPTER 6

Chromatography of macromolecules of biological origin

R. L. Munier

Institut Pasteur,
Paris, France

In recent years much progress has been made in the chromatography of naturally occurring macromolecular substances, for example, enzymes,

hormones, proteins of biological media, peptides, nucleic acids, oligonucleotides, neutral or ionic polysaccharides, viruses, bacteriophages, etc.
The widespread and growing interest in these methods of separation may
be explained by the fact that they have increased to a considerable extent the
scope of the investigations in molecular biology.

Methods for the chromatographic separation of these substances are
based in most cases either upon differences in ionic charge (ion exchange
chromatography on ion exchange resins, celluloses, and cross-linked dextrans, or on methylated albumin kieselguhr), or upon differences in molecular weight (exclusion–diffusion chromatography on cross-linked dextran
gels, gelose, agarose, cross-linked polyacrylamide gels). Besides ion exchange
and molecular sieving phenomena, many molecular interactions may be
involved in these chromatographic separations, for example hydrogen
bonds, van der Waals forces, electrostatic interactions, polar interactions,
and ionic interactions due to polyvalent metallic ions. Sometimes separations
originate in the solubility of macromolecules (salting-out chromatography,
salting-out adsorption chromatography, salting-in chromatography).

Our purpose is to review advances both in the establishment of basic
principles and in the more important applications of specific chromatography, exclusion–diffusion chromatography (gel filtration), salting-out
chromatography, adsorption chromatography and ion exchange chromatography for the separation of macromolecular substances from each other
and from substances of low molecular weight.

I. SPECIFIC CHROMATOGRAPHY

Specific chromatography is the adsorption chromatography in which a
substance is initially linked through characteristic non-covalent bonds to a
specific adsorbent. This specificity may be as stringent as that which governs
the formation of complexes between an enzyme and its substrates or its
inhibitors. The elution of a substance from its specific adsorbent is sometimes
also specific and, in this case, related to properties of the complex formed
between specific adsorbent and substance.

Specific chromatography is mainly applied to the separation of enzymes,
antibodies, antigens, nucleic acids and oligonucleotides; since the separation
of each of these groups poses its particular problems we shall review each
chromatographic method individually.

A. Enzymes

The ability of some powdered materials to retain the very enzymes which
can partly hydrolyse them has been known for many years; for example,
starch[1,2] was often used by enzymologists to specifically retain amylolytic

enzymes (α-amylases).[3–7] However, these workers did not carry out a true chromatographic separation; they separated starch grains from slurry and by means of alcoholic or aqueous salt solutions recovered the free α-amylase.

In 1953, Thayer[8] fixed extracellular α-amylase from *Pseudomonas saccharophilla* on a column packed with a (1:1) mixture of potato starch (the 'particulate' substrate) and infusory earth (Celite). After washing the column, the elution of the enzyme was initiated by 0·5–1 per cent starch solution ('liquid' substrate) or preferably by enzymatically partly-hydrolysed soluble starch (dextrins). As a result, 62 per cent of the original α-amylolytic activity was recovered and a 100-fold increase in concentration obtained. Elution specificity was demonstrated, since the enzyme could be eluted by substrate solutions (for example soluble starch, partly hydrolysed soluble starch and dextrins), and could not be eluted by a substance of the same chemical nature, for example maltose, which is not a product of enzymatic reaction of α-amylase and amylose. It was the first specific chromatography.

Other specific chromatographies of enzymes were made by Hash and King[9] and by Myers and Northcoth.[10] Cellulolytic enzymes of *Myrothecium verrucaria* and *Helix pomatia* (Figure 1) are adsorbed onto a cellulose

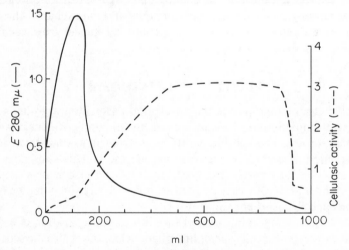

Figure 1. Separation of proteins (–) and cellulases (– – –) of *Helix pomatia* on cellulose powder column (adapted from Myers and Northcoth[10])

column; most of the proteins of the extract are not retained (at pH 6·0 for *H. pomatia* proteins, at pH 3·0 for *M. verrucaria* proteins) but cellulases are firmly retained. Their elution requires large quantities of eluent (at pH 6·0 for *H. pomatia*, at pH 4–8·5 for *M. verrucaria*). In *M. verrucaria* cellulases, the presence of several cellulases (specifically hydrolysing high molecular weight or mean molecular weight carboxymethylcelluloses) and of two β-glucosidases was demonstrated; one of these glucosidases is retained on the column and the other is not effectively retained. The spreading of the cellulase zone during column elution is related to the fact that the authors did not use a specific eluent. It would have been of interest to investigate the eluting power of solutions of several carboxymethylcelluloses with substances of different molecular weight. Cellulases were displaced by a diluted salt solution at a pH close to the pH at which the enzymatic activity of these cellulases was measured (pH 5·6). The very slow elutions were only made possible by the low affinity of cellulases for cellulose (see curves of cellulasic activity against pH by Myers and Northcoth).[10]

Specific chromatography of an enzyme is only possible when the active site of this enzyme (E) can interact with a substrate (S), which has been fixed on a specific adsorbent, in order to form enzyme–substrate complexes (ES)

$$E + S \rightleftharpoons ES \rightarrow P$$

So, this chromatographic method is usable for a limited number of enzymes only.

It is obvious that for cellulases or α-amylases, which can hydrolyse 'particulate' substances such as cellulose or amylose, materials such as cellulose fibres or starch granules may be used as specifically adsorbing supports. The elution of an enzyme fixed on a 'particulate' substrate may be theoretically performed either by means of a solution of its natural substrate (soluble substrate) or preferably by a solution of a competitive inhibitor of the enzymatic reaction, or by means of a salt solution at a pH very different from pH range in which the enzyme is active.

Another example of specific chromatography is the procedure for the isolation of rabbit muscle crystalline phosphorylase a by substrate adsorption (starch granules) and substrate elution[11] (glycogen solution). This enzyme catalyses the reaction:

(n) glycogen + H_3PO_4 → (n − 1) glycogen + glucose-6-phosphate

G. de la Haba[11] adsorbed the enzyme from crude extracts on a starch column (3 °C); other proteins were removed by washing and phosphorylase was eluted (14 °C) by glycogen solution. The pure enzyme is isolated from the phosphorylase–glycogen complex after the complete hydrolysis of glycogen. This process results in a 33-fold purification with an overall recovery of 100 per cent, and an enzyme preparation of specific activity at least as high as that of the twice-crystallized enzyme.

Another type of specific enzyme adsorbent may be prepared by fixing a substrate on inert granules by means of chemical covalent bonds, or preferably by fixing a competitive inhibitor of the reaction catalysed by the enzyme to be adsorbed. In the latter case, the specific bonds created between

Table 1. Adsorption specificity of tyrosinase from *Psalliota Campestris* by several modified celluloses (see Lerman[12])

Table 1 *(cont.)*

Modified celluloses non-adsorbing tyrosinase	Non-adsorbents

cellulose — CH_2 — O — ⟨benzene⟩(OH) — N=N — ⟨benzene⟩ — COOH S_5

cellulose — CH_2 — O — CH_2 — ⟨benzene⟩ — N=N — ⟨benzene⟩(CH_3)(HO) S_6

cellulose — CH_2 — O — CH_2 — ⟨benzene⟩ — N=N — ⟨benzene⟩(OH) — N=N — ⟨benzene⟩ — AsO_3H_2 S_7

enzyme and specific adsorbent are of the same type as those (EI) which exist between competitive inhibitor (I) and enzyme (E)

$$E + I \rightleftharpoons EI$$

As some p-azophenol derivatives are competitive inhibitors of the reaction catalysed by tyrosinase (derived from *Psalliota campestris*)

$$\text{Phenols} \xrightarrow{\text{Tyrosinase}} O\text{-diphenols}$$

Lerman[12] prepared celluloses bearing p-azophenol groups and used them as specific adsorbents. He could specifically adsorb tyrosinase on columns of different 'p-azophenol celluloses' (adsorbents $S_1 S_2 S_3$ in Table 1) and displace them later. The specificity of tyrosinase adsorption was demonstrated (see Table 1). Lerman, after adsorbing tyrosinase on 'p-azophenol cellulose' at pH 5·8 or 6·0 (malonate or phthalate buffer), washed the column by the same solution in order to remove inactive proteins, and then displaced tyrosinase at pH 8·3 (0·02M pyrophosphate) or 9·6 (0·1M glycine–NaOH). Under these conditions a 60-fold purification with an overall recovery of 56 per cent was obtained. It may be seen that the enzyme is specifically adsorbed in a pH range (5·8–6) at which it is active in catalysing transformations (5·5–7) but it is eluted in a very different pH range, at which it is inactive and can hardly generate complexes with its substrates or competitive inhibitors.

In the same way Arsenis and McCormick selectively purified rat liver flavokinase on a 'flavin cellulose' column (7-cellulose acetamido-6,9-dimethylisoalloxazine), and the apoenzyme from spinach glycolate oxidase on flavine mononucleotide cellulose.[13]

B. Antibodies and Antigens

Lerman[12] and Campbell and Luescher[14] have shown the possibility of applying the specificity of antigen–antibody precipitation[14] or hapten–antibodies reaction[12a] to the chromatographic separation of free antibodies; after specific adsorption of these antibodies on an inert material bearing an antigen[14] ('particulate' antigen) or a hapten[12a] ('particulate' hapten) and the removal of non-specific proteins by washing with salt solution, antibodies were displaced by a specific[12a] or a non-specific[14] eluent. Several workers have developed similar processes.[16–28] With these methods, the greatest difficulty encountered was the elution of antibodies retained on column, since antigen–antibody precipitates were only just separable into their two constituents[20,21,25,28,29,31,32,33] without denaturation. It seems that in a few cases[12,14,15,28] antibodies may have been recovered without too great a loss in biological activity.

1. Chromatography on a 'Particulate' Antigen Column

Isliker,[16] Campbell, Luescher and Lerman,[14] Malley and Campbell,[15] Grubhofer and Schleith,[18] Saha,[24] Mougdal and Porter,[28] Sela and Kaltchalski,[17] and Behrens, Inman and Vannier[69] have prepared powders bearing either different antigenic proteins (serum-albumin, ovalbumin, etc.), enzymes (amylase, ribonuclease, carboxypeptidase, chymotrypsin, etc.) or viruses, erythrocyte stroma, etc. which may be used in chromatographic columns and can specifically adsorb corresponding antibodies. These 'particulate' antigens were obtained by fixing proteins through either their histidine or tyrosine residues on diazotized *p*-aminobenzylcellulose,[14,15,21,28,30] on diazotized aminopolystyrene,[18,19,20,22,25,26] on diazotized *p*-aminophenylalanine and leucine copolymers,[14,17] or through amino groups of their lysine residues on bromoacetylcellulose[27] or on carboxylic acid resin chloride.[16]

When all the antigen contained in the granules is fixed by covalent bonds to the chromatographic support, and when the elution of the corresponding adsorbed antibody is performed without any strong decrease in biological activity, valuable results may be obtained.[12,14,17,28]

2. Chromatography on a 'Particulate' Hapten Column

In this field a valuable study has been carried out by Lerman,[12a] who prepared a cellulose (*m*-hydroxy–phenoxycellulose coupled with a *p*-amino-benzene-*p*-arsonic acid diazonium salt) bearing *p*-azobenzene–arsonic chemical groups identical to those fixed on a protein (by coupling histidine and tyrosine groups with diazotized *p*-aminobenzene arsanilate) used as an antigen. Lerman showed that a specific cellulosic adsorbent retains antibodies contained in the serum of the organism which has received the antigenic protein. After removing inert proteins by salt solution (at pH 8·0), antibody proteins may be eluted by a simple hapten solution (sodium arsanilate) of continuously increasing concentration. Elution is really specific since sulphanilate does not remove the adsorbed antibody (see Table 2, Figure 2). As shown by the elution diagram (Figure 2) the antibody is in fact a heterogeneous mixture of antibody molecules; parts A and B of the elution peak, after new chromatography, have the mean mobilities of the A and B fractions respectively.

3. Chromatography on a 'Particulate' Antibody Column

It is possible to eliminate either one or several particular antigens from a mixture of antigens by chromatography on a column in which the granules bear, through covalent bonds, γ-globulins of a serum corresponding to the antigen(s) which are to be specifically retained.[19,22,27]

Table 2. Specificity of adsorption and elution of antibodies homologue to *p*-azoarsonic acid protein on cellulose bearing *p*-azoarsonic acid groups (see Lerman[12a])

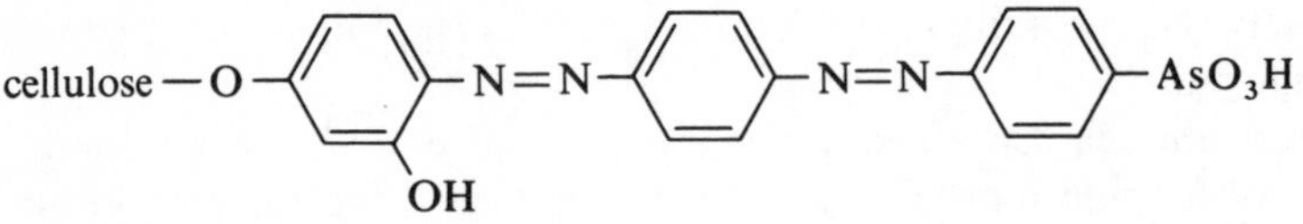	
Antigen	Chromatographied antibodies

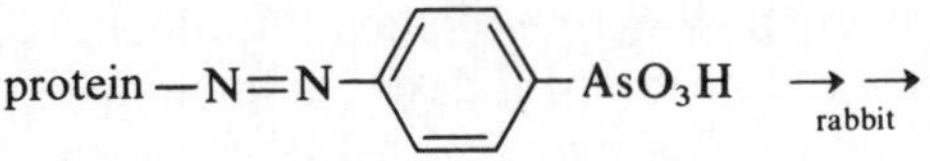

protein —N=N—⟨benzene⟩—AsO$_3$H →rabbit→ Corresponding antibodies

p-azoarsonic
acid-group

Specific adsorbent

cellulose —O—⟨benzene⟩(OH)—N=N—⟨benzene⟩—N=N—⟨benzene⟩—AsO$_3$H

Eluent

H$_2$N—⟨benzene⟩—AsO$_3$Na   Sodium arsanilate

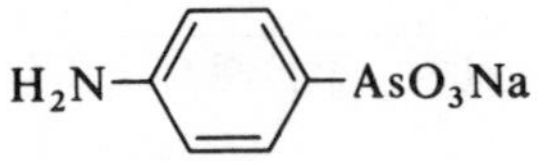

Non-eluent

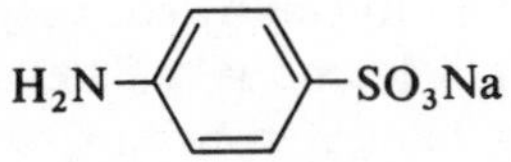

H$_2$N—⟨benzene⟩—SO$_3$Na   Sodium sulphanilate

C. Nucleic Acids and Oligopolynucleotides

For the fractionation of mixtures of oligopolynucleotides or nucleic acids, Gilham,[34] Adler and Rich[35] and Bautz and Hall[36,37] have developed a method using columns packed with granules able to selectively absorb polynucleotides with complementary sequences. They obtained the fixation of polynucleotides or nucleic acids on cellulose through covalent bonds (see Tables 3, 4 and 5). These methods apply the Khorana reaction[39–42] to dicyclohexylcarbodiimide, and may be classified into two groups: in Gilham's method, the phosphate group which binds cellulose to nucleotide is originally present in the terminal ribose (position 5) of the nucleotide (see Table 3) whereas in other methods it is originally present in a glucose residue of a phosphate–cellulose (see Tables 4 and 5).

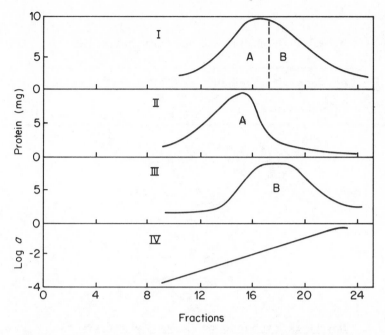

Figure 2. Specific chromatography, on *p*-azobenzene-*p*-arsonic-cellulose of antibodies contained in an antiserum against *p*-azobenzene arsanilate albumin:

I—First chromatography profile

II and III—Diagrams of rechromatography of fractions A and B from first AB peak

IV—Elution of specific antibodies by a sodium arsanilate solution of continuously increasing concentration (a); adapted from Lerman[12a]

These celluloses bearing either polynucleotides (polyadenylate (Adler and Rich[35]) or polythymidylate (Gilham[34]) or polydesoxycytidylate or polydesoxyadenylate (Gilham[38]) or T_4-bacteriophage DNA (Bautz and Hall[36]) may be used as chromatographic adsorbents selectively retaining polynucleotides in which the base sequences are complementary with those of the polynucleotide bound to cellulose. It is well-known that polynucleotides form a strand in which base-bearing ribose (or deoxyribose, or glucose in a few cases) and phosphate residues alternate, and that two strands may associate to form a helix by hydrogen bonding between complementary bases ('adenine–thymine (or uracil)', 'guanine (or inosine)–cytosine') according to Watson and Crick's proposal.[43] These hydrogen bonds between the strands may be split by an increase in the temperature or by a decrease in the ionic strength of the medium. The opposite changes result in re-formation of double strands. These properties may be used to adsorb and elute

Table 3. Polythymidylate cellulose preparation (see Gilham[34])

Table 4. Polyadenylate acetylcellulose preparation (see Adler and Rich[35])

polynucleotides from columns packed with 'polynucleotide-bearing cellulose'. For instance, Adler and Rich[35] compared the 'melting' curve (260 mμ optical density against temperature) of double helices of polyuridylate–polyadenylate in solution with the curve of the polyuridylate elution rate from a 'polyadenylate cellulose' column at different temperatures (Figure 3); the analogy is striking. Figure 4 shows the 'thermal' elution for deoxyribo-polyadenylate (d(pA)$_6$) adsorbed on 'thymidylate cellulose' (Gilham[34]). Polynucleotides are applied at the top of the column of specific adsorbent as a concentrated salt solution (i.e. buffer pH 7, containing 1M NaCl) at 4 °C; the polythymidylate not retained by hydrogen bonds is removed (at very low flow rate); the deoxypolyadenylate selectively retained is later

Table 5. Preparation of acetylcellulose bearing T_4-phage DNA (see Bautz and Hall[36,37])

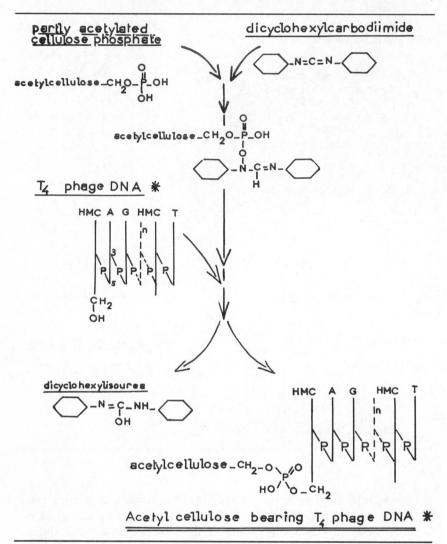

* Scheme[41]; abbreviations: A (adenine), G (guanine), HMC (hydroxymethylcytosine), T (thymine)

displaced by increasing the temperature (24 to 35 °C). Bautz and Hall[36,37] and Gilham[34,38,44] eluted adsorbed polynucleotides by increasing the temperature; Edmonds and Abrams[45] eluted polynucleotides by decreasing the ionic strength of the eluent (0·01M phosphate buffer, pH 6·7, containing NaCl (0·1M), is replaced by pure phosphate buffer). In all cases adsorption is

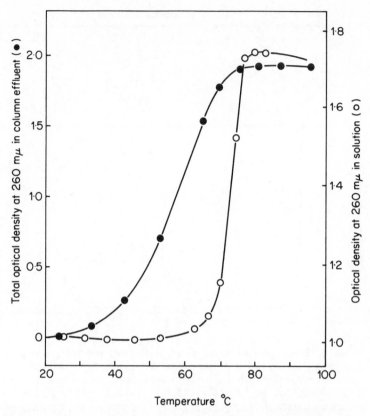

Figure 3. Thermal denaturation of double helices polyuridylate–polyadenylate in solution (260 mμ optical density *vs.* temperature (◯)) compared to elution rate at different temperatures of polyuridylate from polyadenylate cellulose column (●); reprinted from Adler and Rich,[35] *J. Am. Chem. Soc.*, **84**, 3977 (1962). Copyright 1962 by the American Chemical Society. Reprinted by permission of the copyright owner

carried out at low temperature (i.e. 4 °C) and in a medium of high ionic strength (0·1–1·0м NaCl). Gilham and Robinson[38,44] showed that a similar process using a thermally controlled column, with either continuous or stepwise process using a thermally controlled column, with either continuous or stepwise temperature variations, may be used for the fractionation of oligonucleotide mixtures found in enzymatic hydrolysates of nucleic acids. These processes are highly specific.

In the same way Bautz and Hall[36,37] and Bolton and McCarthy[46–48] have developed the fractionation of nucleic acids with base sequences complementary to DNA, i.e. the fractionation of messenger RNA. Bolton, McCarthy and Cowie[46–48] used agar granules to retain single-stranded

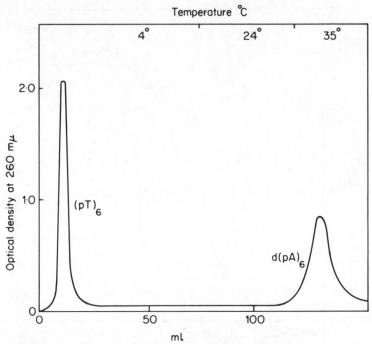

Figure 4. Specific chromatography of polythymidylate $(pT)_6$ and poly-deoxyadenylate $d(pA)_6$ on 'polyadenylate cellulose' column; adsorption at low temperature (4 °C); elution by stepwise increasing temperature; adapted from Gilham,[34] *J. Am. Chem. Soc.*, **84**, 1311 (1962). Copyright 1962 by the American Chemical Society. Reprinted by permission of the copyright owner

DNA, absorbed nucleic acid on agar in a separated flask (hybridization) and then applied the agar suspension at the top of the column. Bautz and Hall used cellulose powder, which covalently retained single-stranded DNA, and adsorbed nucleic acids directly onto the column (see Figure 5). As might be expected, the adsorption and elution of nucleic acids on nucleic acid-bearing cellulose requires higher temperatures than the adsorption and elution of oligonucleotides (see references 36, 37, 46, 47, 48).

Bautz and Hall[36] have obtained a preparation of total RNA, synthesized *de novo* by *E. coli* (the host bacteria) after infection by bacteriophage, and, by using cellulose bearing T_4-phage DNA, they enriched this preparation in messenger RNA corresponding to the phage DNA (see Figure 5). In the same way, Bolton and MacCarthy[48] showed, by means of chromatography on an agar column retaining λ-phage DNA, that in *E. coli* infected by bacteriophage, an RNA was present that was able to hybridize with the infecting phage DNA. So, from these experiments it can be seen that it is possible to enrich a sample in a specific RNA.

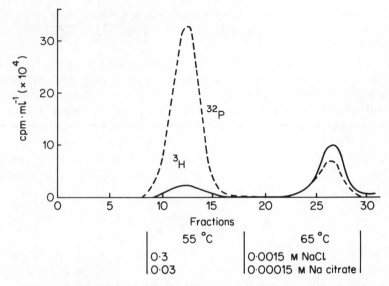

Figure 5. Chromatography of RNA from *E. coli* infected by bacteriophage T_4 on cellulose bearing T_4 phage DNA column. *E. coli* RNA was uniformly labelled with ^{32}P orthophosphate before bacteriophage infection (– – – –) or specifically labelled with ^3H- uracil after bacteriophage infection (——). First peak represents non-retained RNA. Second peak represents *E. coli* RNA enriched in RNA having base sequence complementary to T_4-phage DNA; adapted from Bautz and Hall[36]

D. Particular Cases of Specific Chromatography

1. Chromatography on Macromolecular Substrates

Some substances specifically adsorbed on macromolecular supports can be eluted under suitable conditions, for example animal viruses adsorbed on erythrocytes,[49–54] bacteriophages on bacteria,[55] lytic substances on bacterial bodies,[56] glycogen synthetase on 'particulate' glycogen,[57] synthetase starch on starch granules,[58] etc. can all be treated in this way. It is sometimes possible to pack a chromatographic column with macromolecular substrates; Isliker[59] used an ion exchanger bearing the stroma of red cells in order to specifically fix isoagglutinins.

2. Chromatography in the Presence of Divalent Metallic Ions

Crampton and coworkers[60,61] have developed a nucleic acid chromatography based on the fact that magnesium may link the phosphate groups of nucleic acids and the carboxylic groups of an ion exchanger. On a column of carboxylic acid resin (Mg^{2+} form) Amberlite IRC 50 (polymethacrylic acid cross-linked with divinylbenzene), with magnesium acetate solution in continuously decreasing concentration, they separated *E. coli* DNA from T$_2$-phage DNA.

On a column of magnesium polymethacrylate, using a solution of constant magnesium concentration and continuously decreasing sodium acetate concentration, Mindich, Hotchkiss and Hall[62] have separated native DNA from thermally denatured DNA and *Micrococcus lysodeikticus* DNA from *E. coli* (or pneumococcus) DNA. Fractionation seems to be related to the number of phosphate groups in the molecules,[62] but the base composition of nucleic acids also plays an important role; higher amounts of guanine and cytosine in DNA lead to a lower affinity for magnesium polymethacrylate and to easier elution.[62,63]

3. Specific Elution

The elution of some substances is made possible by the new properties they get when complexing with another substance contained in the washing liquid; non-complexed substances remain fixed on the column.

A good example is the specific elution, as a haemoglobin–haptoglobin complex, of the haptoglobin previously adsorbed on a diethylaminoethylcellulose column. By progressively increasing the ionic strength of the medium and by using a diethylaminoethylcellulose column, proteins are displaced in order of their decreasing isoelectric pH.[64] Thus Jayle, Moretti and Dobryozcka,[65] taking into account the isoelectric pH differences exhibited by haptoglobin (pH$_i$ = 4·15) and the haptoglobin–haemoglobin complex (pH$_i$ = 4·85), have separated haptoglobin from all the other proteins (i.e. α_1, α_2-globulins) retained by DEAE-cellulose; they added to the eluent (at suitable salt concentration) a suitable quantity of haemoglobin, and so induced the formation of the haptoglobin–haemoglobin complex in the column, which, because of its higher isoelectric pH, was eluted while the other proteins (pH$_i$ ≤ 4·15) remained adsorbed.

Some enzymes adsorbed on ion-exchange celluloses may be specifically eluted by solutions of their substrates. It is probable that this elution is caused by variation in the isoelectric pH when an enzyme becomes complexed with a substrate. Examples are the elution of aldolases[66,67] and of rabbit liver diphosphatase[66] by fructose diphosphate from carboxymethylcellulose columns, and the elution of malic dehydrogenase of *Ascaris suum*[68] by a combination of malate and diphosphopyridine nucleotide from diethylaminoethylcellulose column.

In the case of the elution of rabbit liver aldolase by 2·5 mM sodium fructose diphosphate,[66] a 30-fold purification with an overall recovery of 90 per cent was obtained. The specificity of the elution of rabbit liver fructose diphosphatase by fructose diphosphate was shown by the fact that hexose monophosphates, at a 10-fold higher concentration, only partially eluted the enzyme and gave enzymatic preparations with much lower specific activities. As high as a 300-fold purification has been obtained by adsorption on carboxymethylcellulose and elution by sodium fructose diphosphate.

II. ADSORPTION CHROMATOGRAPHY

A. Proteins

Numerous materials can adsorb proteins.[70–76] This more or less specific adsorption is used to fractionate natural mixtures of proteins, i.e. adsorption of ribonuclease[77,78] or egg-white lysozyme[79] on bentonite.

Two adsorbents are commonly used for enzyme purification: C_γ-alumina gel (for the preparation, see references 72, 80, 81) and calcium phosphate gel (for the preparation, see references 72, 81, 82). In general, proteins are adsorbed on these gels in a diluted, slightly acidic salted medium (pH 5–6), then eluted in a concentrated salted medium (for example phosphate) made slightly alkaline (pH 7–8);[72] all adsorbed proteins are wholly eluted by phosphate buffer containing 10 per cent ammonium sulphate.[83] However calcium phosphate, C_γ-alumina gels and bentonite cannot be used as packing materials for chromatographic columns.

For the chromatography of proteins, columns may be packed with a mixture of kieselguhr (Supercel, Hyflosupercel) and calcium phosphate,[84,85] with either the calcium phosphate precipitated on cellulose powder,[84,86–88] or the alumina precipitated on kieselguhr,[89] but the best adsorbents are two microcrystalline calcium phosphates: brushite, $CaHPO_4$, $2H_2O$[90] and hydroxylapatite, $Ca_5(PO_4)_3OH$ (for the preparation, see references 91, 92, 93; for modified preparations see references 94, 95, 96). The separation of all proteins may be obtained under almost identical conditions, namely adsorption at low ionic strength (i.e. 0·005–0·01M phosphate buffer, pH 6·8) and elution by increasing salt concentration (i.e. 0·005M to 0·25M, or more, phosphate). Brushite is not very stable at a pH > 7 and is progressively changed into hydroxylapatite.[90] Hydroxylapatite is stable in the pH range 5·5–10[96] and is the most suitable absorbent for protein chromatography.[91,97] In spite of its low adsorption capacity (1 mg protein per gram of wet adsorbent) valuable separations have been obtained, such as the A 1, A 2 and A 3 components in ovalbumin[90,91] or the monomer and dimer in serumalbumin.[91] Tiselius, Hjerten and Levin[91] investigated the chromatographic behaviour of various proteins on a hydroxylapatite column, using both gradient elution and stepwise elution. Most of the proteins gave multiple zones,[91] but there is no proof that in every case each zone corresponded to one component. Hjerten gives an excellent example of the application of chromatography on hydroxylapatite to the separation of proteins of whole human serum.[94] Of special interest is the fact that the order of emergence of the components shows no correlation with the order observed in ion-exchange chromatography or electrophoresis.

The fractionation of mixtures of enzymes has often been performed on hydroxylapatite columns; the chromatographic purification of levane sucrase from *Bacillus subtilis*[98] and of nitrate reductase from wheat leaves[99] are good examples. In a similar way, a variety of hormones have been purified; for example, pituitary follicle stimulating hormone,[100] pituitary luteinizing hormone[101] and prolactin.[102] The purifications of the enzymes[70, 103–109] was made on alumina[103,105] or bauxite columns.

B. Nucleic Acids

Nucleic acids (animal DNA,[110,111] viral RNA,[112,113] animal RNA[114] and transfer RNA[115–118]) have been investigated by chromatography on hydroxylapatite,*,[110–112,114,119] on brushite[120] and on hydromagnesite,†[113] $4 \, Mg \, CO_3 \cdot Mg(OH)_2 \cdot 4 \, H_2O$. In every case, elution profiles must be interpreted with the utmost care.

C. Viruses

Adsorption chromatography has sometimes been used to purify virus preparations by means of calcium phosphate[90,122–127] or aluminium phosphate precipitated on silica gel.[115,128–131]

III. EXCLUSION–DIFFUSION CHROMATOGRAPHY

A new method for the analysis of mixtures of hydrophilic macromolecular substances, sometimes called 'gel filtration',[132,133] has arisen from observations made by Deuel and Neukon,[134] Lindqvist and Stogards[135] and Lathe and Ruthven.[136] These workers observed molecular sieve effects when solutions of salts, peptides or polysaccharides flow through a column packed with uncharged granular materials swollen in water (granules of galactomannan cross-linked with epichlorhydrin; starch granules swollen at room temperature or at higher temperature). Deuel and Neukon[134] separated polysaccharides from salts on cross-linked locust bean gum granules. Lindqvist and Storgards[135] and Lathe and Ruthven[136] have shown that peptides with molecular weights ranging from 100 to 1,000 can be separated on starch swollen in water at room temperature, and that peptides with molecular weight ranging from 1,300 to 150,000 can be separated on starch swollen in hot water.

---

* Preparation after Tiselius, Hjerten and Levin[91] or special preparation[121]
† Mallinckrodt Chemical Works, Saint-Louis, U.S.A.

This new fractionation method has been applied to the separation of proteins, nucleic acids, polysaccharides and viruses, the exact technique depending on the molecular volumes involved. The method is in common use, since hydrophilic granules are readily available; amongst those that are used are: cross-linked dextran gel beads (Flodin and Ingelman,[137] Flodin and Porath,[138,139] Flodin and Killander,[140] Pedersen[141]), agar gel granules or beads (Polsen,[142] Andrews,[143] Bengtsson and coworkers[144]), agarose granules or beads (Hjerten[145,146]), cross-linked polyacrylamide granules (Hjerten[147]).

In this process, the chromatographic mobility of the substances moving along the column, and consequently the separation of these substances, is related to their more-or-less easily reversible diffusion between an aqueous phase inside the gel granules and an aqueous phase outside the gel granules. This type of fractionation simultaneously involves diffusion, permeation and molecular sieving phenomena. After being swollen in water or salt solutions, gel grains are used for packing the column; applied substances diffuse more or less freely into the aqueous phases inside and outside the gel grains, according to their molecular weight. If the distribution coefficients, $K(K = C_s/C_m$ with C_s = concentration in stationary phase and C_m = concentration in mobile phase) are sufficiently different, in agreement with general chromatographic equation (Wilson, de Vault*,[148-151])

$$dl = \frac{dV}{A_m + KA_s} \qquad (1)$$

a chromatographic separation occurs. Substances with higher molecular weight have higher mobilities; a gel grain column behaves as a molecular sieve.

This new method is most valuable when trying to separate naturally-occurring molecular substances from each other (proteins,[140,143,152]) nucleic acids,[153] polysaccharides,[154] and viruses[144]) or when separating low molecular weight substances (inorganic salts, monosaccharides, amino acids, excess of reagents, etc.) from macromolecular substances (polysaccharides, proteins, nucleic acids, viruses).[133,156,157]

With regard to low molecular weight substances, gel grain columns are sometimes endowed with secondary chromatographic properties,[156-158] in addition to their molecular sieving ability. These are: adsorption, salting-out adsorption, slight exclusion and ion exchange on ionizable substances adsorbed on the gel matrix. The higher the gel cross-linking and the lower the molecular weight of the substances, the stronger are the chromatographic secondary effects.

---

* dl is the infinitesimal displacement of an infinitely narrow band of substance moving as a result of the flow of eluate of volume dV; A_s and A_m are the cross-sectional area occupied by stationary and mobile phases respectively

Exclusion–diffusion chromatography may be performed on columns or on thin layers.[159–164]

A. The Determination of Column Characteristics and Chromatographic Mobilities

The determination of column characteristics (i.e. the volume of the stationary phase V_s and the volume of the mobile phase V_m) and of the mobilities of chemical substances (elution volumes V_R, R_f values and distribution coefficients, K) are essential if exclusion–diffusion chromatography is to produce valuable results.

In this process elution the profile of each substance is symmetrical (linear sorption isotherm); as a result, the distribution coefficient of each substance is independent of the substance concentration and equation (1) becomes

$$l = \frac{V}{A_m + KA_s} \tag{2}$$

in which l = the displacement along the column of a narrow band of substance by the flow of the eluate of volume V.

The chromatographic mobility of each substance may be expressed either by the elution volume*

$$V_R = V_m + KV_s \tag{3}$$

or by the distribution coefficient[165]

$$K = \frac{C_s}{C_m} = \frac{V_R - V_m}{V_s} \tag{4}$$

or by R_f value

$$R_f = \frac{V_m}{V_R} \tag{5}$$

R_f values are very useful to characterize the mobilities of substances, since the values always range from 0 to 1. In exclusion–diffusion chromatography, the use of K values is equally convenient and they are independent of the size of the column.

---

* The volume of eluate necessary to displace a narrow band of substance from the top to the bottom of the column is the elution volume of this substance. In this case, $l = L$ = the length of the column; $V = V_R$ = the elution volume of the substance, and equation (2) becomes equation (3) with: $V_m = L \cdot A_m$ = the void volume of the column = the volume outside the gel grains, and $V_s = L \cdot A_s$ = the volume inside the gel grains.

B. Chromatographic Behaviour of Diverse Substances

1. Possible Values for Distribution Coefficients of Substances

The K value of a substance depends essentially on its properties.

(a) If its molecular volume is sufficiently large, the substance cannot penetrate into the gel grains (exclusion); in this case $C_s = 0$, $C_m \neq 0$, $K = 0$ and $V_R = V_m$; the elution volume of the excluded substance is equal to the void volume of the column.

(b) If a substance is low in molecular weight and not adsorbed by the gel matrix, it undergoes *free diffusion* between intragranular and intergranular phases; in this case $C_s = C_m$, $K = 1$ and $V_R = V_m + V_s$. Tritiated water and potassium chloride ($K = 0.98$) behave like this.

(c) If a substance is low in molecular weight and is adsorbed by the gel matrix (for example, aromatic derivatives, pyridine derivatives, purines and purine nucleosides[155,157,158]) or forms complexes with the gel matrix (for example borate, soda, potash[157]) it moves more slowly than a substance showing free diffusion; in this case, $C_s > C_m$, $K > 1$ and $V_R = V_m + KV_s$.

(d) If a substance is low in molecular weight and slightly excluded from the gel grains (for example, acids), its K value is slightly lower than 1.

(e) If a substance is of medium molecular volume and if the size of its molecules approximately equals the size of the pores and the cavities of

Table 6. K values, after gel filtration on a Sephadex G 25 column, for amino acids, neutral salts, buffer salts, acids, amines and aromatic substances[a] (see Gelotte[157])

Substance	Water	Substance	Water	0.05 M NaCl
NaOH	2.20	γ-collidine	>5	0.9
KOH	2.30	Picric acid	0.40	2.5
NH_4OH, $NaHCO_3$, Na_2CO_3	0.80	Phthalic acid	1.10	—
KCl, NaCl	0.98	Phenol	0.70	1.70
NaH_2PO_4, CH_3COONa	0.70	Alanine, leucine, methionine, oxyproline, proline, serine	0.8	
$Na_2B_4O_7$	1.70			
Sodium citrate	0.50	Phenylalanine	1.0	—
HCl	0.80	Tyrosine	1.1	—
HCOOH, CH_3COOH, citric acid	0.90	Tryptophan	1.9	—
Diethylmalonyl urea	1.20	Aspartic acid, glutamic acid	0.2	0.8
Glycine	0.90	Lysine, chlorhydrate	>3.30	1.00
Triethylamine	1.00	Arginine, chlorhydrate	>13.00	1.00
Pyridine	1.20	Histidine, chlorhydrate	>3.60	0.90

[a] In distilled water are observed slight exclusion ($K < 1$) of some salts (sodium citrate) and acids (HCl, phenol, picric acid, etc.) and adsorption ($K > 1$) of bases and aromatic substances

Table 7. K values, after gel filtration on a Sephadex G 25 column, for various nucleic acids bases, nucleosides, nucleotides, ribonucleic acid and phenol (see Gelotte[157])

Substance	Water	0·05M NaCl	Phosphate buffer pH 7·00 μ = 0·05	0·01M NH$_4$OH (pH 10·6)
Ribonucleic acid	0	—	—	—
Adenosine triphosphate	0	0·6	—	—
Adenylic acid	0·1	1·2	—	0·1
Guanylic acid	0·4	1·3	0·9	0·1
Cytidylic acid	0·1	0·8	0·7	0·1
Uridylic acid	0·1	0·8	0·7	0·1
Adenosine	1·7	1·8	—	1·8
Guanosine	1·6	—	1·8	—
Cytidine	1·2	—	1·2	—
Uridine	1·0	—	1·0	—
Inosine	1·2	1·3	—	0·3
Adenine	2·2	2·4	—	1·2
Cytosine	1·6	—	1·4	1·3
Uracil	1·1	—	1·2	—
Hypoxanthine	1·6	1·6	—	—
Xanthine	1·8	—	—	—
Phenol	0·7	1·7	—	—

the gel grains, the rate of its diffusion from the extragranular to the intragranular phase will decrease when the molecular volume increases; in this case $C_s < C_m$ and $K < 1$ or, more exactly: $0 \leqslant K < 1$.*

The K values of numerous substances in chromatography on cross-linked dextran gels are given in Tables 6, 7 and 8.

2. The Fields of Application of Column Chromatography on Cross-linked Gel Grains

Figure 6 shows various fields of application of column chromatography on cross-linked gel grains. In field I, two macromolecular substances can be separated if their molecular weights are sufficiently different. In field II, a macromolecular substance can be separated from a low molecular weight substance.

In current practice, highly cross-linked gels are used to separate large from small molecules, and slightly cross-linked gels to separate macromole-

---

* From this discussion it is easy to draw a method for determination of characteristics (V_s, V_m) of a column packed with gel grains

Table 8. The relations between K values, the molecular weight of proteins and the porosity of the gel granules used for column chromatography. (According to data given by Determann and Gelotte[166] and by Fritz, Trautschold and Werle[167])

Protein		K value on cross-linked dextran gels Sephadex			Molecular weight
	G 50	G 75	G 100	G 200	
Blue dextran[a,b]	—	—	0	0	2×10^6(average)
Glutamate-dehydrogenase	—	—	—	0	1×10^6
Thyroglobulin[b]	0	0	0	—	670,000
Fibrinogen	—	0	0	0	330,000
Phycoerythrin	—	0	0	0·10	290,000
Phycocyanin	—	0	0	0·20	—
γ-globulins (human)	—	—	—	0·26	150,000
Serumalbumin (dimer)	—	—	—	0·26	140,000
Aldolase (yeast)	—	—	—	0·27	140,000
Glyceraldehyde phosphate dehydrogenase	—	—	—	0·31	99,000
Transferrin	—	—	0·11	0·32	90,000
Lactate dehydrogenase	—	—	—	0·38	72,000
Serumalbumin (monomer)	0	—	0·16	0·43	70,000
Haemoglobin	—	0·08	0·30	0·52	68,000
Ovalbumin	—	0·20	—	—	45,000
Pepsin	—	0·25	—	0·59	35,000
Chymotrypsinogen A	0·12	0·41	—	—	25,000
Trypsin	—	—	0·54	0·71	23,800
Chymotrypsin	—	—	0·55	0·72	22,500
Ribonuclease	0·26	0·61	—	0·76	13,900
Cytochrome c	—	0·63	0·70	—	13,000
Trypsin inhibitor (Soya bean)	0·42	—	—	—	8,400
Trypsin inhibitor of Kunitz (pancreas)	0·49	—	—	—	6,500

[a] Pharmacia, Uppsala, Sweden
[b] Excluded substances for void volume determinations (V_m)

cular substances from each other by the molecular sieving process. In the latter case, it is necessary to have several grades of cross-linked gels available. As a result, each gel type allows the separation of substances in a specific range of molecular weights.

C. Gel Grains for Chromatography

The hydrophilic gels used for the chromatography of naturally-occurring macromolecular substances are of three types: they are either cross-linked water-soluble natural polysaccharides (dextrans), or natural polysaccharides

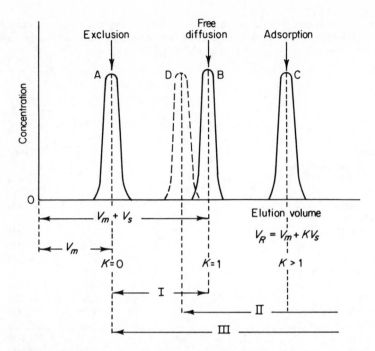

Figure 6. Relationship between elution volumes and K-values in chromatography on hydrophilic cross-linked gels columns:
(I): separation field of macromolecular substances according to molecular weights
(II): separation field of low molecular weight substances according to secondary chromatographic properties of gels (adsorption, 'salting-out' adsorption, 'salting-out' chromatography) adding to free diffusion
(III): separation field of high and low molecular weight substances; elution profile of a substance excluded from intragranular phase (A; $K = 0$), of a substance freely diffusing between phases (B; $K = 1$), of a substance adsorbed on gel matrix (C; $K > 1$), of a substance slightly excluded from gel grains (D; $K \neq 1$ but < 1)

soluble in hot water and forming a gel on cooling (agar, agarose), or synthetic cross-linked copolymers (polyacrylamides).

1. Cross-linked Dextran Gels

These gels are industrial products.* Dextrans of suitable molecular weight are cross-linked by epichlorhydrin[137] and the resulting structure is shown in Figure 7. Each grade is characterized by its degree of cross-linking, which is expressed by the water regain of the gel (i.e. the number of grams of water

* Pharmacia, Uppsala, Sweden

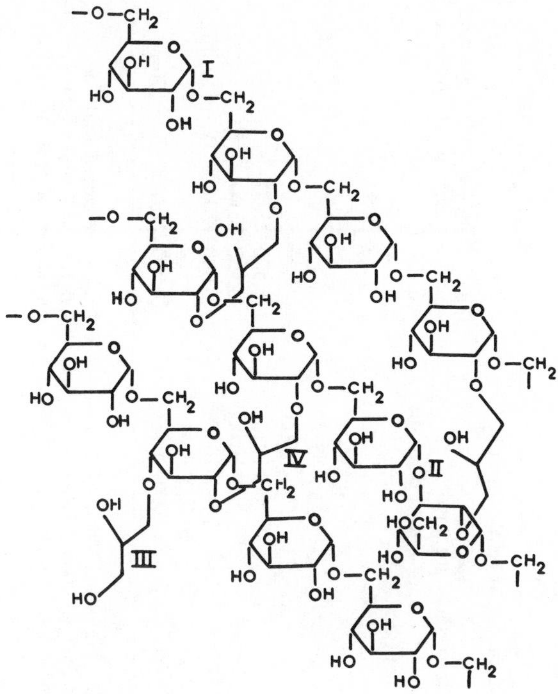

Figure 7. Structural scheme showing dextran gel cross-linked with epichlorhydrin; glucan containing about 90–95 per cent alpha-1,6-glucosidic bonds (I), 5–10 per cent 1,3-glucosidic bonds (II) and bearing some glycerol side chains (III) (side reaction), in which glucose-residues are connected through glycerol bridges (IV)

fixed by 1 gram of dry gel), that is by a number inversely proportional to degree of cross-linking. Eight grades (Sephadex G 10, G 15, G 25, G 50, G 75, G 100, G 150, G 200) allow the application of exclusion–diffusion chromatography to water soluble molecules with molecular weights up to 300,000. The characteristics of these gels are given in Table 9.

Using data from Andrews and Folley,[172,173] Andrews,[162] Whitaker,[152] Determan and Michel,[171] Leach and O'Shea,[170] Fritz, Trautschold and Werle,[167] Flodin and Porath[168] and Eaker[169] one can draw the curves of Figure 8, which shows fields of the fractionation of macromolecular substances according to their molecular weights. It is not surprising to note discrepancies between authors about the limits of the use of gels; these discrepancies are partly related to the form of the curves V_R against log M

Table 9. Cross-linked dextran gels in bead form

Trade Name Sephadex[a]	Water regain (ml H_2O per gram dry Sephadex)	Column volume (ml per gram dry Sephadex)	Approximative range of molecular sieving (molecular weight) with globular proteins and peptides	with dextrans	Range for molecular weight estimation	Reference authors
G 10	1·0 ± 0·1	2·5	50–700	50–700	—	—
G 15	1·5 ± 0·1	3	50–1,500	50–1,500	—	—
G 25	2·5 ± 0·2	5	1,000–20,000	500–5,000	—	—
G 50	5·0 ± 0·3	10	2,000–45,000	1,000–10,000	6,500–25,000	167, 171
G 75	7·5 ± 0·5	13	3,000–70,000	1,500–70,000	3,500–70,000	152, 162
G 100	10·0 ± 1·0	15	10,000–125,000	5,000–100,000	5,000–120,000	152, 162
G 150	15·0 ± 1·5	25	15,000–200,000	10,000–150,000	—	—
G 200	20·0 ± 2·0	35	25,000–300,000	15,000–200,000	20,000–225,000	170

[a] Pharmacia Fine Chemicals, Uppsala, Sweden; Sephadex G 10, G 15, G 75, G 100, G 150 and G 200 are available in one standard grade (40–120μ; 400–140 mesh); Sephadex G 25 and G 50 are available in three grades: coarse (100–300μ; 150–48 mesh), medium (50–150μ; 270–100 mesh); fine 20–80μ; bead preparations (super fine grade: 10–40μ; minus 400 mesh) of all Sephadex types are also available for thin layer chromatography

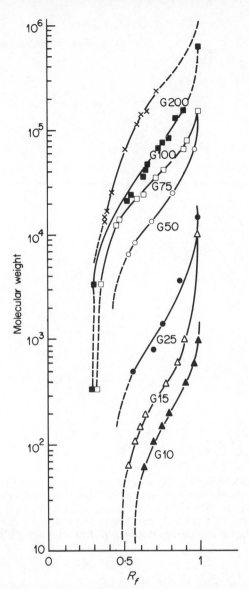

Figure 8. Relationship between molecular weights of substances and mobilities (R_f-values) on cross-linked dextran gels columns (Sephadex G 10, G 15, G 25, G 50, G 100, G 200); on Sephadex G 10, G 15: polyethylene-glycols in water;[168] on Sephadex G 25: polypeptides in 0·2 M acetic acid;[169] on Sephadex G 50, G 75, G 100, G 200: proteins in 0·02 M tris-hydrochloride buffer, pH 7·8, containing NaCl (0·1 M), or 0·05 M triethanol-amine-hydrochloride buffer, pH 7·0, containing NaCl (0·1 M),[167] or 0·1 M acetate buffer, pH 6, containing NaCl (0·1–0·4 M),[152] or in tris-hydrochloride buffer, pH 7·5, containing KCl (0·1 M),[162] or phosphate buffer, pH 7·2, containing NaCl (0·5 M),[171] or 0·2 M citrate buffer, pH 5·0[170] (according to data given by Leach and O'Shea,[170] Determann and Michel,[171] Fritz *et al.*,[167] Whitaker,[152] Andrews[162] for Sephadex G 200, G 100, G 75, G 50, by Eaker[169] for Sephadex G 25, by Flodin and Porath[168] for Sephadex G 15 and G 10

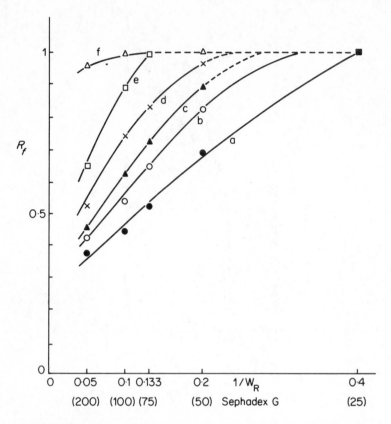

Figure 9. Relationship between chromatographic mobilities (R_f-values) of proteins of various molecular weights and degrees of crosslinking (1/water regain: $1/W_R$) in dextran gel (Sephadex); (a) ribonuclease (molecular weight: 13,900), (b) trypsin (24,000), (c) pepsin (35,000), (d) serumalbumin (67,000), (e) human γ-globulins (150,000), (f) thyroglobulin (670,000)

or R_f against log M. In Figure 9, the curves show the relationship between the chromatographic mobility of a given substance and the degree of cross-linking of the gel used. Gel granules allow the separation of peptides and proteins according to their molecular weights,[139,152,162,167,170,171,174] and also the separation of transfer RNA from ribosomal RNA.[153]

2. Agar and Agarose Gels

It has been known for many years that proteins could diffuse through agar gels (Oudin,[175,176] Ouchterlon,[177] Grabar and Williams[178]). Polson[142] proposed the use of agar gel columns for the separation of macromolecular substances of different molecular weight. However, these granules are made slightly adsorbing[142,143] by the addition of small quantities of an acidic polysaccharide (carboxylic and sulphate groups) existing in agar and known as agaropectine (Araki[179]). This adsorption may be reduced by using eluents containing high salt concentration.* Hjerten[145,180] proposed the use of agarose gel grains instead of agar, since this neutral polysaccharide can be easily prepared from agar,[181-185] of which it is the main component (Araki[179]). This gel is not adsorbing.[186]

Agar and agarose gels are prepared in bead form[144,146] or in pellet form[142,145,180] and are now readily available (see Tables 10 and 11). Their

Table 10. Agar gel granules

Trade name Super Ago Gels[a] (per cent)	Dry weight/ column volume $(g\,ml^{-1})$	Approximative range of molecular sieving (molecular weight)
10	0·1	15,000–400,000
9	0·09	25,000–500,000
8	0·08	35,000–900,000
7	0·07	$50,000-1·5 \times 10^6$
6	0·06	$100,000-2·5 \times 10^6$
5	0·05	$200,000-10 \times 10^6$
4	0·04	$300,000-15 \times 10^6$
3	0·03	$500,000-100 \times 10^6$
2	0·02	$10^6-180 \times 10^6$

[a] Mann research laboratories, 136 Liberty Street, New York, N.Y.; agar gel granules in suspension (0·001 M.E.D.T.A. and 0·02 per cent sodium azide) available in two grades: A grade (65–140μ; 240–100 mesh), B grade (140–250μ; 100–60 mesh); the gel granules are prepared by a method in which the gels are physically disrupted without alteration of the pore sizes

main advantage on cross-linked dextrans (Sephadex G 200) is to allow the use of exclusion–diffusion chromatography for the separation of higher molecular weight substances[143,144-146,187,188] (see Tables 9, 10 and 11). Using this method, nucleic acids[189-191] and proteins[142,192] may be separated according to their molecular weight, and viruses[144,193] and cell particles according to their particular weights.[145]

In Figures 10 and 11, the relationship between chromatographic mobilities and per cent by weight of agar or agarose in gel grains is shown.

* The adsorption of proteins by ion exchangers decreases as the ionic strength of the medium increases

Table 11. Agarose gel grains

Trade name	Grain characteristics			Approximative range of molecular sieving (molecular weight)	Approximative molecular weight of excluded particles
	Dry weight / Wet weight (per cent)	Grain sizes (swollen state) (μ)	Grain shapes		
Sepharose 6B[a]	6	40–190	beads	$100,000–2 \times 10^6$	4×10^6
Sepharose 4B	4	40–190		$300,000–3 \times 10^6$	20×10^6
Sepharose 2B	2	60–250		$2 \times 10^6–25 \times 10^6$	40×10^6
Sagarose 10[b]	10	70–140		$10,000–250,000$	$420,000$
Sagarose 8	8	70–140	granules	$25,000–700,000$	$3 \cdot 4 \times 10^6$
Sagarose 6	6	70–140		$50,000–2 \times 10^6$	11×10^6
Sagarose 4	4	70–140		$200,000–15 \times 10^6$	27×10^6
Sagarose 2	2	70–140		$500,000–150 \times 10^6$	290×10^6
Biogel A-0·5 m[c]	—	—	beads	$10,000–500,000$	$0 \cdot 5 \times 10^6$
Biogel A-1·5 m	—	—		$20,000–1 \cdot 5 \times 10^6$	$1 \cdot 5 \times 10^6$
Biogel A-5 m	—	—		$40,000–5 \times 10^6$	5×10^6
Biogel A-15 m	—	—		$60,000–15 \times 10^6$	15×10^6
Biogel A-50 m	—	—		$200,000–50 \times 10^6$	50×10^6
Biogel A-150 m	—	—		$1 \times 10^6–150 \times 10^6$	150×10^6

[a] Sepharose: Pharmacia fine chemicals, Uppsala, Sweden; spherical agarose gel beads in suspension, prepared by modified bead-gelling method of Hjerten[146]

[b] Sagarose: Seravac Laboratories Ltd, Holyport Maidenhead, Berkshire, England; agarose gel granules obtained by method in which the gels are physically disrupted

[c] Biogel A: Bio Rad Laboratories Inc., 32nd and Griffin, Richmond, Calif., U.S.A.

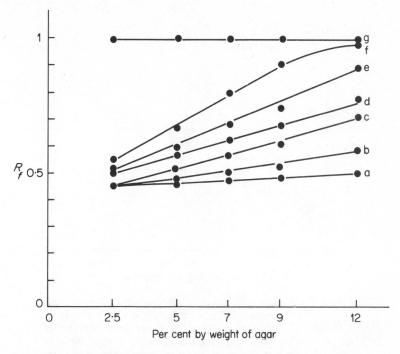

Figure 10. Relationship between chromatographic mobilities (R_f-values) of proteins of various molecular weights and per cent by weight of agar in gel grains; (a) cytochrome c (molecular weight: 13,000), (b) α-lactalbumin (16,000), (c) β-lactoglobulin (36,000), (d) serumalbumin (68,000), (e) human γ-globulins (150,000), (f) thyroglobulin (ox) (670,000), (g) gastric mucoid (excluded substance); (see Andrews[143])

3. Cross-linked Polyacrylamide Gels

Cross-linked polyacrylamide gels have been proposed as supports for zone electrophoresis of proteins[194-198] and, in granule form, for exclusion–diffusion chromatography (Hjerten[147]).

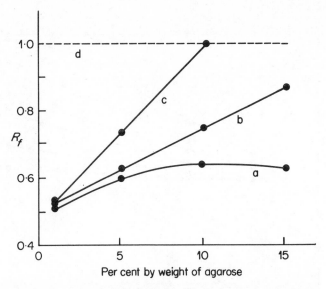

Figure 11. Relationship between chromatographic mobilities (R_f-values) of proteins of various molecular weights and percent by weight of agarose in gel grains; (a) cytochrome c (molecular weight: 13,000), (b) haemoglobin (68,000), (c) R-phycoerythrin (290,000), (d) excluded substance (Indian ink); adapted from Hjerten[145]

Table 12. Cross-linked polyacrylamide-gel in bead form

Trade name Biogel[a]	Water regain (ml/g dry gel)	Column volume (ml/g dry gel)	Approximative range of molecular sieving (molecular weight) (peptides and globular proteins)
P 2	1·6	3·8	200– 2,000
P 4	2·6	6·1	500– 4,000
P 6	3·2	7·4	1,000– 5,000
P 10	5·1	12	5,000– 17,000
P 30	6·2	14	20,000– 50,000
P 60	6·8	18	30,000– 70,000
P 100	7·5	22	40,000–100,000
P 150	9·0	27	50,000–150,000
P 200	13·5	47	80,000–300,000
P 300	22·0	70	100,000–400,000

[a] Bio-Rad Laboratories, 32nd and Griffin Avenue, Richmond, California USA; all types of Gel are available in 50–100, 100–200, minus 400 mesh (US-Standard) grades; P_2 and P_4 type available only in 200–400 mesh grade; minus 400 mesh grade is used in thin layer chromatography

These granules (**4**) are obtained in aqueous medium (water or buffer, pH > 6·0) by the polymerization of a monomer (acrylamide) (**1**) with N,N'-methylene bis-acrylamide (**2**) as a cross-linking agent. The reaction catalysers are β-dimethylaminopropionitrile (**3**) and ammonium persulphate. Gels prepared by Hjerten[147] contain 4–16 per cent monomer (in which the cross-linking agent reaches 5 per cent of the dry weight). Table 12 shows the physical and chromatographic properties of the commercial products. They allow the separation of proteins and nucleic acids according to their molecular

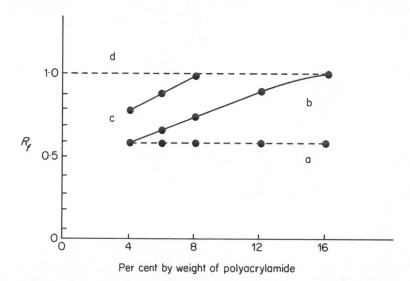

Figure 12. Relationship between chromatographic mobilities (R_f-values) of proteins of various molecular weights and percent by weight of polyacrylamide in gel grains; (a) cytochrome c (molecular weight: 13,000), (b) human haemoglobin (68,000), (c) R-phycoerythrin (290,000), (d) excluded substance (Indian ink); all gels are made with 5 g NN' methylene bis-acrylamide, as cross-linking agent, for 95 g acrylamide; adapted from Hjerten[147]

weights.[147,199–202] Figure 12 shows the relationship between chromatographic mobility and the percentage of the polymer in the gel; Figure 13 shows some fields of application of various commercial gels grouped according to their molecular weights. It can be seen that it is not possible to fractionate mixtures of proteins of molecular weight higher than 350,000.

D. A Comparison of the Properties of Available Gels

It is obvious that the choice of gel depends on the nature of the substances to be separated. Cross-linked dextran gels (Sephadex G 50 to G 200),

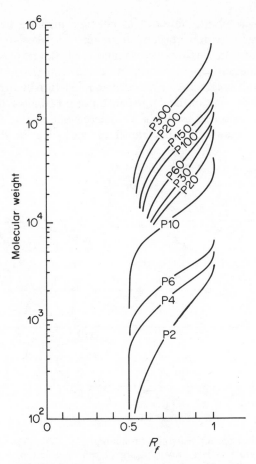

Figure 13. Relationship between molecular weights of substances and mobilities (R_f-values) on cross-linked polyacrylamide gel grains columns (Biogel P2, P4, P6, P10, P20, P30, P60, P100, P150, P200, P300) (after *Bio-Rad Laboratories*: see *Chem. News*, **36**, 107 A (No. 13) (1964); *Bio-Rad List R*, 44 (1966))

polyacrylamide gels (dry weight: 4–16 per cent; Biogel P 10 to P 300) agar and agarose gels (dry weight: 9–14 per cent) allow the separation of proteins according to their molecular weights ($0 < K < 1$) for molecular weights ranging from 5000 to 300,000.

Agar and agarose gels of low percentage dry weight (3–9 per cent) permit the separation of proteins of higher molecular weight,[143,192] nucleic acids (ribosomal RNA, DNA, phage RNA, virus RNA) and most neutral polysaccharides according to their molecular weights. Agar and agarose gels (dry weight: 5–9 per cent) are suitable for separation of proteins with molecular weights ranging from 50,000 to 700,000.[143]

Slightly cross-linked dextran gels (Sephadex G 200) allow a separation of transfer RNA from ribosomal RNA[153] due to the exclusion of ribosomal RNA ($K = 0$) and reduced diffusion of transfer RNA ($0 < K < 1$).

Agar and agarose gels of very low percentage dry weight (1·5–3 per cent) permit the chromatography of viruses and native nucleic acids.

Very highly cross-linked gels (dextran: Sephadex G 10, G 15, G 25; polyacrylamide: Biogel P 2 to P 10) may be used to separate low molecular weight substances (m.w. ranging from 50 to 5,000) according to their molecular volumes.

E. Eluents

For the chromatographic separation of proteins and nucleic acids, the composition of eluent is not of great importance; in most cases buffer solutions (pH 5·0–9·0) containing a neutral salt (i.e. 0·1–1·0 M NaCl) are used.[166,203]

On cross-linked dextran gel columns several volatile eluents may be used: 0·01–1·0 M acetic acid,[204–208] 5 M acetic acid,[209] 0·5 M propionic acid,[210] 1·0 M formic acid,[211] 0·05 M ammonia,[161,212,213] ammonium bicarbonate solution,[214] pyridine acetic acid buffers,[212] triethylamine acetic acid buffers.[215] Peptide chromatography[206,169,216,217] is generally performed in acidic media (40 per cent formic acid, 30 per cent acetic acid, 0·1 M acetic acid, 0·5 M propionic acid); the separation of protein chains may be achieved in acidic media (i.e. 0·5 M propionic acid)[210] or in media containing urea (2–8 M).[218]

F. Applications

We intend to show only some typical applications of exclusion–diffusion chromatography.

1. Separation of Low from High Molecular Weight Substances

This type of separation is very valuable in molecular biology; it allows the elimination of a salt, of an excess of reagent or of an extraneous substance from a macromolecular substance (for example from a protein, polysaccharide, nucleic acid, virus, bacteriophage or bacteria). These separations are obtained on columns of grains having a rather highly cross-linked hydrophilic matrix (i.e. Sephadex G 25, G 50, Biogel P 10). High molecular weight substances (proteins, polysaccharides, nucleic acids, viruses, bacteriophages, bacteria) are excluded from the gel grains ($K = 0$)

whereas low molecular weight substances (salts, amino acids, monosaccharides, etc.) diffuse reversibly between the two phases ($K = 1$) or are adsorbed ($K > 1$) on the gel matrix (borate on Sephadex grain, aromatic amino acids, purines, etc.); low molecular weight substances are delayed relative to the macromolecular substances (Flodin and Porath[133]).

For the separation of small molecules which are non-adsorbed on the gel matrix (salts, monosaccharides, amino acids; $K = 1$) from macromolecular substances (excluded materials; $K = 0$) the sample volume must not be larger than the total volume of the stationary phase in the column (V_s). A good separation is obtained in practice with a sample volume as large as 33 per cent of the total column volume, for example, with Sephadex G 25.[156,219]

The desalting of aqueous solutions of macromolecular substances (for example, proteins,[133,156,220,221] nucleic acids,[222] polysaccharides,[223,224] viruses[225] and bacteria[145]) is performed by applying a solution or suspension to a column of gel grains swollen in water (Sephadex G 25, G 50, Biogel P 10). For desalting solutions of proteins on Sephadex G 25, Flodin[156,219,226] advises the use of a sample volume less than 15–33 per cent of the total volume of the gel column, a column with a diameter: height ratio ranging from 1:5 to 1:20 and a flow rate of 0.3 ml min^{-1} cm^{-2}. A typical example of a separation between protein and ammonium sulphate is shown in Figure 14. The same method has been used for the separation of radioactive macromolecules from an excess of radioactive ions; good examples are the separation of ^{131}INa from iodized proteins,[213,227] ^{65}Zn^{2+} from a ^{65}Zn^{2+}–insulin complex[213] and from a ^{65}Zn^{2+}–carboxypeptidase complex[228] and ^{35}SO$_4$$^{2-}$ from ^{35}S-labelled bacteria.[145] A similar process, using a column of gel grains, soaked in a buffer different from that contained in the sample, allows replacement of one salt by another in a solution of a macromolecular substance.[156]

In the same way, other low molecular weight substances can be separated from macromolecular substances; for example, amino acids,[133,229] peptides,[132,155,211,219] monosaccharides,[133,229] nucleotides, nucleosides, purines and pyrimidines[157] have all been separated in this way. The columns used are packed with, for instance, Sephadex G 25 or G 50 soaked in either water, a volatile buffer, 0.2 M acetic acid, ammonium bicarbonate solution, or a pyridine–acetic acid mixture. Thus amino acids, peptides and glucose can be separated from blood serum proteins,[229] peptides from protease,[155] lactose from milk proteins,[204] vitamin B12 from vitamin B12 bound to macromolecules,[230–234] vasopressin or ocytocin from proteins,[211] nucleosides, bases or oligonucleotides from nucleic acids,[157,235,236] adenosine monophosphate from transfer RNA or from ribosomal RNA,[201] adenosine triphosphate from nucleic acids,[237] ^{14}C-amino acids from ^{14}C-amino acid–transfer RNA complexes,[201] phenol from nucleic acids,[238] glucose from dextrans,[154,133] glucose from inulin,[239] monosaccharides from poly-

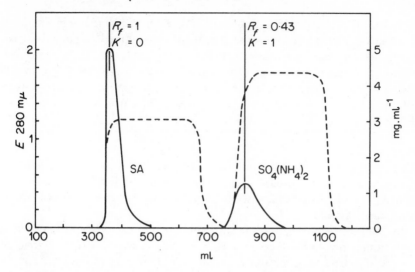

Figure 14. Desalting $(SO_4(NH_4)_2)$ of an aqueous solution of bovine serumalbumin (SA) on cross-linked dextran beads; column: 80 × 4 cm; $V_m = 358$ ml; $V_s = 470$ ml. Cross-linked dextran: Sephadex G 25, 50–150 μ, 143 g. Samples: (– – –) 1·5 g $SO_4(NH_4)_2$ + 0·5 g SA/330 ml, (———) 0·15 g $SO_4(NH_4)_2$ + 0·125 g SA/10 ml

saccharides,[239–242] cofactors from enzymes[243–245] and inorganic ions from enzymes.[228]

These methods have been applied to the structural chemistry of peptides,[155,246–254] glycopeptides,[205,255–262] polysaccharides,[263–265] ferriporphyrin-*c* peptides[155] and in the investigation of naturally iodized substances (iodides, thyroxine and iodotyrosine, either free or bound to macromolecules.)[266–274]

The elimination of side products, of excess reagent or of other low molecular weight substances from a medium containing a macromolecular substance can be effected in the same way, at the end of the reaction, by flowing on a gel grain column (for instance, Sephadex G 25 or G 50); examples are glycolate and sodium diglycolate from carboxymethyldextrin,[275] mercaptoethanol and urea from proteins,[276] sodium triphosphate and urea from proteins,[277] glucose from dextrans,[224,227] colouring matters and excess reagent (fluorescein isothiocyanate) from fluorescent proteins,[278–289] iodine and iodides from iodized proteins,[213,227] hydroxylamine from creatine phosphokinase[290] and so on.

2. Separation of Substances According to Molecular Volumes

Informative data on the grades of gels to be used in order to separate macromolecular substances (proteins, nucleic acids, polysaccharides) or particles (viruses) according to molecular volumes will be found in Figures 8 to 13 and Tables 8 to 12. More or less highly cross-linked gels may be used. In this case chromatographic separations result from the restricted molecular diffusion of macromolecular substances through cross-linked gel grains. Two substances with different K values (K_1, K_2) on a specific gel-type column have elution volumes V_{R1}, V_{R2} which differ from δV_R according to the equation

$$\delta V_R = V_{R1} - V_{R2} = (K_1 - K_2)V_s \tag{6}$$

They will be separated if the sample volume applied to the column is lower than δV_R, if the difference $(K_2 - K_1)$ is sufficiently high (that is if molecular weights are sufficiently different), and if V_s (and the volume of the column) is sufficiently large.

In practice, a good separation can be obtained by molecular sieving with a high, uniformly packed column $(1-3 \times 0.02\,\text{m})$, a fine original zone of sample $(0.3-0.6\,\text{ml cm}^{-2})$ and a rather slow flow rate $(3-6\,\text{ml h}^{-1}\,\text{cm}^{-2})$. Before starting a chromatographic experiment, it is advisable to check the

Table 13. Examples of total separation of proteins according to molecular weights on cross-linked gel grain column (two adjoining peaks wholly separated)

Gel used	Length of column (cm)	Separated substances[a]
Sephadex G 75	150	Monomeric (m.w. 14,000) from dimeric (m.w. 28,000) bovine ribonuclease[291]
Sephadex G 100	190	Pepsin (35,500) from bovine serumalbumin (68,000)[152]
Sephadex G 150	110	Monomeric (68,000) from dimeric (136,000) bovine serumalbumin[141]
Sephadex G 200	55	Monomeric serumalbumin (68,000) from γ-globulins (150,000)[140]
9 per cent agar	50	Bovine serumalbumin (68,000) from α-lactalbumin (36,000)[143]

[a] None of these substances is excluded from gel grains

homogeneity of the packing by filtration of a solution of a coloured substance of high molecular weight through the column bed, for example blue dextran,* haemoglobin, catalase, etc.).

It is difficult to state precisely the actual separating power of the process. However it is possible to conceive of this power when considering total separations of substances as shown in Table 13. Under these precise

* Pharmacia, Uppsala (Sweden)

conditions, substances may be wholly separated if their molecular weights are in ratio 1:2.

a. *Peptides*

Naturally occurring peptides (vasopressin, ocytocin,[211] ACTH,[292–294] glucagon,[167] insulin,[295] tyrocidin,[216] bacitracin, plasteins[296]) can be submitted to molecular sieving chromatography. Separations made according to the molecular weights of the peptides from proteasic hydrolysates of proteins[169,217,297] and of peptides resulting from specific chemical cleavage of proteins[206] were obtained using columns packed with highly cross-linked dextran beads; for example Sephadex G 25,[169,206,217,296] G 50,[217] G 75.[206]

b. *Proteins*

The separation of proteins according to their molecular weights (Flodin and Killander[140]) is one of the main fields of application in molecular sieving chromatography. Several hundred scientific papers have already been devoted to it. Molecular sieving chromatographic procedures for the separation of proteins in artificial mixtures have been intensively investigated using columns packed with cross-linked dextran beads (Sephadex G 50, G 75, G 100, G 200),[167,152,170,173,292,298] agar or agarose beads (dry weight: 4–12 per cent).[143,162,187,188,299]

Figure 15 shows an example of the separation of low molecular weight proteins according to their molecular weights; Figure 16 gives another example with medium molecular weight proteins.

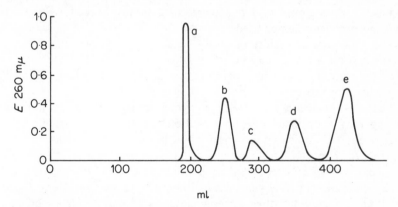

Figure 15. Separation of proteins according to molecular weights on cross-linked dextran beads. Column: 120 × 2·4 cm. Cross-linked dextran: Sephadex G 50, (20 − 80 μ). Eluent: hydrochloric acid, pH 2·5 containing 0·1 M NaCl. Sample: 3·5 mg of each substance/1·25 ml; (a) γ globulins (m.w. 150,000); (b) chymotrypsinogen (25,000); (c) ribonuclease (13,900); (d) Kunitz inhibitor (6500); (e) glucagon (3500); adapted from Fritz, Trautschold and Werle[167]

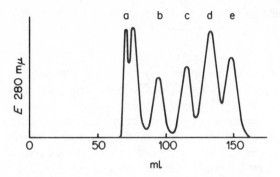

Figure 16. Separation of proteins according to
molecular weights on lightly cross-linked dextran
beads (Sephadex G 100; 40–120 μ); column:
192 × 1·1 cm; eluent: 0·1 M acetate buffer, pH 6·0,
containing NaCl (0·4 M); flow rate 0·4 ml min^{-1}
cm^{-2}; sample: 6 ml of each protein/1 ml; (a) γ
globulins (molecular weight 150,000); (b) serumalbumin (70,000); (c) pepsin (35,000); (d) α chymotrypsin (22,500); (e) cytochrome c (13,000); adapted
from Whitaker.[152] (Reprinted from *Anal. Chem.*, **35**,
1950 (1963). Copyright 1963 by the American
Chemical Society. Reprinted by permission of the
copyright owner)

For proteins of very high molecular weight (for example plasma macroglobulins,[192] thyroglobulin, β-galactosidase) columns packed with 3·5 per
cent agarose beads can be used.

The good reproducibility of results makes this method suitable for the
determination of the molecular weights of proteins[143,152,162,167,170,172,301,302] and for investigations into the affinity of low molecular weight
substances for proteins.[303–305]

These chromatographic methods, which are based upon molecular
sieving, may be applied to the fractionation of naturally-occurring mixtures
of enzymes and other proteins, for example blood serum proteins[140,306–311]
(see Figure 17), proteins from growth media of microorganisms,[312–316] and to
the separation of natural or artificial, monomeric or dimeric protein molecules; examples are: bovine serumalbumin[141] α-amylase from *Bacillus
subtilis*,[302] pancreatic bovine ribonuclease A[291] bovine procarboxypep-
tidase[317] and Bence–Jones proteins.[318] They have also been applied successfully to the separation of protein chains when molecular weights of the
chains were sufficiently different. In this way, Fleischmann, Pain and
Porter,[207] using a column packed with Sephadex G 75 beads soaked in
0·5 M propionic acid, separated chains of 7S γ-globulins after reduction by

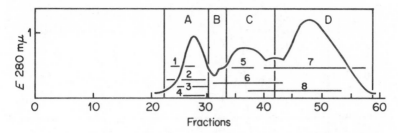

Figure 17. Fractionation by molecular sieving of proteins from human serum on lightly cross-linked dextran beads. (Sephadex G 200; 40–75 μ), column 55 × 4 cm; eluent: 0·1 M tris-hydrochloride buffer, pH 8·0, containing MNaCl; flow rate: 68 ml/h; sample: 10 ml serum = 670 mg of proteins; sedimentation coefficients: (A) 19 S, 10–11 S, (B) 7 S, (C) 7–4 S, (D) 3·5 S; (1) 19 S antiserumalbumin; (2) $\alpha\beta$-lipoproteins; (3) α2-macroglobulin 19 S; (4) β2-macroglobulin 19 S; (5) 7 S antiserumalbumin; (6) 7 S γ-globulins; (7) albumin + α1-globulin; (8) transferrin (β1-globulin); adapted from Flodin and Killander[140]

mercaptoethanol and subsequent alkylation (see also references 210, 319). Similar procedures have been applied to the separation of *S*-carboxymethyl-keratins,[320] chains of fibrin, fibrinogen[321] or insulin[322] after treatment by sodium sulphite.

Pristoupil[218] proposed the use of Sephadex G 200 and of a buffer medium highly concentrated in urea (8 M) and in sodium chloride (1 M) for the determination of molecular weights of constitutive 'monomers' of protein molecules. With the same object, Fritz and coworkers[167] have proposed the use of a Sephadex G 50 column and an acidic medium (HCl, pH 2·5) for low molecular weight proteins (see Figure 15).

c. *Nucleic Acids*

Chromatography on columns filled with grains of a lightly cross-linked dextran gel (Sephadex G 200), agar gel (Superagogel) or agarose gel (Sepharose, Sagarose) leads to excellent procedures for the fractionation of nucleic acid mixtures according to their molecular weights. Dirheimer, Weil and Ebel,[153] and Borman and Hjerten[201] have been able to separate transfer RNA (4*S*; m.w. 27,000) from ribosomal RNA (16*S*, 23*S*, m.w. 2 × 10⁶); the transfer RNA formed from 70 nucleotides undergoes a reversible process of molecular sieving and ribosomal RNA is excluded from the gel grains (see Figure 18). These results have been confirmed by Swensson and coworkers.[323] The same separation has been obtained on polyacrylamide grains.[201] The separation of nucleic acids from polyphosphates has also been investigated.[324,325] Boman and Hjerten[201] have shown that it is possible to separate T₂-phage DNA (60*S*) from *E. coli* total RNA (16*S*,

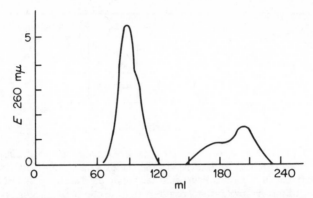

Figure 18. Separation of ribosomal RNA (excluded substances) from transfer RNA of yeast on lightly cross-linked dextran beads (Sephadex G 200); column: $42 \times 3 \cdot 2$ cm; sample: 6 mg RNA; eluent: $0 \cdot 05$ M NaCl; flow rate: 10 ml/h; adapted from Ebel, Dirheimer and Weil[153]

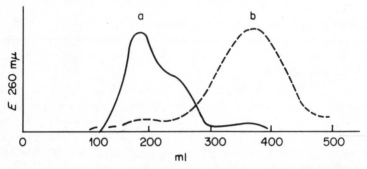

Figure 19. Separation of T_2phage DNA (a) from *E. coli* K 12 total RNA, (b) by molecular sieving on $1 \cdot 5$ per cent agarose grains (average size = 175 μ); column: 95×2 cm; eluent: $0 \cdot 05$ M sodium phosphate buffer, pH $7 \cdot 3$; flow rate: 6 ml/h; adapted from Boman and Hjerten[201]

$23S$, $4S$) on granulated $1 \cdot 5$ per cent agarose gel (see Figure 19). Öberg and Philipson[189] separated KB cell nucleic acids from polio virus RNA (see Figure 20) and several virus nucleic acids on agarose beads (Sepharose 2B).

d. *Polysaccharides*

Molecular sieving chromatography has been applied to the fractionation of mixtures of polysaccharides, for example, dextrans,[153,326] dextrins,[327,328] cellulodextrins,[329,330] acidic polysaccharides,[331] keratosulphates,[263] neutral[332,333] or acidic oligosaccharides[334] have all been separated in this way.

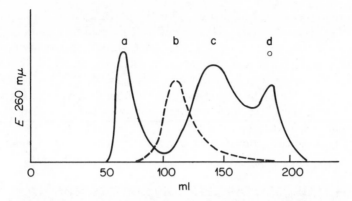

Figure 20. Separation of nucleic acids according to molecular weights on agarose beads (Sepharose 2 B; 60–250 μ); column 60 × 2·1 cm; sample volume: 1 ml; eluent: 0·002 M sodium phosphate buffer, containing $MgCl_2$ (0·001 M); flow rate: $2 \, \text{ml} \, h^{-1} \, \text{cm}^{-2}$; (a) KB cell DNA; (b) poliovirus RNA; (c) KB cell ribosomal RNA; (d) KB cell transfer RNA; adapted from Öberg and Philipson[189]

e. *Viruses, Bacteriophages, Ribosomes*

Bengtsson and Philipson[144] have shown the advantages of columns packed with gel grains of low agar content when separating viruses according to volume. Figure 21 shows the separation of polio virus from influenza virus, the sizes of which are in the ratio 1:3. Molecular sieving

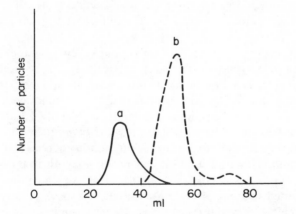

Figure 21. Separation of viruses according to particular volumes on 2·5% agar beads (110–150 μ); column: 30 × 2 cm; sample volume: 2 ml; flow rate: 6 ml/h; (a) *poliovirus type 1* (size: 100 mμ; particular weight: 170 × 10^6); (b) *influenza virus type A* (28 mμ; 6·8 × 10^6); adapted from Bengtsson and Philipson[144]

chromatography has also been used for the purification of virus preparations.[193] Particles of viruses are excluded from 7 per cent agar gel grains[299,335]
and can undergo molecular sieving chromatography on a column packed
with gel grains of low agar or agarose content (1 per cent[185] or 2·5 per
cent[144]): the viruses can originate from animal cells[144,185] or from plant
cells.[193,336,337] On a 6 per cent agar grain column,[145] the phage T_2 (particle
weight: 200×10^6; excluded particle) can be separated from proteins
(molecular weight ranging from 10,000 to 300,000). Hjerten[145] has purified
preparations of yeast ribosomes on 1 per cent agarose grains column.

IV. CHROMATOGRAPHY ON ZONE DISSOLUTION—
PRECIPITATION (SOLUBILITY EQUILIBRIA)

A. 'Salting-out' Chromatography

In general, very good chromatographic separations of ionizable substances of low molecular weight may be obtained by applying them to a
column packed with an inert powdered support and washing the column
with aqueous salt solution: this is known as salting-out chromatography.[338]

Similar processes allow the separation of high molecular weight substances (for example, enzymes, other proteins and viruses). Macromolecular
substances applied to the column as a solution or as a precipitate move
towards the bottom at a rate which depends on whether they precipitate or
pass into solution at the various salt concentrations. Separation between
zones is due to differences in the capacity of substances to be salt-precipitated,
or to be brought into solution when the salt concentration progressively
decreases.[339–341]

In the elution curves obtained in salting out chromatography, the fronts
of the elution peaks generally rise steeply, but the rear parts do not; bands
exhibit 'tailing'. It is very likely that this results from a phenomenon described by Tiselius;[342,343] when proteins come in contact with powdered
substances, either inert (cellulose, kieselguhr) or slightly adsorbing (silica
gel), they tend to fix to them, in the presence of salts, at concentrations
lower than those necessary for their precipitation in aqueous media (salting-
out adsorption). This particular elution profile may also be explained by
non-linear 'sorption isotherm'. The elution peak is symmetrical if a given
protein is eluted in front of or in the middle of a peak corresponding to
several proteins, the mobilities of which are very similar; less mobile proteins, the mobilities of which are very similar; less mobile proteins result in
the easier elution of molecules which are present in rear part of the peak,
i.e., those corresponding to the protein exhibiting the highest mobility.
For instance, the α-hydroxysteroid dehydrogenase of *Pseudomonas testosteroni* by salting out chromatography on cross-linked dextran gel (Sephadex

G 100) gives a symmetrical elution peak,[341] for this protein is eluted in the middle of a peak corresponding to a group of proteins. In this case, although separation may seem of low interest, salting-out chromatography is an excellent purification method for a particular enzyme if specifically active fractions are recovered.

1. Proteins

Three main methods are to be considered:

a. *First Method*

This method involves protein fractionation on the basis of the solubility in aqueous salt solutions. In this case chromatographic processes can give a clean-cut separation of components and may replace precipitation methods.

A narrow band of a mixture of proteins precipitated on a porous inert support at high salt concentration is applied to a column packed with an inert support (for example, a pile of filter paper discs,[339,344] kieselguhr,[345] fine cellulose powder[340] or highly cross-linked dextran gel granules[346]) soaked with a precipitating salt solution. Proteins are eluted successively by washing with the salt solution at progressively decreasing concentration.

In this way Mitchell and coworkers[339,344] separated enzymes from growth medium of *Aspergillus orizae*. They precipitated proteins using an ammonium sulphate solution (60 per cent saturation) at the top of a column made of 500 filter paper discs, then eluted enzymes (at pH 6·5) using an ammonium sulphate solution of progressively decreasing concentration: phosphatase is eluted at 50–40 per cent saturation, amylase at 40–20 per cent and adenosine deaminase at less than 20 per cent.

Figure 22 shows the elution profile of alkaline phosphatase from *E. coli* K 12 on fine cellulose powder (40 μ) column. Protein is precipitated in a slurry of cellulose powder and ammonium sulphate (83 per cent saturation) then applied to the column in the same salt solution (narrow band). By washing with a solution of progressively decreasing ammonium sulphate concentration, phosphatase is eluted by 60 per cent saturation solution. Though homogeneous, the preparation is eluted as a dissymmetrical peak; a non-linear sorption isotherm may explain this.

b. *Second Method*

In a few cases a very simple method, quite similar to salting-out chromatography of low molecular weight substances, is available. Thus Simonart and Chow[347] separated amylase and proteinase from growth medium of *Aspergillus tamarii* using paper chromatography and an aqueous ammonium sulphate

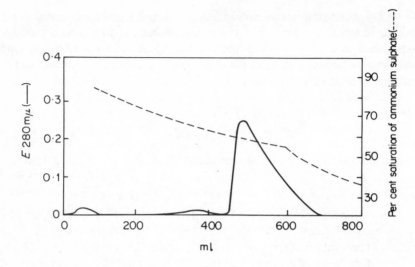

Figure 22. Elution profile of *E. coli* K 12 alkaline phosphatase by salting-out chromatography on cellulose powder column with progressively decreasing concentration in ammonium sulphate; sample: 60 mg phosphatase precipitated in narrow band (3 × 1·3 cm); flow rate: 36 ml/h; column: 3 × 6 cm (Munier[340])

solution (35 per cent (w/v), pH 6·5). In this case separation results from the phenomenon described by Tiselius[342,343] namely salting-out adsorption.

c. *Third Method*

By applying the properties of cross-linked dextran gel granules, Porath[341] proposed a new method of salting-out chromatography. He produced (Figure 23) along a column packed with cross-linked dextran gel (Sephadex

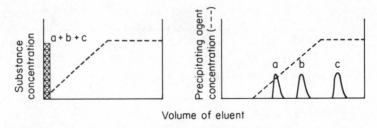

Figure 23. Fractionation of three proteins (a, b, c) by salting-out chromatography on cross-linked dextran column with ammonium sulphate as eluent; left: starting conditions; right: ending pattern; adapted from Porath[341]

G 100) a concentration gradient of the precipitating agent (ammonium sulphate aqueous solution; 85–55 per cent saturation) with a higher concentration at the bottom and lower concentration at the top. A narrow band of a mixture of proteins in salt solution at the lowest concentration is then applied to the column. Progressively diluted salt solutions (55–22 per cent saturation) are used for washing. Since salts move slower than proteins, some of the proteins reach a point where the salt concentration is sufficient to cause their precipitation; these are stopped, whereas the other, non-precipitated proteins continue to move down and, as a result, substances are arranged along the column in form of bands of precipitate. However, flowing solvent displaces the salts sufficiently and bands of precipitated proteins again pass into solution one after the other. Towards the end, the eluate shows several peaks of proteins; their position in the elution pattern results from a precipitation chromatographic process followed by a re-solution chromatographic process. This method is most valuable in protein mixtures analysis field.[341,348]

2. Viruses

Salting-out chromatography has been used to purify preparations of viruses (*Rous sarcoma* virus[350]), bacteriophages,[351] particulates ranging from viruses to mitochondria and bacteria,[352–354] and melanized granules (mouse melanomas[355]).

B. 'Salting-in' Chromatography

'Salting-in' chromatography allows the separation of precipitating from non-precipitating proteins using low concentrated salt solutions. These two types of proteins may be separated by applying a mixture of them in diluted salt solution (i.e. 0·02 M), as a narrow band, to a highly cross-linked dextran gel column (water-soaked Sephadex G 25). Salts are slowed down on such a column ($K \geqslant 1$, free diffusion or slight adsorption) and the band of proteins which cannot penetrate into the gel granules (exclusion) is quickly separated from the salt zone. Water-soluble proteins continue to move down, whereas insoluble proteins precipitate by salt-shortage. After a time, the moving salt zone reaches the precipitated band of proteins and elutes them. They are wholly eluted by washing with NaCl solution (0·5 M). Flodin[349] applied this method to the separation of γ-globulins and albumin from plasma (Figure 24). In the same way pseudoglobulins and euglobulins have been separated from blood serum[306] and C_1' and C_2' components of complement.[356]

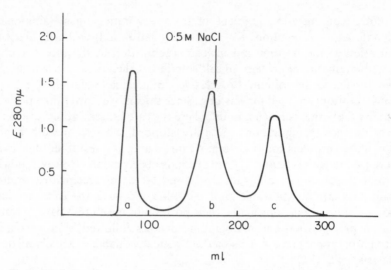

Figure 24. Separation of γ-globulins and serumalbumin by salting-in chromatography; (a) albumin; (b) γ-globulins; (c) merthiolate; adapted from Flodin[349]

V. ION EXCHANGE CHROMATOGRAPHY

Valuable procedures for the separation of components in complex mixtures of macromolecules from biological media have been derived from ion exchange chromatography, due to the high degree of resolution, the large capacity under mild conditions of elution and the quite good recovery of material. Components of mixtures of nucleic acids, enzymes, or other proteins may be easily separated in this way, and ion exchange chromatography may be used either as an analytical method of investigation or as a preparative method of a particular substance present in the mixture. Ion exchange chromatography also provides separation procedures for the components of degradation of enzymes, proteins and nucleic acids, and these are very useful when determining the molecular structure of these biological macromolecules. Binding forces involved in these separations seem to be mainly of an ionic nature although other factors appear to be concerned, namely electrostatic interactions, van der Waals forces and multipolar interactions. In certain chromatographic processes, for instance in the separation of nucleic acids and oligonucleotides, these factors may play a particularly important part.

In this chapter we shall review ion exchangers, processes of elution and applications of ion exchange chromatography to the separation of peptides, enzymes or other proteins, oligonucleotides, nucleic acids and ionic polysaccharides.

A. Ion Exchangers

Ion exchange materials used in the chromatography of biological macromolecules are of four main types: beads of highly cross-linked synthetic resins bearing ionic groups (ion exchange resins), powders of cellulose bearing ionic groups (ion exchange celluloses), beads of cross-linked polysaccharides bearing ionic groups (cross-linked dextran ion exchangers) and finely powdered materials coated with ionic substances (kieselguhr or silica gel coated with methylated serumalbumin or histones). The physical and chemical characteristics of these ion exchangers are given in Table 14–20.

1. Ion Exchange Resins

Among these ion exchangers, carboxylic resins (Amberlite IRC 50, XE 97, XE 64, Zeo-Karb 226–see Table 14) have been mainly used for the chromatographic separation of proteins,[357–361] whereas sulphonic acid resins (Dowex 50, Zeo-Karb 225, Amberlite IR 120)[333–368] and resins bearing quaternary ammonium groups (Dowex 1, Deacidite FF, Amberlite IRA 400)[366–369] have been successfully used for fractionation of mixtures of peptides.

Carboxylic resins (fine particle grades of Amberlite IRC 50: XE 64, XE 97, CG 50 type I, II, III; Zeo–Karb 226 × 2·5 and × 4·5) are polymers of methacrylic acid cross-linked with divinylbenzene. These ion exchangers are mainly used in the sodium salt form,[359,360,362] in the ammonium salt form[357] or sometimes in the barium or calcium salt form.*,[370] Amberlite IRC 50, because of its high percentage of cross-linkages, can only adsorb proteins on the particle surface; consequently, carboxylic resins have a low capacity for fixation of ionic macromolecules (only a few milligrams of protein per gram dry weight). The number of carboxylic groups inside the resin beads is high when compared with number of carboxylic groups at the surface; accordingly, carboxylic ion exchange resins have a very high buffer capacity and the adsorption capacity of ionic macromolecules is little affected by the slight variations of pH or salt concentration in the medium (i.e. in the eluent). The advantage of this high buffer capacity is that dialysis of the protein sample against the first elution buffer before starting the chromatography is sometimes not necessary. Slightly adsorbed substances can be eluted by a medium of constant composition (simple elution).[357–360] More strongly adsorbed substances can be eluted by varying the eluent composition in discrete steps (stepwise elution), or by varying the pH or ionic strength[361] by strongly varying the salt concentration (see Boardman and Partridge).[358]

---

* For the application of the magnesium salt form of these exchangers to chromatography of nucleic acids, see the section on 'specific chromatography'.

Sulphonic resins (Dowex 50 × 4, 50 × 2, Amberlite IR 120, Zeo–Karb 225—see Table 14) are sulphonated beads of polyvinylbenzene cross-linked with divinylbenzene. These resins are ordinarily used in the form of the sodium salt,[365] or the free acid equilibrated with the starting buffer of chromatography (pyridinium salt).[368] Attempts to use them for the chromatography of proteins were unsuccessful, but excellent separations of peptides may be obtained. In the latter case a low degree of cross-linking[365] and the use of beads all of approximately the same diameter will favour a good result.

Resins bearing quaternary ammonium groups (Dowex 1, De-Acidite FF, Amberlite IRA 400—see Table 14) are made of polystyrene cross-linked with divinylbenzene bearing a quaternary base on an aliphatic radical. These exchangers are in most cases used in the acetate form[368,369] and equilibrated with the starting buffer of chromatography. They result in valuable separations in the field of peptides mixtures.[366–369]

2. Ion Exchange Celluloses

Sober and Peterson[371] were the first workers to apply ion exchange celluloses to the separation of biological macromolecules, especially proteins. Ion exchange celluloses which are low in the number of ionic groups are water-insoluble.[371] Ion exchange celluloses possess a loose hydrophilic network; large molecules can penetrate freely and diffuse rapidly. Consequently ion exchange celluloses have high adsorption capacity for proteins;[371] they can sometimes reversibly adsorb their own weight of proteins (see Tables 17 and 18). Proteins can be eluted from ion exchange cellulose columns under mild conditions and with quite a good recovery.[372] Sober and Peterson's discovery resulted in a considerable advance in the fractionation methods available for mixtures of biological macromolecules and resulted in an important improvement in the chromatography of enzymes and other proteins.

Ion exchange celluloses (see Tables 15 and 16) belong to two types: *cation exchange celluloses* (carboxymethylcellulose,[371] oxycellulose,[375,376] sulphomethylcellulose,[377] sulphoethylcellulose,[377] phosphorylcellulose)[371] and *anion exchange celluloses* (diethylaminoethylcellulose,[371] triethylaminoethylcellulose,[389] dimethylaminoethylcellulose,[392] trimethylaminoethylcellulose,[392] dimethylaminoisopropylcellulose,[392] diethylaminohydroxypropylcellulose,[390,391] diisopropylaminoethylcellulose,[392] piperidine-N-ethylcellulose,[392] ECTEOLA-cellulose*,[371]). Table 15 gives the physical and chemical characteristics of ion exchange celluloses; Table 16 gives the formulae of ion exchange celluloses of well-known structure; Tables 17, 18, 19 give data on the commercial ion exchange celluloses.

* Abreviations: see Table 15.

Table 14. Ion exchange resins[a]

Trade name[b]	Active group	Matrix type[b]	Total capacity (mequiv./g dry weight)	Approximative mesh range
Cation exchangers				
Dowex 50 × 2 (200–400 mesh)	$-C_6H_4SO_3H$	Polystyrene	5·1–5·4	80–270[c]
Dowex 50 × 4 (through-400 mesh)	$-C_6H_4SO_3H$	Polystyrene	5·1–5·4	200–400[c]
Dowex 50 × 8 (through-400 mesh)	$-C_6H_4SO_3H$	Polystyrene	5·1–5·4	200–400[c]
Zeo–Karb 225 W.R.1.5. (200–400 mesh)	$-C_6H_4SO_3H$	Polystyrene × 5[e]	4·5–5·0	
Zeo–Karb 225 W.R.1.1 (200–400 mesh)	$-C_6H_4SO_3H$	Polystyrene × 6	4·5–5·0	
Amberlite IR 120 (through-200 mesh)	$-C_6H_4SO_3H$	Polystyrene × 8	4·0	120–400
Zeo–Karb 226 × 2·5	$-COOH$	Polymethacrylic acid	9–10	
Zeo–Karb 226 × 4·5	$-COOH$	Polymethacrylic acid	9–10	
Amberlite IRC 50 (through-200 mesh)	$-COOH$	Polymethacrylic acid × 5	8·5	120–400[c]
Amberlite XE 64[f]	$-COOH$	Polymethacrylic acid × 5	8·5	100–325[d]
Amberlite XE 97	$-COOH$	Polymethacrylic acid × 5	8·5	100–325[d]
Amberlite CG 50, type I	$-COOH$	Polymethacrylic acid × 5	8·5	100–200[d]
Amberlite CG 50, type II	$-COOH$	Polymethacrylic acid × 5	8·5	200–400[d]
Amberlite CG 50, type III	$-COOH$	Polymethacrylic acid × 5	8·5	400–600[d]
Anion exchangers				
Dowex 1 × 2 (200–400 mesh)	$-C_6H_4CH_2\overset{+}{N}(CH_3)_3$	Polystyrene	3·5–3·6	120–280[c]
Dowex 1 × 4 (200–400 mesh)	$-C_6H_4CH_2\overset{+}{N}(CH_3)_3$	Polystyrene	3·5–3·6	140–300[c]

Table 14 (*cont.*)

Trade name[b]	Active Group	Matrix type[b]	Total capacity (mequiv./g dry weight)	Approximate mesh range
Anion exchangers (*contd*).				
Dowex 1 × 8 (through-400 mesh)	—$C_6H_4CH_2\overset{+}{N}(CH_3)_3$	Polystyrene	3·5–3·6	200–400[c]
De-acidite FF	—$C_6H_4CH_2\overset{+}{N}(CH_3)_3$	Polystyrene × 2·5	4	minus 200
	—$C_6H_4CH_2\overset{+}{N}(CH_3)_3$	Polystyrene × 4	4	minus 200
	—$C_6H_4CH_2\overset{+}{N}(CH_3)_3$	Polystyrene × 8	4	minus 200
Amberlite IRA 400 (through-200 mesh)	—$C_6H_4CH_2\overset{+}{N}(CH_3)_3$	Polystyrene × 4	3·0	120–400[c]

[a] Partial list covering currently available fine mesh resins for chromatography of peptides (sulphonated polystyrene, polymethacrylic acid cross-linked with divinylbenzene, polystyrene quaternary ammonium base) or proteins (polymethacrylic acid); Dowex resins manufactured by: Dow Chemical Company, Midland, Michigan, USA; Zeo-Karb and De-acidite resins: Permutit Company Ltd., Gannersbury Avenue, London, W.4. England; Amberlite resins: Rohm and Haas Company, Philadelphia, Pennsylvania, U.S.A.

[b] The percentage cross-linking, indicated by ×, corresponds to the approximate amount of divinylbenzene used in the copolymerization mixture; resins of specified mesh size (dry-screened)

[c] Wet-screened

[d] Dry-screened

[e] W.R. = water regain

[f] Pharmaceutical grade of Amberlite XE 97

Only commercial ion exchange celluloses are currently used by research workers. The main ones are: DEAE-cellulose,* a weak ion exchanger, ECTEOLA-cellulose, a very weak base, CM-cellulose, a weak acid and P and SE-celluloses which are stronger acids (see Table 15). Before use, these exchangers have to be converted either into the free-base form (anion exchange cellulose) or into the free-acid form (cation exchange cellulose), and then in the suitable salt form by equilibration with the starting buffer of chromatography. Most of the ionic groups of the ion exchange celluloses are efficient in adsorbing ionic macromolecules. The number of ionic groups per gram of ion exchange cellulose is low (0·3–1·0 mequiv. per gram dry weight; on average, one ionic group every 50 Å) when compared to ion exchange resins (3·5–10 mequiv. g^{-1} dry weight; on average, one ionic group every 10 Å). Very low variations in the salt composition of the medium (i.e. of eluent) have immediate repercussions on their capacity for fixing charged macromolecules. They can be eluted by continuous variations in the pH (cation exchange celluloses) or in the salt composition of the medium (gradient elution on anion exchange celluloses) or sometimes by variations in both parameters. Ion exchange celluloses have been successfully used for the chromatography of enzymes and other proteins,[371,372,395,396] transfer RNA,[397,398] oligonucleotides,[399,400] peptides,[401] viruses, bacteriophages and ionic polysaccharides.

Two kinds of ion exchange cellulose preparations are suitable for the chromatography of biological macromolecules; those with long fibres for preparative chromatography and those with short fibres for analytical chromatography.

3. Cross-linked Dextran Ion Exchangers

Cross-linked dextran ion exchangers are known under trade name of Sephadex† ion exchangers; the varieties produced are DEAE, CM and SE-Sephadex. Table 20 gives their physical and chemical characteristics. These exchangers may be compared with ion exchange celluloses. The two types of ion exchange polysaccharides differ by certain very specific properties; Sephadex ion exchangers are rather highly cross-linked dextrans bearing a great number of charged groups per gram of dry weight; their total capacity is about three or four times higher than this of currently used ion exchange celluloses, and this high total capacity makes it obligatory to keep them in the salt form (see Table 20); also, cross-linked dextran ion exchangers, in bead form, make the packing of homogeneous columns easier than fibre-shaped ion exchange celluloses.

* Abbreviations: see Table 15.
† Pharmacia, Fine chemicals, Uppsala, Sweden.

Table 15. Ion exchange celluloses

Name	Abbreviation	Total capacity (mequiv./g)	Reference preparation	pK^a in: Water[b]	pK^a in: 0·01M NaCl	pK^a in: 0·5M NaCl	pK^a in: 1·5M NaCl	Author reference
Cation exchange celluloses								
Carboxymethylcellulose	CM-cellulose	0·45–0·50, 0·7	373, 371	3·8–4·2	—	3·6	—	—, 371
Cross-linked CM-cellulose	CM-x-cellulose	0·8–1·90	374	—	—	—	—	—
Oxycellulose[c]	O-cellulose	$1·0^{375}$, $2·7^{376}$	375, 376	4·55	—	3·6	—	375
Sulphomethylcellulose	SM-cellulose	0·4	377	—	—	2·5	—	377
Sulphoethylcellulose	SE-cellulose	$0·42^{377}$	377, 378	—	—	2·2	—	377
Phosphorylcellulose[d]	P-cellulose	2 × 0·6–0·9, 2 × $3·5^e$	371, 379, 380	4·5, 6·55	2·7, 7·4	—; 6·2, —; 6·0	—	381, 371
Succinic half ester of cellulose	SA-cellulose	1·0	382	—	—	—	—	—
Citric half ester of cellulose	CA-cellulose	2·0	383	—	—	—	—	—
Anion exchange celluloses								
Diethylaminoethyl-cellulose	DEAE-cellulose	$0·7–0·9^{371}$	371, 384, 385	6·5–8·5	—	9·1–9·2	—	371
Cross-linked DEAE-cellulose	DEAE-x-cellulose	1·40–4·40	374	—	—	—	—	—
Triethylaminoethyl-cellulose	TEAE-cellulose	0·8	389^f	quaternary base[g]		—	—	389
Methyl diethylaminoethyl-cellulose	MDEAE-cellulose	0·8	389^f	quaternary base[g]		—	—	389
Diethylaminohydroxy-propylcellulose	DEAOP-cellulose	1·30–4·50	390, 391	—	—	—	—	—
Dimethylaminoethyl-cellulose (low capacity)[h]	DMAE-cellulose	0·3	392	—	—	—	8·9	392

Ion exchanger	Abbreviation	Capacity (mequiv. g^{-1})	Reference	pK'			Reference
Trimethylaminoethyl-cellulose (low capacity)[i]	TMAE-cellulose	0·3	392	quaternary base[g]	—		392
Dimethylamino, iso-propylcellulose[j]	DMAIP-cellulose	0·26	392	—	—	9·1	392
Diisopropylaminoethyl-cellulose[k]	DIPAE-cellulose	0·81–1·36	392	—	—	9·05	392
Piperidine-N-ethylcellulose[l]	Pipe-NE-cellulose	0·66–1·02	392	—	—	9·05	392
Guanidoethylcellulose	GE-cellulose	0·38–0·73	393, 371, 394	—	—		393
ECTEOLA-cellulose[m]		0·4, 0·05–0·78		5·8–6·0	7·4–7·6		371

[a] Glycolic acid, pK'(water) 3·7, pK'(0·5M NaCl) 3·6; methoxyacetic acid (water) 3·6; diethylaminoethanol, pK'(water) 3·8; dihydroxyethylaminoethanol, pK'(water) 7·8, pK'(0·5M NaCl) 8·0

[b] The value of pK measured in water depends upon the nature of the cellulose (wood-cellulose, cotton-cellulose) used to form the ion exchanger[371]

[c] Preparation: see also references 386, 387

[d] Preparation: see also reference 388

[e] Whatman ion exchange cellulose (particular preparation)

[f] The alkylation of DEAE-cellulose in an anhydrous medium[389], in the presence of water quaternization does not occur[377]; in the last case, DEAE-cellulose with alkylated carboxylic groups is often commercially designated as 'TEAE'-cellulose

[g] The conversion of a weak-base cellulosic exchanger to a quaternary cellulose anion exchanger is shown by the characteristic differences between titration curves before and after alkylation[389,392]

[h] The ion exchanger is made by reacting dimethylaminochlorethane and cellulose in presence of sodium hydroxide[392]; low capacity products are ion exchangers with tertiary amino groups; above 0·35 mequiv. g^{-1}, cellulose bears tertiary amine groups and quaternary amine groups (i.e.: ion exchanger at 0·55 mequiv. g^{-1})[392]

[i] Alkylation (ICH$_3$) in anhydrous alcoholic medium of low capacity DMAE-cellulose (lower than 0·3 mequiv. g^{-1})[392]

[j] The ion exchanger is made by reacting 2-chloro-N,N-dimethylpropylamine and cellulose in presence of sodium hydroxide[392]

[k] The product is made by reacting β-diisopropylaminoethylchloride and cellulose in the presence of sodium hydroxide[392]

[l] The product is made by reacting N-(β-chloroethyl)-piperidine and cellulose in the presence of sodium hydroxide[392]

[m] The ion exchanger is made by reacting epichlorhydrin, triethanolamine and cellulose in the presence of sodium hydroxide[371]

Table 16. Modified celluloses with definite structure suitable for ion exchange chromatography[a]

Structure	Name	Reference
$COOH-CH_2-O-cellulose$	CM-cellulose	Sober and Peterson[371]
$SO_3H-CH_2-O-cellulose$	SM-cellulose	Porath[377]
$SO_3H-CH_2-CH_2-O-cellulose$	SE-cellulose	Porath[377]
$PO_3H_2-O-cellulose$	P-cellulose	Sober and Peterson[371]
$COOH-CH_2-CH_2-CO-O-cellulose$	SA-cellulose	McIntire and Schenck[382]
$\begin{array}{c} C_2H_4 \\ \diagdown \\ N-CH_2-CH_2-O-cellulose \\ \diagup \\ C_2H_5 \end{array}$	DEAE-cellulose	Sober and Peterson[371]
$\begin{array}{c} C_2H_5 \\ \diagdown \\ C_2H_5 - \overset{+}{N}-CH_2-CH_2-O-cellulose \\ \diagup \\ C_2H_5 \end{array}$	TEAE-cellulose*	Guthrie and coworkers[389]
$\begin{array}{c} C_2H_5 \\ \diagdown \\ N-CH_2-CHOH-CH_2-O-cellulose \\ \diagup \\ C_2H_5 \end{array}$	DEAOP-cellulose*	Champetier and coworkers[390,391]
$\begin{array}{c} CH_3 \\ \diagdown \\ N-CH_2-CH_2-O-cellulose \\ \diagup \\ CH_3 \end{array}$	DMAE-cellulose* (low capacity)	Munier and coworkers[392]
$\begin{array}{c} CH_3 \\ \diagdown \\ CH_3 - \overset{+}{N}-CH_2-CH_2-O-cellulose \\ \diagup \\ CH_3 \end{array}$	TMAE-cellulose* (low capacity)	Munier and coworkers[392]

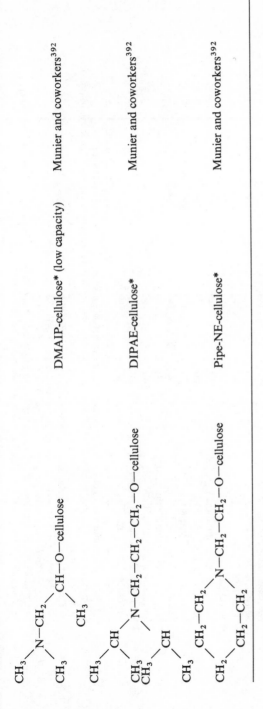

$$CH_3 \diagdown N-CH_2 \diagdown CH-O-cellulose$$
$$CH_3 \diagup \qquad \diagdown CH_3$$

DMAIP-cellulose* (low capacity)   Munier and coworkers[392]

$$CH_3 \diagdown CH \diagdown N-CH_2-CH_2-CH_2-O-cellulose$$
$$CH_3 \diagup \qquad \diagup$$
$$CH_3 \diagdown CH$$
$$CH_3 \diagup$$

DIPAE-cellulose*   Munier and coworkers[392]

$$CH_2-CH_2 \diagdown N-CH_2-CH_2-O-cellulose$$
$$CH_2 \qquad \diagup$$
$$CH_2-CH_2 \diagup$$

Pipe-NE-cellulose*   Munier and coworkers[392]

[a] Asterisked celluloses are not sold commercially

Table 17. Data on Selectacel ion exchange celluloses[a]

Trade name/grade[b]	Average fibre length (microns) (approximate fibre diameter 12–24 microns)	Total capacity (mequiv. g^{-1})	Approximate protein capacity[c] (mg g^{-1} dry exchanger)	Column volume[a] (ml g^{-1})
Selectacel DEAE, Standard	150	0.9 ± 0.1	65–75[e]	14–16
Selectacel DEAE, Type 20	100	0.9 ± 0.1	65–75[e]	11–13
Selectacel DEAE, Type 40	80	0.9 ± 0.1	65–75[e]	10–12
Selectacel DEAE, TLC	40	0.9 ± 0.1	65–75[e]	—
Selectacel ECTEOLA, Standard	150	0.3 ± 0.1	5[e]	7–10
Selectacel ECTEOLA, Type 20	100	0.3 ± 0.1	5[e]	5–6
Selectacel ECTEOLA, Type 40	80	0.3 ± 0.1	5[e]	4–5
Selectacel ECTEOLA, TLC	40	0.3 ± 0.1	5[e]	—
Selectacel CM, Standard	150	0.7 ± 0.1	98[f]	12–17
Selectacel CM, Type 20	100	0.7 ± 0.1	98[f]	9–12
Selectacel CM, Type 40	80	0.7 ± 0.1	98[f]	7–9
Selectacel CM, TLC	40	0.7 ± 0.1	98[f]	—
Selectacel Phosphate, Standard	150	0.9 ± 0.1		9–13
Selectacel Phosphate, Type 20	100	0.9 ± 0.1		7–9
Selectacel Phosphate, Type 40	80	0.9 ± 0.1		5–8
Selectacel Phosphate, TLC	40	0.9 ± 0.1		—

[a] Personal communication of Dr. L. A. Lepage (Brown Co., 650 Main Street, New Hampshire, U.S.A.)

[b] Special grades consisting of non-standard particle sizes and/or higher or lower capacities are available or can be prepared on special order

[c] Protein capacity is not included in the manufacturing specifications; therefore no control tests defining its range are stated here. The numbers listed were taken from the original paper by Peterson and Sober[371]

[d] ml (wet, non-compressed) per gram of dry exchanger

[e] Crystalline bovine plasma albumin in 0·01 M sodium phosphate at pH 7·0 and 2 mequiv. ml^{-1} final concentration

[f] Horse carbon monoxide haemoglobin in 0·01 M sodium phosphate at pH 6·0

Table 18. Data on Whatman ion exchange celluloses

Trade name[a]	Physical structure	Fibre length (swollen in water)	Capacity (mequiv. g^{-1})	Protein capacity (mg g^{-1})	Column volume (ml g^{-1} dry exchanger)
DEAE-cellulose					
DE 22	Fibrous	12–400μ (20–170μ)[b]	1·0 ± 0·1	450[c]	7·7[e]
DE 23	Fibrous (reduced proportion of short fibres)	18–400μ (50–250μ)[b]	1·0 ± 0·1	450	9·1
DE 32	'Microgranular' (dry powder)	24–63μ	1·0 ± 0·1	660	6·3
DE 52	'Microgranular' (wet powder)	24–63μ	1·0 ± 0·1	660	6·3
CM-cellulose					
CM 22	Fibrous	12–400μ (20–170μ)[b]	0·6 ± 0·06	150[d]	7·7[f]
CM 23	Fibrous (reduced proportion of short fibres)	18–400μ (50–250μ)[b]	0·6 ± 0·06	150	9·1
CM 32	'Microgranular' (dry powder)	24–63μ	1·0 ± 0·1	400	6·8
CM 52	'Microgranular' (wet powder)	24–63μ	1·0 ± 0·1	400	6·8
P-cellulose					
P 11	Fibrous	—	2 × 3·7	—	—
AE-cellulose					
AE 11	Fibrous	—	1·0	—	—
ECTEOLA-cellulose					
ET 11	Fibrous	—	0·5	—	—

[a] Whatman Technical Bulletin IE 2, second edition and Whatman Technical Notice WF-CC; H. Reeve Angel and Co. Ltd, 9 Bridewell Place, London, EC4, Great Britain

[b] Fibre length in the middle part of particle size distribution curve

[c] Bovine serumalbumin in 0·01M sodium phosphate buffer pH 8·5 after 90 minutes

[d] 7S γ-globulins in 0·08M sodium phosphate buffer pH 3·5 after 90 minutes

[e] Exchanger equilibrated with 0·05M sodium phosphate buffer pH 7·5

[f] Exchanger equilibrated with 0·05M sodium acetate buffer pH 5·0

Table 19. Data on Serva ion exchange celluloses[a]

Ion exchangers	Trade name/grade	Total capacity (mequiv. g^{-1})	Physical structure	Prevailing fibre length (μ)	Column volume (ml g^{-1})[b]
DEAE-cellulose	DEAE-SN	0·40–0·55	Fibres of middle length	50–200	7·5
	DEAE-SS	0·55–0·75	Fibres of middle length	50–200	7·5
	DEAE-SH	0·75–0·90	Fibres of middle length	50–200	7·5
	DEAE-GS	0·40–0·55	Long fibres	100–1000	9·0
	DEAE-TLC	0·20	Short fibres	<28	—
	DEAE–Neocell	1·0	Crystalline cross-linked cellulose[c]	30–100	6·5
CM-cellulose	CM–(45.030)	0·61–0·63	Fibres of middle length	50–200	7·5
	CM–TLC	0·20	Short fibres	<28	—
	CM–Neocell	1·0	Crystalline cross-linked cellulose[c]	30–100	6·5
P-cellulose	P–(45.130)	0·80–0·90	Fibres of middle length	50–200	7·5
SE-cellulose	SE–(45.190)	0·20–0·30	Fibres of middle length	50–200	7·5
ECTEOLA-cellulose	ECTEOLA–(45.070)	0·30–0·40	Fibres of middle length	50–200	7·0
	ECTEOLA-TLC	0·20	Short fibres	<28	—
GE-cellulose	GE–(45.090)	0·20–0·30	Fibres of middle length Short fibres	50–200	7·5
AE-cellulose	AE–(45.010)	0·32–0·34	Fibres of middle length	50–200	7·5

| Benzoylated DEAE-cellulose[d] | BD–(45.020) | 0·75–0·85 | — | 50–200 | 7·0 |
| Benzoylated–Naphthoylated DEAE-cellulose | BND–(45.025) | 0·75–0·85 | — | 50–200 | 7·0 |

[a] Personal communication of Dr. N. Grubhoffer (Serva–Entwicklungslabor, V. Grothe and Co., Heidelberg, Römerstrass 118, West Germany)
[b] ml (wet, non compressed) per gram of dry exchanger
[c] New microcrystalline crosslinked cellulose exchanger which shows less particle swelling, high capacity, better packing and often better separation
[d] See also Sigma Chemical Co., 3500 De Kalb St., St. Louis, Mo. 63118, U.S.A.

Table 20. Data on cross-linked dextran ion-exchangers

Trade name[a]	Active group	Form	Exchange capacity (mequiv. g^{-1})	Fixation capacity of haemoglobin (gg^{-1})
DEAE-Sephadex A-25 A-50	$-O-CH_2-CH_2-N\big\langle{}^{C_2H_5}_{C_2H_5}$	Cl^-	3–4	0.5^b 3^b
QAE-Sephadex A-25 A-50	$-O-CH_2-CH_2-\overset{+}{N}\big\langle{}^{C_2H_5}_{C_2H_5}$ $CH_3-CHOH-CH_2$	Cl^-	3–3.4	0.3^c 6^c
CM-Sephadex C-25 C-50	$-O-CH_2-CH_2-COOH$	Na^+	4–5	$0.4–0.7^d$ $4.7–7.0^d$
SE-Sephadex C-25 C-50	$-O-CH_2-CH_2-SO_3H$	Na^+	2–2.6	$0.2–0.7^d$ $2.4–3.0^d$

[a] Pharmacia Fine chemicals, Uppsala, Sweden; Sephadex ion-exchangers are available in one standard grade: 40–120μ beads (400–140 mesh)

[b] When equilibrated at pH 8·8 in tris-hydrochloric acid buffer ($\mu = 0.01$)

[c] When equilibrated at pH 8·0

[d] When equilibrated at pH 6·5 in sodium phosphate buffer ($\mu = 0.05$)

Working conditions, and in particular elution conditions, are similar with both ion exchangers. The high total capacity of cross-linked dextrans allows in some cases, the absorption of proteins in solutions relatively stronger in salt concentration than would be possible with ion exchange celluloses. Before use, Sephadex ion exchangers can be converted into the free-base form (anion exchangers) or into free-acid form (cation exchangers), then *immediately* into the suitable salt form, and then finally equilibrated with the starting buffer of the chromatography.

4. Special Ion Exchangers

For certain applications, particularly for the chromatographic separation of nucleic acids of high molecular weight, special ion exchangers have been prepared. The separation of nucleic acids requires the use of slightly adsorbing exchangers; they are either very weak anionic exchangers (ion exchange chromatography) or exchangers, charged groups of which cannot come into contact with the ionic groups of macromolecules to be chromatographied (electrostatic chromatography). Mandel and Hershey[402] prepared a very weak ion exchanger : kieselguhr coated by methylated serum–albumin (MA-kieselguhr). Tener, Gilham and coworkers[403] prepared ion exchangers with masked, charged groups from DEAE-cellulose (0·9 mequiv. g^{-1}) by substitution of the hydroxyl groups of the cellulose by an aromatic acid; they produced benzoylated DEAE-cellulose (B-DEAE-cellulose), 1-naphthoylated DEAE-cellulose (1, N-DEAE-cellulose), benzoylated, 1-naphthoylated-DEAE-cellulose (B, 1-N-DEAE-cellulose). These two types of exchangers adsorb all the nucleic acids (4S–RNA, DNA, ribosomal RNA, phage RNA) in a medium at low salt concentration; different nucleic acids are separated progressively, as the salt concentration of the medium increases (on MA–kieselguhr;[402-406] on B, 1-N-DEAE-cellulose.[403,407,408]) In this case, the main separation is due to differences in the number of phosphate groups in the nucleic acid molecules, and separation appears to depend on molecular weight. Moreover, in each peak different molecular species have slightly different mobilities; this fractionation is based on the secondary structure of the nucleic acids.

These exchangers can also be used in the fractionation (in a same general transfer RNA peak) of various specific aminoacyl acceptor RNAs (on MA-kieselguhr,[404] on B,1-N-DEAE cellulose).[403,409,410] These separations depend on the weak forces of electrostatic attraction, rather than on ion exchange itself. Substitution of the hydroxyl groups of DEAE-cellulose by aromatic acids increases the electrostatic interaction and non-ionic attraction between it and the polynucleotides.[403] These secondary forces, which are dependent upon the secondary structure and the base composition of the polynucleotide chains, can explain the separations obtained. Separative power is lowered in presence of agents (for example urea and alcohols) which disrupt hydrogen bonding and hydrophobic interactions.[403]

Some of these ion exchange adsorbents are commercially available, but they can also be easily prepared.[402,403]

B. Chromatography

In Figures 29–36 the characteristic examples of the separation of biological macromolecules (for example proteins and nucleic acids) by ion exchange chromatography are given below. As regards the chromatography of peptides, oligonucleotides, viruses and phages, precise numerical data is given in the section on applications.

1. Preparation of Ion Exchangers before Use

To obtain reproducible separations, it is essential to use an ion exchanger in a particular ionic form (acid, base or salt) and to establish equilibrium between the ion exchanger and the starting buffer (first eluent). It is also essential that the exchanger contains no impurities. Generally the exchanger is allowed to swell in distilled water and is then washed extensively with 1N NaOH (0·5N NaOH with ion exchange cross-linked dextran), with distilled water, with 1N HCl (0·5N HCl with ion exchange cross-linked dextran), and then with distilled water. This pre-cycling is repeated 3 to 5 times with the synthetic ion exchange resins and ion exchange celluloses, once with the ion exchange cross-linked dextrans. This pre-cycling is necessary to remove the impurities strongly adsorbed on the exchanger.

After pre-cycling, the exchanger is washed with either an acid or an alkali solution to establish the ionic form in which exchanger is to be used. After removing the excess of acid or base with distilled water, the ion exchanger is suspended in the starting buffer, packed in a column, and equilibrated with the starting buffer (the effluent pH—or salt concentration—equals that of the input buffer).

2. Sample

In most cases sample to be analysed must be dissolved in a medium, the composition of which is the same as that of the starting buffer (extensively dialysed against the starting buffer in the case of macromolecular substances). However this is sometimes not necessary when using carboxylic ion exchange resins (see the section on ion exchange resins).

3. Elution

The sample dissolved in the starting buffer is layered on the top of ion exchange column, then the packed column is connected to the eluent reservoir. The components of the sample are adsorbed at the top part of the bed according to their affinities for the ion exchanger. It is possible to fractionate them by judicious choice of eluent. Elution can be accomplished either by using the starting buffer (elution with eluent of constant composition) or by changing the pH and/or the salt concentration of the eluting buffer, either steadily (gradient elution), or discontinuously (stepwise elution).

After adsorption on an exchanger column low in salt concentration (anion exchanger) or at a pH as low as possible (cation exchanger), macromolecular substances are eluted by increasing either the salt concentration or pH. Only very occasionally can two macromolecular substances be separated by an eluent of constant composition (simple elution—see example in Figure 29 below). In preparative chromatography macromolecules are often eluted by several steep variations of eluent composition (stepwise elution—see example in Figure 30 below). To be successful, such an elution requires previous investigation of the elution behaviour of the substances to be separated. In this case elution profiles must always be interpreted with prudence (false peaks may appear).

In biological media, components having similar chromatographic properties are often very numerous and, moreover, these mixtures contain ionic macromolecules, the elution of which requires eluents of very different compositions (and eluting power). Therefore continuous variation in eluting power of the washing solution (gradient elution) is advantageous; gradient elution permits an automatic and gradual attainment of the eluting power required for each macromolecular component. The resolution of complex biological mixtures containing greatly different, strongly sorbed ionic macromolecules in a great number of elution peaks is obtained by gradient elution in a single pass through the ion exchanger column. Another advantage of this elution method is to hinder the broadening of zones (tailing).

a. *Gradient Elution*

As we already said, the elution of ionic macromolecules requires variations in salt concentration and/or pH. Hereafter we shall review methods used to obtain concentration[411] or pH gradients.

i *Concentration Gradient*
Simple mixing devices for concentration gradient elution. In Figures 25 and 26 simple devices for obtaining concentration gradients are shown.

The apparatus for gradient elution allowing an exponential increase of concentration in an eluting agent (concentrations N_1 and N_2 in vessel 1

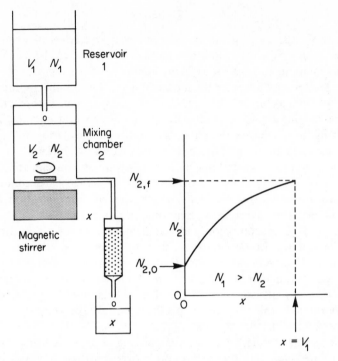

Figure 25. Simple mixing device for exponential concentration
gradient elution [initial values: $V_1, N_{2,0}$; constant values:
N_1, V_2; actual value: N_2; final value: $N_{2,f} = N_1 - (N_1 - N_{2,0})$
e^{-V_1/V_2}]

and vessel 2 respectively) is shown in Figure 25.[412] A flask 2 is fitted onto
a magnetic stirrer and contains the solvent at the lower concentration of
eluting agent (i.e. NaCl: initial concentration $N_{2,0}$, actual concentration
N_2) and is surmounted by a vessel 1 containing the solvent at the higher
concentration of eluting agent (constant concentration N_1). Flask 2 (the
mixing chamber), initially contains the solution of concentration $N_{2,0}$
(constant volume: V_2) used to make the column, and the solution of concentration N_1 is added from vessel 1 (initial volume of solution in reservoir: V_1).
By using the equation

$$N_2 = N_1 - (N_1 - N_{2,0}) \exp\left(-\frac{x}{V_2}\right)$$

in which V_2 is the constant volume of solution in vessel 2, the concentration N_2
of the stronger eluting agent in the vessel 2 can be calculated for any volume x
delivered by the column.

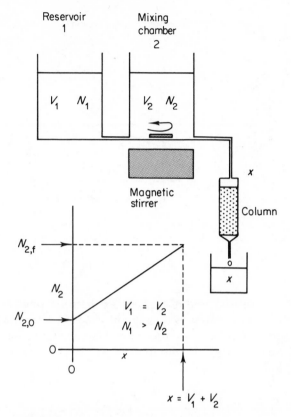

Figure 26. Apparatus for linear concentration gradient elution [initial values: $V_1, V_2, N_{2,0}$; constant value: N_1; actual value: N_2; final value: $N_{2,f} = N_1$]

A mixing device for gradient elution, allowing a linear increase in concentration of an eluting agent (concentrations N_1 and N_2 respectively in vessel 1 and vessel 2) is shown in Figure 26.[413] Two flasks are connected at the bottom by a tube; one of them (the mixing chamber) is mounted on a magnetic stirrer. The mixing chamber 2 initially contains the solution of concentration $N_{2,0}$ (initial volume $V_{2,0}$) used to make the column, and the solution of concentration N_1 ($N_1 > N_{2,0}$) is added from vessel 1 (the reservoir). By using the equation

$$N_2 = N_{2,0} + \frac{N_1 - N_{2,0}}{2V_{2,0}} x$$

in which $V_{2,0} = V_{1,0}$ = the initial volumes of solutions in vessels 2 and 1 respectively, the concentration N_2 of the stronger eluting agent in the vessel 2 can be calculated at any volume x delivered by the column.

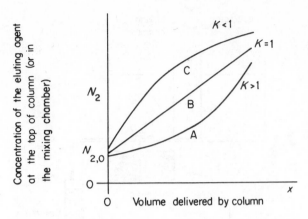

Figure 27. Concentration curves for gradient elution; mixing device shown in Figure 26, with cylindrical vessels of diverse relative diameter sizes; $V_1 > V_2$ (type A curve), $V_1 = V_2$ (type B curve), $V_1 < V_2$ (type C curve)

By altering the relative size of the vessel's diameter (the mixing device shown in Figure 26) one can vary the ratio of V_1 to V_2. The ratio $K = V_1/V_2$ can be adjusted to give concentration curves similar to those depicted in Figure 27. In every case, concentration curves can be calculated by the general equation

$$N_2 = N_1 - \frac{N_1 - N_{2,0}}{V_{2,0}^K}\left[V_{2,0} - \frac{x}{K+1}\right]^K$$

The type of curves with corresponding values and equations are given in Table 21.

Table 21. Typical concentration curves in concentration gradient elution

Type curves	Corresponding K values	K value	$V_1 = KV_2$	Examples Equation
A	$K > 1$	2	$V_1 = 2V_2$	$N_2 = N_1 - \dfrac{N_1 - N_{2,0}}{V_{2,0}^2}\left(V_{2,0} - \dfrac{x}{3}\right)^2$
B	$K = 1$	1	$V_1 = V_2$	$N_2 = N_{2,0} + \dfrac{N_1 - N_{2,0}}{2V_{2,0}}x$
C	$K < 1$	1/2	$V_1 = V_2/2$	$N_2 = N_1 - \dfrac{N_1 - N_{2,0}}{V_{2,0}^{1/2}}\left(V_{2,0} - \dfrac{2x}{3}\right)^{1/2}$

Concentration curves similar to these depicted in Figure 27 can also be obtained with two peristaltic pumps, two vessels, and a magnetic stirrer. This mixing device is shown in Figure 28. Vessel 2 is fitted on a magnetic stirrer and contains initially the lower concentrated solvent ($N_{2,0}$) used to make the column; the more highly concentrated solvent (N_1) is added from vessel 1 at flow rate d_1. If d_2 is the flow rate at which the mixture is added to the column, then the flow rates d_1 and d_2 can be adjusted by peristaltic pumps to give similar curves to those depicted in Figure 28. In the case $d_2 = 2 d_1$, a linear concentration gradient is obtained (type B curve).

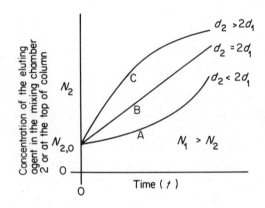

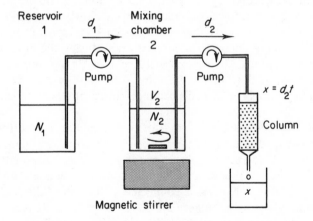

Figure 28. Concentration gradient curves and mixing device including pumps [initial values: $N_{2,0}$, $V_{2,0}$; constant values: N_1, d_1, d_2; actual values: V_2, N_2]

By using the general equation

$$N_2 = N_1 - \frac{N_1 - N_{2,0}}{V_{2,0}^{d_1/d_2 - d_1}}[V_{2,0} - (d_2 - d_1)t]^{d_1/d_2 - d_1}$$

the concentration N_2 of the stronger eluting agent in mixing chamber 2 (see Figure 28) or at the top of the column, can be calculated at any time t. In the case of the linear concentration gradient ($d_2 = 2d_1$) the concentration N_2 of the eluting agent can be calculated by the equation

$$N_2 = N_{2,0} + \frac{N_1 - N_{2,0}}{V_{2,0}}d_1 t$$

or

$$N_2 = N_{2,0} + \frac{N_1 - N_{2,0}}{2V_{2,0}}d_2 t$$

These experimental devices (see Figures 25, 26 and 28) are currently used in the ion exchange chromatography of proteins and nucleic acids with concentration gradients of neutral salts (e.g. NaCl) or buffers (e.g. tris-HCl, tris-CH_3COOH) at constant pH. The type A curve (see Figures 27 and 28) corresponds to separations with minimum tailing.

Other mixing devices for concentration gradient elution. In order to obtain concentration curves with more complicated profiles, Peterson and co-workers[414–416] have described special mixing devices. In their work on the chromatography of serum proteins,[415] these workers used a 'cone-sphere' arrangement (reservoir: Erlenmeyer's flask; mixing chamber: balloon flask) providing a concentration gradient rising slowly in an initial period, then more rapidly towards the end. A 'step–cone–sphere' arrangement gives a more gradual increase in concentration during the first half of the chromatogram, followed by a rapid increase towards the end. The effect of this gradient shape is to elute successively the slightly adsorbed proteins then the more strongly adsorbed substances and at the same time to reduce tailing.

Peterson and coworkers[416–419] also used a mixing device ('varigrad') which makes it possible to effect any desired adjustment in concentration or pH gradient. One example of this apparatus consists essentially of nine identical chambers connected at the bottom by channels and stirred individually. The last chamber is connected at the top of the column. The chambers are filled with buffer or salt mixtures of increasing molarity and/or pH. As the eluent is drawn, the levels in all the chambers fall simultaneously and a concentration and/or pH gradient is produced. This gradient can be controlled by varying the buffer and/or the salt composition of the chambers.[416,417] The salt concentration can be calculated from the theoretical contribution of the individual chambers of the mixing device to the total gradient. The pH curve is empirically determined or calculated from the buffer salt contribution of individual chambers to the gradient.

ii. *pH Gradient*

Usable pH gradients are more difficult to obtain than concentration gradients, since the profile of the titration curves greatly depends on the nature of the base and acid contained in the elution buffer. It is sometimes possible to obtain the pH variation as a linear function of the volume of the eluate;[420] increasing pH gradients are used in chromatography on cation exchange cellulose (P-cellulose;[381] CM-cellulose[421]); decreasing pH gradients are used simultaneously with concentration gradients in the fractionation of certain mixtures of proteins.[415,422]

b. *Constant Components in Eluent*

Besides components whose concentration varies continuously in the course of elution, the eluent often contains components at constant concentration, for instance salts in a buffer when using a concentration gradient in a neutral salt (for example NaCl).

In ion exchange chromatography, the recovery of added material has, in general, been quite good. In some cases where loss of activity has occurred, it may be necessary to include in the eluting buffer either a cofactor or protecting agents (for example, sulphhydryl compounds: cysteine, mercaptoethanol, mercaptoacetic acid; chelating agents: ethylene diamine tetraacetate, glycine, glycylglycine; inorganic ions: manganese, magnesium, phosphate).

C. Applications

1. Peptides

Ion exchange chromatography has provided valuable separation procedures for the peptides of tryptic, chymotryptic and peptic hydrolysates of proteins. The first successful application in this field was the determination of the structure of bovine pancreatic ribonuclease[363,364] by chromatographic fractionation on sulphonated polystyrene resin (150 × 1·8 cm; Dowex 50 × 2; see Table 14), with sodium citrate or acetate buffers of gradually increasing pH and molarity, starting at 0·2N and pH 3·1 and approaching 2N and pH 5·1.[363–365] These procedures have permitted the separation of peptides from the tryptic or chymotryptic hydrolysate of oxidized ribonuclease[363,364] and also of synthetic oligopeptides (di- to tetrapeptides[423]). In spite of the high resolving power of these procedures, in view of certain structural considerations they are definitely at a disadvantage when used to isolate salt-free peptides. The effluent fractions must be desalted by passing them over a column of an ion exchange resin.[364,424,425] Polymethacrylic acid resin (Amberlite IRC 50, see Table 14) has been sometimes used for the fractionation of peptide mixtures[426] under similar conditions. The elution of peptides from Dowex 50 has also been obtained using ammonium acetate or formiate buffers.[427,428] The ammonium salts must be removed by sublimation or evaporation from each effluent fraction but this is time-consuming, and this procedure is rarely used in a laboratory engaged in structural studies of proteins.

Using collidine or pyridine acetate buffer and quarternary ammonium base resin (Dowex 1 × 2, acetate), Braunitzer and coworkers[366,367,369] have succeeded in separating peptides from the tryptic hydrolysate of haemoglobin. For instance Rudloff and Braunitzer[369] have eluted peptides from a Dowex 1 × 2 (acetate) column, using a decreasing pH gradient (collidine or pyridine acetate, starting at pH 8–7·5, and lowered by progressively increasing the concentration of acetic acid from 0·1 M to 2 M). Schroeder and coworkers[429] used either pyridine + α-picoline, collidine, or N-ethylmorpholine acetate buffers (initial pH 8·0, 8·3 and 9·3 respectively) and eluted peptides at progressively decreasing pH by the addition of acetic acid (0·1–2 M). On a Dowex 50 × 2 column with pyridine acetate buffer (0·2 M in pyridine) and with a progressively increasing pH (3·1 to 5·0), Schroeder and coworkers[368,429] have succeeded in separating peptides from haemoglobin. With these procedures, all of which use a volatile buffer and in which none of the components reacts with ninhydrin, peptides of the effluent fractions can be rapidly isolated by lyophilization or evaporation at low temperature (< 40 °C). These procedures, or similar ones, have been used for the fractionation of mixtures of peptides on Dowex 1 × 2;[366,367,369,430–432] on Dowex 50 × 2[368,429,433,437] or Zeo-Karb 225 × 2[438] and on Dowex 50 × 4.[439]

Ion exchange celluloses and cross-linked dextran ion exchangers have rarely been used for the chromatography of peptides; examples of the columns used are DEAE-cellulose[440] and DEAE-Sephadex A 25.[441,442] Chromatographic separation procedures for peptides using a volatile buffer (for example, ammonium acetate buffer[401,443] or pyridine acetate buffer[401,444]) on a P-cellulose column gave excellent results; for example, Canfield and Anfinsen[401] obtained a separation, on P-cellulose column, of peptides produced by the chymotrypsin digestion of carboxymethylated lysozyme using ammonium acetate buffer as eluent (0·02 M to 0·2 M in acetic acid, pH 3·95 to 5·06; then 0·2 M to 0·3 M in acetic acid, pH 4·91 to 6·3; finally 0·3 M in acetic acid, pH 9·0) and also a separation of peptides produced by pepsin digestion of reduced, carboxymethylated lysozyme with pyridine acetate buffer (0·05 M to 0·45 M in acetate, pH 3·93 to 6·12). It may be predicted that the use of a SE-Sephadex column for peptide separations under chromatographic conditions similar to those used with P-cellulose

$(2 \times 1.7\text{–}3.5 \text{ mequiv. g}^{-1})$ and sulphonated polystyrene resin (Dowex 50 × 2 or × 4) will produce interesting results.

2. Enzymes and Other Proteins

The separation of proteins from each other has been performed on columns of polycarboxylic acid ion exchange resins (see Table 14), of ion exchange celluloses (DEAE-cellulose, CM-cellulose, P-cellulose, SE-cellulose; see Tables 15–19), and of cross-linked dextran ion exchangers (DEAE-Sephadex, CM-Sephadex, SE-Sephadex; see Table 20).

a. *Chromatography on Polycarboxylic Acid Resins*

The successful chromatography of enzymes on polycarboxylic acid ion exchange resins,[357,358,365,373,445] has been mainly limited to stable, basic proteins of low molecular weight such as cytochrome c,[357,358,446] lysozymes from chicken egg white,[359] from dog spleen and kidney[447] and from rabbit spleen,[448] pancreatic ribonuclease,[360] modified ribonucleases,[449,450] trypsinogen,[396] α-chymotrypsinogen,[360,396] the least alkaline histones[370] and proteins with an isoelectric pH of about 7–8, such as haemoglobins.[358,361,373,451] Boardman and Partridge[358] investigated quantitatively the effect of variations in the cation concentration and pH of the elution media on the chromatographic behaviour of cytochrome c on Amberlite IRC 50. These authors found that the elution of proteins having a high isoelectric pH ($pH_i = 9\text{–}10$), for example cytochrome c ($pH_i = 10.1$), can be effected by increasing the salt concentration of the medium or increasing the pH of the eluent. For protein chromatography, polymethacrylic acids are used in the pH range 6.5–9. For a given protein, at constant salt concentration, there are only small variations in the elution volume in the pH range 6.8–8.3.[358,446] For proteins of lower isoelectric pH (for example, haemoglobin $pH_i = 8.0$ in citrate buffer[361]) adsorption and elution take place at a lower pH (pH 6.5–8.0) and at relatively high salt concentration (sodium concentration: $0.35 \text{ g ion } l^{-1}$).[358,446] The affinities of proteins for synthetic polycarboxylic acid resins are dependent upon pH and salt concentration, and the equilibration of the resin with the starting buffer is essential for reproducible results.[362] In Table 22, operative conditions in general use with this type of exchanger for chromatography of proteins are given. Figure 29 shows a typical example of chromatographic separation.

b. *Chromatography on Ion Exchange Celluloses*

Ion exchange celluloses are extensively used for the purification of enzymes and other biologically active proteins. The use of ion exchange celluloses as introduced by Sober and Peterson[371] has revolutionized the separation of

Table 22. Chromatography of proteins on polymethacrylic acid resin

Ionic form of ion exchanger	Medium of equilibration of the resin	Chromatography of:	Effluent used
NH_4^+	0·1 M CH_3COONH_4–NH_4OH, pH 9·0	Ox heart cytochrome c (m.w. 12,000; pH_i = 10·1)	Equilibration buffer[357]
Na^+	0·2 M sodium phosphate buffer, pH 7·04, + NaCl (0·34 g ions of Na^+ l^{-1})	Ox heart cytochrome c	Equilibration buffer[358]
Na^+	0·2 M sodium phosphate buffer, pH 7·18	Egg white lysozyme (m.w. 14,000; pH_i = 10)	Equilibration buffer[359]
Na^+	0·2M sodium phosphate buffer, pH 6·02 + NaCl (0·24 g ions of Na^+ l^{-1})	α-chymotrypsinogen (m.w. 22,500; pH_i = 9)	Equilibration buffer[360]
Na^+	0·2M sodium phosphate buffer, pH 6·47 + NaCl (0·24 g ions of Na^+ l^{-1})	Ox pancreatic ribonuclease (m.w. 13,700; pH_i = 8)	Equilibration buffer[360]
Na^+	0·2M sodium citrate, pH 6·5	Foetal haemoglobin, A, B, C-haemoglobin (m.w. 68,000; pH_i ≈ 8)	0·2 to 0·3M sodium citrate, pH = 6·5[361]
Na^+	0·2M potassium phosphate pH 6·0	Cationic proteins from pancreatic juice (non-adsorbed on DEAE-cellulose): trypsinogen, α-chymotrypsinogen	Equilibration buffer[396]

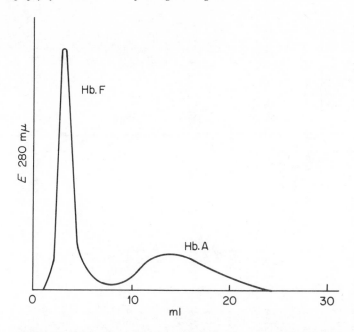

Figure 29. Separation of human adult carbon monoxide haemoglobin (Hb. A) from foetal carbonmonoxide haemoglobin (Hb. F) on a 0·9 × 7 cm. Amberlite İRC-50 column (Na$^+$ salt) sample: 2·5 mg of each haemoglobin; eluent: 0·0476 M sodium citrate buffer, pH 6·5, Na$^+$ concentration made up to 0·20 g ions/l. by addition of NaCl; flow rate: 1·2 ml/h; temperature: 10 °C; adapted from Prins[373]

proteins from biological media. Many complex mixtures of proteins have been fractionated from natural media by ion exchange cellulose chromatography; examples are: egg white proteins,[64,395] proteins of blood serum,[372,452] urinary proteins,[453] proteins of bovine pancreatic juice,[396] and proteins from yeast,[454] from *Bacillus cereus* spores[455] and from *Escherichia coli.*[64]

The chromatographic mobilities of proteins on ion exchange celluloses depend mainly on the number of charged groups that the exchanger has and on the number and nature of the charged groups of the proteins. The effect of the number of charged groups of proteins on their separation has been applied to the separation on CM cellulose[395] and on DEAE-cellulose[64] of three ovalbumins, differing only by absence or presence of 1 or 2 phosphate groups.

i. *Chromatography on Cation-exchange Celluloses*

In a medium in which the pH is as low as possible, proteins can be adsorbed on a *CM-cellulose* column and then displaced by increasing the

Table 23. The relation between pH_i of proteins and the order of their elution after chromatography on CM-cellulose and DEAE-cellulose[a]

Number[b]	Protein	Isoelectric pH of proteins (pH_i)	CM-cellulose pH of eluting buffer	Chromatography on: DEAE-cellulose salt concentrations in eluate (M)[c]:				
				PO_4HK_2	PO_4H_2K	PO_4^{3-} [d]	NaCl	HCl
1	Ovomucoïd A	3.9–4.3	4.3	—	—	—	—	—
2	Flavoprotein	3.9–4.3	4.3	0.0038	0.078	0.0818	0.074	0.0176
3	Ovomucoïd B	—	4.4	—	—	—	—	—
4	Ovalbumin A_1	4.58	4.5	0.0059	0.066	0.0719	0.0585	0.0177
5	Ovalbumin A_2	4.65	4.65	0.0078	0.054	0.0618	0.046	0.014
6	Ovalbumin A_3	4.75	4.85	—	—	—	—	—
7	Globulin F	—	5.2	—	—	—	—	—
8	Conalbumin 1	6.5–6.8	5.8	0.0132	0.0132	0.0264	0	0
9	Conalbumin 2	6.5–6.8	6.1	0.0126	0.0126	0.0252	0	0
10	Globulin I	—	8.0	—	—	—	—	—
11	Globulin J	—	9.3	—	—	—	—	—
12	Globulin K	—	9.4	—	—	—	—	—
13	Avidin	>10	9.5	0.0074	0.0074	0.0148	0	0
14	Globulin M	—	9.9	—	—	—	—	—
15	Lysozyme	10.7–11.3	10.0	0	0	0	0	0

[a] Adapted from Rhodes and coworkers[395] (Chromatography on CM-cellulose) and from Mandeles[64] (Chromatography on DEAE-cellulose)

[b] See Figures 30 and 32

[c] Eluting solution contains also glycin (0.02M) at constant concentration

[d] From PO_4H_2K and PO_4HK_2

pH of the eluent.[371,395] They flow out of the column in the order of their isoelectric pH, the protein with the lowest isoelectric pH flowing first (see Table 23). Besides increasing the pH gradient, increasing the ionic strength of the buffer or increasing the concentration of a neutral salt may also be helpful. Figure 30 shows a typical example of the chromatographic fractionation of proteins from egg-white. Rhodes, Azari and Feney[395] separately eluted the proteins of a mixture, fixed on a CM-cellulose column, by using a

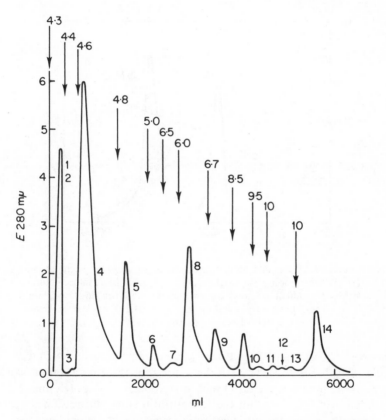

Figure 30. Separation of egg-white proteins of chicken on CM-cellulose column (10 × 2 cm; 0·6 mequiv./g) equilibrated with starting buffer pH 4·3; sample: 30 ml of extensively dialyzed egg white against the starting buffer; elution, primarily, by increasing the pH of the eluting buffer; the proteins having the lowest isoelectric points are eluted first; fractionation with a stepwise elution procedure; numbers and arrows show the pH and point of change of eluting buffers (acetic acid–ammonium hydroxide, 0·1 M with respect to acetate; with the further addition of 0·025 M Na_2CO_3 and 0·2 M Na_2CO_3 to the buffer at 4650 ml and 5285 ml respectively); flow rate: 2·5 ml/min; adapted from Rhodes, Azari and Feeney[395]

medium of pH near that of the isoelectric pH of each protein (see also Table 23 above). The use of pH gradients for the elution of proteins from CM-cellulose columns is particularly valuable.[421,456]

CM-cellulose has uses in chromatography similar to those of polymethacrylic acid resin. Figure 31 shows an example of the separation of haemoglobins from donkey and mouse.[456] CM-cellulose has also been used successfully for the chromatography of trypsin,[457,458] trypsinogen,[458] chymotrypsinogen,[459] pancreatic ribonuclease,[460,461] haemoglobins,[421,456,462-465]

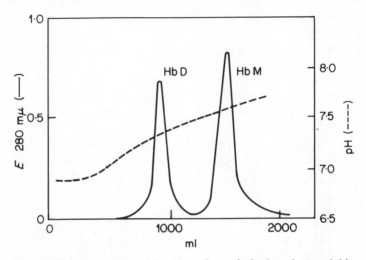

Figure 31. Chromatographic separation of donkey haemoglobin (Hb. D) from mouse haemoglobin (Hb. M) on CM-cellulose column with pH gradient elution; constant flow: 5 ml per hour; column: 50 × 1·8 cm, previously equilibrated with 0·01 M phosphate buffer pH 6·8; see Riggs and Herner[456]

interferon,[466] and D(−) and L(+) lactic dehydrogenase from *Lactobacillus arabinosus*;[467] it has also been used for the separation of proteins having a very high isoelectric pH such as histones,[468-470] polylysines,[471-473] ribosomal proteins,[474,475] for the concentration and purification of specific γ-globulins[415] or for the purification of fragments of γ-globulins.[476]

P-cellulose columns are also used for protein chromatography; elution is effected either by increasing the pH of the eluent (for example, pH 2 to 7[381]) or by increasing the ionic strength of the medium (for example, 0·05 M to 0·5 M phosphate buffer, pH 6·8[477]), or by simultaneously increasing the pH and ionic strength of the medium[452] (for example, 0·04 M phosphate buffer, pH 5·8, to 0·1 M phosphate, pH.7·2); the choice of the range of variations of the pH and ionic strength greatly depends on the nature of the substances to be separated.

Given the same pH and the same salt concentration in medium, a protein will be more strongly fixed on P or SE-cellulose than on CM-cellulose.

ii. *Chromatography on Anion-exchange Celluloses*

DEAE-Cellulose is very frequently used for protein chromatography, often as a first step in the preparation of an enzyme in pure form. On a DEAE-cellulose column, proteins are adsorbed in a medium of very low ionic strength (for example, 0·005 M to 0·02 M) and then displaced by increasing the ionic strength of the medium (0·005 M to 0·5 M of a neutral salt (e.g. NaCl) or a salt buffer) in the order of their isoelectric pH (see Table 23 above). The protein with the lowest isoelectric pH is eluted by the medium of the highest salt concentration, as shown by Mandeles.[64]

In general, chromatography is performed at a constant pH in the range 6–8·8 but sometimes at a decreasing pH, for example, using tris-phosphate (0·005 M to 0·5 M in phosphate) with pH 8·6 to 4·0[415]; various media and concentration gradients are used, including tris-(hydroxymethyl)-amino-methane acetic acid,[478] tris-(hydroxymethyl)-aminomethane hydrochloric acid,[479] tris-(hydroxymethyl)-aminomethane phosphoric acid (0·005 M to 0·5 M in phosphoric acid),[415,478,480] sodium phosphate buffer (0·005 M to 0·5 M)[372,481] potassium phosphate buffer (0·005 M to 0·4 M)[396,482] sodium chloride[372] or lithium chloride[483] (0·005 M to 0·5 M) in various buffers. For the three first buffers, at equal concentrations, the eluting power depends on the nature of the acid and increases in the order: acetic, hydrochloric, phosphoric acid. CO_2 has also been used as an eluent;[484] and eluents containing urea[485–488] or dimethyl formamide[489–491] may be used.

Chromatography on DEAE-cellulose was very successfully used for separation of proteins in complex mixtures, such as serum proteins,[372,415,419,422,452,480,492,493] urinary proteins,[453] egg-white proteins,[64] proteins of pancreatic juice,[396] and proteins from yeast extracts[454] or from spores of *Bacillus cereus*,[455] or *Escherichia coli*.[64] The separation of 7S γ-globulins from 19 S γ-globulins is also very useful.[372,481,494–497] Figure 32 shows the elution profile obtained from separation of proteins of chicken egg-white (for the operative conditions, see legend of Figure 32 and Table 23 above). It is noticeable that the separating power of chromatography on a DEAE or CM-cellulose column (see Figure 30) is higher than the separating power of zone electrophoresis.

Chromatography on DEAE-cellulose has been extensively used for fractionation of mixtures of enzymes and purification of these enzymes. Numerous examples may be given: β-chymotrypsinogen from oxen,[459] leucineaminopeptidase from pig's kidney,[498] pepsinogen from ox's pancreas,[499] malic dehydrogenase from rat's liver,[500] bovine erythrocyte carbonic anhydrase,[501] D(−) and L(+) lactic dehydrogenase from *Lactobacillus*

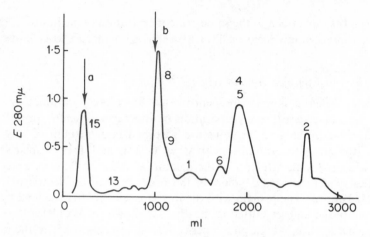

Figure 32. Separation of egg-white proteins of chicken on DEAE-cellulose column (40 × 2·5 cm; 0·7 mequiv./g; equilibrated with 0·02 M glycine); sample: 600 mg of protein extensively dialysed against 0·02 M glycine: elution primarily by increasing concentration of salts; proteins having the highest isoelectric points are eluted first; first gradient (a): mixing chamber—750 ml 0·02 M glycine, reservoir—1 l solution containing 0·02 mole of glycine, 0·02 mole of K_2HPO_4 and 0·02 mole KH_2PO_4; second gradient (b): reservoir volume changed to 1 l solution containing 0·02 mole of glycine, 0·1 mole of KH_2PO_4, 0·1 mole of NaCl and 0·03 mole of HCl; flow rate: 2 ml/min.; for significance of numbers, see Table 23; adapted from Mandeles[64]

arabinosus,[467] and 20-β-hydroxysteroid dehydrogenase from *Pseudomonas testosteroni.*[502]

An example of the chromatographic separation of enzyme is given in Figure 33, namely the separation of the alkaline phosphatase of *Escherichia coli* K 12 from other heat-stable proteins in an extract of this micro-organism.[478]

Certain enzymes have been resolved into several equally active forms by chromatography on DEAE-cellulose;[503–506] this is very valuable in the investigation of the properties and structures of isoenzymes.

DEAE-cellulose was found to be very useful for concentration of proteins from dilute solutions.[372,395,507] As the adsorption capacity is greater at low salt concentrations (0·01–0·005 M of salt), the protein solution is, after dialysis, applied to a DEAE-cellulose column, previously equilibrated at low salt concentration, and then proteins are eluted by a medium at high salt concentration (0·5–2 M of salt).

To obtain good separations on DEAE-cellulose, it is necessary to extensively dialyse the sample against the starting buffer. However, it is sometimes expedient to avoid a dialysis step by diluting the sample solution,

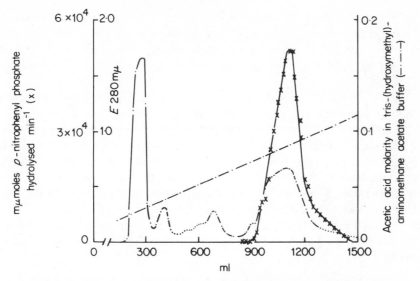

Figure 33. Separation of various heat stable proteins[a] from *Escherichia coli* K_{12}, by chromatography on DEAE-cellulose column (3 × 44 cm); 32 g DEAE-cellulose, 0·88 mequiv./g; elution with increasing concentration gradient: 0·02 M to 0·2 M tris-(hydroxymethyl)-aminomethane acetate buffer, pH 7·7[b], and 0·002 to 0·01 M $MgCl_2$; flow rate: 51 ml/h.; proteins •; alkaline phosphatase (×); Munier[478]
[a] after treatment: 20 min at 70 °C and 5 min at 80 °C
[b] buffer concentration as acetic acid concentration

so that the ionic strength is sufficiently reduced to permit the adsorption of the anionic protein on DEAE-cellulose.[508] Before chromatography, most of the salts can be removed in a preliminary step, in which the solution of proteins is treated with dry Sephadex G 25.[509]

ECTEOLA-cellulose is in practice not used in protein chromatography, since this weak ion exchanger has little affinity for these substances. Because of this property and because of its high adsorption capacity for nucleic acids,[510–511] a column of ECTEOLA-cellulose can be used to remove nucleic acids* from protein solutions.[513]

3. Protein Chains

As we have already said, it is possible to separate protein chains by using molecular sieving chromatography after treatment by both a disulphide splitting reagent (sulphitolyse or reduction and *S*-carboxymethylation) and

---

* Transfer RNA is only eluted by a neutral salt solution (NaCl, KCl etc.) of concentration 0·3M or higher;[514] ribosomal RNA or DNA can only be eluted by a sodium hydroxide solution.[510–512]

by a hydrogen bond splitting medium (urea solution or acid solutions), provided that the molecular weights of the proteins are sufficiently different. For instance, the H-chain (m.w. 55,000) and the L-chain (m.w. 20,000) of human 7S-γ-globulins can be separated on Sephadex G 75 or Sephadex G 100.[207]

When the molecular weights of the chains are similar, their separation can be effected on ion exchangers; carboxylic acid ion exchange resin,[515] CM-cellulose[516-519] and DEAE-cellulose[520,521] can all be used for this. For example, the alpha and beta rabbit haemoglobin chains can be separated by chromatography at low pH on CM-cellulose (0·47 mequiv.g^{-1}), using a gradient from 0·2 M formic acid—0·02 M pyridine to 0·2 M formic acid—0·2 M pyridine as eluent.[516] The B-chain (130 amino acid residues) and C-chain (100 residues) of α-chymotrypsinogen (after sulphitolysis) can be separated by chromatography in 8 M urea medium on DEAE-cellulose (0·91 mequiv. g^{-1}) using 0·01–0·06 M linear gradient of phosphate buffer, pH 8·0.[520]

4. Oligonucleotides

The use of DEAE-cellulose resulted in an important advance in the field of ion exchange chromatography of oligonucleotides. At first, chromatography on DEAE-cellulose only permitted the separation of oligonucleotides up to the pentanucleotide stage on the analytical scale,[522,523] and up to dodecanucleotide stage (in the most favourable cases) on the preparative scale.[524] The first improvement was the use of volatile buffers: triethylamine bicarbonate[525] and ammonium bicarbonate;[399,526] in this way Stahelin, Peterson and Sober[399] separated oligonucleotides with a concentration gradient of 0·01 M to 1 M ammonium carbonate and bicarbonate buffer, pH 8·6.

Tomlinson and Tener[527] showed that the chromatographic separation of oligonucleotides is greatly improved by the inclusion of either urea, formamide or ethylene glycol in the eluent. Secondary binding forces affecting the resolving power of the column chromatography of oligo-nucleotides on DEAE-cellulose are eliminated by these additions in the eluting salt solution. Using this technique Tomlinson and Tener separated oligonucleotides resulting from the action of pancreatic deoxyribonuclease on salmon testis DNA, on DEAE-cellulose column (carbonate form) with a linear gradient of either a sodium acetate buffer (0 to 0·3–0·4 M), pH 7·5, or 5·0, or 4·7, potassium bicarbonate buffer (0 to 0·12 M),[400] sodium chloride (0 to 0·2 M),[528] or ammonium carbonate buffer (0 to 0·4 M), pH 8·4[400,529] in 7 M urea. The nucleotide material which is adsorbed very tightly on DEAE-cellulose, can be eluted with 1 M sodium chloride in 7 M urea.[530] Results of the chromatography of RNA digests in the presence of 7 M urea show that the separation depends on the purine–pyrimidine ratio as well as on the net charge of the oligonucleotide.[531] These processes have found numerous applications,[532-537] particularly in the chromatographic isolation of oligonucleotides from pancreatic DNAse-I digests of salmon testis DNA,[400] from alkaline and enzymatic digests of RNA from MS 2 bacteriophage,[532] from hydrolysates of transfer RNA[530] in pure form for investigation of the structure of nucleic acids.

5. Nucleic Acids

As regards nucleic acids, three kinds of separation can be considered: the chromatographic isolation of low molecular weight RNA (4S transfer RNA, 5S ribosomal RNA), the fractionation of mixtures of amino acid specific acceptor RNA and the separation of various cellular nucleic acids (transfer RNA, DNA, ribosomal RNA).

All *transfer RNA* can be isolated by the chromatography of cellular RNA on a DEAE-cellulose column with 0·02 M potassium phosphate (pH 7·7) and progressively increasing sodium chloride concentration (0·2 to 1·2 M). At first, mononucleotides flow out of the column, then transfer RNA is eluted by sodium chloride solution (concentration ranging from 0·40–0·56 M); ribosomal RNA can only be eluted by 0·1 M sodium hydroxide.[397] Holley and coworkers[538] eluted transfer RNA in 0·1 M tris-(hydroxymethyl)-aminomethane hydrochloride buffer, pH 7·5, containing 1 M sodium chloride. In a similar procedure, transfer RNA can be eluted by lithium chloride.[539] An ECTEOLA-cellulose column can also be used and transfer RNA eluted by 0·3 M sodium perchlorate, pH 7·1.[514]

The *various species of aminoacyl acceptor RNA* can be separated on ion exchangers: MA-kieselguhr,[404] DEAE-cellulose,[398,538-541] DEAE-Sephadex A 50[541-544] and benzoylated and naphthoylated—DEAE-cellulose[403] can all be separated in this way. Early experiments by Stephenson and Zamecnik,[544] Niohiyama and coworkers,[539] Sueoka and Yamane[404] and Kawade and coworkers[543] have shown that various specific aminoacyl acceptor RNAs have different chromatographic mobilities in the elution peak of transfer RNA. Later, the fractionation processes were greatly improved.[398,403,404,540-542] Sueoka and Yamane investigated the first procedure using high separation power: with a MA-kieselguhr column and an increasing concentration gradient of sodium chloride (0·2 M to 0·6 M) at pH 6·7, they obtained several distinct aminoacyl transfer RNA peaks in the elution peak of transfer RNA.[404] Baguley and coworkers investigated the effect of temperature on the separation of various aminoacyl transfer RNAs on a DEAE-cellulose column[541] and on a DEAE-Sephadex A 50 column.[542] Using a DEAE-cellulose column and with the simultaneous application of linear decreasing gradients of both temperature (75 to 30 °C) and sodium chloride concentration (0·75 M to 0·5 M in 0·02 M tris-(hydroxymethyl)-aminomethane hydrochloric acid buffer, pH 8·0), Mirzabekov and

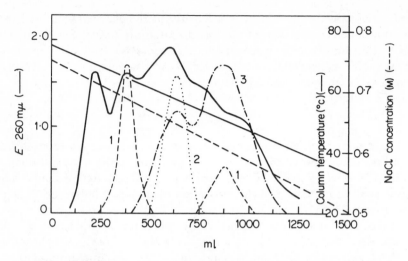

Figure 34. Separation of various aminoacyl transfer RNA by use of a combined decreasing temperature–concentration gradient on DEAE-cellulose column (1·5 × 40 cm); elution with decreasing NaCl gradient 0·75 to 0·5 M in 0·02 M tris-HCl, pH 8·0, and decreasing temperature gradient 75° to 33 °C; flow rate: 100 ml/h; (1) Alanine acceptor activity; (2) Tyrosine acceptor activity; (3) Valine acceptor activity; adapted from Mirzabekov, Krutilina and Bayev[398]

coworkers[398] improved the separating power of the procedure. Figure 34 shows an example of the application of this procedure. Cherayil and Bock[540] found that the relative positions of the aminoacyl transfer RNA depends on pH, on urea concentration in the eluent and on the nature of the chromatographic support (DEAE-cellulose: 0·4 mequiv. g^{-1}; DEAE-Sephadex A 50: 3·3 mequiv. g^{-1}) in the absence of urea. It is also possible to use successively various conditions of chromatography for purification of a particular aminoacyl transfer RNA. Tener, Gilham and coworkers[403] have shown that the substitution of the hydroxyl groups of DEAE-cellulose by aromatic acids increases non-ionic interactions and electrostatic attractions between this chromatographic support and nucleic acids. In this manner, a benzoylated and naphthoylated DEAE-cellulose column can be used to separate individual species of aminoacyl transfer RNA from crude transfer RNA. The various specific aminoacyl transfer RNAs can be eluted from the column by sodium chloride solutions of increasing concentration (0·45 M to 1 M). Certain aminoacyl transfer RNAs (for example, phenylalanine transfer RNA) were eluted with 1 M sodium chloride in 10 per cent 2-methoxyethanol. Under these conditions, the total recovery of the transfer RNA applied to the column was better than 95 per cent. These separations depend not only on ion exchange but also on the weaker forces of attraction (electrostatic attractions) and non-ionic interactions relating to the secondary structure of the various transfer RNAs and to the base composition of the polynucleotide chains. As a result, these separations are strongly disturbed in the presence of urea or alcohols which split hydrogen bonds and prevent hydrophobic interactions.

The efficiency of these methods for the chromatographic preparation of specific aminoacyl acceptor transfer RNAs has been demonstrated by the successful separation of several transfer RNAs in sufficiently pure form to allow the elucidation of their chemical structures.[530,545-548]

The separation of various *cellular nucleic acids* (4S transfer RNA, DNA, 16S and 23S ribosomal RNA) has been achieved on a MA-kieselguhr column[402,404-406,549-551] and on benzoylated and naphthoylated DEAE-cellulose.[403,407,408] Unpurified nucleic acid preparations from bacterial cells[404,405,407,408,549] or mammalian cells[406] and various viral nucleic

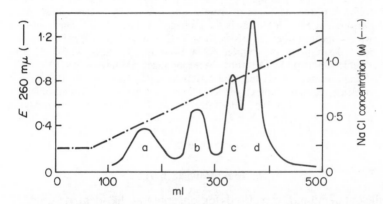

Figure 35. Fractionation of nucleic acids from *E. coli* by methylated serumalbumin column chromatography. Column (1·8 × 11 cm) of methylated serum albumin kieselguhr;[402] sample: 3·3 mg RNA in the starting buffer; elution with NaCl linear gradient 0·2 to 1·2 M in 0·05 M sodium phosphate, pH 6·7; flow rate: 24 ml/h; (a) 4S transfer RNA; (b) DNA; (c) 16 S-ribosomal RNA; (d) 22S ribosomal RNA; see: Sueoka and Yamane[404] and Otaka, Mitsui and Osawa[405]

acids[407,408,552] could be fractionated into their components. Figures 35 and 36 show that nucleic acids are eluted according to their molecular weight as the neutral salt concentration increases in the medium. Four peaks of ultraviolet absorbance, corresponding to transfer RNA, DNA, 16S and 23S ribosomal DNA were resolved. Ribosomal RNA (5S, 16S and 23S RNA) and DNA were separated on a MA-kieselguhr column.[553-555] Three main rapidly renewed RNAs can be detected on the chromatographic elution pattern of cellular nucleic acid on MA-kieselguhr column.[405,406,549]

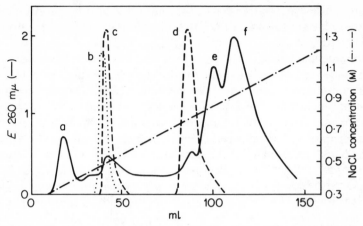

Figure 36. Chromatography on benzoylated naphthoylated DEAE-cellulose of the nucleic acids from *E. coli* infected by bacteriophage and various nucleic acids from bacteriophages; column 1×20 cm; elution with a linear gradient of 0·3 to 1·2 M NaCl, 0·01 M/tris-HCl, pH 7·4; flow rate: 0·5 ml/min.; (a) transfer RNA from *E. coli* (m.w. = 26,000); (b) double-stranded DNA from bacteriophage λ (m.w. = 3×10^7); (c) double-stranded RNA from bacteriophage MS_2 (m.w. = 2×10^6; ribonuclease resistant duplex); (d) RNA infectious material of bacteriophage MS_2 (m.w. = $1·1 \times 10^6$; infectivity in the *E. coli* spheroplast assay); (e) 16S ribosomal RNA from *E. coli* (m.w. = $0·55 \times 10^6$); (f) 23S ribosomal RNA from *E. coli* (m.w. = $1·2 \times 10^6$); adapted from Sedat, Kelly and Sinsheimer[407,408]

6. Ionic Polysaccharides

Efficient fractionation methods for ionic polysaccharides have used DEAE-cellulose,[556–562] ECTEOLA-cellulose,[563–565] DEAE-Sephadex[559,566,567] and Dowex 1.[568] Sodium pectate,[556] sodium hyaluronate,[559,566,567] heparin,[559,560,563,567] keratan sulphate,[564] chondroitin sulphate,[559,567] horatin sulphate,[569] colominic acid,[562,568] fish aminopolysaccharide sulphate,[565] acidic plant polysaccharides,[556,557] acidic polysaccharides from colostrum,[561] and mucopolysaccharides[558,561] have been submitted to ion exchange chromatography. Acidic polysaccharides are eluted in order of increasing molecular weight as hydrochloric ion concentration increases in the eluent (sodium chloride or hydrochloric acid–sodium chloride mixtures).

7. Viruses, Bacteriophages and Rickettsiae

The techniques of column chromatography on ion exchangers have been successfully applied to mammalian viruses,[570–584] plant viruses,[570,585–587] bacteriophages[588–593] and rickettsiae.[571] Chromatographic procedures are very useful for the preparation of purified viruses for further fundamental research. The recovery of viruses and their purification in terms of host material removal (proteins, nucleic acids) and virus degradation products are very good.

For instance, Hoyer and coworkers[571] applied polio virus, type 2, to a DEAE-cellulose column (1 mequiv. g^{-1}) equilibrated with 0·01 M phosphate buffer, and eluted *viruses* by 0·02 M phosphate pH 7·1. Taussig and Creaser[589] purified the basteriophage T_{2r} isolated from *Escherichia coli* by chromatography on an ECTEOLA-cellulose column equilibrated with 0·01 M phosphate buffer of pH 7·0; the *bacteriophage* is eluted by 0·14 M sodium chloride in 0·01 M phosphate buffer.[588,589] Hoyer and coworkers[571] purified a preparation of Q-fever rickettsiae on an ECTEOLA-cellulose column (0·2 mequiv. g^{-1}); yolk sac material was removed by one passage on ion exchange cellulose column and *rickettsiae* eluted by 0·1 M sodium chloride in 0·02 M phosphate buffer.

Various *animal viruses* have been similarly chromatographied on DEAE-cellulose (foot and mouth disease viruses,[584] poliovirus,[570–573,575] adenovirus[576–581] and ECHO virus[571]) or on ECTEOLA-cellulose (Vaccinia virus,[582] Colorado tick fever virus,[571] ECHO virus,[571] Coxsackie virus[571] and poliovirus[571,574]). Various *plant viruses* have been purified on DEAE-cellulose (turnip yellow mosaic virus,[570] tobacco mosaic virus,[585] potato virus X[585]) or on ECTEOLA-cellulose (tobacco mosaic virus[587]). Viruses are adsorbed on ion exchange cellulose at low salt concentration and eluted by increasing concentration in a neutral salt (for example, sodium chloride) or in a buffer.

Carboxylic ion exchange resins[583] or carboxymethycellulose[574,586] have sometimes been used for the chromatographic purification of virus preparations; for example, influenza virus has been purified on the calcium form of Amberlite IRC 50[583] and tobacco mosaic virus on the CM-cellulose.[586]

Similarly, preparations of various *bacteriophages* have been purified by chromatography on ECTEOLA-cellulose (bacteriophages T_1, $T_{2,r}$, T_{2,r^+} isolated from *Escherichia coli*[588,589]) and DEAE-cellulose (bacteriophage G from *Bacillus megatherium*,[590] bacteriophage T_2 from *E. coli*[592]). As with viruses, bacteriophages, in media of low ionic strength, are adsorbed on ion exchange celluloses, and eluted by increasing salt concentration (0 to 0·2 M NaCl). Strongly basic anion exchange resins have sometimes been used for the chromatography of bacteriophages; for example, bacteriophages have been isolated from *Pseudomonas aeruginosa*,[591] and the bacteriophage T_2 from *E. coli*.[593]

Rickettsiae have also been purified by chromatography of ECTEOLA-cellulose; Q-fever rickettsiae and epidemic typhus rickettsiae have both been purified in this way.[571] Particles are also adsorbed at low salt concentration (i.e. 0·02 M phosphate buffer) and eluted at high salt concentration (i.e. 0·1 M sodium chloride in 0·02 M phosphate buffer).

The chromatographic behaviour of viruses, bacteriophages and rickettsiae appears to depend upon the chemical nature of the surface of these infectious particles (i.e. the protein shell) rather than upon their size. Confirmation of this is given by the differences in chromatographic behaviour of various strains of the same virus,[572,575] as may be seen on series of polioviruses.

VI. CONCLUSION

From these examples, we hope to have succeeded in showing the power of chromatographic methods in the characterization and investigation of the evolution of biologically active macromolecules, and also in the preparation of some of them at a high degree of purity with the intention, perhaps, of determining their chemical structure. In fact, purifications involving several successive chromatographic methods based upon different physico-chemical phenomena may often lead to highly pure substances.

The methods described here can certainly be applied to numerous other and as yet unexplored fields, but one can predict that most important advances to be made in the future will be closely related to the discovery and investigation of new chromatographic supports.

REFERENCES

1. O. Holmberg, *Biochem. Z.*, **258**, 134 (1933); *Ark. Kemi*, **11-B**, No. 4 (1933)
2. D. French and D. N. Knapp, *J. Biol. Chem.*, **187**, 463 (1950)
3. E. Starkenstein, *Biochem. Z.*, **24**, 210 (1910)
4. Y. Toknoka, *J. Agric. Chem. Soc. Japan*, **13**, 586 (1937)
5. J. D. Hockenhall and D. Herbert, *Biochem. J.*, **39**, 102 (1945)
6. S. Schwimmer and A. K. Ball, *J. Biol. Chem.*, **179**, 1063 (1969)
7. E. Norbert and D. French, *J. Amer. Chem. Soc.*, **72**, 1202 (1950)
8. P. S. Thayer, *J. Bacteriol.*, **66**, 656 (1953)
9. J. H. Hasch and K. W. King, *J. Biol. Chem.*, **232**, 381 (1958)
10. F. L. Myers and O. H. Northcoth, *Biochem. J.*, **71**, 749 (1959)
11. G. de la Haba, *Biochim. Biophys. Acta*, **59**, 672 (1962)
12. L. S. Lerman, *Proc. Nat. Acad. Sci. U.S.*, **39**, 232 (1953)
12a. L. S. Lerman, *Nature*, **172**, 635 (1953)
13. C. Arsenis and D. M. McCormick, *J. Biol. Chem.*, **239**, 3093 (1964); **241**, 330 (1966)
14. D. H. Campbell, E. Luescher and L. S. Lerman, *Proc. Nat. Acad. Sci. U.S.*, **37**, 575 (1951)
15. A. Malley and D. H. Campbell, *J. Amer. Chem. Soc.*, **85**, 488 (1963)
16. H. C. Isliker, *Ann. N.Y. Acad. Sci.*, **57**, 225 (1953)
17. M. Sela and E. Katchalski, *J. Amer. Chem. Soc.*, **76**, 129 (1954)
18. N. Grubhofer and L. Schleith, *Z. physiol. Chem.*, **297**, 108 (1954)
19. G. Manecke and K. E. Gilbert, *Naturwiss*, **42**, 212 (1955)
20. L. Gyenes, B. Rose and A. H. Sehon, *Nature*, **181**, 1465 (1958)
21. A. E. Gurvitch, R. B. Kapner and R. S. Nezlin, *Biokhimya*, **24**, 144 (1959)
22. L. H. Kent and J. H. R. Slade, *Nature*, **183**, 325 (1959); *Biochem. J.*, **77**, 12 (1960)
23. E. Bar, quoted by M. Sela and E. Katchalski, *Adv. Protein. Chem.*, **14**, 457 (1959)

24. S. Saha, *J. Chromatog.*, **7**, 155 and 165 (1962)
25. Y. Yagi, K. Eugel and J. Pressman, *J. Immunol.*, **85**, 375 (1960)
26. M. Richter, P. Delorme, S. Grant and B. Rose, *Can. J. Biochem. Physiol.*, **40**, 471 (1962)
27. A. T. Jagendorf, A. Patchornik and M. Sela, *Biochim. Biophys. Acta*, **78**, 516 (1963)
28. N. R. Moudgal and R. R. Porter, *Biochim. Biophys. Acta*, **71**, 185 (1963)
29. E. A. Kabat and M. M. Mayer, *Experimental Immunochemistry*, second edition, C. C. Thomas, Springfield, 1961, p. 78
30. D. N. Karsanov and P. A. Solodkov, *Zhur. Priklad. Khim.*, **16**, 11 (1953)
31. L. Gyenes and A. H. Sehon, *Canad. J. Biochem. Physiol.*, **38**, 1235 (1960)
32. R. S. Nezlin, *Biokhim*, **24**, 486 (1960)
33. T. Webb and C. Lapresle, *J. Exp. Med.*, **114**, 43 (1961)
34. P. T. Gilham, *J. Amer. Chem. Soc.*, **84**, 1311 (1962)
35. A. Adler and A. Rich, *J. Amer. Chem. Soc.*, **84**, 3977 (1962)
36. E. K. F. Bautz and B. D. Hall, *Proc. Nat. Acad. Sci. U.S.*, **48**, 400 (1962)
37. E. K. F. Bautz and B. D. Hall, *Biochem. Biophys. Res. Comm.*, **9**, 192 (1962)
38. P. T. Gilham, *J. Amer. Chem. Soc.*, **86**, 4982 (1964)
39. H. G. Khorana, *Chem. Revs*, **53**, 145 (1953); *J. Amer. Chem. Soc.*, **76**, 3517 (1954)
40. H. G. Khorana, *Some Recent Developments in the Chemistry of Phosphate Esters of Biological Interest*, J. Wiley & Sons Inc., New York, 1961, pp. 115–116
41. H. G. Khorana and J. P. Vizzolyi, *J. Amer. Chem. Soc.*, **83**, 675 (1961)
42. R. K. Ralph and H. G. Khorana, *J. Amer. Chem. Soc.*, **83**, 2926 (1961)
43. J. D. Watson and F. H. C. Crick, 9ème Conseil International de Chimie Solvay; *Les proteines*, R. Stoops, Brussells, 1953, pp. 110–112
44. P. T. Gilham and W. E. Robinson, *J. Amer. Chem. Soc.*, **86**, 4985 (1964)
45. M. Edmonds and R. Abrams, *J. Biol. Chem.*, **238**, 1186 (1963)
46. E. T. Bolton and B. J. MacCarthy, *J. Mol. Biol.*, **8**, 201 (1964); *Proc. Nat. Acad. Sci. U.S.*, **48**, 1390 (1962)
47. D. B. Cowie and B. J. MacCarthy, *Proc. Nat. Acad. Sci. U.S.*, **50**, 537 (1963)
48. E. T. Bolton and B. J. MacCarthy, *Proc. Nat. Acad. Sci. U.S.*, **50**, 156 (1963)
49. C. A. Knight, *J. Exp. Med.*, **83**, 11 (1946)
50. G. K. Hirst, 'Virus host cell relation', in *Viral and Rickettsial Infections of Man*, 3rd edn., Rivers and Horsfall, Philadelphia, 1959, pp. 96–144
51. R. M. Franklin, H. Rubin and C. A. Davis, *Virology*, **3**, 96 (1957)
52. G. K. Hirst, *Science*, **94**, 22 (1941)
53. O. Lahelle and T. G. Ward, *J. Immunol.*, **67**, 75 (1951)
54. H. D. Matheka and O. Armbruster, *Virology*, **6**, 584 (1958)
55. D. F. Bradley, *Nature*, **195**, 622 (1962)
56. J. M. Ghuysen and M. Welsch, *C.R. Soc. Biol.*, **146**, 1812 (1952)
57. F. L. Leloir and S. H. Goldenberg, *Methods in Enzymology* (Colowick and Kaplan, editors), **V**, 145 (1962), Academic Press, New York
58. F. L. Leloir, M. A. Rougine, R. de Fekite and C. E. Cardini, *J. Biol. Chem.*, **236**, 636 (1961)
59. H. C. Isliker, *Ann. N.Y. Acad. Sci.*, **57**, 225 (1953)
60. C. F. Crampton, A. M. Benson, J. L. Rodeheavor and A. E. Wade, *Federation Proc.*, **17**, 206 (1958)
61. R. Fraenkel and C. F. Crampton, *J. Biol. Chem.*, **237**, 3200 (1962)
62. L. Mindich and R. D. Hotchkiss, *Biochim. Biophys. Acta*, **80**, 93 (1964)
63. J. Shack and S. Bynum, *Nature*, **184**, 635 (1959)
64. S. Mandeles, *J. Chromatog*, **3**, 256 (1960)
65. M. J. Jayle, J. Moretti and W. Dobryocka, *Bull. Soc. chim. Biol.*, **45**, 301 (1963)

66. B. M. Pogell, *Biochem. Biophys. Res. Commun.*, **7**, 225 (1962)
67. R. Blostein and W. J. Rutter, *J. Biol. Chem.*, **238**, 3280 (1963)
68. M. B. Rhodes, C. L. Marsh and G. W. Kelley, *Exptl. Parasitol*, **15**, 403 (1964)
69. M. M. Behrens, J. K. Inman and W. E. Vannier, *Arch. Biochem. Biophys.*, **119**, 411 (1967)
70. L. Zeichmeister, *Progress in Chromatography (1938–47)*, Chapman and Hall, London, 1951, 2nd edition, pp. 254–262
71. Ch. Grundmann, 'Chromatographie und verwandte Methoden in der Enzymchemie', in *Die Methoden der Enzymforschung* (E. Bamann and M. Myrbäck, editors), G. Thieme, Leipzig, 1940, pp. 1452–1466
72. S. P. Colowick, 'Separation of proteins by use of adsorbents', in *Methods in Enzymology* (S. P. Colowick and N. O. Kaplan, editors), Academic Press, New York, 1955, Vol. 1, p. 90
73. C. A. Zittle, *Adv. Enzymol*, **14**, 319 (1953)
74. F. Turba, *Adv. Enzymol*, **22**, 445 (1960)
75. S. Keller and R. J. Block, 'Adsorption chromatography' in *A Laboratory Manual of Analytical Methods of Proteins Chemistry* (P. Alexander and R. J. Block, editors), Pergamon, London, 1960, Vol. 1, p. 70
76. R. L. M. Synge, *Arch. Biochem. Biophys.*, **Suppl. 1**, 1 (1962)
77. H. Fraenkel-Conrat, B. Singer and A. Tsugita, *Virology*, **14**, 54 (1961)
78. T. J. Brownhill, A. S. Jones and M. Stacey, *Biochem. J.*, **73**, 434 (1959)
79. G. Alderton, W. R. Ward and H. L. Fevold, *J. Biol. Chem.*, **157**, 43 (1945)
80. R. Willstätter and H. Kraut, *Ber*, **56**, 1117 (1923)
81. M. Dixon and E. C. Webbs, *Enzymes*, Longmans, London, 1964, 2nd edition, p. 42
82. D. Keilin and E. F. Hartree, *Proc. Roy. Soc.*, **B-124**, 397 (1938)
83. S. Bach, M. Dixon and L. G. Zerfas, *Biochem. J.*, **40**, 229 (1946)
84. S. M. Swingle and A. Tiselius, *Biochem. J.*, **48**, 171 (1951)
85. K. Wakabayashi, S. Kamei, H. Murakami and N. Shimazono, *J. Biochem. (Japan)*, **52**, 464 (1962)
86. S. Block and N. G. Wright, *J. Biol. Chem.*, **213**, 51 (1955)
87. V. E. Price and R. E. Greenfield, *J. Biol. Chem.*, **209**, 363 (1954)
88. S. S. Korde, *Arch. Biochem. Biophys.*, **101**, 278 (1963)
89. Ch. Rivière, G. Gautron and M. Thely, *Bull. Soc. chim. Biol.*, **29**, 600 (1947)
90. A. Tiselius, *Arkiv. Kemi*, **7**, 443 (1954)
91. A. Tiselius, S. Hjerten and O. Levin, *Arch. Biochem. Biophys.*, **65**, 132 (1956)
92. M. J. Coon, *Biochem. prep.*, **9**, 83 (1962)
93. O. Levin, 'Column chromatography of proteins: calcium phosphate', in *Methods in Enzymol* (S. P. Colowick and N. O. Kaplan, editors), Academic Press, New York, 1962, Vol. 5, p. 27
94. S. Hjerten, *Biochim. Biophys. Acta*, **31**, 216 (1959)
95. L. Broman and K. Kjellin, *Biochim. Biophys. Acta*, **82**, 105 (1964)
96. H. A. Sober and E. A. Peterson, 'Chromatography of the plasma proteins', in *Plasma proteins* (F. W. Putnam, editor), Academic Press, New York, 1960, Vol. 1, p. 107
97. R. L. Munier and P. Chovin, 'L'analyse chromatographique', in *Techniques de Laboratoire*, J. Loiseleur, Masson, Paris, 3rd edition, 1963, p. 447 and 466
98. R. Dedonder, E. Jozon, G. Rapoport, Y. Joyeux and A. Fitsch, *Bull. Soc. Chim. Biol.*, **45**, 483 (1963)
99. W. F. Anacker and V. Stoy, *Biochem. Z.*, **330**, 141 (1958)
100. S. L. Steelman, *Biochim. Biophys. Acta*, **27**, 405 (1958)

101. D. N. Ward, R. F. MacGregor and A. C. Griffin, *Biochim. Biophys. Acta*, **32**, 305 (1959)
102. C. H. Li, *Adv. Protein. Chem.*, **12**, 270 (1957)
103. J. E. Courtois, F. Petek and T. Dong, *Bull. Soc. Chim. Biol.*, **45**, 95 (1963)
104. H. Euler and A. Fono, *Ark. Kemi*, **25A**, 15 (1947)
105. A. Schoeberl and P. Rambacher, *Biochem. Z.*, **305**, 223 (1940)
106. L. Zeichmeister, G. Toth and M. Balint, *Enzymologia*, **5**, 302 (1938)
107. L. Zeichmeister and G. Toth, *Enzymologia*, **7**, 165 (1939); *Naturwiss*, **27**, 367 (1939)
108. L. Zeichmeister, G. Toth, P. Furth and J. Barsony, *Enzymologia*, **9**, 155 (1940)
109. L. Zeichmeister, G. Toth and E. Vajda, *Enzymologia*, **7**, 170 (1939)
110. G. Bernardi, *Biochem. Biophys. Res. Comm.*, **6**, 54 (1961)
111. G. Semenza, *Ark. Kemi*, **11**, 89 (1957)
112. F. Brown, J. F. E. Newman and D. L. Stewart, *Nature*, **197**, 590 (1963)
113. R. D. Lamb and G. R. Dubes, *Anal. Biochem.*, **7**, 152 (1964)
114. G. Bernardi and S. N. Timasheff, *Biochem. Biophys. Res. Commun.*, **6**, 58 (1961)
115. C. A. Knight, 'Chemistry of viruses', in *Protoplasmatologie Handbuch der Protoplasmaforschung* (L. V. Heilbrunn, editor), Springer Verlag, Vienna, 1963, **IV**, 2
116. K. H. Muench and P. Berg, *Biochemistry*, **5**, 982 (1966)
117. V. Harding, H. Schauer and G. Hartmann, *Biochem. Z.*, **346**, 212 (1966)
118. R. L. Pearson and A. D. Kelmers, *J. Biol. chem.*, **241**, 767 (1966)
119. R. Main and L. J. Cole, *Arch. Biochem. Biophys.*, **68**, 181 (1957)
120. G. L. Cantoni, *Colloque intern. C.N.R.S.*, Paris, **106**, 201 (1962)
121. R. Main, M. J. Wilkins and L. G. Cole, *J. Amer. Chem. Soc.*, **81**, 6490 (1959)
122. P. Faulkner, E. M. Martin, S. Sved, R. C. Valentine and T. S. Work, *Biochem. J.*, **80**, 597 (1961)
123. J. Taverne and P. Wildy, *Nature*, **184**, 1655 (1959)
124. J. Taverne, J. H. Marshall and F. Fulton, *J. Gen. Microbiol.*, **19**, 451 (1958)
125. P. Wildy, M. J. P. Stoker, I. A. MacPherrson and R. W. Horne, *Virology*, **11**, 444 (1960)
126. V. Riley, *J. Nat. Cancer Inst.*, **11**, 199 (1950)
127. M. Simons, *Acta Virol*, **6**, 302 (1962)
128. H. K. Miller and R. W. Schlesinger, *J. Immunol.*, **75**, 155 (1955)
129. L. H. Fromhagen and C. A. Knight, *Virology*, **8**, 198 (1959)
130. W. A. Woods and F. C. Robbins, *Proc. Nat. Acad. Sci. U.S.*, **47**, 1501 (1961)
131. A. B. Sabin, *J. Expt. Med.*, **56**, 307 (1932)
132. B. Gelotte and A. B. Krantz, *Acta Chem. Scand.*, **13**, 2127 (1959)
133. J. Porath and P. Flodin, *Nature*, **183**, 1657 (1959)
134. H. Deuel and H. Neukon, 'Natural plant hydrocolloïdes', *Adv. Chem.*, **11**, 51 (1954)
135. B. Lindqvist and T. Stogards, *Nature*, **175**, 511 (1955)
136. G. H. Lathe and C. R. J. Ruthven, *Biochem. J.*, **62**, 665 (1956)
137. P. Flodin and B. Ingelman, *British Patent*, No. 854,715 (1960); *Swedish Patent*, No. 169,293 (1959)
138. P. Flodin and J. Porath, *Biochim. Biophys. Acta*, **63**, 403 (1962)
139. P. Flodin and J. Porath, *U.S. Patent*, No. 3,002,823 (1961)
140. P. Flodin and J. Killander, *Biochim. Biophys. Acta*, **63**, 403 (1962)
141. K. O. Pedersen, *Arch. Biochem. Biophys.*, **Suppl. 1**, 157 (1962)
142. A. Polson, *Biochim. Biophys. Acta*, **50**, 565 (1961)
143. P. Andrews, *Nature*, **196**, 36 (1962)
144. S. Bengtsson and L. Philipson, *Biochim. Biophys. Acta*, **79**, 399 (1964)

145. S. Hjerten, *Arch. Biochem. Biophys.*, **99**, 466 (1962)
146. S. Hjerten, *Biochim. Biophys. Acta*, **79**, 393 (1964)
147. S. Hjerten, *Arch. Biochem. Biophys.*, **Suppl. 1**, 147 (1962)
148. J. N. Wilson, *J. Amer. Chem. Soc.*, **62**, 1583 (1940)
149. D. de Vault, *J. Amer. Chem. Soc.*, **65**, 532 (1943)
150. A. A. Levi, *Biochem. J.*, **43**, 257 (1948)
151. E. Glueckauf, *Nature*, **160**, 301 (1947)
152. J. R. Whitaker, *Anal. Chem.*, **35**, 1950 (1963)
153. J. P. Ebel, G. Dirheimer and J. H. Weil, *C.R. Ac. Sci.*, **255**, 2312 (1962)
154. K. Granath and P. Flodin, *Makromol. chem.*, **48**, 160 (1961)
155. J. Porath, *Biochim. Biophys. Acta*, **39**, 193 (1960)
156. P. Flodin, *J. Chromatog.*, **5**, 103 (1961)
157. B. Gelotte, *J. Chromatog.*, **3**, 330 (1960)
158. R. L. Munier and G. Sarrazin, *Bull. Soc. Chim. France*, 2940, 2941 (1963)
159. C. J. O. R. Morris, *J. Chromatog.*, **16**, 167 (1964)
160. B. Johansson and L. Rymo, *Acta Chem. Scand.*, **16**, 2067 (1962)
161. H. Determann, *Experientia*, **18**, 430 (1962)
162. P. Andrews, *Biochem. J.*, **91**, 222 (1964)
163. B. Johansson and L. Rymo, *Acta Chem. Scand.*, **18**, 217 (1964)
164. P. Fasella, A. Ginstosio and C. Turano, 'Applications of thin layer chromato-
     graphy on Sephadex to the study of proteins', in *Thin Layer Chromatography*
     (G. B. Marini-Bettolo, editor), Elsevier, Amsterdam, 1964, p. 205
165. P. Flodin, *Thesis, University of Uppsala*, Sweden, 47 (1962)
166. H. Determann and B. Gelotte, 'Gel filtration', in *Biochemisches Taschenbuch*,
     H. M. Rauen, Springer Verlag, Berlin (1964), **2**, 905
167. H. Fritz, I. Trautschold and E. Werle, *Z. physiol. Chem.*, **342**, 253 (1965)
168. P. Flodin and J. Porath, 'Molecular sieve processes', in *Chromatography* (Ed.
     E. Heftmann), Reinhold Publ. Corp., New York, 1961, pp. 328–343
169. D. Eaker, quoted by J. Porath, *Adv. Protein Chem.*, **17**, 209 (1962)
170. A. A. Leach and P. C. O'Shea, *J. Chromatog.*, **17**, 245 (1965)
171. H. Determann and W. Michel, *J. Chromatog.*, **25**, 303 (1966)
172. P. Andrews and S. J. Folley, *Biochem. J.*, **87**, 3 P (1963)
173. P. Andrews, *Biochem. J.*, **96**, 595 (1965)
174. J. Porath, *Clin. Chim. Acta*, **4**, 776 (1959); *Biochim. Biophys. Acta*, **39**, 193 (1960)
175. J. Oudin, *C.R. Acad. Sci.*, **222**, 115 (1946); *Bull. Soc. Chim. Biol.*, **29**, 140 (1947)
176. J. Oudin, *Methods in Med. Research*, **5**, 335 (1951)
177. O. Ouchterlony, *Acta path. microbiol. Scand.*, **25**, 189 (1948); **26**, 516 (1949)
178. P. Grabar and C. A. Williams, *Biochim. Biophys. Acta*, **17**, 67 (1955)
179. C. Araki, *Bull. Soc. Chem. (Japan)*, **29**, 543 (1956)
180. S. Hjerten and J. Porath, *Methods Biochem. Anal.*, **9**, 193 (1962)
181. S. Hjerten, *Biochim. Biophys. Acta*, **53**, 514 (1961)
182. S. Hjerten, *Biochim. Biophys. Acta*, **62**, 445 (1962)
183. C. Araki, *4th Intern. Congress Biochem.*, Vienna, 1958; **1**, 15 (1959)
184. C. Araki, *J. Chem. Soc. (Japan)*, **58**, 1338 (1937)
185. B. Russell, T. H. Mead and A. Polson, *Biochim. Biophys. Acta*, **86**, 169 (1964)
186. S. Hjerten, *J. Chromatog.*, **12**, 510 (1963)
187. F. Miller and H. Metzger, *J. Biol. Chem.*, **240**, 3327 (1965)
188. G. Salvatore, G. Vecchio, M. Salvatore, H. J. Cahnmann and J. Robbins, *J. Biol.
     Chem.*, **240**, 2942 (1965)
189. B. Öberg and L. Philipson, *Arch. Biochem. Biophys.*, **119**, 504 (1967)
190. R. L. Erikson and J. A. Gordon, *Biochem. Biophys. Res. Commun.*, **23**, 422 (1966)

191. D. W. Kingsburg, *J. Mol. Biol.*, **18**, 195 (1966)
192. J. Killander, S. Bengtsson and L. Philipson, *Proc. Soc. Exptl. Biol. Med.*, **115**, 861 (1964)
193. K. Fridborg, S. H. Hjerten and A. Höglund, *Proc. Nat. Acad. Sci.*, *U.S.*, **54**, 513 (1965)
194. S. Raymond and L. Weintraub, *Science*, **130**, 711 (1959)
195. L. Ornstein and B. J. Davis, *Meeting of the Blood Soc.*, April 1959
196. H. Ott, *Med. Welt*, **51**, 2697 (1960)
197. P. E. Hermans, W. F. MacGuckin, B. F. McKenzie and E. D. Baynard, *Proc. Staff. Meet Mayo Clin.*, **35**, 792 (1960)
198. S. Raymond and Y. J. Wang, *Anal. Biochem.*, **1**, 1391 (1960)
199. D. J. Lea and A. H. Sehon, *Can. J. Chem.*, **40**, 159 (1962)
200. S. Hjerten and R. Mosbach, *Anal. Biochem.*, **3**, 109 (1962)
201. H. G. Boman and S. Hjerten, *Arch. Biochem. Biophys.*, **Suppl. 1**, 276 (1962)
202. C. G. O. R. Morris and P. Morris, *Separation Methods in Biochemistry*, Interscience Publ., New York (1963), pp. 392, 393
203. W. G. Laver, J. Pye and G. L. Ada, *Biochim. Biophys. Acta*, **81**, 177 (1964)
204. V. Rabek and V. Mansfield, *Experientia*, **19**, 151 (1963)
205. J. S. Marks, R. D. Marshall, A. Neuberger and H. Poppkoff, *Biochim. Biophys. Acta*, **63**, 340 (1962)
206. A. B. Edmundson, *Nature*, **198**, 354 (1963)
207. J. B. Fleischman, R. H. Pain and R. R. Porter, *Arch. Biochem. Biophys.*, **Suppl. 1**, 174 (1962)
208. B. Van Hoften and J. Porath, *Biochim. Biophys. Acta*, **64**, 1 (1962)
209. B. Humbel, *Biochem. Biophys. Res. Comm.*, **12**, 333 (1963)
210. D. E. Olins and G. M. Edelman, *J. Exptl. Med.*, **116**, 635 (1962)
211. E. B. Lindner, A. Elmquist and J. Porath, *Nature*, **184**, 1565 (1959)
212. J. Porath, *Svensk Kem. Tidskr*, **74**, 306 (1962), (C.A. **58**, 8220 h, 1963)
213. K. Brunfeldt, *Science Tools* (L.K.B. Inst. J.), **12**, 5 (1965)
214. C. J. Hepstein and C. B. Anfinsen, *Biochemistry*, **2**, 461 (1963)
215. I. Swensson, H. G. Boman, H. G. Eriksson and K. Kjellin, *J. Mol. Biol.*, **7**, 266 (1963)
216. M. A. Ruthenberg, T. P. King and L. C. Craig, *Biochemistry*, **4**, 11 (1965)
217. F. S. Markland, B. Ribadeau-Dumas and E. L. Smith, *J. Biol. Chem.*, **242**, 5174 (1967)
218. T. I. Pristoupil, *J. Chromatog.*, **19**, 64 (1965)
219. P. Flodin, *Thesis, University of Uppsala*, 49 (1962)
220. B. Lindqvist, *Acta Chem. Scand.*, **16**, 1794 (1962)
221. G. E. Cornell and R. N. Shaw, *Can. J. Biochem. Physiol.*, **39**, 1013 (1961)
222. L. Bosch and H. Bloemendal, quoted by H. Bloemendal, *J. Chromatog.*, **3**, 509 (1960)
223. N. R. Ringertz and P. Richard, *Acta Chem. Scand.*, **14**, 303 (1960)
224. N. R. Ringertz, *Acta Chem. Scand.*, **14**, 312 (1960)
225. H. D. Matheka and G. Wittmann, *Zbl. Bakt.*, **182**, 169 (1961)
226. P. Flodin, *Thesis, University of Uppsala*, 50 (1962)
227. A. Bill, W. Marsden and N. Ulfvendahl, *Scand. J. Chem. Lab. Invest.*, **12**, 392 (1960) [C.A. **59**, 4265 e (1963)]
228. J. Coleman and B. Vallee, *Biochemistry*, **1**, 1083 (1962)
229. J. Porath, *Clin. Chim. Acta*, **4**, 771 (1959)
230. M. Kakei and G. B. J. Glass, *Proc. Soc. Exptl. Biol. Med.*, **111**, 270 (1962)
231. K. Lindstrand, *Acta Med. Scand.*, **173**, 605 (1963)

232. R. Gräsbeck, K. Simons and I. Sinkkonen, *Ann. Med. Exp. Biol. Fenniae*, **40**, Suppl. 6 (1962)
233. R. M. Donaldson and J. H. Katz, *J. Clin. Invest.*, **42**, 534 (1963)
234. K. W. Daisley, *Nature*, **191**, 868 (1961)
235. B. Gelotte, *Naturwiss*, **48**, 554 (1961)
236. S. Zadrazil, Z. Šormova and F. Šórm, *Coll. Czeck. Chem. Comm.*, **26**, 2643 (1961)
237. S. E. Bresler, K. M. Rubina, R. A. Graevskaya and N. N. Vasileva, *Biokhimiya*, **26**, 745 (1961)
238. G. R. Shepherd and D. F. Pedersen, *J. Chromatog.*, **9**, 445 (1962)
239. W. D. Davidson, M. A. Sackner and M. H. Davidson, *J. Lab. Clin. Med.*, **62**, 501 (1963)
240. L. Jacobson, *Clin. Chim. Acta*, **7**, 180 (1962)
241. G. Ostling, *Acta Soc. Med. Uppsaliensis*, **64**, 222 (1960)
242. A. Lundblad and I. Berggard, *Biochim. Biophys. Acta*, **57**, 129 (1962)
243. R. A. Mitchell, H. H. Kang and L. N. Henderson, *J. Biol. Chem.*, **238**, 1151 (1963)
244. R. J. Kisling, *Biochim. Biophys. Acta*, **40**, 531 (1963)
245. E. T. Bucovaz and J. W. Davis, *J. Biol. Chem.*, **236**, 2015 (1961)
246. R. L. M. Synge and N. A. Youngson, *Biochem. J.*, **78**, 31 P (1961)
247. R. J. Hill and W. Köningsberg, *J. Biol. Chem.*, **235**, PC 21 (1960)
248. H. Bennich, *Biochim. Biophys. Acta*, **51** 265 (1960)
249. A. W. Phillips and P. A. Gibbs, *Biochem. J.*, **81**, 551 (1961)
250. V. Stepanov, D. Handschuch and F. A. Anderer, *Z. Naturforsch*, **16 B**, 626 (1961)
251. B. Lindqvist, *Proc. 16th Intern. Dairy Congr*, 673 (1962)
252. Y. Ito, T. Fugii and M. Otakae, *J. Biochem.* (Japan), **52**, 223 (1962)
253. J. H. Glick, *Arch. Biochem. Biophys.*, **100**, 192 (1963)
254. G. Guidotti, R. J. Hill and W. Köningsberg, *J. Biol. Chem.*, **237**, 2184 (1962)
255. J. Montreuil, G. Biserte and A. Chosson, *C.R. Acad. Sci.*, **256**, 3372 (1963)
256. V. V. Mosolov and M. D. Loginova, *Dokl. Acad. Nauk USSR*, **146**, 1209 (1962) (C.A. **58**, 3664 a, 1963)
257. L. Colobert and G. Dirheimer, *Biochim. Biophys. Acta*, **54**, 455 (1961)
258. Y. C. Lee and R. Montgomery, *Arch. Biochim. Biophys.*, **97**, 9 (1962)
259. C. Nolan and E. L. Smith, *J. Biol. Chem.*, **237**, 446 (1962)
260. C. Nolan and E. L. Smith, *J. Biol. Chem.*, **237**, 453 (1962)
261. S. Hakamori and T. Ishimoda, *J. Biochem.* (Japan), **52**, 250 (1962)
262. R. Montgomery and Y. C. Wu, *Biochem. Biophys. Res. Comm.*, **11**, 249 (1963)
263. J. D. Gregory and L. Roden, *Biochem. Biophys. Res. Comm.*, **5**, 430, (1961)
264. A. R. Sanderson, J. L. Strominger and S. G. Nathenson, *J. Biol. Chem.*, **237**, 3603 (1962)
265. J. M. Ghuissen and J. L. Strominger, *Biochemistry*, **2**, 1110 (1963)
266. E. H. Graul and W. Stumpf, *Dtsch. Med. Wochenschr.* **88**, 1886 (1963)
267. E. H. Mougey and J. W. Mason, *Anal. Biochem.*, **6**, 223 (1963)
268. Š. Lissitzky, J. Bismuth and M. Rolland, *Clin. Chim. Acta*, **7**, 183 (1962)
269. S. Lissitzky and J. Bismuth, *Clin. Chim. Acta*, **8**, 269 (1963)
270. S. Lissitzky, J. Bismuth and C. Simon, *Nature*, **199**, 1002 (1963)
271. L. Jacobsson and G. Widström, *J. Clin. Lab. Invest.*, **14**, 285 (1962)
272. W. Stumpf and E. H. Graul, *Med. Klin.*, **58**, 192 (1963)
273. J. L. Rabinowitz, B. Shapiro and P. Johnson, *J. Nuclear. Med.*, **4**, 139 (1963)
274. B. Schapiro and J. L. Rabinowitz, *J. Nuclear. Med.*, **3**, 417 (1962)
275. P. Flodin, *Thesis, University of Uppsala*, 52 (1962)
276. C. B. Anfinsen and E. Haber, *J. Biol. Chem.*, **236**, 1361 (1961)
277. H. Neumann, R. F. Goldberger and M. Sela, *J. Biol. Chem.*, **239**, 1536 (1964)

278. M. A. Gordon, M. R. Edwards and V. Tompkins, *N.Y. State Dept. Health. Ann. Rept. Div. Lab. Res.*, **82** (1961)
279. C. C. Curtain, *J. Histochem. Cytochem.*, **9**, 484 (1961)
280. J. E. Fothergill and R. C. Nairn, *Nature*, **192**, 1073 (1961)
281. W. George and K. W. Walton, *Nature*, **192**, 1188 (1961)
282. H. Rinderkneckt, *Nature*, **193**, 167 (1962)
283. M. A. Gordon, M. R. Edwards and V. N. Tompkins, *Proc. Soc. Exptl. Biol. Med.*, **109**, 96 (1962)
284. M. Wagner, *Zbl. Bakt*, **185**, 124 (1962)
285. W. Lip, *J. Histochem. Cytochem.*, **9**, 458 (1961)
286. J. Zwaan and A. F. Van Dam., *Acta Histochem.*, **11**, 306 (1961)
287. T. Tokumaru, *J. Immunol.*, **89**, 195 (1962)
288. W. Nultsch, *Biochim. Biophys. Acta*, **59**, 213 (1962)
289. J. Killander, K. Ponten and L. Roden, *Nature*, **192**, 182 (1961)
290. N. V. Thoai, R. Kassab and L. A. Pradel, *Biochim. Biophys. Acta*, **73**, 574 (1963)
291. A. M. Crestfield, W. H. Stein and S. Moore, *Arch. Biochem. Biophys.*, **Suppl. 1**, 217 (1962)
292. P. M. Sanfelippo and J. G. Surak, *J. Chromatog.*, **13**, 148 (1964)
293. J. Porath, *J. Appl. Chem.*, **6**, 233 (1963)
294. J. Porath and A. V. Schally, *Endocrinology*, **70**, 738 (1962)
295. C. J. Epstein and C. B. Anfinsen, *Biochemistry*, **2**, 461 (1963)
296. H. Determann and R. Köhler, *Liebig's Ann. Chem.*, **690**, 197 (1965)
297. L. Tantoni, G. Vivaldi and S. Carta, *J. Chromatog.*, **28**, 55 (1967)
298. G. Lundblad and M. Lundblad, *Arkiv. Kemi.*, **20**, 137 (1962)
299. J. F. Largier and A. Polson, *Biochim. Biophys. Acta*, **79**, 626 (1964)
300. A. M. D. C. Battle, *J. Chromatog.*, **28**, 82 (1967)
301. Th. Wieland, P. Duesberg and H. Determann, *Biochem. Z.*, **337**, 303 (1963)
302. K. Kakiuchi, S. Kato, A. İmanishi and T. İsemura, *J. Biochem.* (Japan), **55**, 102 (1964)
303. J. P. Hummel and W. J. Dreyer, *Biochim. Biophys. Acta*, **63**, 530 (1962)
304. A. Ullmann, P. R. Vagelos and J. Monod, *Biochem. Biophys. Res. Comm.*, **17**, 86 (1964)
305. M. Lee and J. R. Debro, *J. Chromatog.*, **10**, 68 (1963)
306. W. V. Epstein and M. Tam., *J. Chromatog.*, **6**, 258 (1961)
307. B. Gelotte, P. Flodin and S. Killander, *Arch. Biochem. Biophys.*, **Suppl. 1**, 319 (1962)
308. L. Hana and B. Styk, *Acta Virol.* (Prague), **6**, 479 (1962)
309. J. Killander and P. Flodin, *Vox. Sang.*, **7**, 113 (1962)
310. G. Lundblad, *Acta. Chem. Scand.*, **15**, 212 (1961)
311. M. Hvatum, J. Jonsen and E. Kass, *Acta. Rheum. Scand.*, **8**, 289 (1962)
312. J. Porath, *Adv. Protein. Chem.*, **17**, 209 (1962)
313. D. R. Whitaker, R. K. Hanson and P. K. Datta, *Canad. J. Biochem.*, **41**, 671 (1963)
314. G. Peterson and J. Porath, *Biochim. Biophys. Acta*, **67**, 9 (1963)
315. G. Peterson, *Biochim. Biophys. Acta*, **77**, 665 (1963)
316. G. Peterson, E. B. Croling and J. Porath, *Biochim. Biophys. Acta*, **67**, 1 (1963)
317. J. R. Brown, R. N. Greenshields, M. Yamasaki and H. Neurath, *Biochemistry*, **2**, 867 (1963)
318. G. M. Bernier and F. W. Putnam, *Nature*, **200**, 223 (1963)
319. D. V. Hoang, M. Rovery and P. Desnuelle, *Arch. Biochem. Biophys.*, **Suppl. 1**, 232 (1962)
320. G. Biserte and Y. Moschetto, *C.R. Acad. Sci.*, **255**, 3263 (1962)
321. A. Henschen, *Acta Chem. Scand.*, **16**, 1037 (1962)

322. S. J. Leach, J. M. Swan and L. A. Holt, *Biochim. Biophys. Acta*, **78**, 196 (1963)
323. I. Swensson, H. G. Boman, K. G. Erikson and K. Kjellin, *J. Mol. Biol.*, **7**, 254 (1963)
324. G. Dirheimer, S. Muller-Felter, J. H. Weil, M. Yacoub and J. P. Ebel, *Bull. Soc. chim. Biol.*, **45**, 875 (1963)
325. G. Dirheimer and J. P. Ebel, *Bull. Soc. chim. Biol.*, **46**, 399 (1964)
326. K. A. Granath, *J. Coll. Sci.*, **13**, 308 (1958)
327. J. Heller and S. Schramm, *Biochim. Biophys. Acta*, **81**, 97 (1964)
328. P. Nordin, *Arch. Biochem. Biophys.*, **99**, 101 (1962)
329. G. L. Miller, *Anal. Biochem.*, **2**, 133 (1960)
330. F. E. Cole and K. W. King, *Biochim. Biophys. Acta*, **81**, 122 (1964)
331. M. Takenchi and Y. Tanaka, *Kurume Med. J.*, **10**, 37 (1963)
332. P. Flodin and K. Aspberg, *Biological Structure and Function*, Academic Press, New York, (1961), Vol. I, p. 345
333. P. Flodin, *Thesis, University of Uppsala*, 56 (1962)
334. P. Flodin, J. D. Gregory and J. Roden, Quoted by P. Flodin—*Thesis, University of Uppsala*, 57 (1962)
335. U. Beiss and R. Marx, *Naturwiss.*, **49**, 142 (1962)
336. R. L. Steere and C. K. Ackers, *Nature*, **194**, 114 (1962)
337. H. Stegmann and V. Loeschcke, *Z. Naturforsch*, **18 B**, 195 (1963)
338. R. L. Munier, 'Chromatographie de relargage sur colonne', in *Techniques de Laboratoire*—J. Loiseleur, Masson, Paris, 1963, 3ème édition, Vol. I(1), p. 572
339. H. K. Mitchell, M. Gordon and F. A. Haskins, *J. Biol. Chem.*, **180**, 1071 (1959)
340. R. L. Munier, *Mises au point Chim. Anal.* (Masson, Paris), **15**, 278 (1967)
341. J. Porath, *Nature*, **196**, 47 (1962)
342. A. Tiselius, *Arkiv. Kemi*, **26 B**, No. 1 (1948)
343. C. C. Shepard and A. Tiselius, *Discuss. Faraday Soc.*, **7**, 275 (1949)
344. H. K. Mitchell and F. A. Haskins, *Science*, **110**, 278 (1949)
345. R. Schwimmer, *Nature*, **171**, 442 (1953)
346. R. N. Sargent and D. L. Graham, *Anal. Chim. Acta*, **30**, 101 (1964)
347. P. Simonart and K. V. Chow, *Experientia*, **14**, 356 (1951)
348. J. Porath and S. Delin, quoted by J. Porath, *Nature*, **196**, 47 (1962)
349. P. Flodin, *Thesis, University of Uppsala*, pp. 53, 54 (1962)
350. V. T. Riley, *Science*, **107**, 573 (1948)
351. C. C. Shepard, *J. Immunol.*, **68**, 179 (1952)
352. V. T. Riley, *J. Natl. Cancer Inst.*, **2**, 199 (1950) (C.A., **45**, 3019 i, 1951)
353. V. T. Riley, *J. Natl. Cancer Inst.*, **2**, 215 (1950) (C.A., **45**, 3019 d, 1951)
354. V. T. Riley and M. W. Woods, *Proc. Soc. Exptl. Biol. Med.*, **73**, 92 (1950)
355. V. T. Riley, M. L. Hasselbach, S. Fiala, M. W. Woods and D. Burk, *Science*, **109**, 361 (1949)
356. C. E. Fjellström, *Acta Path. Microb. Scand.*, **54**, 439 (1962)
357. S. Paleus and J. B. Neilands, *Acta Chem. Scand.*, **4**, 1024 (1950)
358. N. K. Boardman and S. M. Partridge, *Biochem. J.*, **59**, 543 (1955); *Nature*, **171**, 208 (1953)
359. H. H. Tallan and W. H. Stein, *J. Biol. Chem.*, **200**, 507 (1953)
360. C. H. W. Hirs, *J. Biol. Chem.*, **205**, 93 (1953)
361. T. H. S. Huisman and H. K. Prins, *J. Lab. Clin. Med.*, **46**, 255 (1955); *Nature*, **175**, 903 (1955)
362. C. H. W. Hirs, S. Moore and W. H. Stein, *J. Biol. Chem.*, **200**, 493 (1953); *J. Amer. Chem. Soc.*, **73**, 1893 (1951)
363. C. H. W. Hirs, W. H. Stein and S. Moore, *J. Biol. Chem.*, **221**, 151 (1956)

364. C. H. W. Hirs, S. Moore and W. H. Stein, *J. Biol. Chem.*, **219**, 623 (1956)
365. S. Moore and W. H. Stein, *Adv. Protein. Chem.*, **11**, 191 (1956)
366. K. Hilse and G. Braunitzer, *Z. Naturforsch.*, **14**, 603 (1959)
367. B. Liebold, K. Hilse, K. Simon and G. Braunitzer, *Z. physiol. Chem.*, **315**, 271 (1959)
368. W. A. Schroeder, R. T. Jones, J. Cormick and K. MacCalla, *Anal. Chem.*, **34**, 1570 (1962)
369. V. Rudloff and G. Braunitzer, *Z. physiol. Chem.*, **323**, 129 (1961)
370. C. F. Crampton, S. Moore and W. H. Stein, *J. Biol. Chem.*, **215**, 789 (1955); *Federation Proc.*, **15**, 237 (1956)
371. H. A. Sober and E. A. Peterson, *J. Amer. Chem. Soc.*, **76**, 1711 (1954); **78**, 751 (1956)
372. H. A. Sober, F. J. Gutter, M. M. Wyckoff and E. A. Peterson, *J. Amer. Chem. Soc.*, **78**, 756 (1956)
373. H. K. Prins, *J. Chromatog.*, **2**, 471 (1959)
374. J. D. Guthrie and A. L. Bullock, *Ind. Eng. Chem.*, **52**, 935 (1960)
375. D. D. Gilboe and R. M. Bock, *J. Chromatog.*, **17**, 149 (1965)
376. G. F. Davidson and T. P. Nevell, *J. Textile Inst.*, **46**, 407 (1955); **39**, 87 (1948)
377. J. Porath, *Arkiv. Kemi.*, **11**, 97 (1957)
378. P. Karrer, H. König and E. Usteri, *Helv. Chim. Acta*, **26**, 1301 (1943)
379. A. J. Head, N. F. Komber, R. P. Miller and R. A. Wells, *J. Chem. Soc.*, 3418 (1958)
380. K. Katmura and S. Nonaka, *J. Soc. Textile and Cellulose Industries* (Japan), **13**, 24 (1957)
381. J. H. McClendon and J. H. Kreisler, *Anal. Biochem.*, **5**, 295 (1963)
382. F. C. MacIntire and J. R. Schenck, *J. Amer. Chem. Soc.*, **70**, 1193 (1948)
383. G. P. Toucy and J. E. Kiefer (Eastman Kodak C°), *U.S. Patent* 2,759, 787 (1956)
384. M. Hartmann, *U.S. Patent* 1,777,970 (1930)
385. R. Hoffpauir and J. D. Guthrie, *Textile Research J.*, **20**, 617 (1950)
386. E. C. Yackel and W. O. Kenyon, *J. Amer. Chem. Soc.*, **64**, 121 (1942)
387. T. Wieland and A. Berg, *Angew. Chem.*, **64**, 418 (1952)
388. J. Walravens, *Arch. Intern. Physiol.*, **60**, 191 (1952)
389. R. R. Benerito, B. B. Woodward and J. D. Guthrie, *Anal. Chem.*, **37**, 1693 (1965)
390. G. Champetier, E. Kelecsenyi-Dumesnil, G. Montegudet and J. Petit, *C.R. Ac. Sci.*, **243**, 269 (1956)
391. G. Champetier, G. Montegudet and J. Petit, *C.R. Ac. Sci.*, **240**, 1896 (1955)
392. R. L. Munier, A. M. Drapier and Ch. Thommegay, to be published
393. G. Semenza, *Helv. Chim. Acta*, **43**, 1057 (1960)
394. H. Veder, *J. Chromatog.*, **10**, 507 (1963)
395. M. B. Rhodes, P. R. Azari and R. E. Feeney, *J. Biol. Chem.*, **230**, 399 (1958)
396. P. J. Keller, E. Cohen and N. Neurath, *J. Biol. Chem.*, **233**, 344 (1958)
397. R. Monier, M. L. Stephenson and P. C. Zamecnik, *Biochim. Biophys. Acta*, **43**, 1 (1960)
398. A. D. Mirzabekov, A. I. Krutilina and A. A. Bayev, *Biochim. Biophys. Acta*, **129**, 426 (1966)
399. M. Staehelin, E. A. Peterson and H. A. Sober, *Arch. Biochem. Biophys.*, **85**, 289 (1959)
400. R. V. Tomlinson and G. M. Tener, *Biochemistry*, **2**, 697 (1963)
401. R. E. Canfield and C. Anfinsen, *J. Biol. Chem.*, **238**, 2684 (1963)
402. J. D. Mandell and A. D. Hershey, *Anal. Biochem.*, **1**, 66 (1960)

403. G. M. Tener, I. Gilham, S. Millward, D. Blew, M. von Tigerstrom and E. Wimmer, *Biochemistry*, **6**, 3043 (1967)
404. N. Sueoka and T. Yamane, *Proc. Nat. Acad. Sci. U.S.*, **48**, 1454 (1962)
405. E. Otaka, H. Mitsui and S. Osawa, *Proc. Nat. Acad. Sci. U.S.*, **48**, 426 (1962)
406. R. Monier, *Bull. Soc. Chim. Biol.*, **44**, 109 (1962)
407. J. W. Sedat, R. B. Kelly and R. L. Sinsheimer, *J. Mol. Biol.*, **26**, 537 (1967)
408. R. B. Kelly and R. L. Sinsheimer, *J. Mol. Biol.*, **29**, 229 (1967)
409. E. Wimmer, I. H. Maxwell and G. M. Tener, *Biochemistry*, **7**, 2623 (1968)
410. I. H. Maxwell, E. Wimmer and G. M. Tener, *Biochemistry*, **7**, 2629 (1968)
411. P. Lebreton, *Bull. Soc. Chim. France*, 2188 (1960)
412. A. Cherkin, F. E. Martinez and M. S. Dunn, *J. Amer. Chem. Soc.*, **75**, 1244 (1953)
413. C. W. Parr, *Proc. Biochem. Soc. 324th meeting*, **XXVII**, 56 (1954)
414. H. A. Sober and E. A. Peterson, 'Chromatography of proteins and nucleic acids', In *Ion Exchangers in Organic Chemistry and Biochemistry* (C. Calmon and T. R. E. Kressman, editors), Interscience Publishing Co., New York (1957)
415. H. A. Sober and E. A. Peterson, *Federation Proc.*, **17**, 1116 (1958)
416. E. A. Peterson and J. Rowland, *J. Chromatog.*, **5**, 330 (1961)
417. E. A. Peterson and H. A. Sober, *Federation Proc.*, **17**, 288 (1958)
418. E. A. Peterson and H. A. Sober, *Anal. Chem.*, **31**, 857 (1959)
419. E. A. Peterson and E. A. Chiazze, *Arch. Biochem. Biophys.*, **99**, 137 (1962)
420. B. F. Horton, *J. Chromatog.*, **27**, 263 (1967)
421. T. H. J. Huisman, E. A. Martin and A. Dozy, *J. Lab. Clin. Med.*, **52**, 312 (1958)
422. J. L. Fahey, P. F. MacCoy and M. Goulian, *J. Clin. Invest.* **37**, 272 (1958)
423. Y. P. Dowmont and J. S. Fruton, *J. Biol. Chem.*, **197**, 271 (1952)
424. A. Dreze, S. Moore and E. J. Bigwood, *Anal. Chim. Acta*, **11**, 554 (1954)
425. R. B. Merifield and D. W. Woolley, *J. Amer. Chem. Soc.*, **78**, 358 (1956)
426. W. J. Dreyer and H. Neurath, *J. Biol. Chem.*, **217**, 527 (1955)
427. A. R. Thompson, *Biochem. J.*, **61**, 253 (1955)
428. B. A. Askonas, P. N. Campbell, C. Godin and T. S. Work, *Biochem. J.*, **61**, 105 (1955)
429. W. A. Schroeder, R. T. Jones, J. R. Shleton, J. B. Shleton, J. Cormick and K. MacCalla, *Proc. Nat. Acad. Sci. U.S.*, **47**, 811 (1961)
430. F. A. Anderer, H. Uhlig, E. Webber and G. Schramm, *Nature*, **186**, 922 (1960)
431. C. M. Tsung, G. Funatsu and J. D. Young, *Arch. Biochem. Biophys.*, **105**, 42 (1964)
432. G. Funatsu, A. Tsugita and H. Fraenkel-Conrat, *Arch. Biochem. Biophys.*, **105**, 25 (1964)
433. H. Matsubara and E. L. Smith, *J. Biol. Chem.*, **238**, 2732 (1963)
434. M. Dautrevaux, F. Crouwy, Y. Moschetto and G. Biserte, *Bull. Soc. Chim. Biol.*, **48**, 1111 (1966)
435. F. S. Markland, B. Ribadeau-Dumas and E. L. Smith, *J. Biol. Chem.*, **242**, 5174 (1967)
436. J. Vanacek, B. Meloun, V. Kostka, B. Keil and F. Šorm, *Biochim. Biophys. Acta*, **37**, 169 (1960)
437. G. Kreil and H. Tuppy, *Nature*, **192**, 1121 (1961)
438. V. Tomasek, V. Holeysovsky, O. Mikes and F. Šorm, *Biochim. Biophys. Acta*, **38**, 570 (1960)
439. R. E. Canfield, *J. Biol. Chem.*, **238**, 2691 (1963)
440. R. Got, J. Font, R. Bourillon and P. Cornillot, *Biochim. Biophys. Acta*, **74**, 247 (1963)
441. P. R. Carnegie, *Nature*, **192**, 658 (1961)
442. M. J. Crumpton and J. M. Wilkinson, *Biochem. J.*, **94**, 545 (1965)

443. T. C. Merigan, J. W. Dreyer and A. Berger, *Biochim. Biophys. Acta*, **62**, 122 (1962)
444. W. R. Holmquist and W. A. Schroeder, *J. Chromatog.*, **26**, 465 (1967)
445. S. Moore and W. H. Stein, *Ann. Rev. Biochem.*, **21**, 521 (1952)
446. N. R. Boardman and S. M. Partridge, *J. Polymer. Sci.*, **12**, 281 (1954)
447. P. Jollès and C. Fromageot, *Biochim. Biophys. Acta*, **19**, 91 (1956)
448. G. Jollès and C. Fromageot, *Biochim. Biophys. Acta*, **11**, 95 (1953)
449. C. B. Anfinsen, *Biochim. Biophys. Acta*, **17**, 593 (1955); *J. Biol. Chem.*, **221**, 405 (1956)
450. F. M. Richards, *Compt. Rend. Trav. Lab. Carlsberg*, Ser. Chim., **29**, 329 (1955)
451. M. Morrison and J. L. Cook, *Science*, **122**, 920 (1955)
452. K. B. Cooke, M. P. Tombs, R. D. Weston, F. Souter and N. F. MacLagan, *Clin. Chim. Acta*, **4**, 779 (1959)
453. S. H. Jackson, A. W. Farmer, R. J. Farmer, R. J. Slater and M. S. De Wolfe, *Cand. J. Biochem.*, **39**, 881 (1961)
454. J. D. Duerksen and H. Halvorson, *J. Biol. Chem.*, **233**, 1112 (1958)
455. R. Doi, H. Halvorson and B. Church, *J. Bacteriol.*, **77**, 43 (1959)
456. A. Riggs and A. E. Hermer, *Proc. Nat. Acad. Sci. U.S.*, **48**, 1664 (1962)
457. G. Krampitz and K. Knappen, *J. Chromatog.*, **5**, 174 (1961)
458. I. E. Liener, *Arch. Biochem. Biophys.*, **88**, 216 (1960)
459. M. Rovery, O. Guy and P. Desnuelle, *Biochim. Biophys. Acta*, **42**, 554 (1960)
460. G. Taborsky, *J. Biol. Chem.*, **234**, 2652 (1959)
461. R. Shapira, *Anal. Biochem.*, **4**, 322 (1962)
462. F. J. Gutter, E. A. Peterson and H. A. Sober, *Arch. Biochem. Biophys.*, **80**, 353 (1959)
463. A. Saha, *Biochim. Biophys. Acta*, **32**, 259 (1959)
464. M. F. Perutz, L. K. Steinrauf, A. Stockwell and A. D. Bangham, *J. Mol. Biol.*, **1**, 402 (1959)
465. T. H. Huisman and A. Dozy, *J. Chromatog.*, **7**, 180 (1962)
466. G. P. Lampson, A. A. Tytell, M. Nemes and M. R. Hilleman, *Proc. Soc. Exp. Biol. NY*, **112**, 468 (1963)
467. A. M. Snoswell, *Biochim. Biophys. Acta*, **35**, 574 (1959)
468. L. D. Johnson, A. Driedger and A. M. Marks, *Canad. J. Biochem.*, **42**, 795 (1964)
469. E. W. Johns, D. M. P. Phillips, P. Simon and J. A. V. Butler, *Biochem. J.*, **77**, 631 (1960)
470. E. W. Johns, *Biochem. J.*, **92**, 55 (1964)
471. J. W. Stewart and M. A. Stahmann, *J. Chromatog.*, **9**, 232 (1962)
472. R. Aron, M. Sela, A. Yaron and H. A. Sober, *Biochemistry*, **4**, 948 (1965)
473. I. H. Lima, M. A. Smith and R. Stáhmann, *Anal. Biochem.*, **10**, 318 (1965)
474. J. P. Waller and J. I. Harris, *Proc. Nat. Acad. Sci. U.S.*, **47**, 22 (1961)
475. P. Spitnik-Elson, *Biochim. Biophys. Acta*, **80**, 594 (1964)
476. A. L. Grossberg, P. Stelas and D. Pressman, *Proc. Nat. Acad. Sci. U.S.*, **48**, 1203 (1962)
477. H. G. Wood, S. H. G. Allen, R. Stjernholm and B. Jacobson, *J. Biol. Chem.*, **238**, 547 (1963)
478. R. L. Munier, 'Chromatographie d'échange d'ions', in *Techniques de Laboratoire en Chimie Biologique et en Chimie-Physique*, J. Loiseleur, p. 543 and 544, Masson, Paris (1963)
479. C. Levinthal and A. Garen, *Biochim. Biophys. Acta*, **38**, 472 (1960)
480. H. A. Sober and E. A. Peterson, 'Protein chromatography', in *Aminoacids, Proteins and Cancer Biochemistry* (J. P. Greenstein Memorial Symposium) Academic Press, New York, p. 61 (1960)

481. V. Blaton, *Whatman Techn. Bull.*, **IE 2** (second edition), 21 (1967)
482. H. Neurath, J. A. Rupley and B. L. Vallee, *Aminoacids, Proteins and Cancer Biochemistry* (J. P. Greenstein Memorial Symposium), Academic Press, New York, p. 43 (1960)
483. J. E. Silbert, Y. Nagai and J. Gross, *J. Biol. Chem.*, **240**, 1509 (1965)
484. M. A. Mitz and S. S. Yanari, *J. Amer. Chem. Soc.*, **78**, 2649 (1956)
485. E. O. P. Thompson and I. J. O'Donnell, *Aust. J. Biol. Sci.*, **13**, 393 (1960); **14**, 461 (1961)
486. B. R. Dumas, J. L. Maubois, G. Mocquot and J. Garnier, *Biochim. Biophys. Acta*, **82**, 494 (1964)
487. A. Veis and J. Aneshey, *J. Biol. Chem.*, **240**, 3899 (1965)
488. S. Duraiswami, W. H. MacShun and R. K. Meyer, *Biochim. Biophys. Acta*, **86**, 156 (1964)
489. D. H. Simmonds and D. J. Winzor, *Nature*, **189**, 306 (1961)
490. Y. H. Oh and C. W. Gehrke, *Anal. Biochem.*, **10**, 148 (1965)
491. C. W. Wrigley, *Aust. J. Biol. Sci.*, **18**, 193 (1965)
492. S. H. Lawrence and D. C. Benjamin, *Clin. Chim. Acta*, **6**, 398 (1961)
493. O. W. Neuhaus, R. Havez and G. Biserte, *C.R. Soc. Biol.* (Paris), **156**, 1105 (1962)
494. H. B. Levy and H. A. Sober, *Proc. Soc. Exptl. Biol. Med.*, **103**, 250 (1960)
495. D. R. Stanworth, *Nature*, **188**, 156 (1960)
496. J. L. Fahey and E. G. Morrison, *J. Lab. Clin. Med.*, **55**, 912 (1960)
497. J. L. Fahey, *J. Biol. Chem.*, **237**, 440 (1962)
498. J. E. Folk, J. A. Gladner and T. Viswanatha, *Biochim. Biophys. Acta*, **36**, 256 (1959)
499. I. E. Liener, *Biochim. Biophys. Acta*, **37**, 522 (1960)
500. C. J. R. Thorne, *Biochim. Biophys. Acta*, **42**, 175 (1960)
501. S. Lindskog, *Biochim. Biophys. Acta*, **39**, 218 (1960)
502. H. J. Hübener, F. G. Sahrholz, J. Schmidt, G. Nesemann and J. Junk, *Biochim. Biophys. Acta*, **35**, 270 (1959)
503. B. Foltmann, *Acta Chem. Scand.*, **14**, 2059 (1960)
504. K. A. Trayser and S. P. Colowick, *Arch. Biochem. Biophys.*, **94**, 177 (1961)
505. W. J. Reeves and G. M. Fimognari, *J. Biol. Chem.*, **238**, 3853 (1963)
506. T. Wieland and G. Pfleiderer, *Angew. Chem.*, **69**, 199 (1957); *Liebigs Ann. Chem.*, **651**, 172 (1962)
507. R. L. Munier, 'Chromatographie d'échange d'ions', in *Techniques de Laboratoire* (J. Loiseleur), p. 545, Masson, Paris (1963)
508. J. L. Greene, C. H. W. Hirs and G. E. Palade, *J. Biol. Chem.*, **238**, 2054 (1963)
509. J. H. Kycia, M. Elzinga, N. Alonzo and C. H. W. Hirs, *Arch. Biochem. Biophys.*, **123**, 336 (1968)
510. A. Bendich, H. B. Pahl, G. C. Korgold, H. S. Rosenberanz and J. R. Fresco, *J. Amer. Chem. Soc.*, **80**, 3949 (1958)
511. D. F. Bradley and A. Rich, *J. Amer. Chem. Soc.*, **78**, 5893 (1956)
512. L. Astrachan and E. Wolkin, *J. Amer. Chem. Soc.*, **79**, 130 (1957)
513. E. Mihalgi, D. F. Bradley and M. I. Knoller, *J. Amer. Chem. Soc.*, **79**, 6387 (1957)
514. S. Osawa, *Biochim. Biophys. Acta*, **43**, 110 (1960)
515. J. A. Hunt, *Nature*, **183**, 1373 (1959)
516. H. M. Dintzis, *Proc. Nat. Acad. Sci. U.S.*, **47**, 247 (1961)
517. J. B. Clegg and K. Bailey, *Biochim. Biophys. Acta*, **63**, 525 (1962)
518. I. Berggard and G. M. Edelman, *Proc. Nat. Acad. Sci. U.S.*, **49**, 330 (1963)
519. R. E. Humbel and A. M. Crestfield, *Biochemistry*, **4**, 1044 (1965)
520. D. Va Hoang, M. Rovery, A. Guidoni and P. Desnuelle, *Biochim. Biophys. Acta*, **69**, 188 (1963)

521. Y. Moschetto and G. Biserte, *Bull. Soc. Chim. Biol.*, **45**, 75 (1963)
522. G. M. Tener, H. G. Khorana, R. Marklan and E. H. Pol, *J. Amer. Chem. Soc.*, **81**, 6224 (1959)
523. G. M. Tener, P. T. Gilham, W. E. Razzell, A. F. Turner and H. G. Khorana, *Ann. N.Y. Acad. Sci.*, **81**, 757 (1959)
524. H. G. Khorana and J. P. Vizsolyi, *J. Amer. Chem. Soc.*, **83**, 675 (1961)
525. J. Porath, *Nature*, **175**, 478 (1955)
526. M. Staehelin, *Biochim. Biophys. Acta*, **49**, 11 (1961)
527. R. V. Tomlinson and G. M. Tener, *J. Amer. Chem. Soc.*, **84**, 2644 (1962)
528. R. V. Tomlinson and G. M. Tener, *Biochemistry*, **2**, 703 (1963)
529. M. N. Lipsett and L. A. Heppel, *J. Amer. Chem. Soc.*, **85**, 118 (1962)
530. H. G. Zachau, D. Dütting and H. Feldmann, *Z. physiol. Chem.*, **347**, 212, 236 and 249 (1966)
531. E. M. Bartos, G. W. Rushizky and H. A. Sober, *Biochemistry*, **2**, 1179 (1963)
532. D. G. Glitz, *Biochemistry*, **7**, 927 (1968)
533. G. W. Rushizky and H. A. Sober, *J. Biol. Chem.*, **237**, 2883 (1962); *Biochem. Biophys. Res. Comm.*, **14**, 276 (1964)
534. W. M. Stanley and R. M. Bock, *Anal. Biochem.*, **13**, 43 (1965)
535. M. Staehelin, *J. Mol. Biol.*, **8**, 470 (1964)
536. N. B. Furlong, *Anal. Biochem.*, **12**, 349 (1965)
537. B. F. C. Clark and T. M. Jaouni, *J. Biol. Chem.*, **240**, 3379 (1965)
538. R. W. Holley, J. Apgar, B. P. Doctor, J. I. Farrow, M. A. Marini and S. H. Merrill, *J. Biol. Chem.*, **236**, 200 (1961)
539. K. Nishiyama, T. Okamoto, I. Watanabe and M. Takanami, *Biochim. Biophys. Acta*, **47**, 193 (1961)
540. J. D. Cherayil and R. M. Bock, *Biochemistry*, **4**, 1174 (1965)
541. B. C. Baguley, P. L. Bergquist and R. K. Ralph, *Biochim. Biophys. Acta*, **95**, 510 (1965)
542. B. C. Baguley, P. L. Bergquist and R. K. Ralph, *Biochim. Biophys. Acta*, **108**, 139 (1965)
543. Y. Kawade, T. Okamoto and Y. Yamamoto, *Biochem. Biophys. Res. Comm.*, **10**, 200 (1963)
544. M. L. Stephenson and P. C. Zamecnik, *Proc. Nat. Acad. Sci., U.S.*, **47**, 1627 (1961)
545. R. W. Holley, J. Apgar, G. A. Everett, J. T. Madison, M. Marquisec, S. H. Merrill, J. R. Penswick and A. Zamir, *Science*, **147**, 1462 (1965)
546. J. T. Madison, G. A. Everett and H. Kung, *Science*, **153**, 531 (1966)
547. H. G. Zachau, D. Dutting and H. Feldmann, *Angew. Chem. Intern. Ed. Engl.*, **5**, 422 (1966)
548. U. L. Raj Bhandary, S. H. Chung, A. Stuart, R. D. Faulkner, R. M. Hoskinson and H. G. Khorana, *Proc. Nat. Acad. Sci. U.S.*, **57**, 751 (1957)
549. R. Monier, S. Naono, D. Hayes, F. Hayes and F. Gros, *J. Mol. Biol.*, **5**, 311 (1962)
550. N. Sueoka and T. Y. Cheng, *J. Mol. Biol.*, **4**, 161 (1962)
551. T. Kano-Sueoka and S. Spiegelman, *Proc. Nat. Acad. Sci. U.S.*, **48**, 1942 (1962)
552. P. H. Pouwels, G. Veldhuisen, H. S. Jansz and J. A. Cohen, *Biochim. Biophys. Res. Comm.*, **13**, 83 (1963)
553. R. Rosset, R. Monier and J. Jullien, *Bull. Soc. Chim. Biol.*, **46**, 87 (1964)
554. D. G. Comb and S. Katz, *J. Mol. Biol.*, **8**, 790 (1964)
555. D. G. Comb and Z. Willmer, *J. Mol. Biol.*, **23**, 441 (1967)
556. W. Heri, H. Neukon and H. Deuel, *Helv. Chim. Acta*, **44**, 1939 (1961)
557. M. A. Jermyn, *Aust. J. Biol. Sci.*, **15**, 787 (1962)

558. A. Pusztai and W. T. J. Morgan, *Biochem. J.*, **93**, 363 (1964)
559. R. L. Cleland, M. C. Cleland, J. J. Lipsky and V. E. Lyn, *J. Amer. Chem. Soc.*, **90**, 3141 (1968)
560. T. C. Laurent, *Arch. Biochem. Biophys.*, **58**, 120 (1962)
561. R. Got, J. Font and R. Bourrillon, *Biochim. Biophys. Acta*, **78**, 367 (1963)
562. W. F. Goobel, *Proc. Nat. Acad. Sci. U.S.*, **49**, 464 (1963)
563. N. R. Ringertz and P. Reichard, *Acta Chem. Scand.*, **14**, 303 (1960)
564. T. C. Laurent and A. Anseth, *Exptl. Eye Res.*, **1**, 99 (1961)
565. A. G. Lloyd, K. S. Dodgson and R. G. Price, *Biochim. Biophys. Acta*, **69**, 496 (1963)
566. E. R. Berman, *Biochim. Biophys. Acta*, **58**, 120 (1962)
567. M. Schmidt, *Biochim. Biophys. Acta*, **63**, 346 (1962)
568. A. Kimura and K. Turuni, *J. Biochem.* (Japan), **52**, 63 (1962)
569. S. Inoue, *Biochim. Biophys. Acta*, **101**, 16 (1965)
570. A. Polson, *J. Chromatog.*, **5**, 116 (1961)
571. B. H. Hoyer, E. T. Bolton, R. A. Ormsbee, G. Le Bouvier, D. B. Ritter and C. L. Larson, *Science*, **127**, 859 (1958)
572. V. I. Agol, S. V. Maslova and M. Y. Chumahova, *Biokhimiya*, **27**, 1071 (1962)
573. B. H. Hoyer, E. T. Bolton, D. B. Ritter and E. Ribi, *Virology*, **7**, 462 (1959)
574. R. A. Ormsbee and B. H. Hoyer, *Federation Proc.*, **17**, 529 (1958)
575. J. L. Delsal, P. Lepine and V. Sauter, *C.R. Acad. Sci.*, **251**, 290 (1960)
576. L. Philipson, *Virology*, **10**, 459 (1960)
577. V. M. Zhdanov and L. B. Mekler, *Nature*, **195**, 924 (1962)
578. H. Gelderblom, R. Wigand and H. Bauer, *Nature*, **205**, 625 (1965)
579. I. Haruna, H. Yaoi, R. Kono and I. Watanabe, *Virology*, **13**, 264 (1961)
580. H. G. Klemperer and H. G. Pereira, *Virology*, **9**, 536 (1959)
581. W. C. Wilcox and H. S. Ginsberg, *Proc. Nat. Acad. Sci. U.S.*, **47**, 512 (1961)
582. B. H. Hoyer, R. A. Ormsbee, E. Bolton and D. B. Ritter, *Federation Proc.*, **17**, 517 (1958)
583. A. R. Neurath, B. A. Rubin and W. A. Pierzchala, *Arch. Biochem. Biophys.* **120**, 238 (1968)
584. F. Brown and B. Cartwright, *Biochim. Biophys. Acta*, **33**, 343 (1959)
585. O. Levin, *Arch. Biochem. Biophys.*, **33**, 78 (1958)
586. G. W. Cochran, J. L. Chidester, D. L. Stocks, *Nature*, **180**, 1281 (1957)
587. B. Commoner, J. A. Lippincott, G. B. Shearer, E. E. Richman and J. H. Wu, *Nature*, **178**, 767 (1956)
588. A. Taussig and E. H. Creaser, *Biochim. Biophys. Acta*, **24**, 448 (1957)
589. A. Taussig and E. H. Creaser, *Virology*, **4**, 200 (1957)
590. J. S. Murphy and L. Philipson, *J. Gen. Physiol.*, **45**, 155 (1962)
591. J. Yuzura Homma, N. Hamamura and Y. Ashizawa, *Japan, J. Microbiol.*, **5**, 149 (1961)
592. A. S. Tikhonenko and B. F. Poglazov, *Biokhimiya*, **28**, 340 (1963)
593. I. V. Chernyakhovskaya, *Biokhimiya*, **28**, 253 (1963)

IV
Electronic Properties of Biomolecules

CHAPTER 7

Semiconductivity in proteins and nucleic acids

B. Rosenberg

Biophysics Department,
Michigan State University,
East Lansing, Michigan, U.S.A.

E. Postow

Department of the Navy,
Office of Naval Research,
Arlington, Virginia, U.S.A.

I. INTRODUCTION

Solid state physics adds extra dimensions to models of biological mechanisms. It suggests among other things the presence of free, mobile, electronic charge carriers in organized structures. Such structures, which need not be crystalline but may be amorphous, are well documented as present in many biological organelles such as chloroplasts, rods and cones, mitochondria, membranes, both bare and myelinated, and finally, in individual biomacromolecules such as nucleic acids and proteins.

Electronic charge carriers may be delocalized over relatively large distances in such structures and will contribute to the electrical conductivity of the structures, energy transfer, information transfer, spatial separation of oxidizing and reducing entities, generation of free radicals, and localized pH charges. These all represent important mechanisms in biology.

A free electron, with kinetic energy of $\frac{3}{2}kT$, may have thermal velocities at room temperature about 10^7 cm sec^{-1}, and therefore, depending on the lifetime, τ, of the free carrier, may wander over large distances on a molecular scale. In the presence of an electric field, a small component of drift in the direction of the electric field is added to the random thermal motions, and this makes a contribution to the electrical conductivity of the structure. This contribution to the velocity v in the field direction is proportional to the strength of the field, E, or $v = \mu E$, and the constant of proportionality, μ, is called the drift mobility of the carrier. The electrical current per unit cross-sectional area, j, which flows in a material is also proportional to the strength of the field (Ohms Law) and $j = \sigma E$. The constant of proportionality σ is the conductivity of the structure. The conductivity is related to the mobility by the simple formula $\sigma = ne\mu$, where n is the density of free charge carriers and e is the electronic charge ($e = 1\cdot6 \times 10^{-19}$ coulombs).

Generally in any structure, there are many kinds of charge carriers, ions of both signs, protons, electrons and holes (a deficiency in a filled electron level). For sufficiently low charge carrier densities, which usually occurs in non-metallic materials, the charges move independently of all other charges and the total conductivity is the sum of the contributions of each charge carrier type. Therefore we have

$$\frac{j}{E} = \sigma = \sum_i n_i e \mu_i \tag{1}$$

where the sum is taken over all charge carrier types and signs (for multiply charged ions, e must be multiplied by the number of charges per ion). It is therefore necessary to know the charge carrier density n and the mobility μ of each charge carrier type in order to fully understand the conduction processes. The charge carrier density will be the product of the rate of generation of charge carriers, dn/dt multiplied by the lifetime of the carriers, τ. The lifetime reflects the processes of loss of carriers either by recombination or localization (trapping). Thus measurements of the electrical properties of a substance can give significant information on the mechanisms and rates of generation of charge carriers, the equilibrium densities, the lifetimes (loss processes), the mobilities (and also their diffusion constant D, since these are related by the Einstein equation; $D = \mu kT/e$).

Substances whose electrical conductivity as a function of temperature follow the relation

$$\sigma(T) = \sigma_0 \exp(-E_D/2kT) \tag{2}$$

(where E_D is the activation energy for semiconduction; k is the Boltzmann constant; and T the temperature in degrees Kelvin) are operationally defined as semiconductors. Representative molecules of four categories of biochemicals, carbohydrates, proteins, lipids, and nucleic acids, several complex organelles, as well as all organic substances tested have been demonstrated to be semiconductors in the solid state. A recent review by Gutmann and Lyons[1] lists 116 different biochemicals and organelles upon which conductivity measurements have been made. Since equation (2) does not specify the mechanism of conductivity, this mechanism, as well as any of the parameters of semiconductivity, could conceivably differ from one class of biochemicals to another, or perhaps even within a class. In protein, lipid and nucleic acid structures, the consistency of results, however, indicates at least some basic similarities in the conductivity mechanisms.

In vivo systems are not of high purity. Therefore, if semiconductive models are to have biological significance, the effect of environment on the conductive properties of proteins and nucleic acids must be understood. As a first step in this programme the effect of various adsorbed molecules on the semiconductive properties of proteins and nucleic acids has been investigated.

II. THEORY

At constant temperature the conductivity of various biochemicals will increase with increasing amounts of several different adsorbates. The magnitude of the increase is dependent both on the adsorbate and on the biochemical substrate. Hydration effects have been studied most thoroughly. Over a wide range of hydration states the conductivity varies as

$$\sigma(m) = \sigma_D \exp(\alpha m) \tag{3}$$

where σ_D is the dry state conductivity; m is the weight per cent water adsorbed onto the protein; and α is a constant, as illustrated in Figure 1. The validity of equation (3) has been demonstrated in a variety of other biochemical systems including methyl alcohol adsorbed on wool,[2] methyl alcohol adsorbed on a haemoglobin,[3] formic acid adsorbed on keratin,[4] and several alcohols and esters adsorbed on a variety of carotenes.[5,6]

Hydrated proteins are semiconductors, that is, equation (2) is a valid description of the temperature dependence of the hydrated protein's conductivity. The pre-exponential factor in equation (2) has been demonstrated[23] to be independent of the hydration state of the protein. For any given hydration state the temperature dependence of the conductivity of the hydrated protein is described by

$$\sigma(m, T) = \sigma_0 \exp(-E_m/2kT) \tag{4}$$

where E_m denotes the activation energy for semiconductivity of a protein with m per cent adsorbed water. The data for haemoglobin are shown in

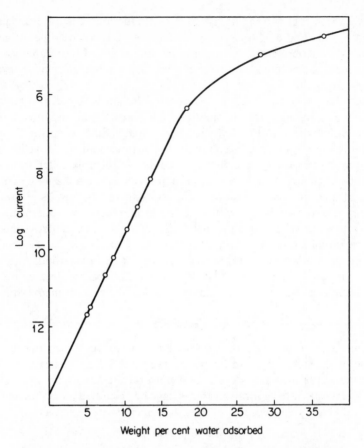

Figure 1. Conductivity of a haemoglobin tablet as a function of the water adsorbed at a constant temperature ($T = 298\ °K$) and constant applied voltage. Up to about a value of 18 per cent, the current increases exponentially with increasing water adsorption. Above this value it approaches saturation

Figure 2. Equation (4) can be verified by determining the dependence of the conductivity, at constant temperature, on the activation energy at several different hydration states. The slope of a plot of the log of the conductivity against the activation energy should and does, as shown in Figure 3, exhibit a value of $1/(2kT)$, where T is the temperature at which the conductivity was measured.[8]

If we assume the mobility is independent of hydration, we may then combine equations (2) and (3)

$$\sigma(T, m) = \sigma_0 \exp[-(E_D - \gamma m)/2kT] \tag{5}$$

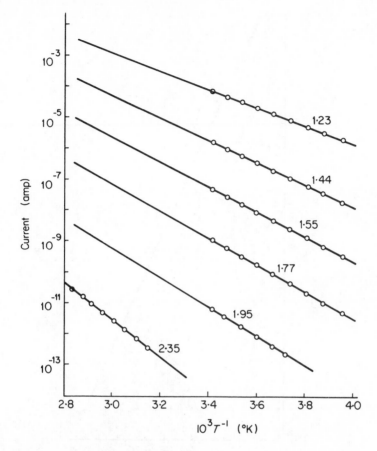

Figure 2. Semilogarithmic plots of the current *vs.* the reciprocal absolute temperature for a haemoglobin tablet with different water absorption values (equilibrated at different P/P_0 of water vapour). The slopes of the straight lines give the values of the semiconduction activation energies, which decrease with increasing water content. All curves intercept the ordinate at $1/T = 0$ at a σ_0 value of 100 (ohm cm)$^{-1}$.[3] (After *Ann. N.Y. Acad. Sci.*, **158**, 161 (1969))

where

$$E_m = E_D - \gamma m \quad \text{and} \quad \gamma = 2kT\alpha \tag{6}$$

The activation energy in a hydrated system is shown to depend on both the intrinsic properties of the system as manifest in E_D, and on the hydration state of the material.

The mechanism of conductivity may be described as the removal of a charge from a neutral portion of the molecule and its relocation on a

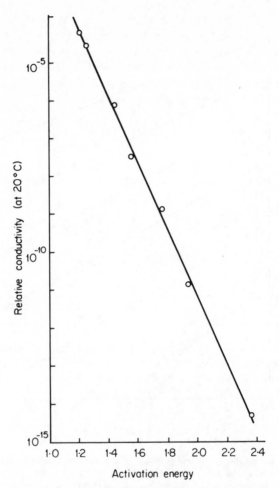

Figure 3. A semilogarithmic plot of the conductivity of a haemoglobin tablet (at the temperature 298 °K) *vs.* the activation energies shown in Figure 2. The slope of this line should, and does, equal $1/2kT$ at this temperature.[3] (After *Ann. N.Y. Acad. Sci.*, **158**, 161 (1969))

previously neutral site of another, or distant portion of the same, molecule. If the charge is moved a considerable distance then the coulomb interaction between charges may be neglected and the charges are essentially free. The energy required for such a process is

$$E = I_g - A_g - 2P \tag{7}$$

where I_g is the gas state ionization potential of the molecule, A_g is the gas state electron affinity and P is the stabilization energy due to polarization relaxation of the dipoles in the medium around each site of ionization. Polarization stabilizes the solid state system, as compared to a gaseous system, because of this relaxation of the dielectric medium in a spherical region around each of the two newly created charges. The polarization stabilization energy for this case is given by

$$P = \frac{e^2}{2R}(1 - 1/\kappa) \tag{8}$$

where R is the radius of the spherical cavity in which the charge is embedded and κ is the effective dielectric constant of the substance. Combining equations (7) and (8) we have

$$E_D = I_g - A_g - \frac{e^2}{R}(1 - 1/\kappa) \tag{9}$$

where κ denotes the dielectric constant of the dry protein. For the hydrated system we can assume the effective dielectric constant increases and hence we can write

$$E_m = I_g - A_g - \frac{e^2}{R}(1 - 1/\kappa') \tag{10}$$

where κ' is the dielectric constant of the appropriately hydrated protein $(\kappa' > \kappa)$. Hydration cannot alter either the ionization potential or the electron affinity of the molecule in the gas state. We may then eliminate I_g and A_g from equations (9) and (10), obtaining

$$E_m = E_D - \frac{e^2}{R}\left(\frac{1}{\kappa} - \frac{1}{\kappa'}\right) \tag{11}$$

Substituting equation (11) in equation (4) yields the final equation

$$\sigma(T, \kappa') = \sigma_0 \exp(-E_D/2kT) \exp\left[\frac{e^2}{2kTR}\left(\frac{1}{\kappa} - \frac{1}{\kappa'}\right)\right] \tag{12}$$

It is thus predicted that the adsorption of water, or other vapour, increases the conductivity of the system by decreasing the activation energy for semiconduction. The decreased activation energy is the result of an increased effective dielectric constant when such molecules are adsorbed on the protein.

From the above development we can see which parameters of the semiconductive process must be measured. The effect of temperature on conductivity will indicate the activation energy for semiconduction. Conductivity must also be measured as a function of the adsorbate concentration, from which the relation between the activation energy and adsorbate concentration may be calculated (see equation (6)). The dielectric constant must be measured as a function of adsorbate concentration. The above development has presupposed that the variation of dielectric constant with temperature is negligible. It has also assumed that experimental verification of equation (12), where the mobility is included in the pre-exponential factor, indicates that the mobility is independent of the adsorption process. Explicit experimental verification of this independence should be obtained. The temperature dependence of mobility may, but need not necessarily,[9] indicate the mechanism of conductivity i.e. the applicability of either the band model or the hopping model of semiconductivity.

III. THE PRE-EXPONENTIAL FACTOR IN SEMICONDUCTION

The temperature dependence of a semiconductor is given by equation (2). If the logarithm of the conductivity $\sigma(T)$ is plotted on the ordinate against the reciprocal of the absolute temperature on the abscissa, a straight line is generated whose slope is $E/2k$ (this is sometimes given as E'/k; the factor of 2 in the equation arises from entropic consideration of the distributions of the positive and negative charges; if they are independent $E/2k$ is required, if they are dependent E'/k is required). The straight line may be extrapolated to intercept the ordinate at $1/T = 0$. This intercept point is the value of σ_0. The temperature dependence of the conductivity is therefore characterized by the two constants, E and σ_0. Since we have described above a method for varying the activation energy, it is interesting to investigate the simultaneous behaviour of σ_0. For the hydrated proteins, since all the lines of Figure 2 intercept at a common point (within a factor of 3) we conclude that σ_0 is invariant, either $T \to \infty$ or $E \to 0$ yields the same value of σ_0. This result is not always true for other substances, Rosenberg and coworkers[8] and Eley[10] have both given experimental evidence that in a given substance σ_0 varies with a variation in E. Generally σ_0 increases as E increases, tending to maintain a constant current. This is a compensation effect. This has been shown to hold for a number of substances such as retinal, oxidized cholesterol, benzoic acid and, more to the point for this paper, nucleic acids. In these compensation effect cases, when the plots of the logarithm of the conductivity against $1/T$ for different activation energies are made, the extrapolated lines intercept the ordinate at larger values of σ_0 as E increases. All the lines do however approximately pass through a common point prior to hitting the ordinate. The implication of this effect is that the simple two constant relation of equation (2) is inadequate to account for these data. A new relation has been postulated by Rosenberg and coworkers[8] which does agree well with the data

$$\sigma(T) = \sigma_0' \, e^{E/2kT_0} \, e^{-E/2kT} \qquad (13)$$

where σ'_0 and $1/T_0$ are the coordinates of the common point through which all the lines pass. A comparison of this equation with equation (2) shows that

$$\sigma_0 = \sigma'_0 \, e^{E/2kT_0} \qquad (14)$$

The conductivity is now completely described by the three constants σ'_0, T_0 and E. T_0 is believed to be characteristic of the molecule involved and not the physical structure or electrode contacts. It is called the 'characteristic temperature' and kT_0 is the 'characteristic energy' of the substance. For the protein haemoglobin, $T_0 = \infty$ and $\sigma'_0 = 1 \times 10^2$ [ohm cm]$^{-1}$; for both the nucleic acids RNA and DNA, $T_0 = 338\,°K$, and $\sigma'_0 = 4 \times 10^{-16}$ [ohm cm]$^{-1}$. Table 1 gives data from Snart[11] for the nucleic acids. We have

Table 1. Semiconduction parameters for the nucleic acids.[a] $[1/(2kT_0) = 17.5\,\text{eV}^{-1}$; $\sigma'_0 = 4 \times 10^{-16}\,(\text{ohm cm})^{-1}]$

Material	E (eV)	σ_0 (ohm cm)$^{-1}$	$\sigma'_0 \exp(E/2kT_0)$ (ohm cm)$^{-1}$
RNA (yeast) in vacuum	3.00	1×10^8	3×10^7
RNA (yeast) in air	1.06	8×10^{-8}	5×10^{-8}
DNA (calf thymus) in vacuum	2.44	6×10^2	1×10^3
DNA (calf thymus) oxygen treated	1.56	3×10^{-4}	3×10^{-4}
DNA (calf thymus) in oxygen	1.27	5×10^{-6}	2×10^{-6}
DNA (calf thymus) in air	1.28	5×10^{-6}	2×10^{-6}
DNA (calf thymus) in nitrogen	2.38	1×10^3	5×10^2
DNA (calf thymus) denatured	3.00	1×10^8	3×10^7

[a] After Snart[11] and Rosenberg *et al.*[8]

no reasonable explanation as yet of the meaning of this compensation effect, although Eley[12] has discussed four possible models for this phenomenon. The implication of this effect for research workers is that it is no longer adequate to determine only σ_0 and E values for a given substance. It is necessary to try and vary E by some means and to evaluate σ'_0 and T_0.

Recently, some further theoretical work has been done on the problem of the compensation law effect. Many, Harnik and Gerlich[13] first showed a correlation of the pre-exponential factor, σ_0, and the semiconduction activation energy, E, for a large number of organic semiconductors. This correlation takes the form of

$$\log \sigma_0 \cong \alpha E + \beta$$

They suggested that this could be due to electronic charges tunnelling from activated levels through transparent intermolecular barriers.

This model was further developed by Eley,[14] and by Kemeny and Rosenberg.[15] In this latter paper, they showed, using a triangular barrier which gives a good approximation to such intermolecular barriers, and using the transmissivity calculations of Mott and Sneddon, that they were able to derive the compensation law. From this, they estimated that for a barrier width of 3×10^{-8} cm, and a height of 6 eV (chosen to match available data), the effective mass of the electron in the activated level was of the order of 100 electron masses.

A second, and probably more appropriate, analysis, on the basis of small polaron theory has been developed by Kemeny and Rosenberg.[16] This too is shown to lead to the correct compensation law; to the invariance of the sum of the activation energy plus twice the polaron binding energy for any substance; and finally, to the identification of the 'characteristic temperature', T_0, as one half the 'Debye temperature' for the substance.

IV. SAMPLE PURITY

Some inorganic semiconductors are of 'electronic grade' purity (one impurity molecule for each 10^{10} substrate molecules). Although a few organic crystals can be greatly purified by zone refining techniques, biological substances are not obtainable at purity levels even approaching this. 'Electronic grade' biochemicals, particularly of the polymeric variety, will not be produced in the foreseeable future. But such high levels of purity cannot be required of biological semiconductors. The living cell contains many 'impurities' and to be relevant to biological systems semiconductive models must be relatively insensitive to general impurities.

The difference in impurity sensitivity is not between inorganic and biochemical semiconductors, but between ionic or covalent crystals on the one hand and molecular crystals, polymers or amorphous materials on the other. Tauc[17] has reported that the conductivity of amorphous germanium is influenced very little by the incorporation of impurities at low levels of concentration. This is in sharp contrast to the case of crystalline germanium, where impurities at low concentration levels dominate the semiconducting process. Ionic and covalent crystals possess a long range order which may extend over as great a distance as 1000 unit cells. This provides a long range interaction not found in amorphous materials, polymers, or molecular crystals. In the case of ionic or covalent crystals, the effects of local perturbations propagate significantly over large distances. In molecular crystals and amorphous materials, impurities, until their concentration is high, produce only local perturbations which are damped out over short distances.[1]

Experimental evidence at present indicates that semiconduction in biomacromolecules is not an impurity-dominated process. A schematic example of classic inorganic impurity semiconductivity is illustrated in

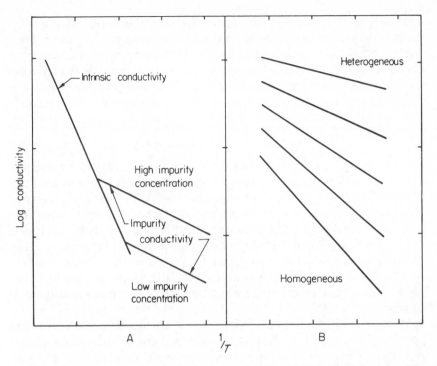

Figure 4. A schematic comparison of the types of semilogarithmic plots of conductivity *vs.* $1/T$ expected for A, classical impurity semiconductivity; B, the haemoglobin–adsorbate semiconductivity.[3] (After *Ann. N.Y. Acad. Sci.*, **158**, 161 (1969))

Figure 4. At low temperatures, and hence low conductivity, impurity conduction will dominate because of the lower activation energy needed to excite charge carriers to the conduction band from an impurity centre, while at higher temperatures, the intrinsic conductivity of the material will dominate because of its higher activation energy and hence its faster increase with temperature. As the impurity concentration is increased the dominance of impurity conductivity is extended to a higher temperature region. Changes in the impurity concentration do not change the activation energy for impurity dominated (or 'extrinsic') conductivity because only the density of carriers is affected by such changes (if impurity band conduction results from a high impurity concentration, then a concomitant change will be found in the semiconduction activation energy). The activation energy curves of the same substance with two different impurity concentrations should be parallel (see Figure 4A), indicating a constant activation energy but a change in the pre-exponential factor. Semiconduction curves for hydrated protein (Figure 4B) are not of this form. The increased concentration of water alters the activation energy of the hydrated protein and according to equation (6) there is a linear relation between the activation energy and the amount of water present in the system. This suggests that water is not behaving as an impurity, but is altering the intrinsic protein semi-conductivity in some more fundamental fashion as discussed in the preceding section. Liang and Scalco[18] have found the same result with hydrated nucleic acids.

V. SAMPLE PREPARATION

As large single crystals of biomacromolecules are usually unavailable, conductivity measurements are generally made on pressed tablets, cast slabs or thin films. Some materials of biological interest (for example, chlorpromazine and β-carotene) form amorphous glasses. For cast slabs, the electrodes which will be employed in the measurements are used to form two faces of the mould. If the remaining faces of the mould are constructed out of Teflon the sample need not be removed from the mould when electrical measurements are being made. When casting a slab, the temperature must not be raised above that temperature at which the substance is likely to decompose or denature.

Most biochemicals can be obtained only in the form of a microcrystalline powder. The powder, preferably of uniform particle size, is placed in a suitable die and compacted to a pressed tablet in a hydraulic press. Electrodes may be included in the die. However, if this is done the sample cannot be checked for denaturation, which may be caused by localized heating when pressure is applied to the die. Pressures in the range of $100 \, kg \, cm^{-2}$ are usually applied to the die at a slow rate to maintain the temperature below the denaturation temperature.

The capacitance of intergranular spaces in compacted powders often causes the dc resistance to be significantly greater than the resistance measured at high frequencies. High frequencies have the effect of short-circuiting the capacitance of the intercrystalline barriers. The dc value of the resistance may be determined by the barrier rather than by the crystallite bulk. Caution must be exercised in the determination of ac resistance, and it should be measured by a bridge. If one merely measures the magnitude of the impedance, by dividing ac voltage by current, then at low frequencies the $1/\omega$ term in the impedance will dominate in $Z = (R^2 + 1/\omega^2 c^2)^{1/2}$ and the sample capacitance will be measured. Siemons, Bierstedt and Kepler[19] have compared the semiconduction activation energy for single crystals and compressed tablets of the highly conducting charge transfer complex $Cs_2(TCNQ)_3$. Compressed tablets exhibit an activation energy of $0.07 \, eV$ while the activation energy of single crystals is $0.01 \, eV$. In the case of the compressed tablet nearly all of the activation energy results from intergranular impedances. If we take this value ($0.07 \, eV$) to be generally indicative of the order of magnitude of the activation energy resulting from inter-granular impedance, then if measured activation energies are large compared to 0·07 eV, as they are for biomacromolecules, we may accept them as being indicative of processes occurring in the bulk material.

Films may be prepared in one of three ways: melt, evaporation from a solution or suspension, or vacuum deposition. In the first case the sample is merely a thin version of a cast slab. Evaporated films often possess non-uniformities and may suffer from the same problems of intergranular contact as compacted powders. Lipid bilayers formed across a hole in a Teflon cup according to the method of Mueller and coworkers[20] results in an evaporated film in which the solvent goes into the solution bathing the membrane. The technique of vacuum sublimation is rarely applicable in biochemical preparations because denaturation usually occurs at tempera-tures lower than those which permit reasonable sublimation rates.

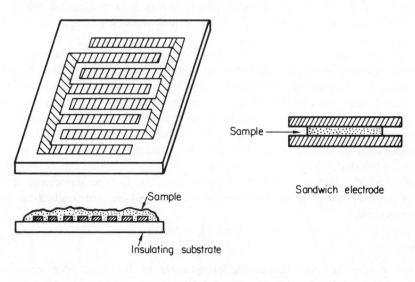

Figure 5. Two alternative types of standard electrode arrangements for conduc-tivity studies on biomacromolecules

Films may be placed between two electrodes, as are compressed tablets, producing a sandwich cell as illustrated in Figure 5. The film may, however, be deposited directly over conducting and insulating areas so as to bridge the gap between the electrodes; this results in a surface cell. When highly resistive biochemical materials are studied, the resistance of the cell may be reduced by increasing the electrode area and decreasing the electrode spacing. This is accomplished when the electrodes are in the configuration of two interpenetrating combs, as illustrated in Figure 5. 'Comb electrodes' are formed by the vacuum deposition of an appropriate metal (for example, nickel or chromium) on a quartz or glass plate.

The discontinuity introduced by the surface of a crystal introduces a series of new energy levels which occur at the crystalline surface but not in the bulk. When the crystal is placed in an electric gradient these states become filled, creating a space charge layer at the crystalline surface. If the charge carrier concentration in the material is small, as it is in biochemical substances, the space charge layer will penetrate deeply into the solid. Films of a thickness of the same order as the space charge layer will exhibit conductive properties indicative of the space charge layer and not of the bulk material. Ioffee[21] has calculated the thickness of this layer to be approximately 10^{-5} cm for organic molecular crystals. Space charge layers play an important, if not dominant role in surface cell conductivity measurements. The importance of space charge layers in a compacted tablet of microcrystalline granules, although not well understood, should not be neglected.

It has been found in our laboratory that altering the metal used for the electrodes did not change the value of either the conductivity or the activation energy for protein crystals. Stainless steel, copper, brass, aluminium and tin oxide-coated glass were used. Eley[12] has reported the same results for gold, platinum, aluminium, and indium electrodes. When one wishes to measure the parameters of the photoconductive process, tin oxide-coated glass electrodes should be used, since they are transparent from 3200 Å to the short infrared. The energy required to transfer an electron from an intrinsic biochemical semiconductor to a metal electrode (p-type semiconduction) or from the metal to the biochemical material (n-type semiconduction) are respectively

$$E = I_c - \phi$$
$$E = \phi - A_c \tag{15}$$

where I_c is the solid state ionization potential of the biochemical material, A_c is the solid state electron affinity of the biochemical material and ϕ is the work function of the metal electrode. The work functions of the various metals used as electrodes differ by several electron volts. Therefore, the measured semiconduction activation energies, which do not vary with the type of metal electrode employed, do not appear to be due to the injection of electrons or holes from the metal electrodes into the biochemical substance. If, however, as is suggested by Eley,[12] the Fermi levels in the metal electrode and in the semiconductor are equal without any bending of the energy bands in the semiconductor, then we have

$$\phi = \frac{I_c + A_c}{2} \tag{16}$$

and the activation energy is independent of ϕ, the work function of the metal electrode. Under these conditions the process of injecting electrons from the electrode into the sample can give rise to the measured activation energy. However, it is unlikely that the Fermi levels of a large number of metals and several biochemical substances are all the same. We therefore believe that the measured activation energy is descriptive of processes in the specimen between the electrodes and not injection of electrons from the electrode.

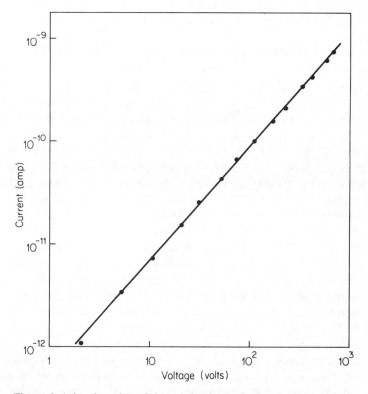

Figure 6. A log–log plot of the conductivity of a haemoglobin tablet (hydrated to equilibrium at a relative humidity of 20 per cent) *vs.* applied voltage. An ohmic relation requires a straight line with a slope of 45°; the measured slope here is 49°

For more than three orders of magnitude change of the applied voltage, the conductivity of hydrated haemoglobin is very closely approximated by Ohm's law as illustrated in Figure 6. Similar results were found for DNA by Burnel, Eley, and Subramanyan.[22] These results indicate an ohmic contact between sample and metal electrodes.

Instead of metal electrodes, held by spring clamps, conductive adhesives can be used to make electrical contact. Aquadag, silver paste and conductive epoxy have all been used to attach the biochemical sample to electrodes. Conductive epoxies and silver paste tend to denature proteins. Adhesive electrodes are not advised for the study of adsorbate effects on conductivity parameters because the adsorbate may affect the adhesive. Solvents in the adhesive may be adsorbed onto the protein or nucleic acid altering its conductivity.

VI. DETERMINATION OF THE NATURE OF THE CHARGE CARRIERS

From equation (1) it is seen that the contribution to the total conductivity of a single species of charge carrier is determined by the product of the carrier species' density and mobility. Conductivity measurements cannot, however, distinguish between concentration and mobility contributions to the product.

Electrolysis will occur if the charge carriers are ionic in nature but will not occur with electronic currents if electrodes which inject electrons but not ions are used. Solid state electrolysis can, therefore, be used to distinguish between the two species of charge carrier. A current passing through a hydrated sample for an extended period of time will decrease the amount of adsorbed water if ionic carriers are dominant, because electrolysis will have converted some of the water into hydrogen and oxygen. Conductivity, a very sensitive measure of a protein's hydration state, can be used to monitor the amount of water adsorbed onto the protein. Thus if ionic charge carriers pass through the sample, the amount of adsorbed water decreases and the conductivity of the sample will be seen to decrease. Rosenberg[7] observed haemoglobin with 7·5 per cent adsorbed water in such an experiment. The conductivity remained constant over a period of time in which the conductivity, if it were ionic, would have dropped by an order of magnitude, as illustrated in Figure 7. It was concluded that at 7·5 per cent adsorbed water the conductivity of haemoglobin is at least 95 per cent electronic.

Maricic, Pifat and Pravdic[24] impressed 150 V across a sample of crystalline haemoglobin with 9·17 per cent adsorbed water for seven days. No evolution of gas could be determined, thus verifying electronic conductivity. However, when 15 per cent water is adsorbed onto haemoglobin, Maricic and Pifat[25] found sufficient hydrogen evolution to assign approximately 90 per cent of the conductivity to ionic charge carriers.

Keratin films containing greater than 15 per cent water were examined by King and Medley.[4] They found the production of hydrogen sufficient to account for the conductivity as 100 per cent ionic. Oxygen evolution was not

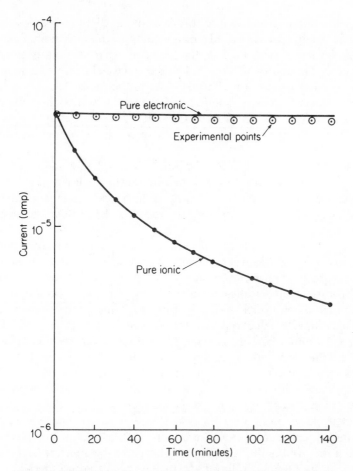

Figure 7. The results of a test of the nature of the charge carriers in haemoglobin with 7·5 per cent adsorbed water. The results indicate that the dominant charge carriers are electronic (either electrons or positive holes) and not ionic at this hydration state[7]

detected but oxygen may have been chemically adsorbed onto the haemoglobin.

The results described above were all based on the observation of evolved hydrogen. Tritiated water (30 per cent by weight) was adsorbed onto haemoglobin in one experiment.[26] The evolution of tritium, as monitored with a proportional counter, in the solid state electrolysis of the hydrated haemoglobin was sufficient to account for only 44 per cent of the conductivity by an ionic mechanism.

The electrolytic deposition of a metal at an electrode indicates the conversion from electronic conductivity in the substance to ionic conductivity in the salt solution which is the electrolysis medium. Digby[27] has demonstrated electronic conductivity by this means in wet quinolin-tanned proteins. Crustacean shells were used as the negative electrode in this electrolysis experiment. Deposition of the metal on the surface of the shell demonstrated the electronic nature of the charge carrier in the shell.

Electronic conductivity is found in Na–DNA with less than 50 per cent adsorbed water.[25] When greater amounts of water are adsorbed on the DNA, the evolution of some hydrogen indicates a contribution of ionic conductivity as well.

It is now evident that at low hydration states, below the adsorption of 15 per cent water, or about 2 B.E.T. monolayers (see below), the conductivity of proteins is predominantly, if not entirely, electronic in nature. The contribution of ionic conductivity increases as the protein becomes more highly hydrated.

Ionic conductivity becomes significant at a level of adsorption on the protein or nucleic acid where water–water interactions become significant when compared with water–protein (or nucleic acid) interactions. At these high hydration states sufficient water is adsorbed on the protein to form a continuous water bridge over the surface of the protein (haemoglobin) molecule. Ionic conduction may then proceed through a path of water molecules without passing through the protein. Estimates of the quantity of water bound to proteins in aqueous solution differ somewhat with the method of determination. The reported values cluster around either 0·2 gram water per gram dry protein for some albumins or 0·3 gram water per gram of dry protein for several other proteins. Included in the first category are estimates based on: (1) an analysis of the nuclear magnetic resonance spectra of water bound to serum albumin[28]; (2) self-diffusion measurements of water in ovalbumin[29]; and (3) dielectric dispersion measurements of human serum albumin solutions.[30] In the second group there are (1) the dielectric dispersion studies of Schwan[31] on haemoglobin; (2) estimates based on the conductivity of hydrated haemoglobin[32]; (3) two studies of the dielectric dispersion of serum albumin; a study of aqueous solutions[33] and one of hydrated protein crystals[34]; (4) estimates based on X-ray scattering data for serum albumin[35]; and (5) calorimetric studies on serum albumin, egg albumin and haemoglobin.[36] A limiting law of protein structure has been developed by Fisher.[37] This theory predicts the hydration of a protein in solution from the amino acid composition of the protein. Values between 0·2–0·3 gram of water per gram of protein are obtained for a variety of proteins.

Ionic conductivity has been shown to be significant only when the protein or nucleic acid has associated with it about the same amount of water as is found to be strongly interacting with the protein, or nuclei acid, when in aqueous solution. This value is about twice as much (on a weight per cent basis) for nucleic acids as it is for proteins. At high hydration states the ionic charge carriers or protons may move only in the hydration shell and never penetrate the protein or nucleic acid.

The critical hydration for nucleic acids appears to correspond to the transition from 'A' form to 'B' form.[25] The 'A' form is characterized by a tilt of the base pairs with respect to the axis of the helix. In the 'B' form, found *in vivo*, the base pairs are perpendicular to the helix-axis. Protonic conduction is indicated for DNA in the 'B' form but not in the 'A' form.

Powell and Rosenberg[38] have extended the solid state electrolysis measurements to a greater degree of sensitivity using a 'palladium window' to measure the total amount of hydrogen evolved. They have reported on the relative protonic and electronic currents as a function of hydration and solvation in DNA, haemoglobin, cytochrome *c*, collagen, lecithin, and melanin. Their conclusions were summarized as:

(1) All the biomolecules tested appear to be mixed semiconductors (i.e., both electronic and protonic charge carriers make significant contributions to the currents).

(2) While the conductivity of each substance increases exponentially with solvation, the ratio of protonic to electronic conductivity increases linearly with hydration for the globular proteins, haemoglobin and cytochrome *c*.

(3) The fibrous protein, collagen, may be a protonic semiconductor in the dry state, with an electronic component that increases linearly with hydration.

(4) Melanin is a mixed semiconductor whose protonic to electronic ratio (65:35) does not change over a hydration range of 10 per cent to 35 per cent.

(5) The DNA–water and the haemoglobin–methanol systems both show dominantly electronic conductivity below a threshhold solvation state, and a fairly sharp transition to a dominantly protonic conductivity above this threshhold. (In DNA-water, the transition occurs between 2-3 BET monolayers. This is the region where DNA undergoes the 'disorder-A form' transition, and not the 'A form-B form' transition.[39])

(6) The protonic to electronic ratio may be a function of the applied dc voltage; being mainly electronic below a certain voltage, and mixed above this voltage, in the haemoglobin–water system.

VII. MOBILITY DETERMINATION

Mobility is the mean velocity of a charge carrier in the direction of an electric field of unit strength. It may be considered either in a microscopic or in a macroscopic context. Drift mobility, a macroscopic concept, depends on both the density and depth of trapping sites for the charge carriers. Increasing the density of traps or the trap depth will serve to decrease the drift mobility.

There have been three different approaches to a determination of charge carrier mobility in semiconductors. Estimates of the mobility can be made from an equation of the form

$$\sigma = e\mu N_0 \exp(-E/2kT) \tag{17}$$

where N_0 is the effective density of conducting states; N_0 is, however, not well-known for biochemical materials. Mobilities can be determined from experiments in which a short pulse of charge carriers is created at a point in the material, which then drifts in an electric field and is monitored when it passes a distant point. The charge carriers may be photocreated, injected via an injecting electrode, or injected via high (approximately 2 kV) energy electron bombardment. In all cases, the experiment is in the form schematically illustrated in Figure 8. The carriers are detected a fixed distance from

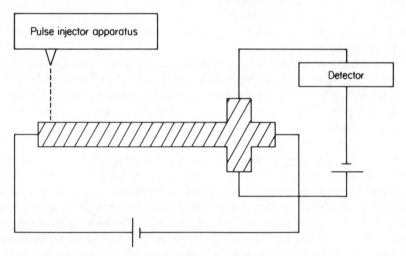

Figure 8. A schematic diagram of the apparatus in general use for measuring the mobility of the charge carriers in many substances

their creation, or introduction with an appropriate impedance matched amplifier circuit and displayed on an oscilloscope which is triggered by the injection pulse. The path length and time required (and hence the velocity) as well as the field strength are known. Therefore the mobility can be calculated. None of these methods has been successfully employed in the determination of the mobility of charge carriers in proteins or nucleic acids, although they have been used with varying degrees of success in other organic materials.

Another method of mobility determination is by use of the Hall effect. This effect occurs when a transverse magnetic field is impressed across a

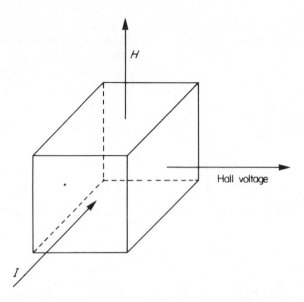

Figure 9. A schematic illustration of the configuration usually used to determine a Hall voltage in a solid. The current flow and the applied magnetic field are along two perpendicular axes, the resulting Hall voltage is along the third axis

conductor or semiconductor. A potential difference is generated which is perpendicular to both the direction of the current and the direction of the impressed magnetic field. This is illustrated in Figure 9. Hall mobilities may not be exactly the same as drift mobilities because traps do not reduce the Hall mobility but they do reduce the drift mobility. The mobility of ionic carriers cannot usually be detected by the Hall effect because they are too small, but the absence of a Hall effect does not necessarily indicate the presence of ionic carriers, since electronic carriers may also possess mobilities too small to be measured. The sign of the Hall effect for a single dominant charge carrier indicates whether the carrier is an electron or a hole. Although the simple development of the Hall coefficient assumes only one carrier, a more complicated analysis[40] will provide information if charge carriers of both signs are present in the material. Although attempts have been made,[41] dc Hall voltages, as discussed above, have not been observed in proteins. Hermann and Ham[42] have developed a system where both the magnetic field and the applied potential are time dependent. The sample is rotated in a static magnetic field at 20 Hz. This simulates an alternating magnetic field. An electric field of $13\frac{1}{3}$ Hz is applied to the sample. The Hall signal appears at $20 \pm 13\frac{1}{3}$ Hz and is detected at the higher frequency, pre-amplified by a field effect transistor mounted on the rotating shaft near the sample and synchronously detected with a phase sensitive detector. Recently Hermann and Rembaum[43] have used this technique to determine an ac Hall mobility for poly-(n-vinylcarbazole)-iodine, a donor acceptor complex, as 0.5 cm^2/volt sec. This method has not yet furnished mobilities for proteins or nucleic acids.

In Trukhan's method,[44] the alternating electric field impressed across the sample is in the microwave region, 10^{10} Hz. At high frequencies the capacitative effect of the intergranular spaces is of no consequence. Trukhan investigated both dry and hydrated DNA and haemoglobin in light and darkness. His results are summarized in Table 2. Illumination does not

Table 2. Microwave Hall mobilities of haemoglobin and DNA

Substance	Illumination	Hydration state	Mobility (cm^2/volt sec)	Carrier species
Haemoglobin	Dark	0.8 moles/g	2 ± 50 per cent	Electrons
Haemoglobin	Light	0.8 moles/g	3.2 ± 40 per cent	Electrons
Haemoglobin	Dark	Heated to 90°C	2 ± 60 per cent	Holes
Denatured haemoglobin	Dark	Heated to 90°C	0	—
DNA	Dark	Normal	0.5 ± 40 per cent	Holes
DNA	Weak ultraviolet	Normal	0.85 ± 50 per cent	Holes
Denatured DNA	Dark	Dry	0	—

appreciably change the mobility of either haemoglobin, which is about 2 cm^2/volt sec or of DNA, which is less than 1 cm^2/volt sec. These values are somewhat higher than would have been expected. Trukhan finds that hydrating the haemoglobin changes it from a p-type semiconductor, dominated by hole conduction, to an n-type semiconductor, in which a majority of the charge carriers are electrons. Denaturation, in both haemoglobin and DNA, appears to decrease the mobility. Mobilities of denatured samples were below the experimental limit of the apparatus. As the data of Trukhan is the first report of protein or nucleic acid charge carrier mobilities, this important result must be verified.

Verification has now been reported by Eley and Pethig.[45] They have corroborated Trukhan's method as applied to known mobility inorganic semiconductors, and report an electronic mobility (n type) in bovine plasma albumin of 40 cm^2/volt sec at a resistivity of about 10^8 ohm cm, which corresponds to a hydration state where only strongly and irrotationally bound water is present. Such a large mobility suggests to them that the band model for conduction is applicable in this system. More recently,[46] they have extended this technique to measurements of electronic conductivity in rat liver mitochondria, spinach chloroplasts, and cytochrome *c*. In the mitochondria, they estimate a mobility value (*n* type) of 50–80 cm^2/volt sec. Addition of respiratory inhibitors to the mitochondria markedly reduced the Hall mobility. This effect suggested to them that they were indeed measuring electronic conduction through the electron transport chain.

VIII. MEASUREMENTS OF SEMICONDUCTION ACTIVATION ENERGIES

Activation energies may be determined by measuring the current flow through a sample as the temperature is varied. In order to test the semiconduction theory discussed above, the measurements should be made with a variety of ambient atmospheres. The sample, clamped between two metal electrodes, is placed in chamber which may either be sealed, evacuated, or may allow the passage of vapour through it. For low conductivity materials the chamber must be shielded, and therefore if the chamber is not constructed of metal it should be placed in a Faraday cage. If photoconductive processes are of interest, an optical port must be included in the chamber.

The sample is placed in contact with a heat source or sink; a mercury pool can be used to establish both electrical and thermal contact. Heat conducting grease may be used to make thermal contact between an electrode and a metal bar if this type of heat sink is used. The metal bar extends outside the chamber and is heated or cooled as required; the sample temperature is monitored with a thermocouple placed as near to the specimen as possible. If a shielded thermocouple probe is used it can be inserted into a hole in the electrode. Such an apparatus is illustrated in Figure 10. Alternatively, a thermoelectric device may be used to heat and cool the sample; it can be preprogrammed to automatically change the sample temperature over a given temperature range at a constant rate.

A battery or dc power supply is used to impress a voltage across the sample. Proteins and nucleic acids in the dry state possess a low conductivity ($\sigma \approx 10^{-12}$ (ohm cm)$^{-1}$), therefore very small currents have to be measured. Currents as low as 10^{-14} amperes can be measured with electrometers as a matter of routine. Two varieties of electrometer are readily available: vibrating reed electrometers, although more sensitive, are unable to make dynamic measurements of the current; they can be used, therefore, only for equilibrium measurements. Electrometers which are built around electrometer tubes (a triode or pentode with very high grid resistance) can follow a dynamically changing current; at optimal conditions (low gain), the Keithley 610BR,* a solid state electrometer, can respond to a 25×10^3 Hz signal.

* Keithley Instrument Co., Cleveland, Ohio, U.S.A.

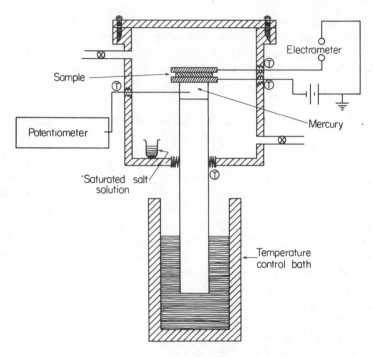

Figure 10. A schematic diagram of the apparatus used in this laboratory
to measure conductivity and semiconduction activation energies.
All insulation indicated by the symbol, T, is of Teflon; the chamber is
metal and is at ground potential[60]

    The desired atmosphere is established in the environment of the sample
either by placing an appropriate saturated salt solution inside the sealed
chamber, by flowing the desired vapours through the chamber or, if the
chamber is vacuum tight, by introducing measured pressures of a given
vapour. Saturated salt solutions come into equilibrium with a static
atmosphere and maintain that atmosphere at a constant relative humidity.
The temperature dependence of the equilibrium relative humidity of several
saturated salt solutions is given in the International Critical Tables. Salts
can be found for which the equilibrium relative humidity is insensitive to
small variations of temperature. With salt solutions and a rate of temperature
change of approximately 1·5 °C per minute, the hydration state of the sample
will not undergo any significant change, as indicated by the reproducibility
with which the system can be recycled. The desired adsorbate may be
introduced into the chamber with the aid of an inert carrier gas, which may
be helium, argon or in some cases nitrogen. If the adsorbate is a liquid, the
carrier gas is bubbled through the liquid, which is maintained at a constant

temperature, and allowed to pass through the chamber. Raising the temperature of the liquid increases the partial pressure of the vapour in the sample chamber. The amount of vapour adsorbed varies with the partial pressure as discussed below. Adsorbates in the gaseous state can be diluted with a carriers gas before introduction into the sample chamber.

Conductivity is a very sensitive measure of the amount of vapour adsorbed. It can therefore be used to indicate the establishment of equilibrium in the absorption process at a given partial pressure of adsorbate vapour. After a constant amount of vapour has been absorbed, the temperature is varied and the conductivity is measured as a function of the sample temperature. When the log of the conductivity is plotted as a function of reciprocal absolute temperature a straight line results whose slope, according to equation (2), is $E/2k$, from which the activation energy for semiconduction may be calculated.

IX. ADSORPTION STUDIES

To understand more clearly the effect of adsorbed vapours on the electrical properties of proteins and nucleic acids, the adsorption process itself must be studied. Both proteins and nucleic acids contain two types of hydrophilic groups which are capable of binding water through hydrogen bond formation. In proteins, these are firstly, the oxygen and nitrogen atoms of the peptide bonds in the peptide chain, and secondly, the polar side-chains of several amino acids, for example tryptophan, tyrosine, histidine, aspartic acid, glutamic acid, serine, threonine, hydroxyproline, aspartamine, glutamine, lysine and arginine. In nucleic acids, water molecules may attach at the phosphate groups and oxygen atoms of the backbone structure as well as at the nitrogen and oxygen atoms of the purine and pyrimidine bases. The number and availability of these groups will in general determine the amount of water held by the protein or nucleic acid. Bull's[47] results on nylon indicate that the peptide carbonyl and imido groups have little attraction for water. The hydrogen bond-forming capacity of the carbonyl and imido groups of the peptide linkage is probably saturated to such a degree that their attraction to water is negligible. Pauling[48] suggested that this saturation of the hydrogen bond-forming capacity of carbonyl and imido groups is a result of the extensive hydrogen bonding between peptide groups: $N-H\cdots O=C$. The stability of the secondary structure in proteins, α-helix or pleated sheet, is a result of hydrogen bonds between the carbonyl and imido groups of peptide bonds. In nucleic acids, the first molecules of water to be adsorbed are hydrogen bonded to the phosphate group.

To measure the quantity of adsorbed water, the sample is equilibrated with an atmosphere at a constant relative humidity in a sealed chamber. A constant relative humidity can be established, at a constant temperature, by placing a saturated salt solution in a sealed chamber. Different salts will equilibrate at different relative humidities and an annotated collection of saturated salt solutions is given by O'Brien.[49] Sulphuric acid solutions can also be used to establish a constant relative humidity. Cardew and Eley[50] equilibrated protein samples over sulphuric acid solutions, in sealed weighing bottles; after seven months the samples were removed and weighed, and then dried by heating in vacuum and reweighed. Titration of the sulphuric acid in the weighing bottle establishes its concentration from which, with the aid of standard curves, the equilibrium relative humidity can be determined. The last increment of water on highly hydrated samples is very loosely bound and therefore the sample will equilibrate very rapidly with an atmosphere of a lower relative humidity. Highly hydrated samples may then lose some water before the completion of the weighing process; similarly dry samples will bind water very readily and therefore may increase their hydration while being weighed. These errors are eliminated if the sample is weighed in the hydration chamber.

Dynamic as well as equilibrium information may be obtained when the sample is weighed in the hydration chamber.[51] The microbalance should be situated in a chamber in such a way that the sample may be heated and

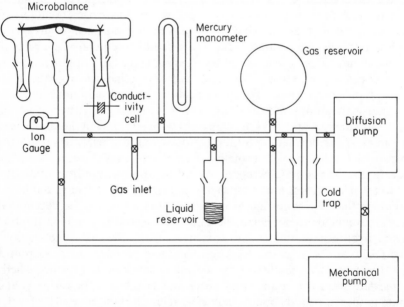

Figure 11. A schematic diagram of the vacuum microbalance apparatus used in this laboratory for measuring adsorption isotherms and simultaneous conductivity changes due to adsorption of various gases and vapours on biomacromolecules in the solid state. An additional connection can be made to a vacuum dielectric cell for simultaneous measurements of the dielectric changes with adsorption[38]

cooled (see Figure 11), and while it is on the balance it is desirable to dry the sample. The chamber itself should be evacuable to approximately 10^{-5} torr. After the sample is dried, a port in the chamber may be opened and an inert gas introduced; additional ports open the chamber to a saturated salt solution (or a sulphuric acid solution) permitting it to equilibrate at a constant relative humidity. The sample weight is continuously monitored on a recorder. A variety of other vapours may be introduced into the micro-balance chamber and their adsorption measured. The partial pressure of the vapour in the chamber can be altered by regulating the temperature of the reservoir of surplus liquid as illustrated in Figure 11 but the temperature of the reservoir must be kept below the temperature of the chamber so that condensation does not occur.

Relatively long periods of time are required to reach equilibrium and therefore it is of the utmost importance that the zero drift of the micro-balance be minimal. If the zero drift is linear in time it may be compensated.

When the increase in weight of the sample is plotted against the relative vapour pressure of water in the equilibrium atmosphere a typical S-shaped curve is obtained. This curve can be separated into three segments as illus-trated in Figure 12. The low vapour pressure portion resembles a Langmuir adsorption isotherm; in this region the amount of vapour adsorbed increases rapidly with an increase in the vapour pressure. In the central region of the curve a more or less linear relation, between the amount of vapour adsorbed and the relative vapour pressure, is approximated. This region has a more gentle slope. The high humidity region is characterized by a sharp upswing of the amount of adsorbed vapour as the relative vapour pressure is increased.

The water adsorption curves can be analysed according to the theory of Brunauer, Emmett and Teller[52] which extended the Langmuir treatment of monolayer gaseous adsorption to cases where more than one layer of molecules are adsorbed. In general $P/V(P_0 - P)$ is plotted against P/P_0, where V is the volume of vapour adsorbed at a vapour pressure P, and P_0 is the vapour pressure at saturation. A straight line is obtained whose y-intercept is $1/V_{\mathrm{M}}C$ and whose slope is $(C - 1/V_{\mathrm{M}}C)$, where V_{M} is the volume of vapour adsorbed in the first layer on 100 grams of dry material and C is a constant related to the heat of adsorption. When the adsorption of water vapour is considered, V can be taken as the weight increase of the sample in grams, and V_{M} is then the weight of the first monolayer of adsorbed water.

It is now possible to calculate the number of water molecules in the first monolayer. This value is comparable with the number of polar groups in the same weight of dry protein for several different proteins.[48] For haemo-globin, our measured value of this number is 430–470 moles per 10^5 gram of haemoglobin compared with 435 moles per 10^5 gram of haemoglobin which is expected.

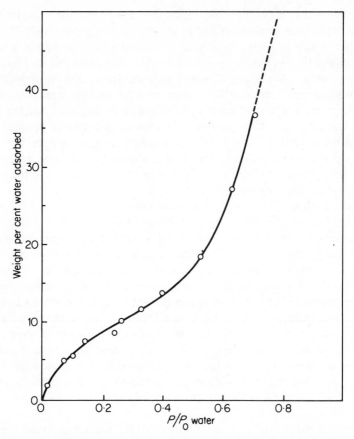

Figure 12. The adsorption isotherm (at $T = 297\,°\text{K}$) for the water–haemoglobin system measured in the balance apparatus shown in Figure 11[60]

X. DIELECTRIC STUDIES

According to the theory of semiconduction discussed above, the effect of adsorbed vapours on the conductivity of proteins and nucleic acids is to increase the effective dielectric constant, which increases the polarization relaxation energy. This decreases the activation energy for semiconduction which in turn increases the conductivity.

Dielectric constants are calculated from measurements of the capacitance according to the equation: $\kappa = C_s/C_A$, where κ is the dielectric constant of the material, C_s is the capacitance of the sample and C_A is the capacitance of an equal geometry of air.

In general, the capacitance, and hence the dielectric constant, of a specimen will vary with frequency. Proteins and nucleic acids in the solid state are most frequently examined in the region from 10^2 to 10^5 Hz. In this frequency region a Schering bridge (for example General Radio type 716-C capacitance bridge) or some modification of it is most often used. Capacitance can be measured at frequencies as low as 0·1 Hz on a bridge manufactured by Ando Electric Company of Tokyo, Japan. Measurements in the radio frequency region can be made on the bridge described by Cole and Curtis[53] or on commercially available models (for example General Radio Model 160 6-A). Ultra high frequency determinations of capacitance utilize transmission lines[54] or wave guides.[55] These measurements are based on a controlled propagation of electromagnetic waves. In the case of bridge measurements, stray and fringe capacitance between electrodes, which become an increasingly large source of error as the sample thickness is decreased, can be eliminated by the use of guard electrodes connected to a Wagner ground circuit. The theory of three terminal measurements of capacitance, employing a guard electrode, is discussed in ASTM specification D 150-65T (1965).

The sample, usually in the form of a compacted tablet is placed between electrodes in the measuring chamber. The thickness of the sample is accurately measured so that the capacitance of an equal thickness of air may be determined. It is therefore convenient to use a chamber in which the moveable electrode is attached to a micrometer. If the effect of adsorbed vapours on the dielectric constant is to be investigated the insulation of the sample chamber should be constructed entirely of Teflon. Such chambers are manufactured by the Balsbaugh Laboratories. This precaution is of added importance in the study of hydrated samples. Several materials often used as insulators will adsorb sufficient quantities of water to produce spurious measurements of dielectric loss.

Compacted tablets contain air filled interstices. The dielectric constant of the solid may be calculated from the dielectric constant of the tablet by extending Böttcher's[56] treatment of a powder to the case of a compacted tablet. One then obtains for the dielectric constant of the crystallite material

$$\kappa_0 = \frac{3\kappa\delta + 2\kappa(\kappa - 1)}{3\kappa\delta - (\kappa - 1)} \tag{18}$$

where κ_0 is the dielectric constant of the crystalline material, κ is the measured dielectric constant of the compacted tablet and δ is the packing fraction, or partial volume of crystallite in the pellet.

Low frequency capacitance measurements often reflect both the sample capacitance and polarization effects. The effect of polarization in different frequency regions can be understood more clearly in terms of the simple model of Figure 13. In this model a lumped polarization capacitance C_p is in series with the specimen conductance G which is then in parallel with

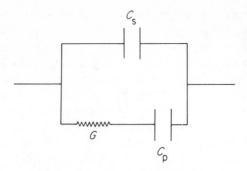

Figure 13. Scheider's (1962) model for the
lumped polarization capacitance in series with
the specimen conductance and both in parallel
with the specimen capacitance

the specimen capacitance C_s. The admittance of this network is given by

$$Y = G + i\omega C_s + \frac{1}{1 + \omega^2 C_p^2 / G^2}(i\omega C_p - G) \tag{19}$$

If $\omega^2 C_p^2 / G^2 \gg 1$, as is often the case in measurements of biomolecules in aqueous solution, then equation (19) reduces to

$$Y = G + i\omega C_s + \frac{i\omega G^2}{\omega^2 C_p} - \frac{G^3}{\omega^2 C_p^2} \tag{20}$$

The measured capacitance is $C_s + G^2/\omega^2 C_p$. Polarization, in this limit, is manifest as an additional term of the form $G^2/\omega^2 C_p$ where C_p is the polarization capacitance. Direct measurements of C_p in aqueous solutions at low frequencies[57] indicate that it varies as the -0.5 to -0.3 power of the frequency. In the approximation $\omega^2 C_p^2 / G^2 \gg 1$, the measured increment resulting from polarization ($G^2/\omega^2 C_p$) will then vary as the -1.5 to -1.7 power of the frequency. This is indeed the case in measurements of aqueous solutions of biomolecules. However, $1 \gg \omega^2 C_p^2 / G^2$, is the condition under which low frequency capacitance measurements are made on compacted tablets. In this approximation the admittance is given by

$$Y = +i\omega(C_s + C_p) \tag{21}$$

Therefore at low frequencies where polarization dominates, the measured capacitance will vary as the -0.3 to -0.5 power of the frequency. That this is indeed the case is shown in Figure 14.

In compacted tablets polarization effects are not limited to the regions around the electrodes but may occur at crystallite boundaries. Polarization

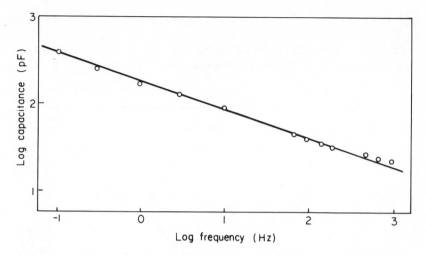

Figure 14. A plot of the log of the capacitance of a haemoglobin tablet at a hydration state of 2·2 mole per cent, *vs.* the log of the applied frequency. The straight line in this low frequency region has a slope of $\omega^{-0.35}$. (From reference 60)

of the Maxwell–Wagner variety will arise at any boundary between two media where

$$\frac{\sigma_1}{\kappa_1} \neq \frac{\sigma_2}{\kappa_2} \tag{22}$$

If a specimen composed of two materials, which do not possess the same ratio of conductivity to dielectric constant, is initially in the completely uncharged state then when a potential is instantaneously impressed across it, the dielectric displacement, D, at the first moment will be constant throughout the specimen. The charges have not yet penetrated into the sample. However, current density is determined by

$$j = \sigma E = \frac{\sigma}{\kappa} D \tag{23}$$

where j is the current density and E is the electric field strength. An accumulation of charge at boundaries which separate regions of different σ/κ must then occur. The build-up of charge at this boundary will continue until j is constant. This process requires time because it depends on a finite conduction through the media on both sides of the boundary being considered. It is then a process which may be characterized by a relaxation time.[58]

Low frequency capacitance measurements of hydrated protein tablets[24,59,60] and nucleic acids[61] show low frequency dispersion which may be the result of polarization. Rosen has also found a discontinuity in the high frequency (10^5–10^7 Hz) dielectric constant of bovine serum albumin at 20 per cent adsorbed water.[34]

Proteins containing less than 20 per cent adsorbed water do not exhibit a frequency dispersion of the dielectric constant in the region of 10^4–10^7 Hz.

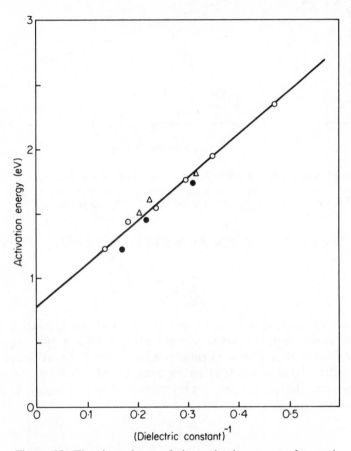

Figure 15. The dependence of the activation energy for semi-conduction in haemoglobin on the reciprocal effective dielectric constant for various amounts of adsorbed water (open circles), ethanol (triangles) and methanol (filled circles)[60]

For this adsorbate concentration the adsorption of water, ethanol, or methanol on protein decreases the activation energy according to equation (11). This can be seen in Figure 15. The decreased activation energy serves to increase conductivity of the sample. Equation (12) relates the conductivity with the effective dielectric constant of the system. The result, a linear

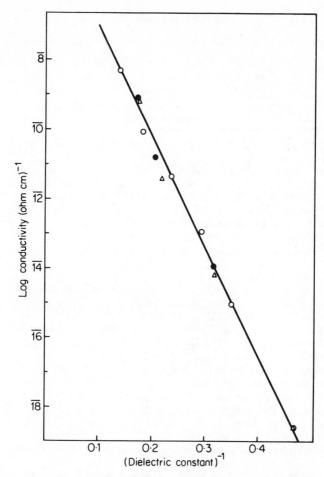

Figure 16. The dependence of the conductivity in haemoglobin on the reciprocal or the effective dielectric constant for various amounts of absorbed water (open circles), ethanol (triangles) and methanol (filled circles)[60]

dependence of the log of the conductivity on the reciprocal of the dielectric constant, is illustrated in Figure 16.

The adsorption of various solvents, for example water, ethanol or methanol on haemoglobin serves to increase the dielectric constant. As the dielectric constant of a medium increases, the activation energy for semiconductivity is decreased and this decreased activation energy then increases the conductivity of the sample at a given temperature. This now appears to be a general phenomenon, which, for biomacromolecules, is the dominant mechanism for the control of the electrical conductivity.

XI. ACKNOWLEDGMENTS

The work of this laboratory reported in this paper has been supported by the Atomic Energy Commission research contract No. AT(11-1)-1714 and the Office of Naval Research contract No. Nonr-2587.

REFERENCES

1. F. Gutmann and L. E. Lyons, *Organic Semiconductors*, John Wiley and Sons, New York, 1967
2. S. Baxter, *Trans. Faraday Soc.*, **39**, 207 (1943)
3. B. Rosenberg and E. Postow, *Ann. N.Y. Acad. Sci.*, **158** (Art. 1), 161 (1969)
4. G. King and J. A. Medley, *J. Colloid Sci.*, **4**, 9 (1949)
5. T. N. Misra, B. Rosenberg and R. Switzer, *J. Chem. Phys.*, **48**, 5 (1968)
6. J. O. Williams and B. Rosenberg. Private communication (1968)
7. B. Rosenberg, *Nature*, **193**, 364 (1962)
8. B. Rosenberg, B. Bhowmik, H. Harder and E. Postow, *J. Chem. Phys.*, **49**, 4108 (1968)
9. S. H. Glarum, *J. Phys. Chem. Solids*, **24**, 1577 (1963)
10. D. D. Eley, A. S. Fawcett and M. R. Willis, *Trans. Faraday, Soc.*, **64**, 1513 (1968)
11. R. Snart, *Trans. Faraday Soc.*, **59**, 754 (1963)
12. D. D. Eley, *J. Polymer Sci.*, C, **17**, 73 (1967)
13. A. Many, E. Harnik and D. Gerlich, *J. Chem. Phys.*, **23**, 1733 (1955)
14. D. D. Eley, *J. Polymer Sci.*, C, **17**, 73 (1967)
15. G. Kemeny and B. Rosenberg, *J. Chem. Phys.*, **52**, 4151 (1970)
16. G. Kemeny and B. Rosenberg, *J. Chem. Phys.*, **53**, 3549 (1970)
17. J. Tauc, *Science*, **158**, 1543 (1967)
18. C. Y. Liang and E. G. Scalco, *J. Chem. Phys.*, **40**, 919 (1964)
19. W. J. Siemons, P. E. Bierstedt and R. C. Kepler, *J. Chem. Phys.*, **39**, 3523 (1963)
20. P. Mueller, D. O. Rudin, H. T. Tien and W. C. Westcott, *Circulation*, **26**, 1167 (1962)
21. A. F. Ioffee, *Physics of Semiconductors*, Information Research Ltd., London, 1960, p. 194
22. M. E. Burnel, D. D. Eley and V. Subramanyan, *Ann. N.Y. Acad. Sci.*, **158** (Art. 1), 191 (1969)
23. B. Rosenberg, *J. Chem. Phys.*, **35**, 816 (1962)
24. S. Maricic, G. Pifat and V. Pravdic, *Ber. Bunsges, Phys. Chem.*, **68**, 787 (1964)
25. S. Maricic and G. Pifat, *Proton conductivity in methemoglobin and Na-DNA* in *Elektrochemische Methoden und Prinzipien in der Molekular Biologie III.* Jenaer Symposium, Akademie-Verlag, Berlin (1966)
26. B. Rosenberg, *Biopolymers Symposia*, **1**, 453 (1964)
27. P. S. B. Digby, *Semiconduction and Electrode Processes in Biological Material.* British Biophysical Society, Biological Semiconductors Conference (1968)
28. M. E. Fuller and W. S. Brey, Jr., *J. Biol. Chem.*, **243**, 274 (1967)
29. J. H. Wang, *J. Am. Chem. Soc.*, **76**, 4755 (1954)
30. J. L. Oncley, G. Scatchard and A. Brown, *J. Phys. Colloid Chem.*, **51**, **184**, 98 (1947)
31. H. P. Schwan, *Ann. N.Y. Acad. Sci.*, **125**, 344 (1965)
32. S. Maricic, G. Pifat and V. Pravdic, *Biochim. Biophys. Acta*, **79**, 293 (1964)
33. V. V. Zhukov and L. D. Stepin, *Biophysics*, **10**, 1081 (1963)
34. D. Rosen, *Trans. Faraday Soc.*, **59**, 2178 (1963)

35. V. Luzzati, J. Witz and A. Nicolaieef, *J. Mol. Biol.*, **3**, 379 (1961)
36. P. L. Privalov and G. M. Mrevlishvili, *Biophysics*, **12**, 19 (1967)
37. H. P. Fisher, *Biochim. Biophys. Acta*, **109**, 544 (1965)
38. M. R. Powell and B. Rosenberg, *Bioenergetics*, **1**, 493 (1970)
39. M. R. Powell and B. Rosenberg, *Biopolymers*, **9**, 1403 (1970)
40. W. Shockley, *Electrons and Holes in Semiconductors*, Von Nostrand, Princeton, New Jersey, 1950, p. 215
41. G. L. Jendrasiak, R. Leffler and B. Rosenberg. Private communication (1967)
42. A. C. Hermann and J. S. Ham, *Rev. Sci. Inst.*, **36**, 1553 (1965)
43. A. C. Hermann and A. Rembaum, *J. Polymer Sci.*, C, **17**, 107 (1967)
44. E. M. Trukhan, *Biophysics*, **11**, 486 (1966)
45. D. D. Eley and R. Pethig, *Bioenergetics*, **1**, 109 (1970)
46. D. D. Eley and R. Pethig, *Bioenergetics*, **2**, 39 (1971)
47. H. B. Bull, *J. Am. Chem. Soc.*, **66**, 1499 (1944)
48. L. Pauling, *J. Am. Chem. Soc.*, **67**, 555 (1945)
49. F. E. M. O'Brien, *J. Sci. Instr.*, **25**, 73 (1948)
50. M. H. Cardew and D. D. Eley, *The Sorption of Water by Haemoglobin* in *Fundamental Aspects in the Dehydration of Foodstuffs*. (Aberdeen Conference) Society of Chemical Industries, London, 1958, p. 24
51. D. D. Eley and R. B. Leslie, *Trans. Faraday Soc.*, **62**, 1002 (1966)
52. S. Brunauer, P. H. Emmett and E. Teller, *J. Am. Chem. Soc.*, **60**, 309 (1938)
53. K. S. Cole and H. S. Curtis, *Rev. Sci. Instr.*, **8**, 333 (1937)
54. E. R. Laird, *Can. J. Phys.*, **30**, 663 (1952)
55. F. E. Harris and C. T. O'Konski, *Rev. Sci. Instr.*, **26**, 482 (1955)
56. C. J. F. Böttcher, *Rec. Trav. Chim.*, *Pays-Bas*, **64**, 49 (1945)
57. I. Wolff, *J. Appl. Phys.*, **7**, 203 (1936)
58. H. P. Schwan, in *Advances in Biological and Medical Physics*. Vol. V (Eds. J. H. Lawrence and C. A. Tobias), Academic Press, New York, 1957.
59. S. Takashima and H. P. Schwan, *J. Phys. Chem.*, **69**, 12 (1965)
60. E. Postow and B. Rosenberg, *Bioenergetics*, **1**, 467 (1970)
61. C. T. O'Konski, P. Moser and M. Shirai, *Biopolymer Symposia*, **1**, 479 (1964)

CHAPTER 8

Magnetochemistry: Methods of measuring static magnetic susceptibility and their applications in biochemistry

W. Haberditzl
Berlin Humboldt University,
G.D.R.

I. PHYSICAL BASIS OF THE MEASUREMENT TECHNIQUES

A. The Magnetic Analogue of Coulomb's Law

Accounts of electrostatics may commence in a very simple manner with a statement of Coulomb's experimental law for the forces between electric charges. Magnetic dipoles, which are analogous to electrostatic dipoles, are the most fundamental units in magnetism, but isolated magnetic poles do not exist. However poles can be introduced in a hypothetical manner as the magnetic analogues of electrostatic charges and the magnetic analogue of Coulomb's law for the force between poles m_1 and m_2 as a function of spacing r

$$F = K\frac{m_1 m_2}{r^2}$$

used to define unit pole in the gaussian cgs system, with $K = 1$. The dimensions of the pole strength are thus $cm\,dyn^{1/2}$, F being in dynes.

B. Magnetic Field

The force on the pole m_2 can be considered to be due to a magnetic field $\mathbf{H}$ originating from m_1 such that $\mathbf{F} = m_2\mathbf{H}$. Thus the (vector) field must be given by

$$\mathbf{H} = \frac{m_1}{r^3}\mathbf{r}$$

The dimensions of the magnetic field, are accordingly $cm^{-1}\,dyn^{1/2} = g^{1/2}\,cm^{-1/2}\,sec^{-1}$. The units of field are termed Oersted (Oe).

C. Dipole in a Uniform Magnetic Field

The magnetic dipole μ may be defined in a similar manner to the electric dipole by

$$\mathbf{\mu} = m\mathbf{l}$$

where $\mathbf{l}$ is the (vector) distance between the constituent poles m. In a uniform field, which may be represented by parallel lines of force with uniform density, a magnetic dipole experiences a torque

$$\mathbf{L} = \mathbf{\mu} \times \mathbf{H}$$

If α is the angle between $\mathbf{\mu}$ and $\mathbf{H}$ the magnitude of the torque is

$$L = \mu H \sin \alpha$$

The potential energy of the dipole in the field is given by integration as

$$E = \int L \, d\alpha = -\mu H \cos \alpha + \text{constant}$$

Taking $E = 0$ when $\alpha = \pi/2$

$$E = -\mu H \cos \alpha = -\mathbf{\mu} \cdot \mathbf{H}$$

This corresponds to the observed tendency of a dipole to become oriented parallel to a field.

D. Magnetic Fields from Electric Currents

According to Maxwell's first law, $\mathbf{H}$ and the current density vector $\mathbf{j}$ are related by

$$\text{rot } \mathbf{H} = \frac{4\pi}{c}\mathbf{j}$$

Using this and Stokes law

$$\int \text{rot } \mathbf{H} \cdot d\mathbf{s} = \oint \mathbf{H} \cdot d\mathbf{l} = \frac{4\pi}{c}\int \mathbf{j} \cdot d\mathbf{s} = \frac{4\pi}{c}i$$

The current i is thus proportional to the field strength integrated along the closed path $\mathbf{l}$. ($\int \mathbf{H} \cdot d\mathbf{s}$ is also termed the magnetic potential ϕ, to which the field is related by $\mathbf{H} = -\text{grad } \phi$.) Accordingly the field inside a toroid of length l with n turns carrying a current i is calculated as

$$H = \frac{4\pi}{c} \cdot \frac{ni}{l}$$

The magnetic field may also be defined directly in terms of the current with which it is associated, and in the GIORGI system

$$\oint \mathbf{H} \cdot d\mathbf{l} = i$$

and for the solenoid

$$H = \frac{ni}{l}$$

If i is measured in amperes (A) then the units of field are amperes per metre, $A\,m^{-1}$.

E. Induction of an Electric Potential by a Changing Magnetic Field

According to Maxwell's second law

$$\mathrm{rot}\,\mathbf{E} = -\frac{1}{c}\frac{\partial \mathbf{B}}{\partial t}$$

$\mathbf{E}$ is the electric field vector and $\mathbf{B}$ the magnetic induction. Using Stokes equation (integrating over an area s contained within the curve l)

$$\int \mathrm{rot}\,\mathbf{E}\,ds = \oint \mathbf{E} \cdot d\mathbf{l} = -\frac{1}{c}\frac{\partial}{\partial t}\int \mathbf{B} \cdot d\mathbf{s} = \frac{1}{c}\frac{\partial \Phi}{\partial t}$$

$\Phi = \int \mathbf{B} \cdot d\mathbf{s}$ is the magnetic flux (induction being flux density or flux per unit area). Since $\mathbf{E} \cdot d\mathbf{l} = V$ is the electric potential this gives the induction law. On carrying out an induction experiment with a coil of area $\mathbf{s}$ with n turns we obtain a potential between the ends of the coil of

$$V = \frac{1}{c}n\frac{\partial \Phi}{\partial t} = -\frac{1}{c}ns\frac{\partial \mathbf{B}}{\partial t}$$

F. Magnetization, Susceptibility and Permeability

The field *in vacuo* produced by a current in a coil is identical to the induction, $\mathbf{B} = \mathbf{H}$. If the coil is filled homogeneously with matter then the relationship between $\mathbf{B}$ and $\mathbf{H}$ depends on the parameters of the material. With magnetization $\mathbf{I}$, susceptibility χ and permeability μ the relationships are $\mathbf{B} = \mathbf{H} + 4\pi\mathbf{I}$, $\mathbf{I} = \chi\mathbf{H}$, $\mu = 1 + 4\pi\chi$. The signs and magnitudes of the values of χ can be used to give an approximate classification of materials:

$\chi < 0$: diamagnetic

$\chi > 0$: paramagnetic (or antiferromagnetic)

$\chi \gg 0$: ferromagnetic (or ferrimagnetic)

The term 'magnetization' is related to the magnetic dipole moment (as defined in section 1.3) by

$$\sum \mu = M = \int \mathbf{I} \cdot d\mathbf{v}$$

I is thus the magnetic moment (M) per unit volume. In the gaussian system B, H and I have the same dimensions ($cm^{-1/2} g^{1/2} sec^{-1}$) and the same units are used. The units of H are called Oersted (Oe) whereas for the induction (B) they are gauss (G). The following conversions apply

$$1 \text{ G} = 10^{-8} \text{ V sec cm}^{-2}$$

$$1 \text{ Oe} = \frac{10}{4\pi} \text{A cm}^{-1}$$

Strictly speaking χ is a tensor, the components of which are field dependent. For ligands and gases the tensor can be represented by its character

$$\chi = \tfrac{1}{3}(\chi_{xx} + \chi_{yy} + \chi_{zz})$$

Since B and H are dimensionally equivalent, χ is a dimensionless number which may be related either to unit volume or to unit mass. The following symbols and relaxations apply: χ per cm^3 is χ_v; χ per g is $\chi_g = \chi_v/d$ (density d); χ per mole is $\chi_M = M\chi_g$ (molecular weight M).

The χ values for most materials are of the order 10^{-5} to 10^{-6} (g^{-1}).

II. TECHNIQUES FOR STATIC SUSCEPTIBILITY MEASUREMENT

A. *Production and Measurement of Magnetic Fields*

1. Electromagnets with Ferromagnetic Cores

In most magnetochemical laboratories commercially available electromagnets are used for static susceptibility measurements. The choice of efficient, stabilized, adjustable magnets has improved greatly in recent years owing to the development of nuclear and electron spin resonance spectroscopy. For magnetochemical work the dimensions of the air gap and of the pole pieces must be adjustable because the range of field strengths, the field gradients and space requirements in the gap must be changed to suit the problem. Moreover it is useful, when the problem permits it, to work with experimental times and currents at which the heat generated is so small that problems of cooling and current stabilization are negligible. When using a magnetic balance with electromagnetic compensation it is possible under certain conditions to eliminate fluctuations in supply voltage by feeding the compensator and the electromagnet from the same source. For precision measurements with high fields it is common practice to use n.m.r. techniques for stabilization and control of the electromagnets. Many standard, commercial models depend for their construction and design on older types (classical horseshoe magnets). The principle of the 'pot magnet', in which the yoke forms an iron container around the whole magnet and the air gap is accessible through a slit in the yoke, has certain advantages. Figure 1

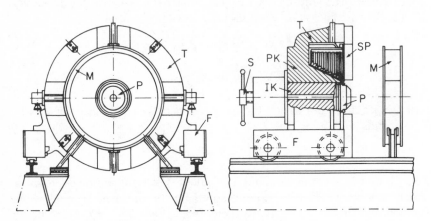

Figure 1. Electromagnet after Kohlhaas and Lange (assembly sketch). T—outer vessel; PK—yoke; IK—pole; P—pole piece; S—spindle; F—drive unit; SP—coil; M—mounting ring

shows the construction of a particularly efficient electromagnet of this type after Kohlhaas and Lange.[1] The magnitude of the magnetic field available in the air gap depends on the relation between the different elements: the poles, pole pieces and the yoke.

Brief description: 290 mm diameter pole pieces, the magnetic circuit consisting of the vessels T (13 mm maximum external diameter) each with a conical pole PK, the pole pieces P and the air gap is shown. The poles each contain an inner piece IK (driven by a spindle drive S) on which the pole pieces can be screwed. The inner pieces have holes for longitudinal magnetic field studies. They may be moved independently of each other by means of the two spindles. Air gaps of between 0 and 170 mm are possible. The coils, cooled with transformer oil, fit around the conical poles. Each vessel has its own coil consisting of 8 individual coils. Each of these units is formed of spun copper tape (25×1.4 mm^2). With a total of 1664 turns the output is 166,400 amp-turns at 100 A. The fields obtained in the air gap as a function of current are shown in Figure 2. The two vessels are each mounted on a separate drive and can be moved hydraulically on rails.

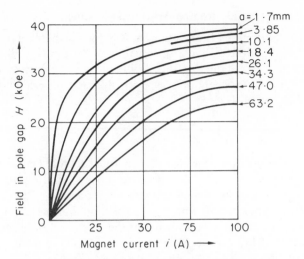

Figure 2. Fields in pole gap as a function of the driving
current (after Kohlhaas and Lange)

Exchangeable mounting rings (W) are used to support the measuring
equipment. In this way the magnet can be used by different groups of
workers at different times. The oil is cooled by two oil coolers each of 20 kW
capacity.

If higher field strengths are required then the saturation magnetization
of the material of the poles must be considered. For example this is 24·5 kG
for a 50:50 Fe–Co alloy. With suitably shaped pole pieces this may be
increased to about 50 kG. The use of pole pieces of rare earth metals has
been proposed. On account of the higher saturation magnetization of, for
example, dysprosium and holmium, fields of 80 kG could be achieved, but
this has not been done in practice.

2. Permanent Magnets

Due to the development of new magnetic materials with very high
remanence and coercivity, and thus of energy product, considerable progress
has occurred in these magnets. However commercial permanent magnets
(e.g. AlNiCo magnets) are of limited use in connection with static susceptibility measurements because in most cases the investigation of field strength
dependence is essential for the full evaluation of the material. For certain
purposes (e.g. investigations requiring measurements over a long period of
time) permanent magnets have proved to be useful. Standard permanent
magnets are especially suited for calibration purposes.

3. Ironfree Water-cooled Coils

If extremely strong fields are required (e.g. for magnetocatalytic studies) then open coils without cores are required. Naturally, unlimited field strengths cannot be obtained in this way because ultimately even the most efficient cooling system cannot cope with the heat generated electrically. The equation due to FABRY

$$H = g\sqrt{\frac{W\lambda}{\rho a}}$$

gives the capacity obtainable. (g = geometrical factor, dependent on the size and shape of the coil, W = electrical power employed, ρ = specific resistance of the coil, a = internal radius of the coil, λ = filling factor giving the fraction of the coil volume occupied by the conducting material.) Using this equation it may be shown that the most favourable construction gives a field of 110 kG per megawatt of power. To obtain fields of this order of magnitude one must cool as efficiently as possible.

The highest field strengths so far obtained with coils are 250 kG continuously (by Montgomery at the Massachusetts Institute of Technology) or 400 kG (Bergles, also at M.I.T.) for 2 sec in pulsed operation only. With such magnets the tensile and compressive forces generated are such as to exceed the yield point even of hard copper. For this reason the principles of construction are still highly specialized.

4. Superconducting Magnets

Superconducting elements (e.g. Hg, Sn, or In) have no electrical resistance below their critical temperatures, T_c (for the most important superconductors this is about 10 °K to 20 °K) and exhibit the phenomenon of the Meissner effect: if they are put into a magnetic field then the magnetic flux only penetrates a thin superficial layer. Above the critical magnetic field H_c the superconductivity disappears and the magnetic flux penetrates all the material. The magnitude of H_c is itself temperature dependent. If a superconductor at a temperature below T_c is put in a magnetic field it may exhibit a linear magnetization curve (Figure 3). Superconductors having this

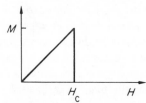

Figure 3. Linear magnetization curve of a soft superconductor

property are called soft superconductors. The magnetization curve of hard superconductors is shown in Figure 4, this depending on the ratio of two quantities H_{c_1} and H_{c_2}. Such hard superconductors are used for superconducting magnets. The most important hard materials are V_3Ga and Nb_3Sn, which are very brittle intermetallic compounds, as well as the ductile

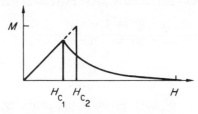

Figure 4. Magnetization curve of a hard superconductor

alloys Ti–Nb and Zr–Nb. Ductibility is of obvious importance for the construction of coils and to date much trouble has been taken to fabricate the least brittle wires. In addition it is found that lattice defects are important in these materials. These cause a very high critical current strength (that is that current strength for which the superconductivity in an external magnetic field breaks down).

Small superconducting magnets for fields of between 50 and 70 kG, will surely play an increasingly important role in magnetochemical laboratories in the future. Naturally these are particularly useful for measurements at low temperatures. The associated problems of refrigeration are readily overcome in laboratories in which low temperature magnetic measurements are carried out.

5. Magnetic Field Measurements

Although in many cases static susceptibilities are measured relative to standard (calibration) materials, it is still very useful to know the absolute field strength and the pattern of the field gradients for various pole piece shapes. A very exact evaluation of the field may not be necessary in most cases. For simple field measurements there are numerous methods available, which are described in greater detail in the literature: flux-meter or ballistic techniques, resistance variation of bismuth spirals, and, above all, Hall probes. Stricter requirements of accuracy are fulfilled by methods in which the force on a current carrying conductor is measured (the principle of the Cotton balance). The measurement of fields by means of proton resonance has achieved great importance. In this method the resonance frequency of the protons in water molecules in the magnetic field, is measured. The resonance frequency, which can be measured with great accuracy, is related to the field strength by the resonance equation. This method is above all suitable for the measurement of strong homogeneous fields.

B. Force Exerted by Magnetic Fields

1. Small Specimen Volume: Gradient Methods

The basic magnetostatic relation for the force which a field with gradient $\partial H/\partial x$ exerts *in vacuo* on a body with the magnetic moment M is:

$$F = M\frac{\partial H}{\partial x}$$

For a small specimen of homogeneous susceptibility χ_v with volume dv we have the following differential equation:

$$dF = \chi_v H\frac{\partial H}{\partial x}\,dv$$

since $M = \chi_v H\,dv$. If the quantity $H(\partial H/\partial x)$ is constant over the volume dv then this relation can be used in practice directly for force measurements on small specimens; otherwise the force is given by

$$F = \chi_v\!\int H\frac{\partial H}{\partial x}\,dv$$

The pattern of the three quantities H, $\partial H/\partial x$ and $H(\partial H/\partial x)$ determines the mode of operation of nearly all methods of measurement which are based on magnetic force effects. Figure 5 shows a schematic representation of the

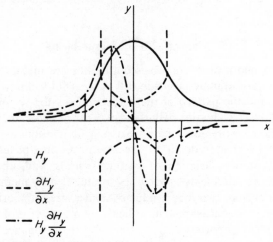

Figure 5. Plots of H, $\partial H/\partial x$ and $H(\partial H/\partial x)$ in a pole gap designed for field gradient methods

pattern of these quantities in the air gap between two pole pieces with 'inhomogeneous' profiles. The shape of the $H(\partial H/\partial x)$ curve is decisive for the force effect (note that the ordinate gives the force effect on the sample in the direction of the abscissa relative to the origin).

It is important, when performing measurements in non-uniform fields, that the sample is always placed in the same position in the field.

2. Cylindrical Specimens: Integral Methods

A general expression, which takes into account all the data relating to the measurement of forces in a magnetic field, allowing for the non-homogeneous character of the density ρ and also of the permeability μ, is

$$F = -\frac{1}{8\pi} \int \int \int \left\{ \text{grad}\left(H^2 \rho \frac{\partial \mu}{\partial \rho}\right) + H^2 \, \text{grad} \, \mu \right\} dx \, dy \, dz$$

In the special case of a uniformly dense and magnetically isotropic sample (i.e. for liquids; this condition is not usually fulfilled by solids) we have

$$\mu = 1 + 4\pi\chi\rho, \qquad \frac{\partial \mu}{\partial \rho} = 4\pi\chi, \qquad \text{grad} \, \mu = 0$$

From this we obtain

$$F = \frac{\chi_v - \chi_{v_0}}{2} \int \int \int \text{grad} \, H^2 \, dx \, dy \, dz$$

χ_{v_0} is the susceptibility of the surrounding medium which has been displaced by the introduction of the sample.

In the Gouy method, or the 'cylinder method' the sample has a uniform cross-section q and is suspended so that one end is in a region of constant maximum field strength, H_{max}, and the other in a field strength H_0. We then have the following relation ($dv = q \, dx$)

$$\int dF(x) = (\chi_v - \chi_{v_0})q \int_{H_0}^{H_{\text{max}}} H(x) \, dH(x)$$

from which we obtain

$$F = \frac{\chi_v - \chi_{v_0}}{2} q(H_{\text{max}}^2 - H_0^2)$$

If it is so arranged that $H_0 = 0$ we have finally

$$\chi_v = \frac{2F}{q H_{\text{max}}^2} + \chi_{v_0}$$

or for the gram susceptibility ($\chi = \chi_v/\rho$)

$$\chi = \frac{2F}{q\rho H^2_{\text{max}}} + \frac{\chi_{v_0}}{\rho}$$

ρ is here the true or apparent density of the substance. If the substance is finely crystalline or a powder then the lack of uniformity of the density is an important factor limiting the accuracy of the results.

For highly accurate results the following points should be noted:

(a) The second term in the last equation can lead to errors, particularly if measurements are carried out in air, since oxygen has a high paramagnetic susceptibility. Therefore measurements by this method are sometimes performed *in vacuo* or in a helium atmosphere.

(b) Measurements on a powder include the susceptibility of the residual gas within the interstices. It is recommended that for diamagnetic measurements the sample tube should be filled in a helium atmosphere.

(c) The dependence of the buoyancy of the sample on the temperature in particular can act as a source of error in some cases.

3. Torsion Methods

Increasing attention is being paid to measurements on magnetically anisotropic substances, particularly single crystals. In a homogeneous magnetic field a freely rotating anisotropic material takes up a position in which the axis of greatest susceptibility is parallel to the field direction. If an anisotropic body is turned out of its stable position it experiences a torque. This torque can be measured if the body is suspended by means of a quartz fibre of known torsion constant.

For the method due to Krishnan[2] the orientation of the crystal is adjusted by turning the torsion head until it is no longer affected when the field is increased (5 to 10 kG). The direction in which the magnitude of the susceptibility is greatest is then parallel to the field direction. Beyond this stage Krishnan[2] has proposed and developed two techniques, the *oscillation* and the *angle of torque* method. The first of these relies upon the fact that when the crystal performs an oscillatory motion the period of oscillation is related to the magnetic anisotropy of the two principal susceptibilities in the plane of oscillation $\Delta\chi$ by the following equation:

$$\Delta\chi = \frac{T^2_0 - T^2}{T^2} \times \frac{cM}{H^2 m}$$

T and T_0 = period of oscillation with and without the field, c = torsion constant of the suspension, m = mass of the crystal, M = molecular weight, $\Delta\chi$ = difference between the maximum and minimum susceptibilities along the magnetic axes lying in the plane of oscillation. In order to determine the susceptibility along the third principal axis the crystal must be resuspended. The various orientations appropriate to the different crystal classes, and the labelling of the magnetic axes for the crystal classes may be found in the specialist literature. The angle of torque method is based on the following reaction: the torque exerted by a magnetic field on an anisotropic crystal, with principal susceptibilities χ_1 and χ_2 in the plane of torque is:

$$D = \frac{H^2 m}{2M}(\chi_1 - \chi_2) \sin 2\phi$$

where ϕ is the angle of torque of the crystal.

At equilibrium this torque must equal that of the suspension fibre:

$$\frac{H^2 m}{2M}(\chi_1 - \chi_2) \sin 2\phi = c(\psi - \phi)$$

where ψ is the angle measured on the torsion head. It is always the case that $\psi > \phi$. If $\phi = \pi/4$, then the crystal takes up a new equilibrium position. The corresponding value of ψ is read off. After some calculation one obtains

$$\Delta\chi = \frac{2M}{m} \frac{c}{H^2}\left(\psi - \frac{\pi}{4}\right)$$

C. Techniques Based on Force Measurements

1. Vertical Cylinder Method (Gouy Method)

A very simple Gouy balance, readily installed in the laboratory, but without great claims for precision (not better than 2 per cent) is shown in Figure 6.

The force exerted on the sample tube in the magnetic field is determined with a commercial semimicro-balance. This rests upon a mechanical stage, adjustable in two directions, which permits adjustment of the sample between the pole shoes. Thermal disturbances of the suspension system are reduced by means of a protecting jacket. The magnetic field (3 to 10 kG) is provided by a watercooled Weiss magnet. The air-gap is 3·8 cm, or 4·8 cm when a heating or cooling jacket is required. Fluctuations in the laboratory direct current voltage frequently call for a special supply for the magnet (see block diagram, Figure 7). In this the alternating current supply is fed via a magnetic stabilizer (constant voltage transformer) to an adjustable transformer and rectified in a two-way rectifier stage. The magnet current can be controlled by the regulating transformer. Figure 6 shows the housing of a heating mantle, for measurements at elevated temperatures.

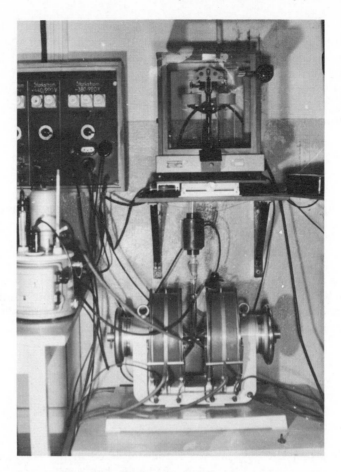

Figure 6. Assembly of a simple Gouy balance

A self-compensating magnetic balance with an automatic temperature programme has been developed by Wuttig and Wagner.[3] It can be used to obtain susceptibility–temperature curves, for which a low rate of change of temperature is required, the total change possibly stretching over several days. Figure 8 shows the block diagram of the whole apparatus. The position of the pendulum balance is also compensated electromagnetically. The balance compensating current is automatically adjusted by means of a servo-amplifier, which is activated photoelectrically by changes in the position of the pendulum. A second amplifier is used for the temperature programme, this controlling the heating-jacket current by the comparison of a programmed voltage with the voltage from a thermocouple attached to the sample. The corresponding values of the compensating

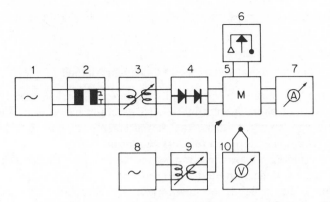

Figure 7. Block diagram of the equipment in Figure 6: (1) a.c. source; (2) magnet constant voltage supply; (3) variable transformer with control; (4) two-way rectifier; (5) magnet; (6) semimicro-balance; (7) ammeter for field current; (8, 9) power supply for the heating jacket; (10) temperature control (thermocouple)

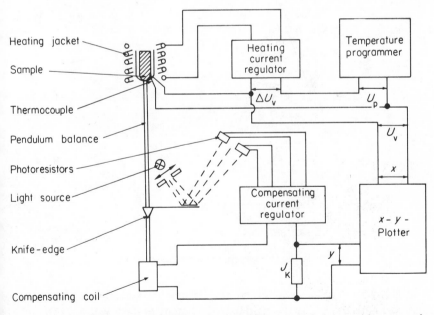

Figure 8. Block diagram of the self-compensating magnetic balance with automatic temperature programme, after Wuttig and Wagner:

U_p—predetermined potential for temperature programmer;
U_v—thermally produced potential; $\Delta U_v = U_p - U_v$;
J_K—balance compensating current

current and temperature are shown on a x–y plotter, whose rate of print-out is adjustable.

2. Horizontal Cylinder and Pendulum Method

Naturally the deflection of a horizontally mounted cylindrical sample can also be utilized to measure susceptibilities. This method has various advantages: for example changes during the measurements (due to deposits of dust or water vapour) are appreciably less significant.

The force exerted axially on the sample after switching on the field can be measured either as a deflection (under certain conditions) or by a compensation technique. The principle of the horizontal deflection was used by Theorell[4] for a highly sensitive magnetic micro method. This is based on compensating the deflection of a pendulum system (the sample hanging on a fibre about 1 m long) by a horizontal displacement of a sliding carriage which carries the pendulum system with it. It is true that this kind of force compensation is associated with the disadvantage that the force measurement is not independent of the sensitivity of the system.

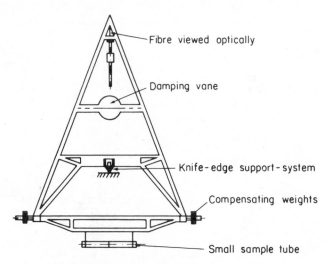

Figure 9. Balance system of a Havemann balance

This disadvantage is avoided by the design of Havemann's[5] magnetic balance. In this case there is a rigid balance system, to the lower end of which the small sample tube is attached in a horizontal position in the magnetic field. Figure 9 shows a balance system constructed of aluminium, which is relatively robust and is suitable for routine measurements. Figure 10 shows improved equipment based on this system, a particularly light balance

Figure 10. Diamagnetism balance

system made of cemented aluminium tubing (as used for instrument pointers) whose torque is electromagnetically compensated by a compensating coil placed in the pole gap of a small subsidiary magnet. The current connections (by means of two small platinum springs contacting platinum surfaces) is so arranged that the additional torque is insignificant.

3. Differential Methods (Faraday Methods)

By referring to the differential equation it is seen that these methods depend on the sample being positioned always in the same part of the field (since the integral is not carried out over a region of field strength). It has been found that specially-shaped pole pieces are most suitable (see Figure 11).

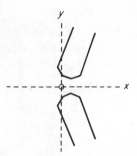

Figure 11. Principle of the Faraday balance

The principle is shown in Figure 11. The sample is placed so that it is in a region of maximum value of $H \, \partial H / \partial x$ (= origin of the coordinate system). The sample can move freely along the x-axis; the suspension can be achieved in various ways. The oldest technique is due to Weiss and Foëx[6] (see Figure 12).

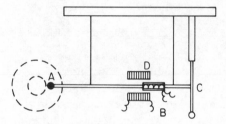

Figure 12. Method of Weiss and Foëx

The sample A is connected rigidly to a rod, supported bifilarly, which can move in the direction AC. The mirror C (which reflects a beam of light) is so attached that it tilts when the rod moves. Another method has been given by Curie and Chéneveau[7] (see Figure 13). It requires a torsion balance and a magnet which can rotate about one axis. The latter is so arranged that no

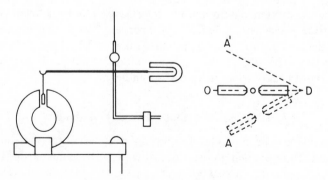

Figure 13. Curie–Chéneveau balance

deflection of the torsion balance occurs (null point). If the magnet is now turned then a maximum deflection will occur for a particular position (A). The system is calibrated by reference to the maximum deflections produced by standard materials.

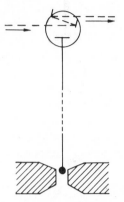

Figure 14. Sucksmith balance

Figure 14 shows Sucksmith's[8] apparatus. The sample is suspended vertically from a phosphor–bronze ring. On this ring two small mirrors are mounted in such a way as to optimize the deflection of a light beam when the ring is moved. Figure 15 shows the Faraday balance after Selwood.[9]

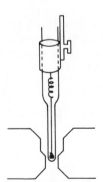

Figure 15. Faraday balance, after Selwood

4. Torsion Methods

Numerous torsion balances are described in the literature; they were developed principally for precision measurements as well as for measurements on very small amounts of materials. Since they involve mainly specialized designs only a few are described here. Rankine[10] used a vertical metal suspension strip hanging from a torsion head; an aluminium beam is supported by the strip and the sample, in a magnetic field, is attached to one end of this beam (optical level system 1). The reading on the torsion head can be obtained very accurately by measuring a light interference system. Bockris and Parsons[11] describe equipment capable of measuring a change in volume susceptibility of 4×10^{-10}. A torsion balance described by Schoffa, Ristau and Mai[12] is suitable for anisotropy measurements over a wide temperature range. The same principle has been used for measuring gas susceptibilities as well as for studies of photochemical reactions. A very sensitive apparatus based on Bitter's[13] method was developed by Evans.[14] The small sample tube, hanging on a quartz fibre between the pole pieces, is illuminated from below. The tube, whose illuminated surface is divided into four sectors (alternate sectors being blackened) rotates as a result of the photochemical reaction (Figure 16). The torsion effect is detected photoelectrically using an optical lever arrangement. The complete suspension

Figure 16. Arrangement by Bitter for the measurement of gas susceptibilities (Evans used the same principle for the study of photochemical equilibria)

system is maintained in a hydrogen atmosphere. Mercier and Bovet[15] have described equipment using a torsion pendulum, in which the change in the frequency of oscillation of the pendulum on applying the magnetic field is measured with a quartz clock. Recently Geist[16] has succeeded in detecting susceptibility differences of 3×10^{-11} (corresponding to 3×10^{15} electron spins at $140\,°K$) by means of a torsion pendulum Faraday method with a permanent magnet. These latter methods may well represent the absolute upper limit of sensitivity of static susceptibility techniques.

D. Induction Methods and Techniques Based on Other Principles

1. Induction Techniques

These depend on the following principle. The sample is placed inside the coil of a high frequency oscillator; the inductance of the coil is altered by changes in the permeability and this alters the frequency of the oscillator. The frequency changes are determined by measuring the beats produced by combining the unknown with a known fixed frequency.

The following relationships exist between the induction L, frequency v and susceptibility χ_v:

$$\frac{dL}{L} = -2\frac{dv}{v}, \qquad L = L_0(1 + \chi_v)$$

Since χ_v is of the order of 10^{-6}, dv is of the order of 1 Hz when $v = 1$ MHz. If the frequency is measured to ± 0.01 Hz we have

$$\chi_v = \frac{\Delta L}{L_0} \doteq \frac{dv}{v_0} \doteq 10^{-8}$$

and χ can be measured to about 2 decimal places ($\times 10^{-6}$).

Original equipment of this kind was built by Schwenkhagen and Gutwill[17] based on a suggestion by Eugen Müller.[18] Figure 17 shows a block diagram of the arrangement.

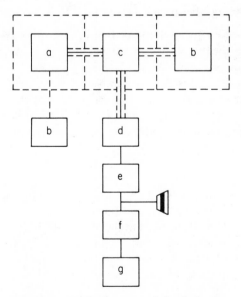

Figure 17. Block diagram of the induction apparatus of Schwenkhagen and Gutwill

(a) is a high frequency oscillator, with (b) the auxiliary oscillators. The beat frequency is amplified (d) after being rectified (c); (e) is a filter chain and (f) a frequency measurer (condenser–load current principle). The load current is compensated in a compensation control.

Further equipment of this type has been described by many workers e.g. Pacault, Lemanceau and Joussot-Dubien,[19] Effemey, Parsons and Bockris[20] as well as by Seidel and Vogel[21] (apparatus shown in Figure 18).

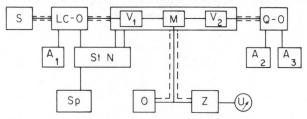

Figure 18. Induction measuring apparatus of Seidel and Vogel. S—coil of the LC-oscillator; LC-O—LC-oscillator; Q-O—standard quartz oscillator; Z—counting equipment; U—switch clock; StN—stabilizer; Sp—constant voltage supply; V_1—frequency multiplier of the LC-O; V_2—frequency multiplier of Q-O; M—mixer; O—oscillograph; A_1—power supply for the heater of the LC-O; A_2—power supply for the heater of the Q-O; A_3—anode battery of the Q-O

2. Susceptibility Measurements with the Aid of Nuclear Resonance

A novel method has been developed and put into practice, especially by H. J. Friedrich,[22] for the purpose of magnetochemical investigations.

In nuclear resonance the resonant frequency is related to the field strength acting at the nucleus. This field is composed of the external field H_0 and that induced at the nucleus by local dia- and paramagnetic ions. The induced field produces a shift of the frequency, for which the following relationship holds under certain definite conditions:

$$\frac{\Delta H}{H} = \frac{2\pi}{3}\Delta\chi_v$$

The practical procedure is as follows: A standard solution is placed in one part of a cylindrical double tube and the same solution containing some of the material to be studied is placed in the other part. If there is no interaction between the reference solution and the unknown substance the (paramagnetic) susceptibility of the unknown may be determined from the nuclear resonance effect using the following equation:

$$\chi_g = \frac{3\Delta v}{2\pi vm} + \chi_{gs} + \chi_{gs}\frac{(\rho_s - \rho_L)}{m}$$

in which v = measuring frequency in Hz, m = weight of unknown substance in 1 ml of solution, ρ_s = density of the solvent, ρ_L = density of the solution, χ_{gs} = gram susceptibility of the solvent.

3. Other Methods

The method of G. Quincke[23] is nowadays of little importance. It is only suitable for liquid materials (possibly for gases) and the accuracy is low compared with that of magnetic balances. It depends on the forces exerted on a column of liquid placed in a capillary between the pole pieces of a magnet. Using the equipment shown in Figure 19, the movement of the liquid in the column is a direct measure of its magnetic susceptibility.

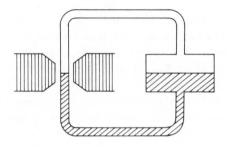

Figure 19. Quincke method (moving meniscus method)

Another method, depending on a different hydrostatic principle, was developed by Grün and Haas[24] using the principle of the cartesian diver. Such a diver, filled with the liquid under investigation, experiences a certain upward thrust. This is compensated by the addition of a paramagnetic solution.

E. Special Investigation Conditions, Preparation and Evaluation of Results

1. Measurement of the Temperature Dependence of the Susceptibility

A knowledge of the temperature dependence of the susceptibility is important for the full interpretation of measurements on paramagnetic materials. Measurements at temperatures down to about 70 °K is relatively straightforward (see Figure 20), e.g. the sample tube may be surrounded with a double-walled glass tube and the connections sealed with a thermostatic material. Heat transfer fluids appropriate to the temperature range under study can be used. A platinum resistance thermometer is sited immediately beneath the sample to record the temperature. The resistance of

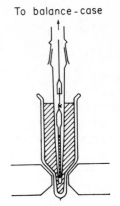

To balance - case

Figure 20. Simple arrangement for measurements at low
temperatures

the platinum thermometer is determined with a bridge circuit. The choice of the intervals between the measurements is determined by the rate of the temperature variation. The system must be allowed to come to thermal equilibrium, and this is recognized when the susceptibility at constant field strength becomes steady. Of course the changes in density of the sample with temperature must also be allowed for. This is best done by following the level of the solution in the sample tube as a function of temperature (e.g. with a travelling microscope). In most cases there is a linear relation between level and temperature.

Measurements at elevated temperatures are best carried out with a small electrical furnace (e.g. non-inductively wound platinum heating coils). The temperature gradient in the furnace must be measured as accurately as possible. At higher temperatures relatively large temperature gradients are found in double-walled quartz tube furnaces with platinum heating coils. An improved procedure is to embed the coils in a ceramic paste. A tube serving as an additional jacket reduces the temperature gradients even at temperatures above 500 °C. The gradients can be measured with two thermocouples and a compensation circuit.

2. Ferromagnetic Impurities: Field Strength Dependence

Ferromagnetic impurities are troublesome, even at the lowest concentrations. They may be recognized by the fact that the susceptibility of the sample appears to be dependent on the field strength. It is well known that ferromagnetic susceptibility, unlike dia- and paramagnetism, is characterized by being dependent on the field strength, particularly in certain regions of field. By measuring this dependence it is possible to allow for the contribution due to ferromagnetic impurities. The susceptibility data are plotted against H^{-1} and the plots, in most cases linear, are extrapolated to obtain the

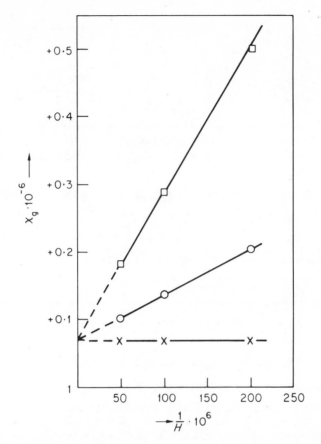

Figure 21. Extrapolation to eliminate ferromagnetic contributions

susceptibility for $H = \infty$. Figure 21 shows such extrapolations for materials with different amounts of ferromagnetic impurities. Knappwost[25] in partic- ular has commented on the uncertainty in this procedure. He showed that, when using the cylinder method, the integrated susceptibility (between H_{max} and H_0) is not exactly proportional to the reciprocal of the field strength if the paramagnetic materials are contaminated with ferromagnetics. The extrapolation procedure then gives susceptibility values which are too high. He proposed the use of a field gradient method instead, based on the use of shaped pole pieces (see Figure 22) producing two different regions of homogeneous field (H_{max} and H_z).

In order to saturate the ferromagnetic components H_z is arranged to be greater than 7 kG. Under this condition we have (F = effective force)

$$F_{ferro} = I_0 q_{(\text{'ferro'})}(H_{max} - H_z)$$

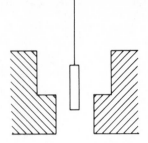

Figure 22. Pole piece shapes for the field difference
method after Knappwost

Together with the paramagnetic contribution (χ_p, q_p) we have $(q_p \approx q)$ for $\chi_{\text{total}} \equiv \chi_M$

$$F_{\text{total}} = \tfrac{1}{2}\chi_M q(H_{\max}^2 - H_z^2) = \tfrac{1}{2}\chi_p q(H_{\max}^2 - H_z^2) + I_0 q_{\text{ferro}}(H_{\max} - H_z)$$

Multiplying this equation by $2/[q(H_{\max}^2 - H_z^2)]$ gives

$$\chi_v = \chi_p + \frac{2I_0 q_{\text{ferro}}(H_{\max} - H_z)}{q(H_{\max}^2 - H_z^2)} = \chi_p + \frac{2I_0 q_{\text{ferro}}}{q} - \frac{1}{H_{\max} + H_z}$$

Extrapolation to $H = \infty$ yields a reliable value for $\chi_v = f/(H_{\max} + H_z)$.

3. Sample Preparation; Sources of Error; Calibration

Measurements on powders demand a painstaking preparation, since any inhomogeneity of packing can represent a considerable source of error. A relatively simple method of obtaining as uniform a packing as possible requires that only a few millimetres of the very finely powdered material are added at a time and then compacted by tamping. A ball mill is useful for grinding to a powder, but one must beware of the introduction of ferromagnetic impurities. A second method is to centrifuge homogeneous suspensions of the powder in organic liquids (benzene or carbon tetrachloride).

Before commencing magnetic measurements based on the cylinder method, it must be established what height of sample is required for one end of the sample to be in a field-free region, i.e. so that increasing the sample length no longer affects the force exerted. This may be accomplished by using $NiCl_2$ solutions of different concentrations and filling the tube to different levels in order to determine the weight changes in different magnetic fields. Figure 23 shows some typical results for the dependence on current I, and on the height of sample in the tube.

Magnetic investigations should take advantage of the most modern techniques of purification (and the estimation of purity). Zone melting and chromatography are the most important of these for solid materials. The

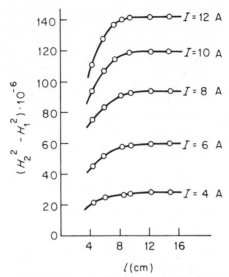

Figure 23. Establishing the correct length of sample

course of purification can be followed, in many cases, by means of a combination of thermal melting point analyses, diamagnetism and spectroscopic measurements. It is frequently the case that beyond a certain stage in the purification process the diamagnetic susceptibility remains constant and this may be used as a criterion of purity. With liquids, such as H_2O and organic solvents (particularly when they are used as calibrants) it is essential that they be free of dissolved oxygen. Otherwise only air-saturated organic liquids should be used as calibrants. According to Juulman[26] the χ_g values of liquids saturated at 20 °C and 760 mm are for H_2O $1\cdot0 \pm 0\cdot1$ per cent, for benzene $7\cdot6 \pm 0\cdot2$ per cent and for nitrobenzene $5\cdot20 \pm 0\cdot02$ per cent below the χ_g values of the degassed liquids.

Susceptibility data taken from the literature are listed below for various calibrants.

(a) H_2O, $\chi_{25\,°C} = -0\cdot721 \times 10^{-6}$, $\partial\chi/\partial T \doteq 0\cdot12$ per cent per degree at 20 °C

(b) Pt, $\chi_{20\,°C} = +0\cdot971 \times 10^{-6}$

(c) $HgCo(CNS)_4$, $\chi_{20\,°C} = +16\cdot44 \times 10^{-6}$; this is particularly suitable for calibration purposes in the strongly paramagnetic region.

(d) $(NH_4)_2Fe(SO_4)_2\cdot6H_2O$, $\chi_{20\,°C} = +32\cdot3 \times 10^{-6}$; for very strongly paramagnetic materials. This should only be used when freshly prepared.

(e) Solutions of nickel salts, $NiCl_2$ and $NiSO_4$ in particular, represent the most commonly used standards for solution work. The nickel content of these solutions must be known very reliably (e.g. by electrogravimetry).

When water is used as a calibrant it should be doubly distilled (preferably in a N_2 atmosphere) and then sealed off in quartz tubes previously filled with N_2. The susceptibility of the (nickel containing) solutions may be calculated from Wiedemann's law of mixtures.

$$\chi_{g\,\text{solution}} = \frac{1}{100}(c\chi_{g\,\text{NiCl}_2} + [100 - p]\chi_{g\,\text{H}_2\text{O}})$$

where c is the concentration of the $NiCl_2$ solution which must be known as accurately as possible. The values of $\chi_{g\,\text{NiCl}_2}$ are calculated after Nettleton and Sugden[27] using the expression:

$$\chi_{g\,\text{NiCl}_2} = \frac{10030}{T}$$

The susceptibilities at room temperature of some standard solutions are given in the following table:

Per cent NiCl$_2$	$\chi_g \times 10^{-6}$
10·01	2·778
15·66	4·753
18·62	5·788
23·27	7·413
29·68	9·657

The following procedure is usually adopted:

(a) 'rough' measurement, to estimate the magnitude of χ_g
(b) calibration, i.e. indirect determination of the field strength
(c) measurement of the (unknown) substance.

If solutions are being studied then Wiedemann's law of mixtures may again be applied. Obviously its validity should be confirmed for the case under study.

III. EXAMPLES OF THE APPLICATION

A. The Magnetochemistry of the Metal–Porphyrin–Protein System

As a result of the pioneering work of Pauling magnetic investigations (static susceptibility and resonance methods) have played an important role in illuminating the chemical bonding and reactions of protein-heavy metal–porphyrin complexes. These three components individually possess very characteristic magnetochemical properties. We shall discuss these in subsequent separate sections in order to amplify their biochemically significant interplay.

1. Paramagnetism of Transition Metal Ions

These metals are paramagnetic due to the unpaired spins of the $d_\varepsilon^n d_\gamma^m$ electrons with or without an orbital contribution. Their behaviour is strongly dependent on the term sequence and term separation, as well as on the orbital contribution, on the symmetry and strength of the ligand field, which is greatly influenced by changes in valence (oxidation state) and on the nature of the bonding. Let us consider as an example Fe^{3+} $(d_\varepsilon^3 d_\gamma^2)$, as it exists in for example methaemoglobin. The free ion is in a $^6S_{5/2}$ ground state, whose resultant orbital moment is zero. Nevertheless a small zero field splitting is found even for weak crystal fields. Its origin has been explained by van Vleck and Penney[28] for cubic crystal fields and by Abragam and Pryce[29] for trigonal and tetragonal fields. It depends on the energy separation (*ca.* $2\cdot5 \times 10^4\,\mathrm{cm}^{-1}$) between the ^6S ground term and the $3d^4 4s\,^6D$ or ^4P term, respectively, of the free ion as well as from the spin–orbit coupling constant λ (*ca.* $300\,\mathrm{cm}^{-1}$) and the crystal field potential. The splitting was calculated to be *ca.* $0\cdot1\,\mathrm{cm}^{-1}$, which is in agreement with the experimental e.s.r. data.

The fundamental study of the splitting of the d^5 configuration in strong crystal fields, as found in numerous Fe^{3+} complexes such as cyanides, is due to Howard.[30] The work has been amplified and extended by Stevens.[31] The implication of the molecular orbital technique in crystal field theory of octahedral d^5 complexes was examined by Owen[32] in particular. In addition this gave the most practical current method of investigating the relationship of the bonding in complexes, even in complicated ligand fields. For example in a MX_6 complex the important d-molecular orbitals (for d_ε/d_γ splitting $\Delta = 10\,Dq$) are formed from the following linear combination between metal d and ligand $p\sigma$ orbitals.

$$\sigma_{z^2}{}^* = \alpha d_{z^2} - \sqrt{1 - \alpha^2}\,\frac{1}{\sqrt{12}}(2p_6 - 2p_3 + p_1 + p_2 - p_4 - p_5)\sigma$$

$$\sigma_{x^2-y^2}{}^* = \alpha d_{x^2-y^2} - \sqrt{1 - \alpha^2}\,\tfrac{1}{2}(p_2 + p_4 - p_1 - p_5)\sigma$$

$$\sigma_{z^2} = \sqrt{1 - \alpha^2}\,d_{z^2} + \frac{\alpha}{\sqrt{12}}(2p_6 - 2p_3 + p_1 + p_2 - p_4 - p_5)\sigma$$

$$\sigma_{x^2-y^2} = \sqrt{1 - \alpha^2}\,d_{x^2-y^2} + \alpha\tfrac{1}{2}(p_2 + p_4 - p_1 - p_5)\sigma$$

The subscripts 1 to 6 relate to the ligand atoms on x, y, z, $-x$, $-y$, $-z$ axes, respectively. For purely ionic interactions the overlap coefficient $\alpha = 1$, and for complete covalency α^2 is $0\cdot5$. When Δ and λ are known, α may be determined by e.s.r. measurements. We have, following Owen $g = 2\cdot0023 - \alpha^2 8\lambda/\Delta$.

The magnitude of Δ is, in addition, affected by π-bonding due to linear combinations of the metal d_ε and ligand p_π atomic orbitals, thus:

$$\pi_{xy}{}^* = \beta d_{xy} - \sqrt{1 - \beta^2}\tfrac{1}{2}(p_1 + p_2 - p_4 - p_5)\pi$$

$$\pi_{xy} = 1 - \beta^2 d_{xy} + \beta\tfrac{1}{2}(p_1 + p_2 - p_4 - p_5)\pi$$

The corresponding $\pi_{yz}, \pi_{yz}{}^*, \pi_{zx}, \pi_{zx}{}^*$ molecular orbitals are obtained by exchanging the indices in a cyclic fashion. In the absence of π bonding $\beta = 1$. The currently available experimental evidence on Fe^{3+} compounds may be divided into two categories: Δ is so small that the $d_\varepsilon^3 d_\gamma^2$, $^6S_{5/2}$ ground term is unaffected; or Δ is so large that a d_ε^5 ground term with $S' = \tfrac{1}{2}$ results. The majority of hydrated Fe^{3+} salts belong to the first category whereas $K_3[Fe(CN)_6]$ belongs to the second. In the first case we have $\Delta \approx 10{,}000\ cm^{-1}$ and in the second $\Delta \approx 50{,}000\ cm^{-1}$. It should be noted that in the latter case a considerable orbital moment and a lowering of the symmetry (Jahn–Teller effect) appear due to the asymmetric occupation of the d_ε orbital and the consequent threefold degeneracy of the ground state. This increases with increasing stabilization due to the $d_\varepsilon-\pi$ bonding.

The interaction of the d_ε^5-MO's, split into three Kramers-doublets, with a static magnetic field has been treated quantitatively by Kamimura[33] in particular. This is the basis of an analysis of the experimental e.s.r. and susceptibility data. The position of the Kramers doublets is especially dependent on λ and Δ, where λ decides the separation between the d_ε levels. Both quantities are influenced by the extent of covalency; λ because of the change in potential at the metal ion as a result of dative π-bonding. The effective magnetic moment deviates from the spin-only value (μ_s) as follows:

$$\mu_{eff} = \mu_s\left(1 - \frac{\alpha\lambda}{\Delta}\right) \qquad \alpha = 2 \text{ for } D \text{ terms}$$

$$\alpha = 4 \text{ for } F \text{ terms}$$

$$\alpha = 0 \text{ for } S \text{ terms}$$

2. Diamagnetism of the Porphyrin System

The porphyrin system is notable for its magnetochemical properties. Its diamagnetism is the greatest known of all organic compounds. The diamagnetic contribution, corresponding to the London-diamagnetism due to the ring current of the conjugated π-electron system, can be ascertained in two independent ways.[34] On the one hand it may be estimated by means of reliable diamagnetic contributions to the measured susceptibilities, and on the other it may be obtained experimentally by comparison with the structurally analogous bilirubin, in which the extensive conjugated π-system is interrupted. A ring current diamagnetism is obtained in both ways, and from this the radius of the ring current is determined by means of the following equation

$$\chi_{\text{ring}} = \frac{N_L e^2}{4mc^2} \sum_n \overline{r^2} = 4 \cdot 248 n \overline{r^2} \times 10^{10}$$

(n = number of π-electrons)

This gives $r = 3 \cdot 21$ Å. This shows that a ring current of this magnitude coincides with the area of the conjugated double bonding of the porphyrin system. Moreover this result supports the views on the origin of the abnormally large diamagnetism of porphyrin. The anisotropic nature of the diamagnetism is also related to the ring current diamagnetism. Interesting relationships appear from the fact that the diamagnetism of the π-electron assembly in such macrocycles is intermediate in nature between the Langevin diamagnetism of localized electrons and the Landau diamagnetism of the electron gas. The effect of the induced secondary magnetic moment of the π-electrons ring current on the paramagnetic metal ions bonded at the centre of the porphyrin molecule surely has an influence on the special role of the porphyrin system as a ligand.

3. Protein Diamagnetism

The macromolecular portion of the haem-proteins, such as globin, are magnetically normal. It can be shown[35] that the diamagnetism of globin arises as the sum of the diamagnetism of the component amino acids.

The gram susceptibilities of the following amino acids have been measured:

Compound	Gram susceptibility $\chi_g \times 10^6$
glycine	$-0 \cdot 520$
DL-alanine	$-0 \cdot 521$
β-alanine	$-0 \cdot 549$
DL-aminobutyric acid	$-0 \cdot 599$
DL valine	$-0 \cdot 606$
DL-norvaline	$-0 \cdot 593$
DL-lysine-HCl	$-0 \cdot 574$
DL-aspartic acid	$-0 \cdot 463$
glutamic acid	$-0 \cdot 499$
DL-serine	$-0 \cdot 536$
L-cystine	$-0 \cdot 487$
DL-methionine	$-0 \cdot 581$
DL-histidine	$-0 \cdot 552$

The author has calculated an approximate value for globin of $\chi_g =$ -0.56×10^{-6} cgs units from the experimental data using the formula

$$\chi_{g\,\text{Globin}} = \frac{\Sigma p_i \chi_i}{100}$$

in which p_i is the percentage of the ith component amino acid of gram susceptibility χ_i.[36] For globin in oxblood a value of $\chi_g = (-0.532 + 0.007)$ $\times 10^{-6}$ cgs units has been obtained. This corresponds to a susceptibility of $\chi_{\text{mol}} = (-8600 \pm 100) \times 10^{-6}$ cgs units. This value is in good agreement with the author's experimental data for oxy- and carbon-monoxide-haemoglobin of $\chi_{\text{mol}} = -9000 \times 10^{-6}$ ($\chi_g = -0.54 \times 10^{-6}$) cgs units. The discrepancy may be related to the diamagnetic contribution of the porphyrin component.

It is possible to calculate a value of $\chi_g = -0.53 \times 10^{-6}$ cgs units for globin on the basis of the results for anhydro methaemoglobin, oxy- and carbon-monoxide-haemoglobin. The denaturing of methaemoglobin by urea has also been investigated magnetochemically. It was found that the reaction product had a molar susceptibility of -7700×10^{-6} cgs units. The results of an investigation on the denaturing of methaemoglobin by acid and alkali also show the tendency for diamagnetic products to be formed. These results accord with Keilin and Hartree's spectrophotometric investigations on methaemoglobin.[37]

Nevertheless the ligand field of protein component (probably bonded to the metal via imidazole-nitrogen) can be varied by the marked steric variability of the tertiary structure of the protein.

Another example of the use of magnetic measurements is to follow the kinetics of H_2O-diffusion in the reversible methaemoglobin–anhydro-methaemoglobin reaction, since the former is paramagnetic and the latter diamagnetic (or rather, has only one unpaired spin).[38]

The gram susceptibility of anhydro-methaemoglobin is -0.63×10^{-6} cgs units. Taking account of the residual paramagnetism of the Fe, we have a value of $\chi_g = -0.77 \times 10^{-6}$ for the protein–porphyrin component. The magnetic behaviour on taking up H_2O is interesting (this is accompanied by a colour change from pink to dark brown). The susceptibility variation of the methaemoglobin does not parallel the uptake of water from air or water vapour; it proceeds much more slowly and is detectable only when the water content corresponds to about 10 weight per cent (i.e. 100 H_2O molecules per haem unit (m.w. 17,000)). For example the water absorption from a water vapour atmosphere is as follows:

Measurement number	1	2	3	4
Time (days)	1	7	9	12
Weight per cent H_2O	21·2	21·3	32·0	45·0
χ_g based on weight of anhydrous Mhb taken	-0.47	-0.41	-0.15	$+0.12$

If the law of additivity were valid for protein–water systems then the number of hydrated Fe atoms could be directly estimated from the χ_g values (e.g. for measurements 2, 3 and 4 of 4 per cent, 75 per cent and 100 per cent respectively of the Fe atoms are hydrated). However this condition is certainly not fulfilled. For dissolved dry materials the calculated χ_g values (obtained from experimental data by assuming additivity) are higher than those for Mhb with a calculated value for $\mu = 5\cdot 8$ and χ_g (protein–porphyrin) $= -0\cdot 77 \times 10^{-6}$; this may be related to a decrease in the diamagnetism of globin arising from a change in structure on taking up H_2O. The attainment of equilibrium in the reaction

$$N_{GI}\cdot Fe^+\cdot N_{GI} + H_2O \rightarrow N_{GI}\cdot Fe^+\cdot H_2O + N_{GI}$$

is controlled by the rate of formation of the globin-hydrate structure. The latter is itself dependent on the rates of adsorption and diffusion of H_2O and thus on the effective surface area of globin and on the vapour pressure of H_2O. More detailed study of these kinetics has allowed further deductions to be made concerning the protein structure and the position of the prosthetic groups. From the existing evidence it may be said that the new Fe^+–N_{GI} bond probably forms intramolecularly and is accompanied by an extensive rearrangement of the protein molecule as a result of the development of a difficultly accessible, cytochrome-like 'cage structure'.

B. *Magnetochemistry of the Fe–Porphyrin–Protein System*

It has already been noted that the known Fe^{3+} complexes can be classed as either 'strong' ($\Delta \approx 50,000 \text{ cm}^{-1}$) or 'weak' ($\Delta \approx 10,000 \text{ cm}^{-1}$) ligand field types. The difference between the crystal field stabilization energy $\Delta = 10\,Dq$ and the energy of the spin exchange interaction π is decisive for the spin distribution and is in both cases appreciably greater than kT at room temperature (*ca.* 200 cm^{-1}).

The characteristic of the porphyrin–protein ligand field arises partly from the lowering of the symmetry (compared with the cubic symmetry of MX_6) and the related additional splitting and anisotropy of the magnetic properties, but more especially in that the above mentioned Δ–π difference is of the order of 200 cm^{-1} (as was suggested by the author as early as 1957).[39] The following experimentally verifiable properties are apparent.

1. Small variation in the ligand field bring about various ground states $^{2S+1}\Gamma$ with different degeneracies and different spin (e.g. $^2T_2$, $^4T_1$, $^6A_1$).

2. Temperature dependence of the multiplicity of the ground state and thereby of the magnetic moment for such ligands, for which $E(^{2S+1}\Gamma) - E(^{2S+1}\Gamma) \approx kT$.

3. The relatively close lying states $^6A_1$, $^4T_1$ and $^2T_2$ bring about large values of the constants D and a of the tetragonal and cubic terms respectively, of the spin-Hamiltonian operator and thereby the unusually large zero-field splitting ($\sim 10\,\mathrm{cm}^{-1}$), as well as the marked anisotropy of the g factors of the high-spin Fe–porphyrin complexes.

Property 1 has been demonstrated by the author[40] as well as by Scheler, Schoffa and Jung[41] in connection with the complexes of methaemoglobin with N_3^-, NO_2^-, SCN^-, CNO^- etc.

Table 1 shows the interdependence of the magnetism and some spectroscopic and thermodynamic properties of methaemoglobin complexes with F^-, CNO^-, SCN^-, NO_2^-, N_3^-, CN^- as well as with 'acid' Mhb_s and 'alkaline' Mhb_a methaemoglobin.

Property 2 was postulated by the author in 1957 and was confirmed later for absorption spectra by George, Beetlestone and Griffith[42] as well as by Scheler, Blanck and Graf.[43]

Property 3 was found by Bennett, Ingram, George and Griffith[44] and Gibson, Ingram and Schonland[45] in 1955 and 1958, respectively.

The magnetically significant molecular orbitals for all Fe–porphyrin complexes arranged in order of increasing energy are $d_{xy}:d_{xz}:d_{yz}:d_{z^2}*$ and $d_{x^2-y^2}*$.

The magnitude of Δ is, more than all the other factors, influenced by small changes in the strength and symmetry of the ligand field. Going from haemin to methaemoglobin apparently causes two effects.

(a) Δ is increased, because of the new N—(imidazole) bond, into a region where $\Delta - \pi \approx kT$ and the supposition of the above mentioned properties 1 to 3 is satisfied.

(b) The protein ligand field is sterically so variable that supposition of 1 (above) is particularly well satisfied.

In the case of the protein-free Fe porphyrins the suppositions for properties 1 to 3 are apparently not yet completely evident. For example the protohaemin complexes with Cl^-, Br^-, I^-, SCN^-, COO^- all show the theoretical spin–moment corresponding to the $d_\varepsilon^3\,d_\gamma^2$ configuration without any orbital contribution.[46]

With certain methaemoglobin compounds (e.g. with NO_2^- or OH^-) the suppositions are however valid, and Δ may be estimated for these complexes. Since $\pi \approx 30{,}000\,\mathrm{cm}^{-1}$ from the spectroscopic data, then Δ is also approximately $30{,}000\,\mathrm{cm}^{-1}$.

Ligand field theory gives the energies of the $^2T_{2g}$ and $^4T_{1g}$ states, relating to $E(^6A_{1g}) = 0$, as

$$E(^2T_{2g}) = 15B + 10C - 2\Delta$$

$$E(^4T_{1g}) = 10B + 6C - \Delta$$

Table 1. Magnetochemistry of F^-, CNO^-, SCN^-, NO_2^-, N_3^-, CN^- complexes as well as with 'acid' and 'alkaline' methaemoglobin (Mhb$_s$ and Mhb$_a$)

Compound	Redband (Intensity)	Greenband (Intensity)	γ-band (mμ)	γ-side-band	$\log \varepsilon_M$ (γ-band)	pK 20° (Dissociation constant)	Enthalpy of formation H kcal mol^{-1}	Entropy of formation S cal mol^{-1}	χ_M	$\mu B = 2\cdot84\sqrt{(T\cdot\chi_M)}$	Effective spin-moment
Mhb F	strong	—	403·5	—	5·117	2·38	~ −2	~ −1	14,000	5·8	~5
Mhb$_s$			405·7	—	5·193				14,000	5·8	~5
Mhb CNO	Decrease →	Increase →	409·0	—	5·122	3·42	−6·6	−7·5	12,750	5·5	~4·6
Mhb SCN			410·5	—	5·122	2·55 / 3·14	−8·7	−19·7	11,200	5·1	~4·2
Mhb$_a$			408·5	—	5·005		−9·5	−7·9 / −7·0	8,340	4·47	~3·5
Mhb NO$_2$			410·5	+	5·078	2·7	−5·1	−5·0	7,300	4·2	~3·2
Mhb N$_3$			417·5	+	5·095	5·66	−15·8	−28·8	3,530	2·9	~2
Mhb CN	—	strong	420·0	+	5·016	5·5	−18·6	−24	2,300	2·4	~1·4

Myoglobin

The empirical values for the Racah parameters are $B = 1133\,\text{cm}^{-1}$ and $C = 3883\,\text{cm}^{-1}$. The energies of $^2T_{2g}$ and $^6A_{1g}$ and of $^4T_{1g}$ and $^6A_{1g}$ are identical when $\Delta = 27900$ and $34630\,\text{cm}^{-1}$, respectively (see Figure 24).

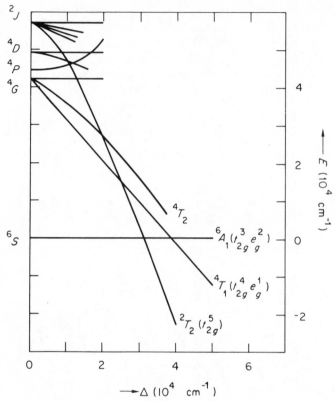

Figure 24. Energies of different states as functions of Δ

For methaemoglobin compounds showing temperature equilibria, Δ has a value for which $E(^6A_1)$ is roughly $10^1\,\text{cm}^{-1}$ above $E(^2T_2)$. Thus with increasing temperature the magnetic moment of these compounds increases. The influence of Δ can be defined more precisely as a result of the semi-empirical molecular orbital calculations performed by Spanjaard and Berthier[47] as well as by Ohno, Tanabe and Sasaki.[48] New possible combinations of Me-d- and ligand symmetry orbitals arise through the ligands at positions 5 and 6: these are $d_{x^2-y^2}(\sigma B_{1g})$, d_{yz}, $d_{xz}(\pi E_{1g})$ where the corresponding ligand orbitals are given in brackets for the D_{4h} symmetry. In consequence the term sequence alters for spin-critical $d_{x^2-y^2}*$ and $d_{z^2}*$ (σ-bonding) as well as d_{yz} and d_{xz} (π-bonding) orbitals. Δ is increased or decreased according to the degree of overlap, especially of the orbitals of symmetry B_{1g} and more particularly $A_{1g}(d_{z^2})$. In this way redox potential (the separation between the highest occupied and lowest unoccupied orbital) and charge distribution between metal and ligand orbitals due to changes in protein conformation can be controlled.

The uniquely high biochemical specificity of the system metal–porphyrin depends therefore on the magnitude of Δ and on the mechanism of the control of Δ by very small changes in conformation of the particular protein matrix, in which the haem system is embedded. Long known effects such as the Bohr effect and the haem–haem interaction are explained by this mechanism of control.

1. Magnetic Susceptibility and Zero-Field Splitting of Ferrihaemoglobin

The theory of the magnetic properties of haemoglobin derivatives has been considerably developed in recent years owing to the work of Kotani,[49] Griffith,[50] Weissbluth[51] and G. Harris.[52] As an illustration of this, consider the relationship between the zero-field splitting parameter D (see Figure 25) and the magnetic moment of methaemoglobin. Neglecting the temperature independent paramagnetism (that is perturbation components of second order) we have according to van Vleck:

$$\chi = N_L \frac{\sum_n [(E_n^1)^2/kT]\,e^{-E_n^0/kT}}{\sum_n e^{-E_n^0/kT}}$$

with $E_n^0 = 0,\ 2D,\ 6D$; $E_n^1 = \pm\beta,\ \pm 3\beta,\ \pm 5\beta$, in the case that H_z is parallel to the porphyrin plane. With $x = D/kT$ we have

$$\chi = N\beta^2 \frac{1 + 9\,e^{-2x} + 25\,e^{-6x}}{kT(1 + e^{-2x} + e^{-6x})}$$

and together with

$$\chi = \frac{N_L \beta^2 \mu_{\text{eff}}^2}{3kT}$$

one obtains the relation

$$\mu_{\text{eff}\parallel}^2 = \frac{3(1 + 9\,e^{-2x} + 25\,e^{-6x})}{1 + e^{-2x} + e^{-6x}}$$

The limiting values for $T \to \infty$ and $T \to 0$ give the magnetic moments of the high and low spin, methaemoglobin derivatives

$$\mu_{\text{eff}}(T \to \infty) = \sqrt{35} = 5{\cdot}92$$

and

$$\mu_{\text{eff}}(T \to 0) = \sqrt{3} = 1{\cdot}73$$

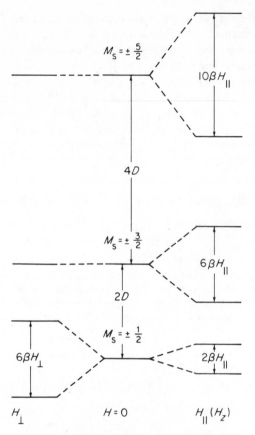

Figure 25. Energy levels for methaemoglobin,
showing the splitting by fields

A similar derivation gives the relationship for H_z normal to the porphyrin plane

$$\mu_{\text{eff}\perp}^2 = \frac{3(9 + 8/x - (11/2x)\,e^{-2x} - (5/2x)\,e^{-6x})}{1 + e^{-2x} + e^{-6x}}$$

with the limiting values $\mu_{\text{eff}}(T \to \infty) = 5.92$ and $\mu_{\text{eff}}(T \to 0) = 5.19$.

The magnetic moment of powder samples has an average value obtained by using the weighting factors $\frac{1}{3}$ and $\frac{2}{3}$ for $\mu_{\text{eff}\,\|}^2$ and $\mu_{\text{eff}\,\perp}^2$, respectively. In this way Beetlestone and George have calculated an exceptionally large value for D of 28 cm^{-1} (for free ions values of D are of the order 0·01–0·1 cm^{-1}). The origin of this large zero-field splitting is again the same phenomenon, as has been described above; the very small energy separation between the $^4T_1$ level and the $^6A_{1g}$ ground state here influences the considerable size of the spin–orbit coupling matrix elements $\langle {}^6A_1|H_{SB}|^4T_1\rangle$ and hence the zero-field splitting.

To conclude we should mention the discussion by Griffith[53] of the diamagnetism of oxyhaemoglobin and the conclusions about the nature of attachment of oxygen.

REFERENCES

1. R. Kohlhaas and H. Lange, *Z. angew. Phys.*, **10**, 461 (1958)
2. K. S. Krishnan, B. C. Guha and S. Banerjee, *Phil. Trans. Roy. Soc. (London)*, A **231**, 235 (1933)
3. H. Wuttig and K. Wagner, *Z. f. Instrumentenkunde*, **71**, 230 (1963)
4. H. Theorell and A. Ehrenberg, *Ark. Fysik*, **3**, Nr. 16, 299 (1950)
5. R. Havemann, *Forschen und Wirken, Festchr. z. 150-Jahrfeier d. Humboldt-Universität*, Akademie–Verlag, Berlin, 1960, p. 243
6. J. Weiss and G. Foëx, *Ann. Physique*, **16**, 174 (1921)
7. C. Chéneveau, *Phil. Mag.*, **20**, 357 (1910)
8. W. Sucksmith, *Phil. Mag.*, **8**, 158 (1929)
9. P. W. Selwood, *J. Amer. chem. Soc.*, **55**, 3161 (1933)
10. A. O. Rankine, *Proc. physic. Soc.*, **46**, 1 u. 391 (1934)
11. J. O. M. Bockris and D. F. Parsons, *J. sci. Instruments*, **30**, 362 (1953)
12. G. Schoffa, O. Ristau and G. Mai, *Exper. Techn. Physik*, **7**, 217 (1959)
13. P. Bitter, *Physic. Rev.*, **33**, 389 (1929)
14. T. Evans, *Nature (London)*, **176**, 777 (1955)
15. R. Mercier and D. Bovet, *Bull. Soc. vandoise Sci. natur.*, **66**, 481–483, Dec. 1957
16. D. Geist, *Z. Physik*, **158**, 359 (1960)
17. E. Schwenkhagen and K. Gutwill, *VDE-Fachberichte*, **10**, 101 (1938)
18. E. Müller, *Fortschr. chem. Forsch.*, **1**, 235
19. A. Pacault, B. Lemanceau and J. Joussot-Dubien, *C.R. hebd. Séances Acad. Sci.*, **237**, 1156 (1953); *J. Chim. physique Physico-Chim. biol.*, **54**, 198 (1956).
20. H. G. Effemey, D. F. Parsons and J. O. M. Bockris, *J. Sci. Instruments*, **32**, 99 (1955)
21. H. Seidel and C. Vogel, *Chem. Tech.*, **15**, 207 (1963)
22. H. J. Friedrich, *Z. f. Naturforsch*, **19 b**, 280 (1964)
23. G. Quincke, *Ann. Physik*, **24**, 347 (1885); **34**, 401 (1888)
24. R. Haas and F. Grün, *Helv. chim. Acta*, **40**, 1299–1304, Basel (1957)
25. A. Knappwost, *Z. Physic. Chem.*, **188**, 246 (1941); *Z. Elektrochem.*, **56**, 840 (1952)
26. D. Juulman, *Z. Physik*, **160**, 109 (1960)
27. H. R. Nettleton and S. Sudgen, *Proc. Roy. Soc. (London)*, Ser. A **173**, 313 (1939)
28. J. H. van Vleck and W. G. Penney, *Phil. Mag.*, **17**, 961 (1934)
29. A. Abragam and M. H. L. Pryce, *Proc. Roy. Soc.*, A **205**, 135 (1951)
30. J. B. Howard, *J. chem. Phys.*, **3**, 207 (1935)
31. D. Stevens, *Proc. Roy. Soc.*, A **219**, 542 (1953)
32. J. Owen, *Proc. Roy. Soc.*, A **227**, 183 (1955)
33. H. Kamimura, *J. Phys. Soc. Jap.*, **11**, 1171 (1956)
34. R. Havemann, W. Haberditzl and P. Grzegorzewski, *Z. Phys. Chem.*, **217**, 217 (1961)
35. R. Havemann, W. Haberditzl and G. Rabe, *Z. Phys. Chem.*, **218**, 417 (1961)
36. G. R. Tristram, *Haemoglobin-Symposium, London*, Butterworths, 1949
37. O. Keilin and R. Hartree, *Nature (London)*, **170**, 161 (1952)
38. R. Havemann and W. Haberditzl, *Z. Phys. Chem.*, **210**, 267 (1959)

39. R. Havemann and W. Haberditzl, *Z. Phys. Chem.*, **209**, 135 (1958)
40. R. Havemann and W. Haberditzl, *Chem. Technik*, **8**, 418 (1956)
41. W. Scheler, G. Schoffa and F. Jung, *Naturwiss.*, **43**, 159 (1956)
42. P. George, J. Beetlestone and J. S. Griffith, *Haematin Enzymes*, p. 105, Oxford, 1961
43. W. Scheler, R. Blanck and K. Graf, *Naturwiss.*, **50**, 500 (1963)
44. J. E. Bennett, D. S. E. Ingram, P. George and J. S. Griffith, *Nature (London)*, **176**, 394 (1955)
45. J. F. Gibson, D. S. E. Ingram and D. Schonland, *Discuss. Faraday Soc.*, **26**, 81 (1958)
46. W. Haberditzl, K. H. Mader und R. Havemann, *Z. Phys. Chem.*, **218**, 71 (1961)
47. C. Spanjaard and G. Berthier, *J. Chim. Phys.*, **1961**, 169
48. K. Ohno, Y. Tanabe and F. Sasaki, *Theoret. Chim. Acta*, **1**, 378 (1963)
49. M. A. Kotani, *Biopolymers Sympos.*, **1**, 67 (1964)
50. J. S. Griffith, *Biopolymers Sympos.*, **1**, 35 (1964)
51. M. Weissbluth, *Structure and Bonding*, Vol. 5, Berlin–Heidelberg–New York, 1966
52. G. Harris, *Theoret. Chim. Acta*, **5**, 379 (1966)
53. J. S. Griffith, *Proc. Roy. Soc. (London)*, **A 235**, 23 (1956)

BIBLIOGRAPHY

Survey of the literary works on magnetic susceptibility of naturally occurring substances and biochemically important molecules.

Amino Acids, Proteins
 1. P. Pascal, *Traité de Chimique de Grignard*, Vol. II, 566 (1936)
 2. R. Perceau, *Compt. rend.*, **236**, 76 (1953)
 3. R. Havemann, W. Haberditzl and G. Rabe, *Chem. Tech.*, **10**, 539 (1958)
 4. G. Schoffa, W. Scheler, O. Ristau and F. Jung, *Acta biol. et med. german*, **3**, 65 (1959)
 5. H. Theorell and A. Ehrenberg, *Acta Chem. Scand.*, **5**, 823 (1951)

Terpenes
 6. W. Haberditzl, *Sitz.-Berichte DAW Bln. Klasse für Chemie*, No. 2 (1964)

Alkaloids
 7. F. Wagenknecht, *Z. f. Naturforsch.*, **16a**, 635 (1961)

Gallic Acids
 8. D. L. Woernley, *Archiv Biochem. Biophys.*, **50**, 199

Porphyrin, Bilirubin
 9. W. Haberditzl, R. Grzegorzewski and R. Havemann, *Z. physik. Chem.*, **217**, 91 (1961)

Metal–porphyrin Systems, Haemin
10. L. Pauling and Ch. D. Corvell, *Z. physik. Chem.*, **22**, 159 (1936)
11. L. Cambi and L. Szegö, *Reale ist. lombardo di scienze*, **67**, 275 (1934)
12. G. Schoffa and W. Scheler, *Naturwiss.*, **44**, 464 (1957)
13. G. Schoffa and coworkers (unpublished)
14. W. A. Rawlinson und P. B. Scutt, *Australian J. Sci. Research, Ser. A*, **5**, 173 (1952)

Haemoglobin
15. A. Gamgee, *Proc. Roy. Soc. (London)*, **68**, 503 (1901)
16. H. Kubo, *Acta Med. Scand.*, **81**, 511 (1934)
17. P. Berthier, *Acta Med. Scand.*, **208**, 943 and 1435 (1939); **209**, 774 (1939)

18. L. Pauling and Ch. D. Coryell, *Acta Med. Scand.*, **22**, 210 (1936)
19. D. S. Taylor, *Acta Med. Scand.*, **61**, 2150 (1939)
20. F. Stitt and Ch. D. Coryell, *J. Am. Chem. Soc.*, **61**, 1263 (1939)
21. D. S. Taylor and Ch. D. Coryell, *J. Am. Chem. Soc.*, **60**, 1177 (1938)

Methaemoglobin
22. Ch. D. Coryell, F. Stitt and L. Pauling, *J. Am. Chem. Soc.*, **59**, 633 (1937)
23. C. D. Russel and L. Pauling, *Proc. Nat. Acad. Sci. U.S.*, **25**, 517 (1939)
24. J. S. Griffith, *Biochim. et Biophys. Acta*, **28**, 439 (1958)
25. W. Haberditzl and R. Havemann, *Chem. Technik*, 7, 418 (1956)
26. R. Havemann and W. Haberditzl, *Z. physik. Chem.*, **209**, 232 (1958)
27. W. Scheler, G. Schoffa and F. Jung, *Biochem. Z.*, **329**, 232 (1957)
28. F. Jung, G. Schoffa, W. Scheler and O. Ristau, *Naturwiss.*, **46**, 317 (1959)
29. W. Scheler, G. Schoffa and F. Jung, *Naturwiss.*, **43**, 159 (1956)
30. G. Schoffa, W. Scheler, O. Ristau and F. Jung, *Acta biol. et med. german.*, **3**, 65 (1959)
31. W. Haberditzl and R. Havemann, *Naturwiss.*, **44**, 31 (1957)
32. G. Schoffa, *Das Dt. Gesundheitswesen*, **14**, 1828 (1959)

Myoglobin
33. D. S. Taylor, *Das Dt. Gesundheitswesen*, **61**, 2150 (1939)
34. H. Theorell and A. Ehrenberg, *Acta Chem. Scand.*, **5**, 823 (1951)
35. W. Scheler, G. Schoffa and F. Jung, *Biochem. Z.*, **329**, 232 (1957)
36. H. Theorell and A. Ehrenberg, *Arch. Biochem. Biophys.*, **41**, 442 (1952)
37. A. S. Brill, A. Ehrenberg and H. den Hartog, *Biochim. Biophys. Acta*, **40**, 313 (1960)

Cytochrome
38. H. Theorell, *J. Am. Chem. Soc.*, **63**, 1820 (1941)
39. H. Theorell, *Ergebn. d. Enzymforschung*, **9**, 232 (1943)
40. S. Paleus, A. Ehrenberg and H. Tuppy, *Acta Chem. Scand.*, **9**, 365 (1955)

Peroxidase
41. H. Theorell, *Arkiv Kemi Mineral. Geol.*, **A 16**, 3 (1942)
42. H. Theorell, *Ergebn. d. Enzymforschung*, **9**, 232 (1943)
43. H. Theorell and A. Ehrenberg, *Acta Chem. Scand.*, **5**, 823 (1951)

Catalase
44. Sh. Tanaka, *J. Biochem. Soc. Japan*, **22**, 141 (1950)
45. H. Theorell, *Ergebn. d. Enzymforschung*, **9**, 232 (1943)
46. H. Theorell and K. Agner, *Ergebn. d. Enzymforschung*, **A 16**, 7 (1942)
47. H. F. Deutsch and A. Ehrenberg, *Acta Chem. Scand.*, **6**, 1 (1952)
48. A. S. Brill, *Lit.*, **6**, 53

Ferritin and other Iron Complexes
49. L. Michaelis, Ch. D. Coryell and S. Granik, *J. Biol. Chem.*, **148**, 463 (1943)
50. H. E. Bayer and K. H. Hausser, *Experientia*, **11**, 254 (1955)
51. M. Kubo, *Monogr. Ser. Res. Inst. of Appl. Electron.*, **5**, 37 (1955)
52. M. Kubo, T. Kobayashi and M. Kishita, *Bull. Chem. Soc. Japan*, **29**, 767 (1956)
53. J. B. Neilands, *J. Am. Chem. Soc.*, **74**, 8648 (1952)
54. R. Chain, A. Towlo and A. Carilli, *Nature*, **176**, 645 (1955)
55. A. Ehrenberg, *Nature*, **178**, 379 (1956)

Vitamin B 12, Cobalt Chelate
56. F. Grün and R. Menasse, *Experientia*, **6**, 263 (1950)
57. H. Diehl, R. W. van der Haar and R. K. Sealak, *J. Am. Chem. Soc.*, **72**, 5312 (1950)
58. H. Diehl and coworkers, *Iowa State Coll. J. Sci.*, **22**, 165 (1948)
59. J. E. Wallmann, B. B. Cunningham and M. Calvin, *Science*, **113**, 55 (1951)
60. L. Michaelis, *Arch. Biochem. Biophys.*, **14**, 17 (1947)
61. J. Hearon, D. Burk and A. Schade, *J. Nat. Cancer Inst.*, **9**, 337 (1949)
62. J. M. White, Th. J. Weismann and N. C. Li, *J. Phys. Chem.*, **61**, 126 (1957)
63. R. H. Bailes and M. Calvin, *J. Am. Chem. Soc.*, **69**, 1886 (1947)
64. J. Gilbert, N. Otey and V. Price, *J. Biol. Chem.*, **190**, 337 (1951)
65. Ch. Tanford, D. C. Kirk, Jr. and M. K. Chautoon, Jr., *J. Am. Chem. Soc.*, **76**, 5325 (1954)

Cancer Cells, Tumour Tissue
66. A. Pacault, *Compt. rend.*, **219**, 186 (1944)
67. A. Pacault, *Experientia*, **10**, 41 (1954)
68. P. Rondini, G. Mayr and E. Gallico, *Experientia*, **5**, 537 (1949)
69. F. E. Senftle and A. Thorpe, *Nature*, **190**, 410 (1961)

Influence of Magnetic Fields on Enzymes
70. W. Haberditzl and K. Müller, *Z. Naturforsch.*, **20b**, 517 (1965); *Angew. Chemie*, **76**, 891 (1964)
71. G. Akoynoglou, *Nature*, **202**, 452 (1964)
72. M. J. Smith and E. S. Cook, *Biological Effects of Magnetic Fields*, p. 246 (Ed. M. F. Barnothy), New York, 1964
73. G. Dorfman, *Biophysics (Russ.)*, **6**, 733 (1962)
74. P. W. Neurath, *Biological Effects of Magnetic Fields*, p. 25 (Ed. M. F. Barnothy), New York, 1964
75. M. Valentinuzzi, *Biological Effects of Magnetic Fields*, p. 63 (Ed. M. F. Barnothy), New York, 1964
76. L. Gross, *Biological Effects of Magnetic Fields*, p. 74 (Ed. M. F. Barnothy), New York, 1964

CHAPTER 9

Polarography of biomolecules

P. Zuman

Department of Chemistry,
Clarkson College of Technology,
Potsdam, New York, U.S.A.

I. INTRODUCTION

Even when numerous applications of polarography in the study of bio-molecules have been reported,[1] the method does not belong to the regularly used techniques in biochemical laboratories. Polarography is not frequently applied even in these cases, when it offers considerable advantages, in particular speed and selectivity, over the commonly used methods. There are, perhaps, two main reasons for such limitation: firstly the application of polarography requires more specially trained personnel than, for example, most optical methods. The handling of the sample and apparatus is usually simple, but the proper intepretation of the curves obtained in the course of polarographic investigation, requires experience. Secondly, many biochemists are not aware of the scope of possible applications that polarography offers in the study of biomolecules. It is the aim of the present contribution to indicate on selected examples various types of applications of polarography, both in the solution of practical, mainly analytical, problems, and in the study of theoretical problems, such as studies of kinetics and mechanisms of biochemically important reactions or studies of relations between the structure, polarographic behaviour and biological activity. But before discussing the various applications, the principles of polarography will be briefly summarized.

II. PRINCIPLES OF POLAROGRAPHY

Applications of polarography are based on the measurement and interpretation of the current–voltage curves. Attempts to record reproducible current–voltage curves have been made since the beginning of this century, but because solid electrodes have been used the activity and surface properties of which were altered in the course of electrolysis, the attempts were unsuccessful. It was the important contribution made by J. Heyrovský from the Charles University at Prague in 1922 that he introduced a dropping mercury electrode for the study of electrolysis. This electrode consists of a glass capillary tube, similar to that used for thermometers (0·05 to 0·1 mm inside diameter and 15 cm long), connected to a mercury reservoir by rubber or plastic tubing (Figure 1). The tip of the capillary tube is immersed in the solution to be analysed and, during operation, a droplet of mercury grows on the tip until it becomes heavy enough to fall. The size of the capillary orifice and mercury pressure—controlled by the height of the mercury reservoir—determine the drop size. While each droplet (which represents the electrode) is attached to the capillary orifice, its surface is exactly the same. The ideally smooth surface reproduced by each new drop is practically unaffected by previous electrolysis.

The electrolysed solution with an immersed capillary electrode is connected to a second reference electrode. This electrode is chosen in such a way that, even though the potential of the capillary electrode is changed, the potential of the reference electrode remains constant. Two types of reference electrodes are most frequently used: a calomel electrode consisting of a mercury layer covered with a thin layer of mercurous chloride above which is a solution of 1 M or saturated potassium chloride, or a mercurous sulphate electrode ($Hg \cdot Hg_2SO_4 \cdot Na_2SO_4 \cdot H^+$) (Figure 2). A regularly increasing voltage is applied to the system consisting of the dropping mercury and reference electrodes using a potentiometer and an e.m.f. source (Figure 3) and the current is measured after each successive increment. The current–voltage curve can then be plotted (Figure 4); modern polarographic instruments record such current voltage curves automatically.

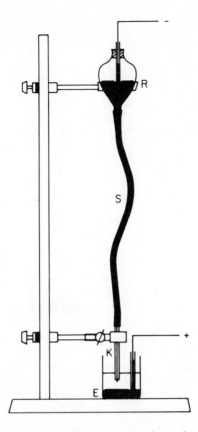

Figure 1. Dropping mercury electrode.
K—glass capillary; E—reference elec-
trode; S—rubber or plastic tubing;
R—mercury reservoir

When electrolysis occurs to a measurable extent in the presence of substances which are either oxidized or reduced at the mercury surface, the
current increases with increasing voltage and produces what is known as a
polarographic wave. The current rises in a given potential range and this
increase is followed by a region of potentials in which the current reaches
its limiting value.

The shape and position of polarographic waves provides us with information on both the quantitative and qualitative composition of the electrolysed
solution. The difference between the current before the wave-rise and the
limiting current, called the wave-height (cf. Figure 4) usually depends on
the concentration of the electroactive substance in the solution. Most of

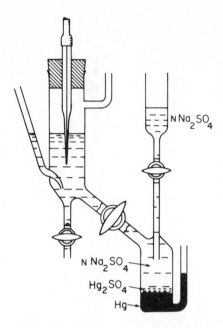

Figure 2. Kalousek vessel for polarographic electrolysis with separated reference electrode. Left compartment—electrolysed solution; right hand compartment —reference electrode (mercurous sulphate shown in the figure)

the analytical applications are based on the increase in wave-height with concentration.

The second measurable quantity is the half-wave potential, i.e. the potential at that point on a polarographic curve at which the current reaches half of its limiting value. Whereas the wave-height depends on concentration, the half-wave potential is practically independent of the concentration of the electroactive species, but its value does depend on the kind of organic compound involved. Because the half-wave potential depends on the nature of the electrolysed substance and on the composition of the solution (called the supporting electrolyte), it is a quantity that can be used for qualitative characterization of organic substances. For systems in which the equilibrium between the oxidized and reduced form at the surface of the electrode is rapidly established (these are called reversible), the half-wave potentials measured polarographically are practically equal to standard oxidation–reduction potentials measured potentiometrically. In these cases the half-wave potentials are a function of the equilibrium constants of the oxidation–reduction equilibrium.

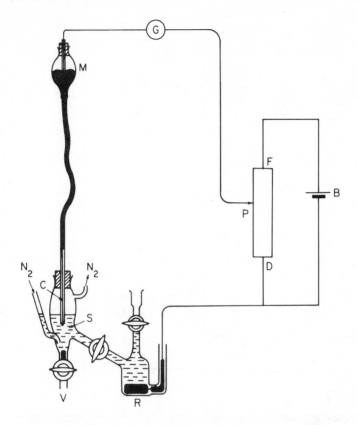

Figure 3. Polarographic circuit (schematically). C—glass capillary of the dropping electrode; V—electrolytic vessel; S—analysed solution; R—reference electrode; N₂—nitrogen inlet and outlet; M—mercury reservoir; G—recording instrument; DF—potentiometer; P—potentiometer sliding contact; B—source e.m.f., battery

However, polarography offers the possibility of characterizing oxidation–reduction properties of numerous systems for which the application of potentiometry is excluded. In such systems, which involve a step with a high activation energy (these are called irreversible), the half-wave potential is a function of the rate constant of the electrode process.

The analogy between polarography and absorptiometry is obvious. The wave-height depends on concentration of the active component similarly as the absorbance at a given wave-length does and the half-wave potential is affected by the structural properties of the studied compound and on the medium used in a similar way as the wave-length of the absorption bond is.

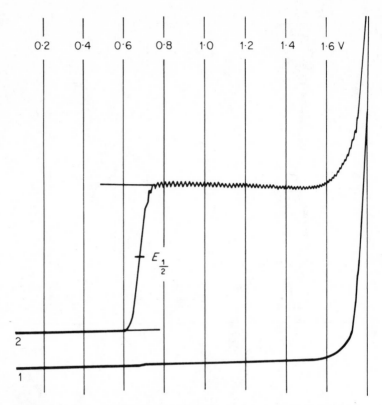

Figure 4. Polarographic current–voltage curve. 1—supporting electrolyte;
2—curve in the presence of an electroactive (reducible) compound. $E_{1/2}$—
half-wave potential

III. LIMITATIONS OF POLAROGRAPHIC METHODS

Before indicating the possible types of applications, limitation of the use of
polarographic methods will be discussed first.

Limitations to the application of polarographic methods are given by the
electroactivity, solubility and the absence of interfering substances. Only
substances that are electroactive, i.e. that give a polarographic wave on the
$i–E$ curve in the available potential range can be electrolysed at the dropping
mercury electrode. The compounds to be studied ought to be soluble in
some solvent that is polar enough to be made conductive. Substances that
prevent measurement of the wave of the studied substance are considered
interfering.

Polarographic waves are given by substances that undergo at the dropping
mercury electrode reduction or oxidation, that form salts with mercury, that

affect catalytically electrode processes and that are surface active. Waves which are recorded above galvanometer zero lines are called cathodic and are observed when reduction takes place as well as for most catalytic processes. Waves below the galvanometer zero line are called anodic. They are observed when oxidations take place and when mercury salts are formed.

Most frequent types of bonds that give reduction waves in organic compounds are shown in Table 1.

Table 1. Some simple reducible groupings

C—C	C—N	C—O	C—S	C—X[a]
C=C	C=N	C=O	C=S	
C≡C	C≡N			
	N—N	N—O	N—S	N—X
	N=N	N=O		
		O—O	O—S	
			S—S	

[a]X = halogen

The reductions of bonds between carbon atoms and carbon and heteroatoms, given in the first three lines (Table 1) take place in the available potential range only when another activating grouping is present in a proper position in the molecule. In other words, a presence of an isolated bond of the type given in the first three lines of Table 1 is often insufficient to produce a polarographic wave. The reduction of bonds between two heteroatoms is often easier and hence the presence of the groups containing bond between nitrogen atoms, oxygen and sulphur atoms is often sufficient to give a polarographic wave. Also condensed aromatic rings and some heterocyclic rings undergo reduction in the ring system.

Due to the high hydrogen over-voltage on mercury the accessible voltage range for reductions at the dropping mercury electrode is relatively wide. On the other hand, the number of compounds oxidizable at the dropping mercury electrode is limited by the fact that the mercury, of which the electrode consists, is dissolved. Under optimum conditions $+0.4$ V can be achieved, which makes it possible to oxidize only strong reducing agents, such as hydroquinones, and their nitrogen analogues, some leucodyestuffs, enediols, phenylhydroxylamino derivatives, etc. For less easily oxidizable substances solid electrodes such as platinum or carbon must be used, with all the intricacies caused by the little defined surface of such electrodes.

Anodic waves are observed in the presence of some organic compounds that react with mercury ions. To this group belong mercaptans and other thiols, dithiocarbamates, urea and thiourea derivatives and some sulphonates, to name only a few. Even when these waves are mostly complicated by adsorption phenomena, it is in most cases possible to find conditions under which their height is a simple function of concentration and hence they can be used for all applications as well as the reduction waves.

Catalytic hydrogen evolution takes place mainly in the presence of ammonium type of acids and of thiol compounds, but examples in which an O-acid or a C-acid can be catalytically active are also described.

Effects of surface-active compounds on polarographic curves can be used for their detection and determination. It is possible usually to estimate only the total content of surface active substance in the polarographed solution.

Whenever it is possible, aqueous and water containing mixtures are preferred for polarographic studies. This is made possible by the fact that most polarographic studies are carried out in solutions containing the examined solution in concentrations of the order of 10^{-4} M. Many compounds considered in preparative chemistry as insoluble, are sufficiently soluble in such dilutions. The reason for this preference is that the chemistry and in particular the equilibria in water containing solutions are at least partially understood, whereas for non-aqueous media information such as on acid-base equilibria, proper buffer solutions, ion-pair formation, etc. are usually not available. Nevertheless, in the future, when more fundamental knowledge on reactions in non-aqueous media will be available, the polarography in non-aqueous solutions seems to be promising. It is already now possible to record waves in non-aqueous solvents such as dioxane, acetonitrile, dimethylformamide and dimethylsulphoxide, but their interpretation is sometimes involved. The control of the water content is essential.

Interfering are those substances that have a half-wave potential differing by less than 0·1 V or 0·2 V (depending on the shape of the wave) or substances present in large excess when they are reduced more positively (or oxidized more negatively) than the studied compound. These interferences can be eliminated either by a change in the composition of the supporting electrolyte, by a modification of the polarographic instrumentation used or by separation preceding the polarographic examination. Either the interfering species or the studied compound can be separated. Most advantageous seems to be a combination of polarography with chromatographic techniques. Paper, thin-layer, gas chromatography or ion-exchange techniques all proved to be useful in combination with polarography, as both these methods enable the handling of small volumes of dilute solutions. Combination of other separation techniques—such as distillation, extraction, dialysis, precipitation or electrolysis—can be used in some cases as well, but one should ensure that the great advantage of polarographic methods, that is their speed, is not lost.

IV. PRACTICAL ANALYTICAL APPLICATIONS

In practical applications the wave-heights are measured, either of the substance to be determined, when it is electroactive (direct methods) or after conversion of the electroinactive substance into an electroactive one by means of a chemical reaction (indirect methods). It is also possible to follow concentration change of an electroactive reagent that reacts with an electroinactive substance. In all these methods the wave-heights are evaluated by means of standards, using the methods of calibration curves or of standard addition.[2]

A. Compounds of Lower Molecular Weight

1. Carbonyl Compounds

Numerous carbonyl compounds, such as aldehydes, ketones, ketoacids, lactones, etc. undergo polarographic reduction and can be polarographically determined. Direct determination is possible[1] for practically all aldehydes (even when for compounds bearing CHO group on a saturated carbon the reduction can be complicated by hydration of the aldehydic group), all ketoacids and those ketones, in which the carbonyl group is adjacent to an electronegative group and/or conjugated with an unsaturated system. Saturated ketones, together with other carbonyl compounds, can also be determined indirectly after transformation into semicarbazones.[3,4] Particularly important is the possibility to determine α-ketoacids, often directly or when present in complex mixtures, after transformation into 2,4-dinitrophenylhydrazones.[5] Their mixture was separated by paper electrophoresis and after elution the waves of the dinitrophenylhydrazones were measured.

Polarography offers possibilities in the study of sugars, particularly of mixtures of aldoses and ketoses, in which most ketoses can be determined in the presence of an excess of aldoses, the waves of which are more affected by formation of cyclic forms.

2. Amino Acids

Naturally occurring amino acids do not undergo electrolysis at the dropping mercury electrode and can be therefore determined only by indirect methods. Some of these methods are general, other allow selective determination of a certain type of amino acids.

General methods are based on complex formation or condensation with carbon disulphide. In the former method amino acids are made to react with insoluble copper salt (usually phosphate). Polarographically the amount of cupric ions brought into solution at a controlled pH as amino acid complexes is determined.[6,7] In the latter amino acids are condensed with carbon disulphide either in aqueous or acetone containing solutions and the anodic waves of dithiocarbamates are measured at pH 11·5. Proline and hydroxy-proline give separate anodic waves.[8]

If individual amino acids should be determined by either of these methods, it is necessary to separate them first chromatographically. The reaction with ninhydrine should be also adaptable for polarographic determinations. To determine histamine and histidine in mixtures, both acids are first converted to dithiocarbamates and then separated by electrophoresis.[9]

Among selective methods for analysis of amino acids it is possible to mention determination of secondary amino acids, such as proline, after treatment with nitrous acid. Waves of resulting nitrosoderivatives were measured. Amino acids bearing a phenyl ring can be determined after nitration, followed by a measurement of the reduction wave of the nitro-benzene derivative formed. For determination of tyrosine[10] a nitration with 0·15 M nitric acid was recommended, for tryptophan[11] 1·5–3·0 M nitric acid and for phenylalanine[12] a mixture of potassium nitrate and sulphuric acid.

Hydroxyamino acids, such as serine, react with periodic acid to yield formaldehyde, which can be determined polarographically.[13] Asparaginic acid and asparagine can be converted by reaction with methyl sulphate and acidification into a mixture of fumaric and maleic acids, reducible polaro-graphically.[14] Histidine can be determined after reaction with 1-fluoro-2,4-dinitrobenzene.[15]

In solutions of divalent cobalt salts in complexing buffer solutions, such as ammonia–ammonium chloride, cysteine and cystine give characteristic catalytic waves (Figure 5). These waves correspond to a catalytic hydrogen

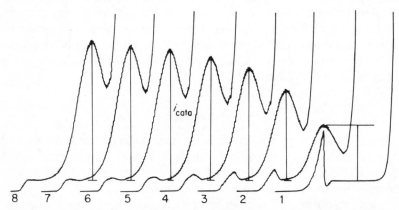

Figure 5. Calibration polarogram of cystine. To 0·002 M cobaltous chloride in 0·1 M ammonia, 0·1 M ammonium chloride solution of cystine was added to give final concentrations: (1) 0; (2) 0·95; (3) 1·8; (4) 2·6; (5) 3·3; (6) 4·0; (7) 4·6; (8) 5·2 × 10^{-5} M cystine. Curves starting at −0·8 V, Hg-pool, open to air, full scale sensitivity 100 μA

evolution.[16,17] The electroreduction of hydrogen is facilitated, the proton-transfer being made easier and faster probably by a complex involving cobalt, cysteine and ammonia. The steric conditions affect reaction rates and equilibria in complex formation, as was shown[18] by the difference in catalytic waves of *threo*- and *erythro*-phenylcysteine (Figure 6). Such catalytic waves in

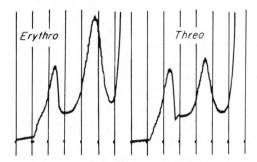

Figure 6. Catalytic waves of phenylcysteine. 0.001 M cobaltous chloride, 0.1 M ammonia, 0.1 M ammonium chloride, 3×10^{-5} M phenyl-cysteine, *erythro*- and *threo*-form. Curves starting at -0.8 V, S.C.E., full scale sensitivity 150 μA

ammoniacal solutions of divalent cobalt are observed for other sulphur-containing complex forming compounds, e.g. thiohydantoines and thio-pyrimidines.[19] The role of the ammoniacal buffer is to participate (by the free ammonia form) in the complex formation and the adjustment of pH. For the usually used excess of buffer, the proton donor ability of the ammonium ion has no direct relation to the height of the catalytic wave.

3. Vitamins[20]

All water-soluble vitamins can be determined polarographically. Thiamine gives on polarographic curves reduction steps, anodic waves corresponding to a mercury salt formation of the thiol form at pH greater than about 9, catalytic waves corresponding to hydrogen evolution in buffer solutions (protonation on the heterocyclic rings) and in ammoniacal cobalt solutions (complex formation and protonation). Contrary to the catalytic waves of cysteine, the catalytic waves of thiamine are observed both in solutions of divalent and trivalent cobalt.[21]

The riboflavin with its leucoform form a reversible oxidation–reduction system.[22] Formation of lumichrome and decomposition products can be followed during the photolysis of riboflavin.

Among further compounds belonging to the vitamin B group it is possible[23] to follow polarographic waves of compounds described by the general name pyridoxine, such as pyridoxol, pyridoxamine, pyridoxamine-5-phosphate, pyridoxthiol, pyridoxal and pyridoxol-5-phosphate. Reactivity of all these compounds is affected strongly by protonation—either on the pyridine ring or on side chains. It is also possible[24] to investigate the properties of the product formed by the reaction of pyridoxal and cysteine, in which a thiazolidine ring is formed.

Also folic acid and other pterins undergo reduction[25] at the dropping mercury electrode. Cobalamin gives in buffered solutions a very pronounced catalytic maximum.[26] This effect is very sensitive and allows determination of 0.04 μg/ml, but is rather unspecific. At higher concentrations the reduction of the central cobalt ion can be followed as well.

Anodic wave of ascorbic acid offers much more specific determination than most conventional analytical techniques. Many substances present in the biological material that interfere in conventional methods do not interfere in polarography. Preparation of samples is thus simplified, e.g. with turbid liquid samples containing proteins and thiols a simple dilution with a buffer solution is sufficient and the analysis is carried out faster. This in turn diminishes the danger of autoxidation of ascorbic acid in samples during preparation of the solution for analysis. Limited application of polarography in ascorbic acid determination is a striking example of the underestimation of polarography.

Whereas naphthoquinone derivatives belonging to the vitamin K group give reduction waves directly, tocophenols are first oxidized to form quinoid species.

4. Hormones

α, β-Unsaturated ketosteroids undergo reduction at the dropping mercury electrode; saturated ketosteroids must be first transformed into compounds with azomethine grouping, which is reduced at more positive potentials than the parent ketone. Betainylhydrazones are, due to their solubility, most frequently used derivatives. The method is faster and above all much more selective for analysis of the biological material than the conventional methods. Oestrogenic hormones can be determined after treatment with nitrous acid,[27] adrenaline and adrenodrome after oxidation to adrenochrome derivatives.[28] Of the thyroid gland hormones it is possible to determine polarographically thyroxine and 3,5-diiodotyrosine, possibly in mixture.[29]

5. Pigments and Dyes

Polarographically active pigments and dyes contain either a naphthoquinone or anthraquinone system, azo groups or pyrilium rings, as flavanoid compounds or anthocyanines. Generally, all substances that absorb in the visible region, give also polarographic waves. Polarography enables acid–base changes of such dyestuffs to be followed. Polarography offers again selectivity for structurally related compounds and possibility to follow changes in dye-concentration even in turbid solutions.

B. Biomacromolecules

Greatest attention has been so far paid to polarographic effects of proteins. Presence, properties and structural changes of proteins can be polarographically studied principally in two ways. The catalytic effect of proteins and hydrogen evolution can be followed, or various reactive species can be added to protein solutions and the concentration of the reagent studied.

Brdička protein reaction[30] is based on a catalytic wave observed when a protein solution is added to a buffered ammoniacal solution of cobalt salts (Figure 7). Whereas low-molecular sulphur compounds give catalytic waves

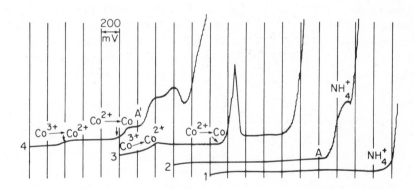

Figure 7. Brdička protein reaction. To 1 M ammonia, 0·1 M ammonium chloride added: (1) 0; (2) 400 times diluted blood serum. To 0·001 M hexaminnecobaltichloride, 1 M ammonia, 0·1 M ammonium chloride added: (3) 0; (4) 400 times diluted blood serum. (A) catalytic 'praesodium' wave; (A') catalytic protein wave. Curves starting at 0·0 V, Hg-pool, open to air, full scale sensitivity 100 μA

only in divalent cobalt solution, proteins show a characteristic double-wave also in solutions of trivalent cobalt. The detailed mechanism of the catalytic process is not yet fully understood, but it is known that condition for the catalytic activity is the presence of a thiol group in the protein molecule and formation of a complex between the protein, cobalt and ammonia. It is assumed that the proton transferred from this complex undergoes reduction more easily than a hydronium ion. There are indications that only molecules of proteins in a given tertiary structure can form such complexes and in the formation of the complex one or more other functional groups are involved.

In Czechoslovakia the method is in clinical use in following changes in various fractions of serum proteins. The following modification was recommended[31] as the most sensitive test to indicate pathological changes in blood serum:

0·4 ml of the serum, obtained by centrifuging the non-haemolysed blood taken from the vein after fasting (in the morning) is added at room temperature to 1 ml of 0·1 M potassium hydroxide. After 45 minutes standing to allow denaturation to take place the solution is deproteinated by addition of 1 ml of 20 per cent solution of sulphosalicylic acid. After 10 minutes precipitated high molecular weight albumin and globulin fractions are separated by filtering through a hardened filter paper. 0·4 ml of the clear filtrate is added to 4 ml of the cobaltic solution, consisting of 0·001 M $Co(NH_3)_6Cl_3$, 0·1 M ammonium chloride and 0·1 M ammonia. The polarographic curve is recorded in a vessel, open to the atmosphere.

The height of the catalytic double wave is measured from the diffusion current corresponding to the reduction of divalent cobalt. For each polarographic equipment, i.e. capillary, recording instrument, mercury pressure, etc., a standardization must be carried out by the analysis of at least 30 sera of clinically healthy individuals. The height of the catalytic wave of a clinical case is compared with the mean value obtained for normal sera and an increase or decrease for the analysed sera is followed.

Pathological sera show frequently an increase in the height of the catalytic wave. It was first assumed that an increase of the catalytic double-wave in this so called 'Brdička filtrate test' indicates cancer diseases. More than 90 per cent of the cancer cases studied showed an elevated catalytic wave and only small non-metastatic skin tumors give non-conclusive results. Later it was nevertheless shown that some other diseases, in particular those accompanied by fever and inflammatory diseases show also an increased reaction. Hence only when such diseases are excluded by clinical observations, the increase in the catalytic wave can be considered as indicating cancer. On the other hand, hepatitis and other liver diseases cause a decrease in the cancer reaction when compared with the normals.

There are three main types of applications of the filtrate test: (1) In general medicine is Brdička reaction used as a screening test: every substantial increase above or decrease below the normal level shows pathological changes in the organism. Its importance for the diagnosis is comparable with the classical sedimentation test. (2) In connection with other tests (usually about 10 different tests are used) it can be used for proof of malignancy. The most important limitation can be seen in the fact that the test is positive only at a more advanced stage of the cancer disease. This prevents detection in early stages, when the patient can be more easily treated. (3) Perhaps the most important application is the control of the effectiveness of the treatment and the control of the course of the disease after an operation or irradiation. A decrease in the height of the catalytic wave is observed when the treatment has been successful.

The increase of the filtrate reaction for pathological sera is, according to our present state of knowledge, due to increase in level of certain fractions of serum proteins, such as mucoproteins and others. The change in individual serum proteins can be followed when paper electrophoresis is combined with polarography,[32] The fractions separated by electrophoresis are evaluated using the polarographic method with a separate calibration curve constructed for each of the separate protein fractions.

Other applications of the catalytic waves enable us to follow the structural changes in proteins by irradiation, heat, changes in acidity addition of solvents and various chemical reagents or by enzymes. Reduction of azoproteins enables study of antigen–antibody reaction.

Behaviour of proteins, studied on the basis of concentration changes of various reagents, was investigated after addition of numerous metals, organomercurials, but also iodine, azodyes or acridine derivatives.

Some of the methods used in the investigation of proteins can be used also for peptides, such as glutathione, oxytocine, vasopressine, etc.

V. STUDY OF BIOCHEMICALLY IMPORTANT REACTIONS

Polarography offers the possibility of following of rates and equilibria for a number of biochemically important reactions. The techniques used depend on time intervals involved and differ for fast and slow reactions. From this point of view we consider as slow those reactions that have half-times of the order of 10 sec or slower.

For slow reactions the wave-height is measured. Changes in wave-height with concentration of one participant of the reaction mixture makes it possible to determine equilibrium constants, time-changes of wave-heights enable us to follow the kinetics of the slow reaction.

For very fast reactions the equilibrium constants can be determined from the shifts of the half-wave potentials with changes in concentration of a reagent participating in the equilibrium, provided that this reagent is present in reaction mixture in excess. For fast reactions so called kinetic waves appear on polarographic curves. From the height of such waves together with an independently determined value of the equilibrium constant it is possible to calculate rate constants of these fast reactions.

A. Examples of Studies of Slow Reactions

An example application of polarography to the study of slowly established equilibria of a biochemically important reaction is the formation of Schiff bases. When to a solution of a carbonyl compound, e.g. a α-keto acid, a solution of a primary amine, for example of an amino acid, is added, two waves are observed on polarographic curves. The more positive wave corresponds to the reduction of the Schiff base, the more negative one to the reduction of the remaining unreacted α-keto acid.[33] With increasing concentration of the amino acid the more positive wave increases (Figure 8),

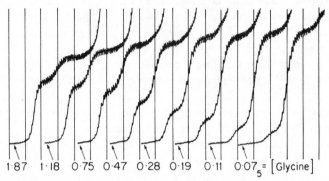

Figure 8. Equilibrium between pyruvic acid and glycine. 5.6×10^{-4} M sodium pyruvate, 0·005 per cent gelatin, glycine buffer pH 9·2. Concentration of amino (NH_2) form of glycine given in the polarogram. Ionic strength kept constant by sodium chloride additions. Curves starting at -0.8 V, S.C.E., full scale sensitivity 4·2 μA

as more of the keto acid is transformed into the Schiff base. From the ratio of the heights of the positive and negative wave it is possible to calculate the equilibrium constant of the overall reaction:

$$RCOCOO^- + H_2NCHR'COO^- \rightleftarrows RC{=}NCHR'COO^- + H_2O$$
$$\underset{COO^-}{|}$$

Some values are given in Table 2.

Table 2. Equilibrium constants of Schiff base formation[33]

Carbonyl compound	Amine	K	$E_{1/2}(C{=}N)$ ($v.$ S.C.E.)
Pyruvic acid	Ammonia	0·25	-1.00
	Glycine	2·47	-1.05
	Alanine	0·93	-1.06
	Colamine	3·11	-1.00
	Histidine	1·99	-0.99
	Histamine	6·65	-0.99
Phenylglyoxalic acid	Ammonia	0·1	-0.84
Ascorbic acid oxidation product }	Ammonia (2 moles)	22	-1.50

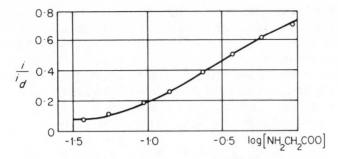

Figure 9. Equilibrium between pyruvic acid and glycine. Dependence of the wave-height of the more positive (ketimine) wave (i) expressed as the ratio i/i_d on the glycine concentration

The plot of the height of the more positive Schiff base wave against the logarithm of the amine (present in excess) concentration (Figure 9) has a shape of a dissociation curve. The slope of this curve allows us to distinguish whether with one mole of the carbonyl compound one mole of amine reacts (as it is in most cases given in Table 2) or two moles react (as it is the case for the oxidation products of ascorbic acid).

The value of the equilibrium constant is a function of pH and this can be interpreted by considering protonation of the amine and of the Schiff base:

$$H_3NR'^+ \rightleftharpoons H_2NR' + H^+$$

$$RCOCOO^- + H_2NR' \rightleftharpoons \underset{\underset{COO^-}{|}}{RC}{=}NR' + H_2O$$

$$\underset{\underset{COO^-}{|}}{RC}{=}NR'H^+ \rightleftharpoons \underset{\underset{COO^-}{|}}{RC}{=}NR' + H^+$$

The equilibrium conditions for all the amines studied in the reaction with α-ketoacids were reached within 1 min, for aldehydes slower reactions were observed. Histidine and histamine show in addition to the formation of Schiff bases another equilibrium, in which a polarographically inactive compound is formed, perhaps of type (**1**)

$$
\begin{array}{c}
\begin{array}{ccc}
& CH_2 & NH \\
\diagup & \diagdown \diagup & \diagdown \\
CH_2 & C & CH \\
| & \| & \| \\
NH & C\text{-----}N \\
\diagdown & \diagup \\
& C \\
\diagup & \diagdown \\
H_3C & COOH
\end{array}
\end{array}
\qquad (\mathbf{1})
$$

From the decrease in the total limiting current it was possible[34] to calculate values of the equilibrium constants of these reactions ($K = 24.5$ for the reaction of histidine at pH 9.17 and 25.5 (pH 9.33) or 28.1 (pH 9.80) for the reaction of histamine).

Changes of limiting currents with time enable us to determine the rate constants of reactions with half-times above 15 sec.[35] As a model reaction for the investigation of transamination reaction, the rates of reactions of pyridoxal with hydroxylamine were followed.[36] As the wave of pyridoxal

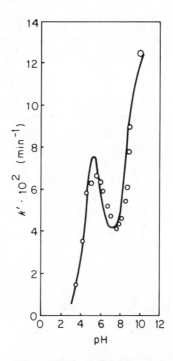

Figure 10. Theoretical curve of the pH-dependence of the rate constant for condensation of pyridoxal and experimental points. Buffers used: pH 3.6–5.6, 0.18 M acetate; pH 5.9–7.8, 0.18 M phosphate; pH 8.2, 0.09 M veronal; pH 8.6, 0.12 M veronal; pH 8.8–9.0, 0.18 M veronal; pH 10, 0.18 M phosphate; ionic strength $\mu = 0.5$ controlled by addition of sodium chloride. Concentration of hydroxylamine, $c = 10^{-2}$ M

corresponds to a two-electron reduction[23] (and is in acid media even smaller due to the hemiacetal ring formation), whereas the reduction of the oxime is a four-electron process, an increase of the wave-height during the oxime formation was followed.[36] The observed dependence of the determined rate constants on pH (Figure 10) can be interpreted by the assumption that all three forms of pyridoxal react with the unprotonized form of hydroxyl-amine, but their rate constants vary: the fully protonized form reacts fastest $(k_{pH_2^+} = 500\ 1\ mol^{-1}\ min^{-1})$, then the fully dissociated form $(k_{p^-} = 13\cdot1\ 1\ mol^{-1}\ min^{-1})$, and slowest the uncharged or zwitterionic form $(k_{pH} = 3\cdot5\ 1\ mol^{-1}\ min^{-1})$. General acid and base catalysis, reported for other azomethine (C=N) bond formations, had negligible effect on reaction rates.

In the previously mentioned reaction of pyruvic acid with imidazole derivatives[34] it was possible to measure the rate of the formation of the adduct which for histamine was eight times faster than for histidine.

Difference between the wave-heights of pyridoxal and pyridoxal-5-phosphate[37] at pH 2-5 enabled us to follow the nonenzymatic hydrolysis of the latter. Monoanion of pyridoxal-5-phosphate undergoes hydrolysis (at 80 °C) considerably faster than the other two forms of pyridoxal-5-phosphate. Measurement of the change in wave-height makes it possible to study the activity of a phosphatase.

Because of the possibility of measurement of concentration changes directly and continuously in the reaction mixture, polarography is suitable for the study of enzymatic reactions. Apart from the examples mentioned in the section on proteins and in the study of the action of phosphatase in preceding paragraph, it is possible to follow the action of peroxidase,[38] and other oxidases,[39,40] tyrosinase[41] and cholinesterase.[42]

B. Fast Reactions

Fast reactions with second order rate constants of the order of 10^4 to 10^{10} 1 mole sec^{-1} can be studied polarographically from the height of the so called kinetic current. Provided that the equilibrium constant of such reactions accompanying the electrode process proper is known, value of the rate constant can be determined from polarographic measurements.

Acid–base reactions preceding the electrode process proper are the most frequently encountered type of such reactions. Polarographic curves give unambiguous evidence of such proton transfer even in cases when they are not expected—e.g. for 4,5-unsaturated-3-ketosteroids,[43] but the calculation of the rate constants can be complicated by adsorption, hydration (in the case of α-ketoacids) and other factors. In some instances, nevertheless, these effects do not play an important role, e.g. for β-diketones.[44]

Complications of the above mentioned type usually do not affect the treatment of systems, in which a dehydration takes place preceding the electrode process. It is hence possible to show which aldehydes or ketones have the carbonyl group hydrated. The calculation of the rate constants is here complicated by the fact that the overall equilibrium constant of the hydration–dehydration equilibria cannot be split into the individual constants K_1 and K_2:

$$>C\begin{matrix} OH \\ \\ OH \end{matrix} + B \underset{}{\overset{K_1}{\rightleftharpoons}} >C\begin{matrix} O^- \\ \\ OH \end{matrix} + BH^+$$

$$>C\begin{matrix} O^- \\ \\ OH \end{matrix} \underset{}{\overset{K_2}{\rightleftharpoons}} >C=O + OH^-$$

The value of the constant K_1 is needed for the calculation of the rate constant of the dehydration reaction.

For sugars polarography enables the study of the ring formation.[45] Among catalytically active systems reactions of hemin[46] and catalase[47] with hydrogen peroxide were studied.

VI. STRUCTURE, POLAROGRAPHY AND BIOLOGICAL ACTIVITY

Because the polarographic half-wave potentials of reversible systems are a simple function of equilibrium constants and those of irreversible systems a function of rate constants, it is not surprising that simple relationships have been found between structure, half-wave potential and reactivity in chemical reactions in homogeneous solutions.[48] It is therefore also understandable that in those cases in which the biological activity depends predominantly on the chemical reactivity rather than on the rate of transport to the reaction centre, it is possible to find qualitative and in some cases quantitative correlations between the biological activity and half-wave potentials. Because in all cases in which we want to compare polarographic data with the chemical reactivity of organic compounds it is necessary that the mechanism of the electrode process is identical for all compounds compared, it is necessary to restrict such comparisons to small groups of closely related organic compounds, bearing the same electroactive group. Effects of quinoid compounds on growth of microorganisms,[49] allergic properties of p-phenylenediamines,[50] bacteriostatic activity of β-aroylacrylic acid derivatives,[51] chemotheropeutic activity of acridine derivatives[52,53] are few examples of systems in which biological activity was more or less satisfactorily related to polarographic half-wave potentials.

VII. CONCLUSIONS

In view of its wide range of possible applications, organic polarography should deserve more attention from scientists working in biological sciences. One inherent limitation is the need of skilled workers needed to gain the optimum, polarography can offer.

REFERENCES

1. M. Březina and P. Zuman, *Polarography in Medicine, Biochemistry and Pharmacy*, Interscience, New York, 1958
2. J. Heyrovský and P. Zuman, *Practical Polarography*, Academic Press, London, 1968
3. P. Souchay and M. Graizon, *Chimie analytique*, **36**, 85 (1954)
4. B. Fleet and P. Zuman, *Collect. Czechoslov. Chem. Comm.*, **32**, 2066 (1967)
5. W. J. P. Neish, *Rec. Trav. Chim.*, **72**, 105, 1098 (1953)
6. A. J. P. Martin and R. Mittelmann, *Biochem. J.*, **43**, 353 (1948)
7. W. J. Blaedel and J. W. Todd, *Anal. Chem.*, **32**, 1018 (1960)
8. R. Zahradník, *Collect. Czechslov. Chem. Comm.*, **21**, 447 (1956)
9. R. Zahradník, V. Mansfeld and B. Souček, *Pharmazie*, **10**, 364 (1955)
10. D. Monnier and Y. Rusconi, *Anal. Chim. Acta*, **7**, 567 (1952)
11. D. Monnier and Z. Besso, *Helv. Chim. Acta*, **35**, 777 (1952)
12. D. Monnier and Y. Rusconi, *Helv. Chim. Acta*, **34**, 1297 (1951)
13. M. J. Boyd and K. Bambach, *Ind. Eng. Chem. Anal. Ed.*, **15**, 314 (1943)
14. B. Warshowsky and M. W. Rice, *Anal. Chem.*, **20**, 341 (1948)
15. P. E. Wenger, D. Monnier and S. Faraggi, *Anal. Chim. Acta*, **13**, 89 (1955)
16. R. Brdička, *Research*, **1**, 25 (1947)
17. R. Brdička, *Collect. Czechoslov. Chem. Commun.*, **5**, 238 (1933)
18. P. Zuman, *Collect. Czechoslov. Chem. Comm.*, **24**, 2027 (1959)
19. P. Zuman, M. Kuik, *Collect. Czechoslov. Chem. Comm.*, **24**, 3861 (1959)
20. P. Zuman, *Acta Chim. Acad. Sci. Hungar.*, **9**, 279 (1956)
21. I. Tachi and S. Koide, *J. Agr. Chem. Soc. Japan*, **26**, 249 (1952)
22. R. Brdička and E. Knobloch, *Z. Elektrochem.*, **47**, 721 (1941)
23. O. Manoušek and P. Zuman, *Collect. Czechoslov. Chem. Commun.*, **29**, 1432 (1964)
24. O. Manoušek and P. Zuman, *Collect. Czechoslov. Chem. Commun.*, **29**, 17 (1964)
25. E. L. Rickes, N. R. Trenner, J. B. Conn and J. C. Keresztesy, *J. Am. Chem. Soc.*, **69**, 2751 (1947)
26. K. H. Fantes, J. E. Page, L. F. J. Parker and E. Lester-Smith, *Proc. Roy. Soc.* (*London*), **B136**, 604 (1950)
27. O. Gry, *Dansk. Tidskr. Farm.*, **23**, 139 (1949)
28. J. Henderson and A. S. Freedberg, *Anal. Chem.* **27**, 1064 (1955)
29. G. K. Simpson and D. Trail, *Biochem. J.*, **40**, 116 (1946)
30. R. Brdička, *Collect. Czechoslov. Chem. Comm.*, **5**, 112, 148 (1933)
31. R. Brdička, quoted in P. Zuman, *Organic Polarographic Analysis*, Pergamon Press, Oxford 1964

32. J. Homolka and V. Krupička *Proc. 1st Internat. Polarograph. Congr.*, Prague 1951, Part III, p. 662, Prague (1952)
33. P. Zuman, *Collect. Czechoslov. Chem. Comm.*, **15**, 839 (1950)
34. P. Zuman, *Chem. listy*, **45**, 40 (1951)
35. P. Zuman, *Adv. Phys. Org. Chem.*, **5**, 1 (1968), Academic Press
36. P. Zuman and O. Manoušek, *Experientia*, **20**, 301 (1964)
37. P. Zuman and O. Manoušek, *Collect. Czechoslov. Chem. Comm.*, **26**, 2314 (1961)
38. J. Doskočil, *Proc. 1st Polarograph. Congr. Prague 1951*, Part I, p. 651, Prague 1952
39. E. Knobloch, *Collect. Czechoslov. Chem. Comm.*, **12**, 407 (1947)
40. O. Pihar and L. Dupalova, *Chem. listy*, **48**, 265 (1954)
41. J. Doskočil, *Collect. Czechoslov. Chem. Comm.*, **15**, 614, 780 (1950)
42. V. Bergerová-Fišerová, *Collect. Czechoslov. Chem. Comm.*, **28**, 3811 (1963)
43. P. Zuman, J. Tenygl and M. Březina, *Collect. Czechoslov. Chem. Comm.*, **19**, 46 (1954)
44. G. Nisli, D. Barnes and P. Zuman, *J. Chem. Soc. (B)*, **1970**, 764, 771, 778.
45. J. Los and K. Wiesner, *J. Am. Chem. Soc.*, **75**, 6346 (1953)
46. R. Brdička and K. Wiesner, *Collect. Czechoslov. Chem. Comm.*, **12**, 39 (1947)
47. J. Koutecky, R. Brdička and V. Hanuš, *Collect. Czechoslov. Chem. Comm.*, **18**, 611 (1953)
48. P. Zuman, *Substituent Effects in Organic Polarography*, Plenum Press, New York, 1967
49. J. A. Page and F. A. Robinson, *J. Chem. Soc.*, **1943**, 133
50. R. L. Bent, *et al.*, *J. Am. Chem. Soc.*, **73**, 3100 (1951)
51. S. Wawzonek and J. H. Fossum, *Proc. 1st Internat. Congr. Polarography*, Prague 1951, Part I, p. 548, Prague 1952
52. R. C. Kaye and H. I. Stonehill, *J. Chem. Soc.*, **1951**, 2638
53. D. L. Hammick and S. F. Mason, *J. Chem. Soc.*, **1950**, 345

V
Action of Physical Agents on Biomolecules

CHAPTER 10

Electron spin resonance of irradiated biomolecules

H. W. Shields
Wake Forest University,
Winston-Salem, North Carolina, U.S.A.

I. INTRODUCTION

The objective of this chapter is to provide an introduction to electron spin resonance (e.s.r.) techniques applied to irradiated biomolecules. Essentially all of the information obtained on radicals stabilized in irradiated biomolecules such as the nucleic acids and proteins has been obtained from e.s.r. spectroscopy. In this context, a radical is defined as a molecule or molecular fragment with an unpaired electron in an outer orbital. If an e.s.r. absorption is to be observed, an unpaired electron is required. Radicals, paramagnetic ions or trapped electrons all satisfy this condition, but the e.s.r. spectra observed in irradiated biomolecules are predominantly from the radicals.

The presence of radicals can also be detected by measurements of static susceptibility or by monitoring a reaction with added radical scavengers. Neither of these methods leads directly to an identification of the radical species. However, e.s.r. spectroscopy may be used to identify radicals, describe their molecular structure, and determine their concentration. In many biological materials a high water content decreases the effectiveness of e.s.r. spectroscopy. Numerous attempts have been made to perfect an e.s.r. spectrometer which performs well when the sample being studied has a high water content, but most of the e.s.r. data on biomolecules have been obtained from dry or frozen samples.

A brief theoretical description of the resonance phenomenon, hyperfine interactions, and g-values is given which the author believes will provide sufficient background for understanding and interpreting e.s.r. spectra. The section on the experimental aspects of e.s.r. describes the methods used in taking data and indicates how principal values for both the hyperfine interactions and g-values are calculated from observed spectra.

It was impractical to cover all of the many applications of e.s.r. to irradiated biomolecules. Two classes of molecules, the nucleic acids and proteins were chosen as examples of the role e.s.r. spectroscopy may play in analysing irradiation effects in biomolecules. Additional information on the application of e.s.r. to irradiated biomolecules may be found in reviews by Andros and Calvin,[1] Boag,[2] Henriksen,[3] Müller[4] and Gordy.[5] A more complete theoretical treatment and discussions on special topics may be found in books by Gordy, Smith, and Trambarulo,[6] Ingram,[7] Slichter,[8] Schoffa,[9] Carrington and McLachlan,[10] Ayscough,[11] Poole,[12] Wyard,[13] and Fehr.[14]

II. THEORY

A. Electron Spin Resonance

Ionizing electromagnetic irradiation interacts with the atoms of an organic solid to yield photoelectrons, Compton electrons, and electron–positron pairs. The classical absorption behaviour is observed with the photoelectric effect dominating at low energies (0·05 Mev), the Compton effect at intermediate energies (1 Mev), and pair production at higher energies (50 Mev).[15] In the photoelectric effect, an electron from an inner shell of an atom is usually ejected. The other electrons in the higher energy levels of this atom rearrange in such a way that the lower energy levels are occupied and the atom is left without one of its valence electrons. If this atom is part of a covalent bond in a molecule, the missing electron may cause the bond to break and the molecule is split into a radical and a positive ion. Radicals and ions are also created by the electrons ejected in the photoelectric effect, Compton scattering, or by positron–electron pairs as they loose their energy by ionization in an organic solid. It is the unpaired electron of the radical or ion which leads to an e.s.r. absorption.

The magnetic moment of a free electron is

$$\mu = -\frac{ge}{2mc} S \tag{1}$$

where g is the spectroscopic splitting factor, e and m are the electronic charge and mass respectively, c the speed of light, and S the intrinsic spin angular momentum. A magnetic moment in a magnetic field, H, has

$$E = -\mu \cdot H \tag{2}$$

ergs of potential energy. When the expression for the magnetic moment is substituted into equation (2) the following equation for the energy is obtained

$$E = g\beta H m_s \tag{3}$$

In equation (3) the Bohr magneton, β, is equal to $eh/4\pi mc$ and has a value of $0·927 \times 10^{-20}$ erg gauss^{-1}. The magnetic quantum number m_s has values of $\pm\frac{1}{2}$ corresponding to projections of S parallel and antiparallel to H. Electron spin resonance occurs when transitions are induced by electromagnetic radiation between these two energy levels as shown in Figure 1. The energy difference of the two states is $g\beta H$, and the frequency of an absorbed or emitted photon is

$$f = \frac{g\beta H}{h}, \text{ where } h \text{ is Planck's constant.} \tag{4}$$

If the frequency is expressed in MHz, and the magnetic field in gauss, the frequency condition for resonance reduces to

$$f = 2·8 H \tag{5}$$

Transitions corresponding to m_s changes of $+1$ and -1 are equally probable. A net absorption occurs because the lower energy level has the

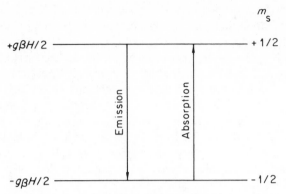

Figure 1. Energy level diagram for a free electron in a
magnetic field

higher population. The number of magnetic moments, $N(m_s)$, in a particular
state is given by Boltzman statistics as

$$N(m_s) = \frac{N \exp(-g\beta H m_s/kT)}{Z} \tag{6}$$

Here N is the total number of magnetic moments and Z is the partition
function. The ratio of $N(-\frac{1}{2})/N(\frac{1}{2})$ reduces to $\exp(g\beta H/kT)$. Thus

$$N(-\tfrac{1}{2}) \simeq N(\tfrac{1}{2})(1 + g\beta H/kT\dots) \tag{7}$$

since $g\beta H/kT \ll 1$. When H is equal to 2,250 G and T is equal to 300 °K,
$N(-\frac{1}{2}) \simeq N(\frac{1}{2})(1 + 10^{-3})$. For every 1000 moments in the upper state,
$N(\frac{1}{2})$, there are 1001 in the lower state. It is this small difference in population
which gives a net e.s.r. absorption.

Transitions between the two energy levels described in Figure 1 are
induced by a linearly polarized alternating magnetic field, H_1, oriented at
right angles to the static field, **H**. A linearly polarized field may be resolved
into components rotating in opposite directions with the same frequency.
The frequency of the rotating components should be the Larmor precession
frequency of the magnetic moment in the static field, **H**. The resolution of the
linearly polarized field into rotating components is illustrated in Figure 2.
The resultant along the x direction of the two vectors in Figure 2 is $H_1 \cos \omega t$.
This is a polarized wave in the x direction with a sinusoidal amplitude.
One of the rotating components rotates in phase with the magnetic moment
which is rotating about **H** at the Larmor frequency and exerts a perturbation
on the magnetic moment which may cause it to 'flip' from the $\frac{1}{2}$ state to the
$-\frac{1}{2}$ state or vice-versa. The effect of the rotating component in the opposite
direction may be ignored.

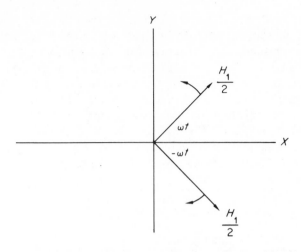

Figure 2. Resolution of a plane wave polarized in the X direction into two circularly polarized waves rotating in opposite directions

The e.s.r. spectrum for a free electron not interacting with its environment is a single absorption. In irradiated biomolecules, the unpaired electron usually interacts with neighbouring hydrogen nuclei to give an absorption which is split into several components. These hyperfine interactions often provide the information needed to identify the radical to which the unpaired electron belongs.

B. Hyperfine Interaction: First Order

The interaction of the unpaired electron with nuclei in the radical or with nuclei of neighbouring molecules is expressed by the spin Hamiltonian

$$\mathcal{H} = \beta \mathbf{S} \cdot \mathbf{g} \cdot \mathbf{H} + \sum_i \mathbf{S} \cdot \mathbf{A}^i \cdot \mathbf{I}^i - \sum_i g_{N_i} \beta_N \mathbf{H} \cdot \mathbf{I}^i \tag{8}$$

where β_N is the nuclear magneton, g_{N_i} is the nuclear g factor and $\mathbf{I}^i$ is the intrinsic spin of the ith nucleus. The hyperfine interaction constant $\mathbf{A}^i$ and the spectroscopic splitting factor g are tensors. The first term in equation (8) represents the Zeeman energy of the electron in the external magnetic field, the second term the hyperfine energy and the third term the nuclear Zeeman energy. In equation (8) the orbital angular momentum $\mathbf{L}$ is assumed to be quenched and terms which normally occur in the Hamiltonian involving $\mathbf{L}$ are neglected.[8]

The main features of e.s.r. spectra from free radicals may be explained in terms of a Hamiltonian reduced from equation (8) by several approximations: (i) The g-tensor of equation (8) is replaced by an isotropic g-value: since in most irradiated biomolecules, the observed g-values are approximately isotropic, and have a value close to that of the free electron, 2·0023. (ii) The nuclear Zeeman energy is considered to be small when compared with the first two terms of equations (8) and it is neglected. (iii) Finally, the direction of quantization for the spin S is taken as that of the external field H. If H is in the Z direction, terms involving S_x and S_y are neglected since they do not have diagonal matrix elements. With these approximations, equation (8) may be written as

$$\mathscr{H} = g\beta H S_z + \sum_i S_z(A^i_{zz}I^i_z + A^i_{zx}I^i_x + A^i_{zy}I^i_y) \qquad (9)$$

The eigenvalues of this equation are[10]

$$E = g\beta H m_s + \sum_i [(A^i)^2_{zz} + (A^i)^2_{zx} + (A^i)^2_{zy}]^{1/2} m_s m_{I^i} \qquad (10)$$

where m_{I^i} is the magnetic quantum number of the ith nucleus. The direction of quantization for the nuclear spin is not along the Z axis, since the field at the nucleus is the resultant of the external field plus the effective field from the electron at the nucleus. For a typical hyperfine interaction, the effective field from the electron is around 5,000 gauss. Since this field is not directed parallel to the external field, the direction of quantization for the nucleus will not be along the field direction.

For a number of radicals found in irradiated biomolecules, the hyperfine interaction constants are nearly isotropic. This means that the off diagonal terms in equation (10) are approximately zero and $A^i_{zz} \simeq A^i_{xx} \simeq A^i_{yy}$. With this approximation, the energy equation becomes

$$E = g\beta H m_s + \sum_i a^i m_s m_{I^i} \qquad (11)$$

when the hyperfine interactions are isotropic and $a^i = A^i_{zz}$. A typical energy level diagram as prescribed by equation (11) is given in Figure 3 below for irradiated chloracetylalanine.[16] In this radical, structure (1),

$$Cl-CH_2-\overset{\overset{\displaystyle O}{\|}}{C}-\underset{\underset{\displaystyle H}{|}}{N}-\overset{\overset{\displaystyle CH_3}{|}}{\underset{\displaystyle \cdot}{C}}-C\overset{\diagup\!\!\!\!O}{\diagdown_{OH}} \qquad (1)$$

the unpaired electron is localized on the carbon adjacent to the methyl group and interacts equally with each of the hydrogen nuclei in the methyl group when it is freely rotating. For equal interactions the hyperfine term

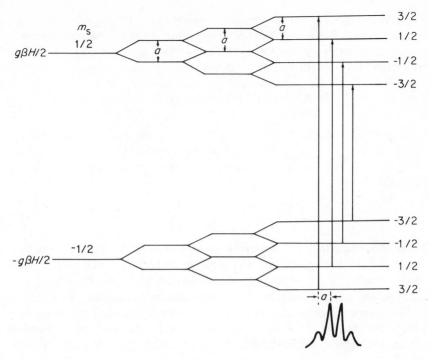

Figure 3. Energy level diagram and spectrum for irradiated chloroacetylalanine

may be written as $am_s M_I$, where M_I is the $\sum_i m_{I^i}$ for the interacting nuclei. The selection rules for first order spectra are $\Delta m_s = \pm 1$ and $\Delta M_I = 0$.

In Figure 3 the energy required for a transition between the $M_I = 3/2$ levels is equal to the difference in energy between the states $(1/2, 3/2)$ and $(-1/2, 3/2)$, where the first number in the parenthesis refers to the value of m_s and the second to M_I. The values of the energy for this transition and that of the other transitions in Figure 3 are

$$\Delta E(3/2) = g\beta H + a(3/2)$$
$$\Delta E(1/2) = g\beta H + a(1/2)$$
$$\Delta E(-1/2) = g\beta H + a(-1/2)$$
$$\Delta E(-3/2) = g\beta H + a(-3/2)$$

$$\tag{12}$$

A general expression for the energy of these transitions is

$$\Delta E(M_I) = g\beta H + aM_I \tag{13}$$

The energy ΔE of an absorbed photon is equal to hf, and also $g\beta H$ according to equation (4). If it is assumed that the g value of the resonance is approximately equal to that of the free electron, g_0, the magnetic field position of a transition may be written from equation (13) as

$$H = H_0 + \frac{aM_I}{g_0\beta}, \quad \text{or} \quad \Delta H = \frac{aM_I}{g_0\beta} \tag{14}$$

where ΔH is the separation in gauss of a transition from H_0 the magnetic field value for the free electron resonance. There will be $(2M_I + 1)$ resonances or hyperfine components. The intensity of these components will follow a gaussian intensity distribution given by

$$1 : n : \frac{n(n-1)}{2} : \ldots : \frac{n!}{(n-k)!k!} : \ldots n : 1 \tag{15}$$

In the above equation n is the number of equivalent nuclei, and k is an integer which goes from 0 to n. The intensity of a hyperfine component is proportional to the number of ways the M_I value of the transition may be formed. For example the relative intensities of the four transitions in Figure 3 are $1 : 3 : 3 : 1$.

The hyperfine tensor A is in general the sum of an isotropic coupling **a** and an anisotropic coupling A'. An expression for the isotropic interaction which is frequently called the Fermi contact interaction[10] is given in equation (16).

$$a = \frac{8\pi}{3} gg_N\beta_N\beta|\psi_e(0)|^2 \tag{16}$$

In this equation $|\psi_e(0)|^2$ is the square of the wave function of the unpaired electron evaluated at the interacting nucleus. For atomic wave functions, only s functions have a non-zero value at the nucleus. The expression for the isotropic coupling does not depend on either the direction or magnitude of the external magnetic field. Since the magnitude of the isotropic hyperfine interaction is directly proportional to the square of the wave function evaluated at the nucleus, the amount of s character in the wave function may be evaluated. When an unpaired electron is 100 per cent in the $1s$ orbital of hydrogen, the isotropic coupling is 508 gauss.[11] The isotropic splitting for the methyl group in Figure 3 is 23 gauss per hydrogen. Therefore, the unpaired electron has 23/508 or 4·5 per cent $1s$ character about each hydrogen nucleus.

The anisotropic part of the hyperfine tensor may be treated as a classical dipole–dipole interaction, if the region containing $r = 0$ is excluded,

$$\mathcal{H}' = gg_N\beta\beta_N\left[\frac{3(\mathbf{I} \cdot \mathbf{r})(\mathbf{S} \cdot \mathbf{r})}{r^5} - \frac{\mathbf{I} \cdot \mathbf{S}}{r^3}\right] \tag{17}$$

where **r** is the vector between the two interacting magnetic moments. Since the electron is not fixed at a point in space, an expectation energy for the interaction which must be calculated using the wave function of the electron is

$$E' = \int \psi_e^* g\, g_N \beta \beta_N \left[\frac{3(\mathbf{I} \cdot \mathbf{r})(\mathbf{S} \cdot \mathbf{r})}{r^5} - \frac{\mathbf{I} \cdot \mathbf{S}}{r^3} \right] \psi_e \, d\tau \tag{18}$$

This integral is zero for s orbitals and has a finite value for an orbital with higher orbital angular momentum. The magnitude of this integral is very sensitive to the direction of the external magnetic field with respect to the direction of **r**. If point dipoles and quantization in the direction of **H** are assumed

$$E' = g\, g_N \beta \beta_N \left(\frac{3 \cos^2 \theta - 1}{r^3} \right) m_s m_I \tag{19}$$

and the hyperfine constant is

$$A' = g\, g_N \beta \beta_N \left(\frac{3 \cos^2 \theta - 1}{r^3} \right) \tag{20}$$

Here θ is the angle between **H** and **r**.

The hyperfine interaction tensor is conveniently represented in a principal axis system having principal values A_x, A_y, and A_z along the x, y and z axes respectively. These principal values may be written as

$$A_x = A_x' + a$$
$$A_y = A_y' + a \tag{21}$$
$$A_z = A_z' + a$$

Here the magnitude of a is $\frac{1}{3}(A_x + A_y + A_z)$ since the sum of the A' values taken along any three mutually perpendicular axes is zero.

C. Hyperfine Interaction: Second Order

In many cases, spectra from irradiated organic solids contain transitions which are not predicted by the first order selection rules, $\Delta m_s = \pm 1$, and $\Delta m_I = 0$. These second order transitions are due to $\Delta m_I = \pm 1$ transitions. The spin Hamiltonian will adequately describe these additional transitions if the nuclear Zeeman term of equation (8) is not neglected.[10] When the nuclear Zeeman term is included, equation (9) may be written as

$$\mathcal{H} = g\beta H S_z + \sum_i [S_z(A_{zz}^i I_z^i + A_{zx}^i I_x^i + A_{zy}^i I_y^i) - g_{N_i}\beta_N H I_z^i] \tag{22}$$

Since S_z is quantized along $\mathbf{H}$ this equation may be written in the form

$$\mathcal{H}(\pm) = \pm\tfrac{1}{2}g\beta H \pm \tfrac{1}{2}\sum_i (\mathbf{Z}\cdot\mathbf{A}^i \mp X_i\mathbf{Z})\cdot\mathbf{I}_i \qquad (23)$$

where $\mathbf{Z}$ is a unit vector in the direction of $\mathbf{H}$ and X_i is $2g_{N_i}\beta_N H$. The expression $(\mathbf{Z}\cdot\mathbf{A}^i \mp X_i\mathbf{Z})$ determines the direction in which I^i is quantized and will be represented by the vector $\mathbf{A}_i(\mp)$. Equation (23) may now be written as

$$\mathcal{H}(\pm) = \pm\,\tfrac{1}{2}g\beta H \pm \tfrac{1}{2}\sum_i [\mathbf{A}_i(\mp)\cdot\mathbf{I}^i] \qquad (24)$$

The eigenvalues of this equation for a single nucleus are

$$E(\pm) = \pm\tfrac{1}{2}g\beta H \pm \tfrac{1}{2}A_{\mp}m_I \qquad (25)$$

In equation (25), $A_{\pm}$ is $[(A_{zz} \pm X)^2 + A_{zx}^2 + A_{zy}^2]^{1/2}$. The Hamiltonian $H(+)$ is not diagonal in the same representation as $H(-)$ because for $H(+)$ the nuclear spin is quantized along the $\mathbf{A}(-)$ direction and for $H(-)$ it is quantized along $\mathbf{A}(+)$. The angle between the directions of quantization is

$$\cos\xi = \mathbf{A}(+)\cdot\mathbf{A}(-)/[A(+)A(-)] \qquad (26)$$

The nuclear wave functions, ψ_{N+}, along $\mathbf{A}(+)$ may be expressed in terms of those, ψ_{N-}, along $\mathbf{A}(-)$ and the angle ξ. The representations of these nuclear wave functions are important since the intensity of a $\Delta m_s = \pm 1$ transition is proportional to

$$|\langle m_s = +\tfrac{1}{2}|S_x|m_s = -\tfrac{1}{2}\rangle|^2|\langle\psi_{N+}|\psi_{N-}\rangle|^2 \qquad (27)$$

Transitions corresponding to $\Delta m_I = \pm 1$ are allowed since ψ_{N-} contains a component of the ψ_{N+} state.

An energy level diagram for the case $I = \tfrac{1}{2}$ is given in Figure 4. The energies of the four levels are

$$E_{+1/2} = \tfrac{1}{2}g\beta H \pm \tfrac{1}{4}A_-$$
$$E_{-1/2} = -\tfrac{1}{2}g\beta H \pm \tfrac{1}{4}A_+ \qquad (28)$$

Transitions represented by the solid lines give absorptions at magnetic fields

$$H_1 = H_0 + \frac{1}{4g\beta}(A_+ + A_-)$$
$$H_2 = H_0 - \frac{1}{4g\beta}(A_+ + A_-) \qquad (29)$$

The doublet separation d_+ of these two transitions is $(A_+ + A_-)/2g\beta$. In Figure 4 the dotted lines represent two additional absorptions with a

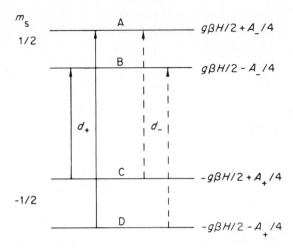

Figure 4. Energy level diagram showing permitted transitions for coupling with a nucleus of spin $1/2$

doublet separation d_- of $(A_+ - A_-)/2g\beta$. If c_1, c_2 and c_3 are the direction cosines of Z in a principal axis system of the tensor A, the intensities of the doublets are[17]

$$I_+ = \cos^2(\xi/2) = \tfrac{1}{2}\left[1 + \sum_j (A_{jj}^2 - X^2)c_j^2/A_+A_-\right]$$

$$I_- = \sin^2(\xi/2) = \tfrac{1}{2}\left[1 - \sum_j (A_{jj}^2 - X^2)c_j^2/A_+A_-\right]$$

(30)

There are some interesting cases where the intensity of one of the doublets is much greater than that of the other: (i) When the external magnetic field is along one of the principal axes $I_+ = 1$. (ii) If the hyperfine tensor A is isotropic, $I_+ = 1$. (iii) When the external magnetic field is so large that X is much greater than any element $|A_{ij}|$, I_- is the dominant intensity. And (iv) when the external field is small such that X may be neglected in comparison with any element $|A_{ij}|$, I_+ predominates. Simultaneous observation of both doublets requires an anisotropic hyperfine tensor and a favourable magnitude and direction of the external magnetic field. The hydrogen hyperfine structure from irradiated organic solids is normally a d_+ doublet when observed at X-band (9 GHz).

Miyagawa and Gordy[18] have given a theoretical description of the variation of I_+ and I_- as a function of magnetic field strength. They found that second order transitions are observable only if the two conditions $A_{zx}^2 + A_{zy}^2 \neq 0$ and $H \simeq A_{zz}/2g_N\beta_N$ are satisfied. In their study of the CH_3CHR radical in L-alanine, they found mainly the d_+ doublet at X band. The d_-

doublet predominated at J band (34 GHz), and at K band (24 GHz) where $A_{zz} \simeq 2g_N\beta_N H$ the two doublets were of comparable intensity.

According to first order theory, the separation of hyperfine absorptions does not vary as a function of the external magnetic field strength. An examination of equation (29) reveals that the separation of hyperfine components does depend on the strength of the magnetic field in second order theory. Miyagawa and Gordy observed this type of variation in the separation of the hyperfine components from the radical in L-alanine.

D. α-Hydrogen Hyperfine Interaction

In e.s.r. of irradiated organic solids two types of hyperfine interactions with hydrogen nuclei are prevalent. These interactions are known as α-hydrogen coupling[19-20] and β-hydrogen coupling.[21-22] The α-hydrogen interaction occurs when the unpaired electron of the radical is in a $2p$ orbital whose axis is perpendicular to a plane containing a hydrogen bonded to the atom on which the $2p$ orbital is localized. A radical of this type, $>\dot{C}-H$, is found in glycylglycine[23-25] structure (2). The observed spectra

$$
\begin{array}{c}
H_3N^+ \quad \overset{H}{\underset{\diagdown}{\big|}} \quad \overset{O}{\overset{\|}{C}} \quad \overset{H_\alpha}{\underset{\diagdown}{\cdot C_\alpha}} \quad O \\
C \quad\quad N \quad\quad C \\
\overset{|}{H} \quad\quad \overset{|}{H} \quad\quad O^-
\end{array}
\tag{2}
$$

at X band frequency from irradiated glycylglycine contain two absorption peaks, d_+, as predicted by Figure 4 for the interaction of an unpaired electron with a single nucleus of spin $\frac{1}{2}$. Principal values for the interaction were found to be 9 G, 19 G, and 28 G. According to equation (21) the isotropic part of the interaction would be 18·6 G which is one-third the sum of these principal values. The principal values may be written as the sum of the isotropic part and the anisotropic part

$$
\begin{pmatrix} -9 & & \\ & -19 & \\ & & -28 \end{pmatrix} = \begin{pmatrix} -18\cdot 6 & & \\ & -18\cdot 6 & \\ & & -18\cdot 6 \end{pmatrix} + \begin{pmatrix} 9\cdot 6 & & \\ & -0\cdot 4 & \\ & & -9\cdot 4 \end{pmatrix}
\tag{31}
$$

Data from e.s.r. measurements do not give the sign of the principal values, but e.s.r. data combined with n.m.r. data show that the coupling constants of an α-hydrogen are negative. The anisotropic terms are attributed to the dipole–dipole interaction of the unpaired electron with the proton of the hydrogen nucleus. If it is assumed that the spin of the electron and the proton are quantized in the same direction, equation (18) becomes

$$E' = g\beta g_N \beta_N \int \frac{\psi_{2p}^*(3\cos^2\theta - 1)\psi_{2p}}{r^3} d\tau m_s m_I \qquad (32)$$

where θ is the angle between $\mathbf{r}$ and $\mathbf{H}$. The principal directions of this interaction are along the $2p$ orbital, the C—H bond, and a direction perpendicular to both of these directions. Figure 5 may be used to obtain a qualitative

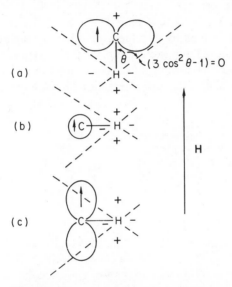

Figure 5. The α-hydrogen anisotropic hyperfine interaction. The regions where $(3\cos^2\theta - 1)$ are positive or negative are indicated for three cases: (a) magnetic field parallel with the CH bond; (b) perpendicular to the bond and the p orbital; (c) parallel to the p orbital and perpendicular to the radical plane

estimate of the coupling expected from the dipole–dipole interaction.[25] The $(3\cos^2\theta - 1)$ part of the integral is positive for angles less than 54° 44' and negative for larger angles. In Figure 5 the regions of positive and negative coupling are indicated, and the cone boundaries make an angle of 54° 44' with respect to the field $\mathbf{H}$. The dipole–dipole coupling is positive when $\mathbf{H}$ is parallel to the C—H bond, negative when $\mathbf{H}$ is perpendicular to the C—H bond, and approximately zero when $\mathbf{H}$ is parallel to the p orbital. The anisotropic coupling constant for a $>\dot{C}$—H radical has been evaluated by McConnell and Strathdee[26] for the applied magnetic field along the p

orbital, the C—H bond, and perpendicular to these directions. The values they obtained using Slater type wave functions are, in gauss, $A_p = -1.6$, $A_{C-H} = 15.4$, and $A_{\perp} = -13.8$. These values agree with the qualitative discussion given with reference to Figure 5.

An isotropic coupling constant would not be expected for a radical with the unpaired electron in a p orbital as pointed out in the discussion of equation (16). Since the α-hydrogen type radical does have an isotropic coupling constant, there must be a contribution to the coupling from an unpaired electron in an s type orbital. An unpaired spin density in the $1s$ orbital of the hydrogen atom in a $>\dot{C}$—H type radical may be understood in the following way.[27-28] The unpaired electron is in a $2p$ orbital perpendicular to the plane formed by the three sp^2 orbitals which are separated by 120°. According to the Pauli exclusion principle, the spin in the σ portion of the C—H bond may have either of two equivalent orientations as shown in Figure 6. Because of an exchange interaction between the electron in the $2p$

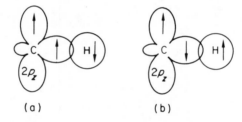

(a)   (b)

Figure 6. Orbitals showing the spin orientation at the hydrogen atom in an α-hydrogen radical

orbital and the carbon electron of the σ bond, Figure 6a is favoured. Therefore, there is a slight unpairing of the electron spin, or spin polarization, in the $1s$ orbit of hydrogen. It should be noted that the direction of the spin on the hydrogen atom is opposite to that in the p orbital, and it is referred to as negative. This negative spin density gives a negative hyperfine coupling constant.

McConnell used the empirical equation[27]

$$a_H = Q\rho_c \tag{33}$$

to describe this isotropic hyperfine interaction. In equation (33) ρ_c is the unpaired spin density localized in the $2p$ orbital of the carbon atom, and Q is an empirical constant which has a value of about -22.5 G. This value is in excellent agreement with the theoretical value of minus 20–25 G obtained from MO theory.[10] The magnitude of the spin density in the $1s$ orbital of hydrogen is about -0.04 for a $>\dot{C}$—H radical, since the coupling constant for atomic hydrogen is 508 G. In the case of glycylglycine the 18·6 G isotropic coupling represents a spin density of -0.037 in the $1s$ orbit of the α-hydrogen. A spin density of 0·83 is found for the unpaired electron in the $2p$ orbital of the α-carbon in glycylglycine when the isotropic coupling is substituted in equation (33) for a_H.

E. β-Hydrogen Coupling

Isotropic coupling from β-hydrogens is often larger than that from α-hydrogens. In the alanine radical[29-30] at room temperature the β-hydrogen coupling is 25·1 G. The β-hydrogen coupling may be explained in terms of valence bond structures or hyperconjugation. In the valence bond approach[31] structures such as (4) are mixed with the ground state structure (3).

(3)   (4)

Structure (4) and the two equivalent forms have a small positive spin density in the $1s$ orbitals of the methyl hydrogens. This results in a positive coupling constant in agreement with observation. The other method[32] assumes hyperconjugation between the $2p$ orbital of the α-carbon and a pseudo-π-system of the methyl hydrogen group orbitals. Coulson and Crawford[33] showed that the three hydrogen orbitals may be represented by three linear combinations of the $1s$ hydrogen orbitals, $\psi_1 + \psi_2 + \psi_3$, $\psi_1 - \frac{1}{2}(\psi_2 + \psi_3)$ and $\psi_2 - \psi_3$. The first of these combinations has σ symmetry about the C—C bond, and the other two have π symmetry. Overlap of either of these π orbitals with the $2p$ orbital of the unpaired electron leads to delocalization of the unpaired electron into the hydrogen $1s$ orbitals. It has been found that the magnitude of the coupling due to this interaction may be represented by the equation[34]

$$A(\theta) = Q(\theta)\rho_c \tag{34}$$

where ρ_c is the spin density in the $2p$ orbital of the α-carbon, and $Q(\theta)$ is expressed as

$$Q(\theta) = B_0 + B \cos^2 \theta \tag{35}$$

The dihedral angle θ is the angle between the axis of the $2p$ orbital and the projection of the C_β—H_β bond on a plane containing the $2p$ axis and perpendicular to the C_α—C_β bond. In equation (35) B_0 is small in comparison to B. Typical values for B_0 range from 0 to 4 gauss and the value of B is about 50 gauss. The B_0 term is probably the result of spin polarization. Methyl groups at room temperature are usually rotating about the C—C bond, thus the β coupling for a methyl group is an average value. The average of $\cos^2 \theta$ is $\frac{1}{2}$ and the coupling for a rotating methyl group is $A_\beta = B_0 + \frac{1}{2}B$, which is about 25 G. Principal values for the methyl coupling in alanine[29] at room temperature are 23·9 G, 24·1 G and 27·3 G.

In some molecular solids at lower temperatures (for example 77 °K) a methyl group may not be rotating, or it may exhibit hindered rotation. When there is no rotation, each β-hydrogen of a methyl group will in general have a different θ and therefore a different coupling constant according to equation (34). If there is no distortion of the hydrogen positions, $\theta_2 = \theta_1 + 120°$ and $\theta_3 = \theta_2 + 120°$. These dihedral angles may be calculated from the measured coupling constants, and their separation is usually very close to 120°. Equation (34) may be used to adequately describe the coupling of an unpaired electron localized in a $2p$ orbital of an α-carbon with one, two, or three hydrogens attached to the β-carbon.

F. Hyperfine Structure from ^{14}N and ^{13}C

Hyperfine structure in the e.s.r. spectra of irradiated biomolecules is mainly due to coupling with hydrogen nuclei. In some cases coupling with the nucleus of ^{14}N or ^{13}C is observed. The unpaired electron is generally localized on ^{14}N or ^{13}C in a $2p$ orbital directed perpendicular to the plane containing the other atoms. The expected coupling results from a Fermi contact interaction and a dipole–dipole interaction between the electron and the ^{14}N or ^{13}C nucleus. The Fermi contact term is due to spin polarization in the s orbitals of the central atom by the unpaired electron in the $2p$ orbital. For nitrogen and carbon, spin polarization occurs in both the $1s$ and $2s$ orbitals.[35-36] The total spin polarization in the orbitals for carbon or nitrogen is positive, and therefore the isotropic Fermi coupling constant is positive. Isotropic coupling constants[37] are generally around 45 G for ^{13}C and 16 G for ^{14}N. A comparison of these values with those given for atoms in Table 1 shows that spin polarization in the s orbitals is very small.

The dipole–dipole part of the hyperfine interaction is obtained by evaluating an integral similar to that in equation (18) for the dipole–dipole interaction with hydrogen in the $>\dot{C}$—H radical. In this case the dipole–dipole interaction is with the nucleus of the atom on which the orbital of the unpaired electron is centred as illustrated in Figure 7. The values obtained for the dipole–dipole coupling have cylindrical symmetry about the axis of the $2p$ orbital, with principal values for the coupling of $2B$ along the $2p$ orbital and $-B$ perpendicular to the $2p$ orbital. Values for B are given by the equation $B = \frac{2}{5}g\beta g_N \beta_N \langle r^{-3} \rangle$ which follows from equation (18) with $\langle r^{-3} \rangle$ the expectation value of $1/r^3$. The total coupling is therefore equal to

Table 1. Calculated hyperfine constants for atoms of interest in biomolecules

Nucleus	Quantum number of unpaired electron n	Nuclear spin	$\psi_{ns}^2(0)$ (a.u.)	$\langle r^{-3}\rangle_{np}$ (a.u.)	Isotropic constant a (gauss)	Anisotropic constant B (gauss)	References
^{1}H	1	$\frac{1}{2}$			508		37
^{2}D	1	1			78		37, 38
^{13}C	2	$\frac{1}{2}$	2·767	1·692	1110	32·5	37, 39, 40
^{14}N	2	1	4·770	3·101	550	17·0	37, 39, 40
^{15}N	2	$\frac{1}{2}$			771	23·9	38, 39, 40
^{17}O	2	$\frac{5}{2}$	7·638	4·974	1653	51·5	37, 39, 41
^{31}P	3	$\frac{1}{2}$	5·625	3·319	3636	102	37, 40, 42
^{33}S	3	$\frac{3}{2}$	7·919	4·814	970	27·9	37, 40, 42

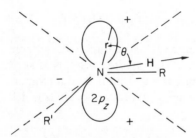

Figure 7. Regions of positive and negative dipolar hyperfine coupling for a R′—Ṅ—R radical. The unpaired $2p$ electron couples with the nitrogen nucleus

$(a + 2B)$ when the field is directed along the p orbital and $(a - B)$ when the field is in the nodal plane of the $2p$ orbital. In Table 1 values for a, B, $|\psi_{ns}(0)|^2$ and $\langle r^{-3}\rangle$ are given for atoms of interest in biomolecules.

G. g-Tensor

In the derivation of the first and second order equations for hyperfine interactions, the spectroscopic splitting factor was considered to be isotropic. This approximation is certainly valid, but there is a small directional dependence of the g factor which may be calculated and measured. A correlation of the principal g-values and their directions with the principal values and directions of the hyperfine coupling constants is often helpful in establishing the identity of a radical. The hyperfine coupling constant of intermediate magnitude and the smallest g-value are in the same direction which is along the p orbital for a $>\dot{C}-H$ type radical.

The g-values for sulphur type radicals, $R-\dot{S}$, which are frequently found in irradiated proteins are very anisotropic in comparison with the $>\dot{C}-H$ radical. Radicals referred to as sulphur type radicals in organic solids are radicals with an unpaired electron localized on a sulphur atom. A typical difference between g (max) and g (min) for the sulphur radical[43] is 0.050. The corresponding variation in the g-values for the $>\dot{C}-H$ radical[44] is an order of magnitude less, or 0.003.

Kurita and Gordy[45] used the theory developed by Pryce[46] for calculating g-values in discussing the structure of the proposed sulphur radical in irradiated L-cystine dihydrochloride, structure (5).

$$
\begin{array}{cccc}
\text{HO} & & \text{H} & \text{H} \\
\diagdown & & | & | \\
& \text{C}-\text{C}\!\!-\!\!-\!\!-\text{C}-\dot{\text{S}} & & \\
\diagup\!\!\diagup & & | & | \\
\text{O} & & \text{NH}_2 & \text{H} \\
\end{array}
\qquad (5)
$$

In Figure 8 it is assumed that the unpaired electron is in a $3p_z$ orbital of sulphur denoted by ϕ_1. The bonding orbitals of sulphur are considered to be hybrids made up of s and p orbitals; and for simplicity it is assumed that they are sp^2 hybrids. The sp^2 hybridized orbitals which lie in the nodal plane

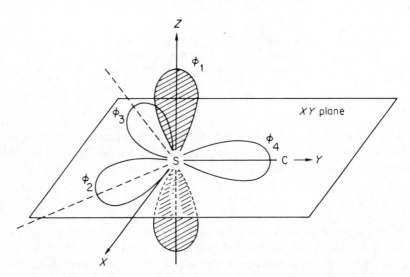

Figure 8. Orbitals in a $R-\dot{S}$ type radical with the unpaired electron localized in the ϕ_1 orbital which is directed perpendicular to the XY plane

of the $3p_z$ orbital are formed from the $3p_x$, $3p_y$ and the $3s$ orbitals of the sulphur atom. Two of the sp^2 hybrids, ϕ_2 and ϕ_3 of equation (36), are non-bonding orbitals.

$$\phi_1 = 3p_z$$
$$\phi_2 = (1/3)^{1/2}(3s) - (1/6)^{1/2}(3p_y) + (1/2)^{1/2}(3p_x)$$
$$\phi_3 = (1/3)^{1/2}(3s) - (1/6)^{1/2}(3p_y) - (1/2)^{1/2}(3p_x)$$
$$\phi_4 = (1/2)^{1/2}[(1/3)^{1/2}(3s) + (2/3)^{1/2}(3p_y) + \psi_c]$$

(36)

The σ bond, ϕ_4 between sulphur and carbon is formed from the third orbital of the sp^2 hybrid and a carbon orbital, ψ_c, which is most likely an sp^3 hybrid. In the g value calculation the exact form of ψ_c is not required. The ground state for the radical is assumed to be $\phi_1^1\phi_2^2\phi_3^2\phi_4^2$ with energy E_1. Spin–orbit coupling is assumed to mix three excited states with the ground state. The configurations of the excited states, in which the unpaired electron is in an sp^2 hybrid, are $\phi_1^2\phi_2^1\phi_3^2\phi_4^2$, $\phi_1^2\phi_2^2\phi_3^1\phi_4^2$, and $\phi_1^2\phi_2^2\phi_3^2\phi_4^1$ with energies E_2, E_3 and E_4, respectively. If electron interactions with neighbouring molecules are neglected E_2 should equal E_3 and the relative magnitude of the energies are $E_1 < E_2 = E_3 < E_4$.

The anisotropy in the g-factor due to spin orbit coupling is calculated from second order perturbation theory according to the method developed by Pryce,[46]

$$g_{ij} = g_f\left\{\delta_{ij} - \lambda \sum_{n=2}^{n=4} \frac{\langle\phi_1|L_i|\phi_n\rangle\langle\phi_n|L_j|\phi_1\rangle}{E_n - E_1}\right\}$$

(37)

where g_f is the free spin value of g, λ is the spin orbit coupling, $\delta_{ij} = 1$ for $i = j$, and 0 if $i \neq j$. In equation (37) L_i and L_j are components of the orbital angular momentum. When equation (37) is evaluated for $i = j = x$,

$$g_{xx} = g_f\left[1 - \lambda \sum_{n=2}^{n=4} \frac{(\phi_1|L_x|\phi_n)(\phi_n|L_x|\phi_1)}{E_n - E_1}\right]$$

(38)

After evaluating the matrix elements in equation (38)

$$g_{xx} = g_f\left\{1 - \frac{\lambda}{6}\left[\frac{1}{E_2 - E_1} + \frac{1}{E_3 - E_1} + \frac{2}{E_4 - E_1}\right]\right\}$$

(39)

Equation (37) also gives

$$g_{yy} = g_f\left\{1 - \frac{\lambda}{2}\left[\frac{1}{E_2 - E_1} + \frac{1}{E_3 - E_1}\right]\right\}$$

(40)

$$g_{zz} = g_f$$

and

$$g_{xy} = g_{xz} = g_{yz} = 0$$

Since the off-diagonal terms all vanish, g_{xx}, g_{yy} and g_{zz} must be principal values of the g-tensor.

The principal values determined experimentally for L-cystine dihydrochloride[45] are $g_u = 2 \cdot 003$, $g_v = 2 \cdot 025$ and $g_w = 2 \cdot 053$. The smallest value is very close to the free electron value, $2 \cdot 0023$, and it is therefore identified with g_{zz}. An examination of equations (39) and (40) shows that $g_{xx} < g_{yy}$; thus g_v is identified as g_{xx} and g_w as g_{yy}. These g-values may be used in equations (39) and (40) with a value of λ equal to -382 cm^{-1} to give

$$E_2 - E_1 = 15,000 \text{ cm}^{-1}$$

$$E_4 - E_1 = 43,000 \text{ cm}^{-1}$$

Because of several approximations made in the calculation these energy values must be regarded as approximate. However, the discussion of this particular radical does show why there is anisotropy in the g-factor, and how the corresponding g-tensor may be determined from theoretical considerations.

III. APPARATUS AND EXPERIMENTAL TECHNIQUES

A. Spectrometers

A brief description of e.s.r. spectrometers will be given since many articles and monographs on their construction and operation are available.[5,7,9,12,47–50] The simplest type of e.s.r. spectrometer consists of a klystron, transmission cavity, crystal detector, connecting waveguide, and magnet. The klystron generates electromagnetic radiation in the microwave region of the spectrum, and most e.s.r. spectrometers operate at a frequency close to 9 GHz ($\lambda = 3$ cm). The radiation passes through a cavity containing the sample in a region where the alternating magnetic field, $\mathbf{H}_1$, of the radiation is directed perpendicular to the static field $\mathbf{H}$, Figure 9. The radiation passing through the cavity is detected by a diode, which converts the microwave radiation into a direct current. Electron spin resonance occurs when equation (5) is satisfied. The presence of resonance is indicated by a decrease in the current at the diode detector; since at resonance, the sample absorbs power from the microwave field in the cavity and the power level at the detector is decreased accordingly.

Resonances have a finite width which range from less than 1 G to more than 100 G depending on the sample being studied. In almost all spectrometers, the frequency is fixed, and the magnetic field $\mathbf{H}$ is varied in order to cover the width of the resonance. The shape of a resonance which depends

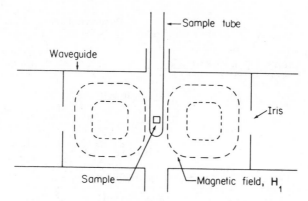

Figure 9. Cross-section of microwave cavity operating in the TE_{102} mode

on the local environment of the electron is usually close to either a Gaussian or Lorentzian curve. This will not be true for a polycrystalline sample whose g-factor is anisotropic and which has a dipole–dipole type hyperfine interaction, since the shape of the resonance results from an average over the random orientations.

The direct current method of detection is relatively insensitive and it is not used in research instruments. Sensitive instruments use alternating current detection. An a.c. component is produced at the diode by modulating the magnetic field **H**. An auxiliary set of coils are placed either in or outside the microwave cavity and oriented so that the magnetic field from the coils is parallel to the field **H**. When an alternating current is passed through the coils, the magnetic field **H** is modulated at the frequency of the current. Modulation frequencies around 100 kHz are preferred since diode noise which is a limiting factor in instrument sensitivity is inversely proportional to the modulation frequency.

The a.c. signal from the diode is amplified and displayed on an oscilloscope or on a graphic recorder. A block diagram of a research instrument with these features is given in Figure 10. In order to display the entire resonance on the oscilloscope the amplitude of the modulation must be slightly larger than the total spread of the resonance. If a high fidelity, or wide band, amplifier is used the shape of the resonance will not be distorted. The sensitivity may be improved further by using a very small modulation with phase sensitive detection. In this method, the static field **H** is swept through the resonance very slowly, i.e. 3 minutes may be required for the field to move through a resonance spread over 100 G. As the field **H** moves through the resonance, the superimposed magnetic modulation produces an alternating component in the crystal current which is proportional to the slope of the absorption curve at the static field value. The phase of the a.c. signal changes

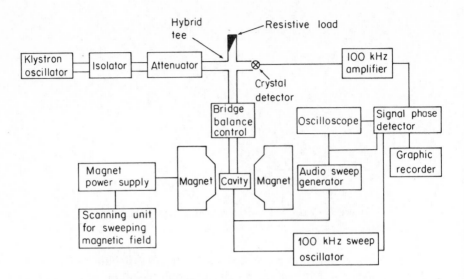

Figure 10. Block diagram of e.s.r. spectrometer

as the static field passes through the peak of the resonance curve. This a.c. signal is amplified with a narrow band amplifier tuned to the frequency of the modulation and it is then passed to a phase sensitive detector. The output of the phase sensitive detector gives the first derivative of the absorption curve when the field **H** is swept through the resonance, Figure 11. In some e.s.r. instruments, the modulation and phase sensitive detection are arranged to give a second derivative output. The point where the first derivative curve crosses the base line corresponds to the centre of the absorption curve. Points of maximum and minimum amplitude correspond respectively to the points of maximum and minimum slope on the absorption curve. The point of minimum amplitude on a second derivative curve corresponds to the centre of the absorption curve. Most research instruments will detect about 10^{11} spins. Sensitivities in this range are needed for the detection of trapped radicals when their concentration is very low.

In applications requiring a low power level and high sensitivity a superheterodyne spectrometer is frequently used. Two klystrons operate at a fixed frequency difference. One klystron provides the power used to excite the cavity and the power from the other, which is referred to as a local oscillator, is mixed with the power from the cavity at the crystal diode. An intermediate frequency (IF) results from this mixing. The frequencies of the klystrons are set to give IF frequencies around 60 MHz. The e.s.r. signal is detected after amplification of the IF carrier wave. Superheterodyne spectrometers are usually about ten times more sensitive than homodyne spectrometers.

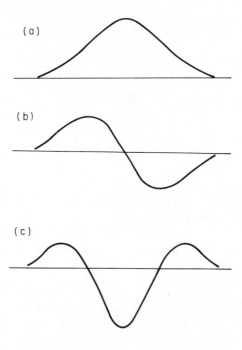

Figure 11. Absorption, first derivative, and second derivative e.s.r. curves. The curve in (a) is a Gaussian absorption, (b) is the first derivative of (a), and (c) is the second derivative of (a)

Box and coworkers have recently applied the ENDOR (electron-nuclear-double-resonance) technique developed by Fehr[51] to radicals in irradiated organic solids.[52-55] In this method an e.s.r. transition is saturated and nuclear resonance is observed indirectly. For example, if the e.s.r. of the A ↔ D transition in Figure 4 is saturated the populations of the two states tend to become equal and the e.s.r. signal disappears or becomes very small. A radio-frequency which will induce transitions between the A and B states is then applied. This causes the population of the A state to decrease and the e.s.r. signal reappears or increases in intensity. A similar increase occurs in the e.s.r. intensity when transitions are induced between C and D states. The nuclear hyperfine spectrum is obtained by monitoring the e.s.r. signal and sweeping the radio-frequency through the values required to induce transitions between the nuclear states.

The ENDOR frequency for resonance is

$$h\nu_i = g_N^i \beta_N^i |\mathbf{H}^i| \tag{41}$$

where the components of the effective field, $\mathbf{H}^i$, are

$$H^i_j = \sum_k B^i_{jk} u_k \tag{42}$$

Hyperfine components may be obtained from the following relation:

$$B^i_{jk} = \frac{m_s}{g^i_N \beta^i_N} A^i_{jk} + H\delta_{jk} \tag{43}$$

Components of the hyperfine coupling tensor A^i_{jk} are determined with respect to axes fixed in the crystal. The u_k are components of a unit vector in the direction of the external field, $\mathbf{H}$.

A high power amplifier driven by a radio-frequency oscillator is commonly used to provide the radio-frequency excitation. Various physical arrangements may be used to place the sample simultaneously in a microwave field and a radio-frequency field. One arrangement is a cylindrical microwave cavity operating in the TE_{011} mode with the radio-frequency carried on a pair of wires which pass through the cavity parallel to its axis.

When an unpaired electron interacts with several nuclei, the e.s.r. spectrum is often poorly resolved. Many of the hyperfine components may fall on top of each other and thus the spectrum may not be interpretable. Even though an e.s.r. spectrum is not resolved the coupling constants of the interacting nuclei may be measured with ENDOR since the resolution in radio-frequency spectroscopy is very high. A disadvantage with ENDOR is the saturation requirement. Experiments are often done at 4 °K in order to achieve saturation.

B. Irradiation

The radiation sources most often used are X-rays, γ-rays, and high energy particles from an accelerator or a reactor. In general the radical species produced in an organic solid do not depend on the type of ionizing radiation. This is because a high energy photon may give its energy to an electron in either of the following processes: photoelectric effect, Compton scattering, and pair production. Most of the damage in the crystal is done by secondary high energy electrons as they lose their energy. In view of this, it is not surprising that different types of ionizing radiation usually produce the same radical species.

Electron spin resonance studies of irradiated biomolecules have been restricted mainly to the solid state. Most of the organic solids will contain radicals trapped in the matrix of the undamaged molecules after exposure to ionizing radiation. The lifetime, or stability, of these trapped radicals varies greatly from sample to sample. Radicals trapped in DL-valine at room temperature are stable for years; whereas e.s.r. signals in norvaline are difficult to detect at room temperature. The lifetime of radicals may be greatly increased at reduced temperatures. Recent experiments have shown interesting conformation changes in radical structure for samples irradiated at nitrogen temperature and then allowed to warm. These experiments will be discussed later.

Samples which are liquids at room temperature are usually frozen and then irradiated. The resulting trapped radicals are very stable in many frozen liquids. If the radiation is from a low intensity source, the liquids must be frozen since the lifetime of a radical in an irradiated liquid is short, about 10^{-2} sec, and the steady state number is not large enough for detection. Fessenden and Schuler[56–57] have observed e.s.r. in organic liquids irradiated with electrons from an accelerator. The spectra from radicals in solution are in many cases better resolved since the dipole–dipole interactions are averaged out.

Radicals may be produced and studied by e.s.r. in molecules which are gases at standard temperature and pressure.[58] The technique most often used is to condense the gas on a cold finger and then irradiate.

Recently interesting e.s.r. spectra have been obtained from biomolecules irradiated with low energy atomic hydrogen or deuterium.[59–61] The sample being studied is subjected to a beam of low energy hydrogen atoms produced by dissociation of molecular hydrogen in an electric discharge. Suitable electric discharges may be obtained with a magnetron at microwave frequencies or with a higher power oscillator at radio frequencies. Care must be taken in the design of the discharge apparatus to prevent strong ultraviolet light and free electrons, also products of a discharge, from reaching the sample. Irradiation with electrons or ultraviolet light may produce radicals whose e.s.r. spectra would tend to obscure the spectra from the radicals formed by hydrogen addition or abstraction. A typical arrangement for irradiating with slow hydrogen atoms is shown in Figure 12. Hydrogen or deuterium is passed through a quartz tube at a pressure of ~ 0.1 torr. The quartz tube in which the discharge occurs is passed either through a waveguide powered with a low power magnetron, or the coil in the tank circuit of a radio-frequency oscillator. The flow rate should be $\sim 10^{-4}$ moles/sec. The magnet and right angle bend are used to prevent charged particles and ultraviolet light from reaching the sample. The tube may be coated on the outside with a black film to further reduce light transmission to the sample. Hydrogen atoms impinge on a powdered sample placed on a watch glass in a chamber. Because of the larger surface area, it is much easier to produce a detectable number of radicals in a powder than a single crystal.

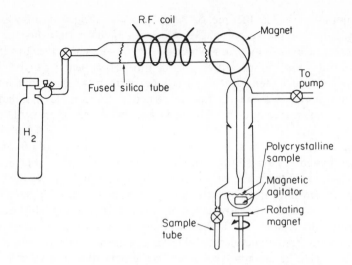

Figure 12. Experimental apparatus for the production of radicals
in samples exposed to thermal hydrogen atoms

C. Low Temperature Studies

It is frequently advantageous to study e.s.r. spectra over a wide range of
temperature. Many interesting experiments are being done by irradiating a
sample at 4 °K and observing its e.s.r. spectra as its temperature is increased.
In many cases the conformation or structure of the trapped radical changes
with the temperature. Interesting changes in the spectra due to a change in
the rate of rotation of methyl groups are sometimes observed. Also, radicals
which are not trapped at room temperature may be observed at lower
temperatures.

Several methods may be used to irradiate and observe spectra at various
temperatures.[7,11] Samples may be cooled from room temperature to close
to liquid nitrogen temperature, 77 °K, by nitrogen gas which has been
circulated through a metallic coil submerged in liquid nitrogen. The cool
nitrogen gas is then passed through a dewar tube to the dewar insert which
holds the sample in the cavity. The sample is cooled by the circulating gas,
and its temperature is measured by a thermocouple placed in the dewar
insert near the sample.

Variable temperatures between 4 °K and 77 °K may be achieved in
several ways. Helium dewars with pressure controls and provisions for
applying heat to the sample may be used. An alternate method uses a dewar
containing a miniature open-cycle Joule–Thomson refrigerator. Temperature
dependent spectra from liquid helium temperature to room temperature
can be investigated without the requirement for liquid helium. High pressure

hydrogen and helium gases are required by the refrigerator. The sample is placed on the end of either a single crystal quartz rod or a copper rod which is covered by a quartz vacuum shroud which may be inserted into the microwave cavity. In some experimental arrangements, the microwave cavity is part of the vacuum shroud surrounding the rod on which the sample is placed. The cavity may be constructed with a slotted wall and a quartz window which permits irradiation of the sample at 4 °K in the cavity.

D. Irradiation and Observation in a Vacuum

Care must be taken to either exclude air from the sample being studied or determine that the trapped radicals do not undergo a change when the sample is exposed to air. This may be done if the sample being studied is kept in an evacuated tube during irradiation and observation of the e.s.r. spectra. Radicals formed in some biomolecules combine with oxygen in the air to form an $R-O_2^-$ type radical.[62] These radicals have a characteristic shape which is slightly asymmetrical with a g value close to that of the free electron. However, radicals in alanine, valine, and many other biomolecules are stable indefinitely in air at room temperature.

Irradiated glass is usually coloured and gives a spectrum which tends to obscure that from an irradiated biomolecule. This difficulty may be avoided by irradiating the sample in one end of the sealed tube and transferring it to the other for observation. The radicals trapped in the irradiated end of the glass tube may be removed by heating this end while the sample is in the other. High purity silica glasses have been developed which do not give an e.s.r. signal at low irradiation doses and at high doses the observed signal is narrow and may not interfere with a strong signal from the sample being studied.

E. Evaluation of the Hyperfine Tensor

The elements of the hyperfine tensor in an x, y and z orthogonal coordinate system may be calculated from equation (10). Insofar as possible the x, y and z axes are normally assigned to the crystal being studied along crystallographic axes. For example, if the crystal is monoclinic, the x axis could be along the **a** crystallographic axis, and y along the **b** crystallographic axis. The separation of adjacent hyperfine transitions as determined by equation (10) for a particular nucleus is

$$\Delta E = (A_{zz}^2 + A_{zx}^2 + A_{zy}^2)^{1/2} \tag{44}$$

or

$$(\Delta E)^2 = (A_{zz}^2 + A_{zx}^2 + A_{zy}^2)$$

These equations were obtained with **H** parallel to z. It is now necessary to take **H** in an arbitrary direction with respect to the x, y and z axes. Let **h** be a unit vector in the direction of **H** with direction cosines **l**, **m** and **n**. For **H** in the xy, xz, and yz planes, $(\Delta E)^2$ is respectively

$$(\Delta E)^2 = \mathbf{l}^2 A_{xx}^2 + 2\mathbf{lm}A_{xy}^2 + \mathbf{m}^2 A_{yy}^2$$

$$(\Delta E)^2 = \mathbf{l}^2 A_{xx}^2 + 2\mathbf{ln}A_{xz}^2 + \mathbf{n}^2 A_{zz}^2 \qquad (45)$$

$$(\Delta E)^2 = \mathbf{m}^2 A_{yy}^2 + 2\mathbf{mn}A_{yz}^2 + \mathbf{n}^2 A_{zz}^2$$

The diagonal components of the squared tensor may be obtained by taking **H** along x, y and z to give $(\Delta E)^2$ values of A_{xx}^2, A_{yy}^2, and A_{zz}^2 respectively. The off-diagonal elements may then be calculated by taking **H** at an angle with respect to the axes in each of the three planes. An off-diagonal element may also be evaluated from a relationship between the values of $\mathbf{A}^2$ along the two axes of the plane, the maximum variation of $(\Delta E)^2$ in the plane denoted as ΔA^2 and the square of the off-diagonal element.[63] The equation for A_{xy}^2 is

$$4(A_{xy}^2)^2 = (\Delta A^2)^2 - [A_{xx}^2 - A_{yy}^2]^2 \qquad (46)$$

The sign of an off-diagonal term is positive for a maximum and negative for a minimum value of $(\Delta E)^2$ in a quadrant bounded by **H** directed along the positive directions of the axes forming the plane. The $\mathbf{A}^2$ tensor is readily diagonalized to give principal values and directions by standard techniques. The principal values of **A** are obtained by taking the square root of the $\mathbf{A}^2$ principal values. They have the same direction as the $\mathbf{A}^2$ values. It should be noted that only the magnitude and not the sign of A is given by this method.

F. Evaluation of the g-Tensor

In order to evaluate the g-tensor, the frequency and magnetic field strength must be known. The hyperfine structure is symmetric about a central magnetic field position in the first order treatment, and the g-value of the resonance is that which is measured for this central position in the hyperfine structure. Rough frequency measurements may be made with a cavity type wavemeter. More accurate measurements are made by comparing the unknown frequency with harmonics from a carefully calibrated and stabilized crystal oscillator. Magnetic field measurements are made from proton resonance observations. The frequency at which the proton resonance occurs may be accurately measured with a crystal controlled oscillator. Since the magnetic moment and g-value of the proton have been accurately measured, the magnetic field may be calculated from these values. In practice, the g-value may be evaluated by making a comparison of the magnetic field position of the resonance to that of a standard whose g-value is known. An expression for the g-value is

$$g = g_s\left(1 + \frac{\Delta H}{H}\right) \tag{47}$$

Here g_s is the g-value of the standard, H is the field position of the centre of the resonance under investigation and ΔH is the field separation between H and the centre of the standard. If H is greater than the field value of the standard, ΔH is negative; when H is less, ΔH is positive.

The equations for the first order hyperfine interaction were derived on the assumption that the g-value of a free radical is isotropic. In practice there is a small anisotropy in the g-value which is measurable. Principal g-values and their directions are evaluated in the same manner as they are for the hyperfine interaction. In an x, y and z coordinate system with $\mathbf{H}$ in the xy plane

$$g^2 = l^2 g_{xx}^2 + 2lm g_{xy}^2 + m^2 g_{yy}^2 \tag{48}$$

The off-diagonal term g_{xy}^2 may be evaluated from the relation

$$4(g_{xy}^2)^2 = [\Delta g^2]^2 - [g_{xx}^2 - g_{yy}^2]^2 \tag{49}$$

where Δg^2 is the maximum variation of g^2 in the xy plane. The sign of g_{xy}^2 is determined in the same manner as the sign of A_{xy}^2 in equation (46). The

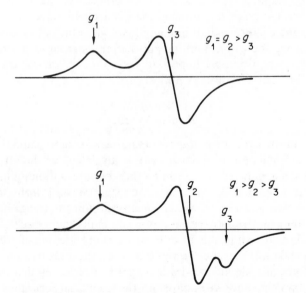

Figure 13. Spectra showing g-values in polycrystalline materials

principal values for g and their directions are obtained from the diagonalized g^2-tensor.

It is frequently difficult to obtain single crystals from which the components of the g-tensor may be determined. However, the expected line shapes for polycrystalline samples containing radicals with two or three different principal g-values have been theoretically calculated.[64] Typical line shapes expected for $g_1 = g_2 > g_3$ and $g_1 > g_2 > g_3$ are especially useful in identifying the sulphur resonance in proteins and other sulphur containing molecules. The positions of the g-values for the two cases are indicated in Figure 13. The data in Table 2 show a close correlation between the principal g-values

Table 2. Comparison of principal g-values determined from polycrystalline samples with those from a single-crystal of cystine dihydrochloride

Sample	g_1	g_2	g_3	Reference
Penicillamine	2·004	2·026	2·053	43
Cysteine	2·006	2·025	2·052	43
Cysteamine hydrochloride	2·005	2·027	2·054	43
Cystine dihydrochloride (single crystal)	2·003	2·025	2·053	45

obtained for sulphur type radicals in the polycrystalline state and those measured for L-cystine dihydrochloride. The close agreement of the g-values in Table 2 indicates that in many cases the principal g-values may be measured from the spectra of polycrystalline samples with reasonable accuracy. Perhaps the most important feature is the characteristic shape which may be used in radical identification.

G. Radical Symmetry

Radicals in an irradiated organic crystal are in most cases formed by the loss of a single atom or functional group from the parent molecule. The radicals in general occupy a limited number of sites and obey the symmetry operations of the undamaged crystal. Spectra from chemically identical radicals at different crystal sites would not in general be identical because of the anisotropy in the g-tensor and hyperfine constants. In many crystals spectra from radicals at different sites are identical for **H** parallel to some special direction, such as a unique axis. The g-tensors and hyperfine coupling constants should be described with a right handed orthogonal coordinate system, and when possible the coordinate system should be identified with the crystallographic axes of the crystal. In a monoclinic crystal, no site splitting is observed for **H** perpendicular to the two-fold crystallographic **b**-axis. Magnetically distinguishable sites may be observed when **H** is rotated in a plane containing **b**. Distinguishable sites have opposite signs for corresponding off-diagonal tensor elements. There is usually no site splitting in an orthorhombic crystal for **H** parallel to a crystallographic axis; although two magnetically non-equivalent sites may be observed for **H** perpendicular to a crystallographic axis and four when **H** is not along an axis.

IV. APPLICATIONS

A. Proteins and Their Constituents

1. Ions, Primary Radicals and Secondary Radicals in the Amino Acids

Many investigations on free radicals in irradiated amino acids, poly-peptides and proteins have been reported since Combrisson and Uebers-feld[65] and Gordy, Ard and Shields[66] first used e.s.r. to show that free radicals are trapped in these substances after exposure to ionizing radiation.[3,9,23,67–71] Most of the early studies were on polycrystalline material and the hyperfine structures were very difficult to interpret because of the aniso-tropies in both the *g*-value and hyperfine coupling constants. When single crystals are used the spectra are usually better resolved and in many cases a definite identification of the trapped radicals may be made. Resonances have been detected in all of the essential amino acids after irradiation, and single crystal studies have been done on more than half of them, Table 3. The fact that a positive identification of the radicals in all of these amino acids has not been made does not rule out the usefulness of e.s.r. in the study of irradiated proteins since the amino acids do not occur as single crystals in polycrystalline proteins.

Most of the constants for the radicals of Table 3 were measured at room temperature. In general the radicals found in the amino acids at room temperature are not the primary radicals. Many of the amino acids have been irradiated and their spectra observed at low temperatures (77 °K or 4·2 °K) without allowing the sample to warm between the time of irradiation and observation. The radicals observed at these low temperatures seem to be either positive or negative ions and in some cases both species are observed simultaneously. The hyperfine structures for a number of polycrystalline amino acids irradiated and observed in a vacuum at 77 °K was not as spread out, nor were they as well resolved as the hyperfine structure obtained for the same amino acids irradiated and observed at room temperature.[82] When the hyperfine structures were resolved, they consisted of fewer com-ponents than obtained at room temperature. Evidence for more than one radical species was also more frequent at the lower temperature. These features made the identification of radicals trapped in irradiated poly-crystalline amino acids at low temperatures difficult. As the samples were

Table 3. Radicals identified in the essential amino acids at room temperature[a]

Name	Radical	α-Hydrogen coupling	β-Hydrogen coupling	References
Glycine	$^+NH_3{-}\overset{\displaystyle }{C}{-}COO^-$ with H; and $\overset{H}{\diagup}\dot{C}{-}COO^-$	35.2 21.9 13.9	19.5 16.4 16.4 -3.6 -3.5 -2.5 nitrogen coupling	[53]
		25 20 17		[72]
Alanine	$CH_3{-}\overset{\displaystyle }{\dot C}{-}COOH$ with H	32.4 17.8 9.0	23.9 24.1 27.3 (CH_3) 41.5 42.1 45.7 (β_1) 26.4 28.2 27.5 (β_2) 77 °K 3.9 5.7 5.0 (β_3)	[29] [30]
Serine	$HO{-}\overset{H}{\underset{H}{C}}{-}\dot C{-}COOH$ with H	23 with H∥b axis	40 (β_1) 40 (β_2)	[73]
Cysteine hydrochloride	$\dot S{-}\overset{H}{\underset{H}{C}}{-}CH(NH_3^+Cl^-){-}COOH$		38 (β_1) Principal g-values 1.99, 1.99, 2.29 77 °K 14 (β_2)	[74]
Aspartic acid	$COOH{-}\overset{H}{\underset{H}{C}}{-}\dot C{-}COOH$ with H	32.1 21.4 13.9	24.3 (β_1) 41.4 (β_2)	[75] [76]
Valine	$\overset{CH_3}{\underset{CH_3}{C}}{-}\dot C{-}COO^-$ with NH_3^+		24.4 23.2 22.4 9.0 7.3 6.9 nitrogen coupling	[77] [78]
Methionine	$\dot S{-}\overset{H}{\underset{H}{C}}{-}CH_2{-}CH(\overset{+}{N}H_3){-}COO^-$; and $CH_3{-}S{-}CH_2{-}\overset{H}{\underset{H}{C}}{-}\dot C{-}COO^-$	29 16 9	β-coupling small 33 (β_1) 18 (β_2)	Principal g-values 2.004, 2.022, 2.062 [79]

Compound	Structure	Constants	Reference
Glutamic acid hydrochloride	COOH—C—C—CH(NH$_3^+$Cl$^-$)—COOH (H, H H)	30 20 10 ; 25 (β_1), 35 (β_2)	[80]
Leucine	(CH$_3$)(CH$_3$)C—CH($^+$NH$_3$)—COO$^-$ (H)	23 for CH$_3$ Protons and (β_1) ; 8 (β_2)	[81] [82]
Isoleucine	(CH$_3$)(CH$_3$—CH$_2$)C—C(H)(NH$_3^+$)—COO$^-$	Radical postulated, no constants reported	[82]
Lysine hydrochloride	NH$_3^+$—CH$_2$—CH$_2$—CH$_2$—C—C—COO$^-$ (H H, H) ; and NH$_3^+$—CH$_2$—C—C—CH(NH$_3^+$Cl$^-$)—COO$^-$ (H, H H H)	30 18·5 12·8 ; 29·8 28·8 27·4 (β_1), 40·0 36·5 36·5 (β_2) ; No constants reported	[83]
Cystine dihydrochloride	$\overset{\text{H}}{\underset{\text{H}}{\text{S—C}}}$—CH($^+NH_3Cl^-$)—COOH	9 (β_1) ; 2 (β_2) ; Principal g-values 2·003, 2·025, 2·053	[45]
Histidine hydrochloride	ring structure with COO$^-$, NH$_3^+$, N—H, Cl$^-$, H (β_1), H (β_2)	53·2 50·7 49·4 (β_1) ; 50·0 47·2 46·5 (β_2) 4·2°K	[54] [84]

a The β-hydrogen constants of the alanine radical and the cysteine data were taken at 77°K; and the histidine constants were measured at 4·2°K warmed to room temperature, in most cases, the 77 °K spectra changed to that observed when the samples were irradiated and observed at room temperature. The changes in the spectra are not reversible. Such experiments show that the free radicals obtained at room temperature are generally not the primary radicals produced by ionizing radiation, but are secondary radicals produced by conversion of the primary radicals after the initial ionization.

Single crystal studies of a few of the amino acids at 77 °K have shown the existence of positive and negative ions. Alaska and coworkers[85] identified the primary radical produced at 77 °K in the L-cystine dihydrochloride molecule with its S—S bond ionized as $[R—CH_2—S \dot{-} S—CH_2—R]^-$. At 201 °K the ionized bond is unstable and is ruptured to form the 'semi-stable' $R—CH_2^*—S\cdot$ radical in which the direction of the C—S bond remains almost unchanged from that of the parent radical. The spectra observed for this 'semistable' radical are characterized by a triplet hyperfine structure resulting from the interaction of the unpaired electron localized on the sulphur atom with the two methylene hydrogens of the adjacent carbon atom. On further warming to room temperature the radical undergoes a conformation change in which the $CH_2—S$ group has rotated to a stable position accompanied by a change in the electronic state. The room temperature spectra give a doublet hyperfine structure because the rotation of the $CH_2—S$ group places one of the methylene hydrogens near the nodal plane of the unpaired electron and its interaction is therefore essentially zero. These results indicate a possible model for the production of the room temperature $RS\cdot$ radical. At low temperatures electrons removed from various atoms of the molecules by the ionizing radiation are trapped at S—S bonds to form the ionized radicals. On warming, the S—S bond of the ionized radical breaks to form the 'semistable' radical. Further warming causes a change in conformation to give the room temperature $R—CH_2—S\cdot$ radical.

Box and coworkers have studied interesting conformation changes in α-aminoisobutyric acid, glycine, and valine. In α-aminoisobutyric[86] acid irradiated and observed at 77 °K a single absorption peak with weak satellites is observed. The strong centre peak is attributed to the negative ion shown in structure (6)

$$CH_3—\overset{\overset{\displaystyle +NH_3}{|}}{\underset{\underset{\displaystyle CH_3}{|}}{C}}—C\!\!\cdot\!\!\overset{\diagup O^-}{\diagdown O^-} \tag{6}$$

and the weak satellites to a positive ion or products related to its dissociation. As the sample is warmed the satellites fade at about 145 °K and the strong

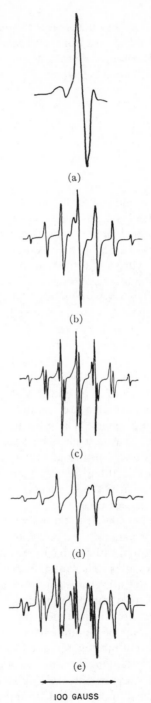

Figure 14. E.s.r. spectra from a single crystal of α-amino-isobutyric acid irradiated at 77 °K and then allowed to warm. The temperature at which the spectra were recorded is as follows: (a) at 77 °K, (b) at 165 °K, (c) at 190 °K and (d) at 300 °K. Spectrum (e) was observed when the crystal was re-cooled. The orientation of the crystal was the same for all of the spectra in the figure (from H. C. Box and H. G. Freund, *J. Chem. Phys.*, **44**, 2345 (1966))

central absorption which fades at about 165 °K is replaced by the spectrum of Figure 14(b). This spectrum contains seven hyperfine lines with the binomial intensity distribution expected from either of the radicals in structure (7) where the unpaired electron

$$
\begin{array}{ccc}
CH_3 & O & CH_3 & O \\
\backslash & \!\!\!\!\!/\!\!/ & \backslash & \!\!\!\!\!/\!\!/ \\
& C\!\!-\!\!C & \text{or} & \dot{C}\!\!-\!\!C \\
/ & \backslash & / & \backslash \\
CH_3 & O^- & CH_3 & OH
\end{array}
\qquad (7)
$$

interacts equally with six β-hydrogens. The first of the radicals could be formed from the zwitterion structure by the loss of the amino group. The second radical is a possibility since protons produced by dissociation of as yet unidentified positive ions may recombine with the negative oxygen. An interaction with the hydroxyl group was not resolved. In solids, interactions with hydrogen in the γ position normally are not detected. When the sample is allowed to warm to 190 °K the spectrum changes to that of Figure 14(c). Here the intensities of the hyperfine lines agree well with that expected from two sets of three equivalent hydrogens. Further warming gives the room temperature spectrum which is also different. The room temperature spectrum is believed to be the average between two conformations. Hyperfine constants and g-value measurements show that the three hyperfine structures belong to different orientations of the $2p$ orbital containing the unpaired electron. In a radical of this type, the $2p$ orbital is expected to be oriented perpendicular to a plane containing the two methyl groups, α-carbon, and β-carbon. The minimum values for the g-factor and hyperfine coupling are measured when the magnetic field is perpendicular to this plane. An analysis of the principal values for the g-factor and hyperfine coupling constants for the three different hyperfine interactions shows that the orientation of the $2p$ orbital with respect to the crystallographic axes of the crystal is different in each case.

E.s.r. studies similar to that of α-aminoisobutyric acid have been made for glycine,[87-88] alanine,[89] valine,[78] and methionine.[90] In these amino acids at 77 °K the initial radicals have been identified as negative ions, or negative and positive ions. The anion radical on warming converts to a neutral radical. The resulting radical may remain stable on warming to room temperature, or as in some cases undergo another conversion before reaching room temperature. The anion radical is believed to be formed when the carboxyl group of an undamaged molecule captures an electron ejected from a different molecule by the irradiation. Structure (6) is a typical anion radical at 77 °K. On warming, the anion radical becomes unstable and the amino acid may decompose to give a neutral radical and ammonia. This neutral radical in some cases disappears by abstracting a hydrogen from a neighbouring molecule, resulting in the transformation of the neighboring molecule into a different radical.

The e.s.r. spectra from valine[78] irradiated at 77 °K indicate an anion radical at 77 °K, an intermediate radical at ~ 225 °K and a different stable radical at room temperature. The anion radical has a structure similar to that shown in structure (6) and the intermediate radical is identified as structure (8)

$$
\begin{array}{c}
CH_3 \qquad\qquad\qquad\quad O \\
\diagdown \qquad\qquad\qquad \diagup\diagup \\
H-\overset{|}{C}(\beta)-\overset{\cdot}{\underset{|}{C}}(\alpha)-C \\
\underset{CH_3}{\vert} \quad \underset{H}{\vert} \qquad\qquad \diagdown OH
\end{array}
\qquad (8)
$$

Observed spectra for this radical are attributed to the hydrogen atoms attached to the α- and β-carbons. The principal values of the coupling constants are 32·8 G, 25·4 G and 8·2 G for the α-hydrogen and 35 G, 26 G, and 23·9 G for the β-hydrogen. On warming from 225 °K to 300 °K, the radical of structure (8) disappears when it abstracts a hydrogen from a neighbouring molecule to give structure (9). At room temperature the radical of structure (9) is stable. Constants for this radical are given in Table 2.

$$
\begin{array}{c}
CH_3 \qquad\quad H \qquad\quad O \\
\diagdown \qquad\quad \vert \qquad\quad \diagup\diagup \\
\overset{\cdot}{C}-\overset{|}{\underset{|}{C}}-C \\
\diagup \qquad \underset{+}{\overset{|}{N}H_3} \qquad \diagdown O^- \\
CH_3
\end{array}
\qquad (9)
$$

A number of intermediate radicals have been found in glycine[87-88] irradiated at 77 °K and warmed to a temperature in the range 110 °K to 140 °K. The intermediate radicals seem to form according to two different paths. One path as described for valine originating from an anion and the other path from a cation, a glycine molecule which has lost an electron. The cation is believed to decompose immediately to give $N^+H_3\dot{C}H_2$ and CO_2. The $\overset{+}{N}H_3\dot{C}H_2$ radical abstracts a hydrogen from a neighbouring molecule at ~ 110 °K to give a $\overset{\cdot\ +}{N}H_2CH_2COO^-$ radical. Near 120 °K, $\dot{N}H_2CH_2COO^-$ decomposes to form $\overset{+}{N}H_3\dot{C}HCOO^-$, one of the stable radicals at room temperature. On the other path, the anion decomposes to give the radical of structure (10a) or (10b).

In either of the structures the unpaired electron interacts with two α-hydrogens. No further conversions take place and the radical of structure (10) is identified at room temperature.

$$
\begin{array}{ccc}
\underset{H}{\overset{H}{\diagdown}}\!\!\overset{\displaystyle .}{C}\!\!-\!\!\overset{\displaystyle O}{\underset{O^-}{C}}\diagup & & \underset{H}{\overset{H}{\diagdown}}\!\!\overset{\displaystyle .}{C}\!\!-\!\!\overset{\displaystyle O}{\underset{OH}{C}}\diagup \\[2mm]
(a) & & (b)
\end{array}
\qquad (10)
$$

L. P. Kayushin and coworkers[91–93] have used a technique of photoconversion to good advantage in identifying intermediate radicals. After irradiation at 77 °K the samples are irradiated with light and in a number of cases photoconversions are observed. Upon exposure to light the anion radical in glycine, $NH_3{}^+CH_2CO^-O^-$ was photoconverted to $N^+H_3\dot{C}H_2$ at 77 °K.

These examples show how e.s.r. may be used to identify different phases in the irradiation damage of organic solids. In these molecules ionic, primary neutral, and final secondary neutral radicals have been observed. The ionic radicals are observed around 77 °K or lower. On warming a primary neutral radical is frequently observed. As the sample is raised to higher temperatures, intermolecular or intramolecular transfer processes often cause the primary neutral radical to change to a secondary neutral radical.

Table 3 and data for other amino acids, such as aminoisobutyric and isovaline show that a number of the stable secondary radicals are formed by breaking the C—N bond to the amino group. Among the smaller molecules, glycine and valine are noticeable exceptions. However, the low temperature studies for these two molecules show that the primary neutral radical in each case results from a rupture of the C—N bond. Box has suggested that a transfer of hydrogen from a neighbouring molecule at higher temperatures to the primary radical may account for the secondary radicals with an amino group.[78]

2. Identification of Secondary Radicals in Proteins

In 1955 Gordy, Ard, and Shields[66] showed that in most proteins a sulphur type resonance and/or a doublet resonance is observed at room temperature. These resonances have been attributed to structures (11) and (12) respectively.

$$
\begin{array}{cc}
\underset{\begin{array}{c}|\\H-\overset{\textstyle|}{C}-H\\ |\\ \underset{\textstyle .}{S}\end{array}}{\cdots\!-\!\!\overset{\textstyle H}{\underset{\textstyle}{N}}\!\!-\!\!\overset{\textstyle H}{\underset{\textstyle}{C}}\!\!-\!\!\overset{\textstyle O}{\underset{\textstyle}{C}}\!\!-\cdots} & \cdots\!-\!\!\overset{\textstyle H}{\underset{\textstyle}{N}}\!\!-\!\!\overset{\textstyle}{\underset{\textstyle H}{\dot{C}}}\!\!-\!\!\overset{\textstyle O}{\underset{\textstyle}{C}}\!\!-\cdots \\[6mm]
(11) & (12)
\end{array}
$$

The sulphur resonance arises from an unpaired electron localized mainly on the sulphur atom with a small β-hydrogen hyperfine interaction. In polycrystalline samples the resonance is very asymmetrical with respect to $g = 2$ and spreads over about 100 G at X-band frequencies. A description of this resonance is given in the section of the theory on g-values. The doublet resonance is attributed to a glycylglycine type resonance which has been described in the section of the theory on α-hydrogen coupling. A typical protein spectrum is shown in Figure 15.

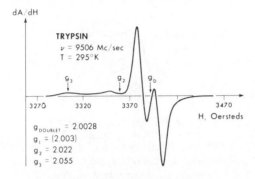

Figure 15. E.s.r. spectrum of irradiated trypsin showing the superposition of doublet and sulphur type resonances found in irradiated proteins (from T. Henriksen, 'Electron spin resonance signals in irradiated proteins', in *Electron Spin Resonance and the Effects of Radiation on Biological Systems* (Ed. W. Snipes), Publication 1355, Committee on Nuclear Science, National Academy of Sciences—National Research Council, Washington, D.C., p. 82, (1966))

Of the two structures (11) is the more firmly established. It is essentially identical to the spectra from polycrystalline cystine or cysteine. A comparison of the spectrum from cystine with spectra from several proteins is given in Figure 16. The radical to which the spectra from cystine or cysteine is attributed has been identified in single crystal studies as structure (11). The assignment of the doublet spectra to structure (12) is not as definite. It is possible to obtain predominantly a doublet spectrum from the superposition of hyperfine components from different radicals, or the doublet could be from a hydrogen not bonded to the backbone of the protein although there are very few sites in a protein where an unpaired electron could interact with a single α-hydrogen because in most of the residues an α-hydrogen would have β-hydrogen neighbours. The protein doublet is similar to the

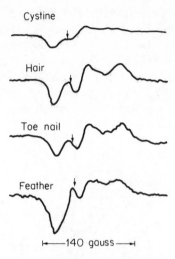

Figure 16. Spectra of cystine and some fibrous proteins. The arrows indicate the free electron g-value (from W. Gordy, W. B. Ard, and H. Shields, *Proc. Nat. Acad. Sci. U.S.*, **41**, 983 (1955))

doublets found in polycrystalline glycylglycine, polyglycine, and acetylglycine.[94] Single crystal studies on glycylglycine[24,95] and acetylglycine[96] indicate that the radical responsible for the doublet is given by structure (**12**). A typical doublet spectrum is shown in Figure 17.

Data obtained by Drew and Gordy[71] for irradiated polyamino acids are helpful in identifying the radicals formed in proteins. A doublet like that of polyglycine was found only in polysarcosine. Since polysarcosine is not a common constituent of proteins, the similarity of the doublet in the proteins to polyglycine seems definite.

One of the factors which makes the identification of the radical in an irradiated protein difficult is the polycrystalline nature of the proteins. Anisotropies which usually exist in both the g-factor and the hyperfine coupling frequently cause the observed spectrum to be poorly resolved. Gordy and Shields[69] studied the resonances in a number of proteins which could be partially oriented with respect to the external magnetic field. In silk[97] doublets of 26 G and 13 G are observed for the field parallel and perpendicular, respectively, to the strands. These spectra are attributed to an α-hydrogen type radical which is assumed to be formed by the loss of the glycine residue from the backbone of the protein, structure (**12**). Observed and calculated spectra, Figure 17, are in close agreement. In calculating the

ESR OF GAMMA-IRRADIATED SILK

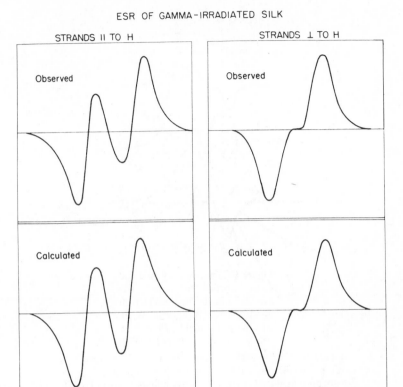

Figure 17. Doublet spectra from gamma-irradiated silk (from W. Gordy and H. Shields, *Proc. Nat. Acad. Sci. U.S.*, **46**, 1124 (1960))

spectra, it is assumed that (i) the NCCH group is planar, (ii) the unpaired electron is in a π orbital perpendicular to the plane of the radical, and (iii) the plane of this group is parallel to the axis of the silk strands and thus the CH bond is perpendicular to the strands. These restraints are in agreement with the antiparallel pleated sheet model of Pauling and Corey. The spin density is mainly in the $2p$ orbital of the α-carbon, with the axis of the $2p$ orbital perpendicular to the plane of the radical and the CH bond. In Figure 18 the X axis is along the $2p$ orbital, the Z axis along the CH bond, and the Y axis perpendicular both to the CH bond and the $2p$ orbital. The X and Z axes are in a plane perpendicular to the strands of silk.

The coupling parallel to the strands, A_y, is equal to 26 G and the spectroscopic splitting factor g_y is 2·0034. When the field is perpendicular to the strands the doublet splitting is an average of A_x and A_z, and the same is true for g_x and g_z. The average A for the coupling in the plane X and Z is

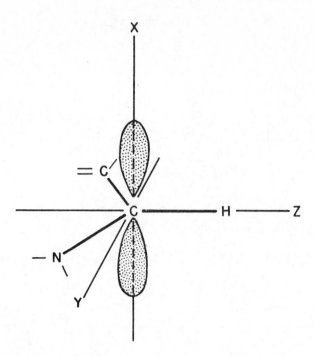

Figure 18. Structural diagram for the α-hydrogen type radical in silk. The 2p orbital in which the unpaired electron is mainly localized is directed perpendicular to the NCHC plane which contains the perpendicular Y and Z axes (from W. Gordy and H. Shields, *Proc. Nat. Acad. Sci. U.S.*, **46**, 1124 (1960))

13 G. From equation (21) the isotropic coupling is found to be $\frac{1}{3}(A_x + A_y + A_z)$ which gives an isotropic coupling of 17 G. The spin density of the unpaired electron on the α-carbon may be estimated from the McConnell relation $a_H = Qp_c$, equation (33). If Q is taken as 25 G, the experimental value of 17 G for the isotropic coupling gives a density of 0·68. The expected anisotropic components are equal to 0·68 of the theoretical values calculated by McConnell,[26] when the unpaired electron is completely localized in a 2p orbital of the α-carbon. The combined isotropic and anisotropic components of the coupling in silk give $A_x = -18\,G$, $A_y = -26\,G$ and $A_z = -7\,G$. These values were used to calculate the curves in Figure 16. The good agreement between the observed spectra and that predicted for an α-hydrogen indicates that the radical in silk is consistent with structure (**12**). It was not possible to recognize nitrogen hyperfine splitting as was done in the case of glycylglycine.[24]

Structure (12) can be formed by breaking away any of the side chains, but it has been suggested that it is most likely formed by removing the H of the glycine residue.[98] A spectrum resulting mainly from the superposition of a doublet from structure (12) and a sulphur resonance from structure (11) may be decomposed to give the relative contributions of each structure to the total spectrum. This is easily done since the shape of the sulphur resonance is known and a significant portion of the resonance on the low field side is not distorted by the superimposed doublet. In experiments on a series of proteins the relative intensities of the doublet and sulphur resonance were found to be related to the glycine and cysteine–cystine content of the protein.[99]

Recently, the free radical distribution in γ-irradiated, dry proteins has been determined by means of the free-radical interceptor technique.[100] This technique relies on the reaction of carbon free radicals with tritiated hydrogen sulphide to form carbon–tritium bonds with the radical. The tritium distribution among the amino acid components is determined by amino acid analysis and a determination of the level of radioactivity in each component due to tritium. This tritium distribution is assumed to be the same as the free radical distribution. The tritium distributions for the proteins studied showed high specific activities for methionine and proline, with relatively low activities for many other residues. These results may seem to disagree with the assumption from e.s.r. data that the stabilized radical in proteins such as silk is due to the loss of a hydrogen from the glycine residue. However, the distributions from the tritium data are given in terms of specific activities, and when the fraction of the total tritium in the protein which is located on glycine in a glycine-rich protein is considered, the radicals from the glycine residue would be expected to make a large contribution to the e.s.r. spectrum.

In addition to structures (11) and (12) Patten and Gordy[101] have observed evidence for contributions to protein spectra from structure (13) and possibly structure (14).

(13)   (14)

Data from the polyamino acids[71] were helpful in identifying structure (13) in some of the irradiated proteins. Polyalanine gave a quartet which was assigned to structure (13). A quartet is expected from the interaction of the unpaired electron localized on the carbon atom of the polypeptide chain and interacting through hyperconjugation with the β-hydrogens of the methyl group which is usually rotating at temperatures as low as 77 °K. The two central components of the polyalanine quartet would fall approximately on the doublet from polyglycine if these two spectra were superimposed. A composite spectrum of this type could be recognized by the two central components. Spectra believed to be the superposition of spectra similar to the polyalanine quartet and the polyglycine doublet have been observed for zein and gelatin.[101]

The radicals in irradiated rat tail tendon and fish fin bone[69] have also been attributed to a superposition of spectra from structures (12) and (13). In rat tail tendon the orientation dependence of the spectra from the α-proton of structure (12) could be studied. As was pointed out in the theory sections on α-hydrogen and β-hydrogen coupling, the α-hydrogen coupling is orientation dependent, whereas β-hydrogen coupling is essentially isotropic.

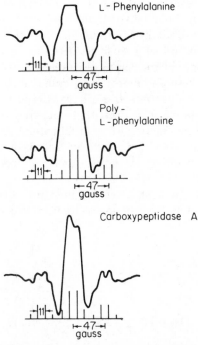

Figure 19. Spectra for radicals formed in L-phenylalanine by H atoms at 300 °K, in poly-L-phenylalanine by H atoms at 77 °K, and in carboxypeptidase A by H atoms at 77 °K (from G. F. Liming and W. Gordy, *Proc. Nat. Acad. Sci. U.S.*, **60**, 794 (1968))

These properties aid in the identification of structures (12) and (13) in an oriented protein, such as collagen in rat-tail tendon.

E.s.r. data from irradiated polyamino acids indicate the possible existence of structure (14) in irradiated proteins. This structure would be formed by the loss of hydrogen from an α-carbon of the protein chain. The unpaired electron localized on this carbon would normally interact with the two hydrogens attached to the β-carbon. It is possible that only one of the hydrogens would give a doublet hyperfine pattern if the other hydrogen happened to be in the nodal plane of the unpaired electron's orbital (see the theory section on β-coupling). In polycrystalline samples a doublet from β-hydrogen coupling would be sharper and better resolved than that from an α-hydrogen since the β-hydrogen coupling is essentially isotropic. If there is a small unresolved coupling with the second β-hydrogen of structure (14) this would broaden the doublet from the first β-hydrogen to give a spectrum similar to that observed from an α-hydrogen. Doublets with an indication of additional doublet splitting observed for polyleucine and polyglutamic acid irradiated at room temperature have been attributed to structure (14). In the proteins a composite spectrum from structures (12), (13), and (14) might be mistaken for a doublet from only structure (12).

Liming and Gordy[102] have recently identified radicals with the unpaired electron localized in the aromatic ring of some amino acids, polyamino acids, and proteins. Radicals of this type were observed in monomers and polymers of phenylalanine, tyrosine, and tryptophan after bombardment with either thermal hydrogen atoms or γ-radiation. Similar resonances were also observed for the proteins; carboxypeptidase A, insulin, and papain which contain these residues. The radicals result from hydrogen-addition at the meta carbons of the rings. Figure 19 shows the spectra observed for phenylalanine and carboxypeptidase A. The hybrid structures to which the phenylalanine, poly-phenylalanine, and carboxypeptidase A spectra are attributed are

(15)   (16)   (17)

The triplet splitting of 47 G is from the two equivalent hydrogens on the 3-carbon and the smaller quartet splitting of 11 G is from equivalent hydrogens on the 2, 4 and 6-carbons. Spectra observed for insulin and papain are similar to that of poly-tyrosine where the unpaired electron couples strongly with the two equivalent hydrogens of the 3-carbon and weakly with the equivalent hydrogens at the 2 and 6-carbons. No hydrogen coupling occurs at the 4-carbon which has an OH group in tyrosine.

3. Production of Secondary Radicals in Proteins

The predominance of the doublet and/or sulphur resonance in irradiated proteins is surprising. Absorptions of ionizing irradiation by a protein is expected to be random, that is, not restricted to one or two residues. Since in most proteins the glycine and sulphur containing residues account for only a small per cent of the total protein structure there must be a migration of irradiation effects to the sites responsible for the doublet and sulphur spectra.

A mechanism for the production of these secondary radicals at room temperature in the proteins has been proposed by Gordy and coworkers.[68,103] They assume that the first phase of radical formation involves the random formation of electron holes by ionization. In the next phase, the electron holes are assumed to migrate by hyperconjugation along the backbone of the protein to the glycine or sulphur residues where the room temperature e.s.r. centres are mainly localized.

An alternate mechanism has been suggested by Henriksen, Sanner and Phil,[104] in which the main features are shown in Figure 20. As in the Gordy

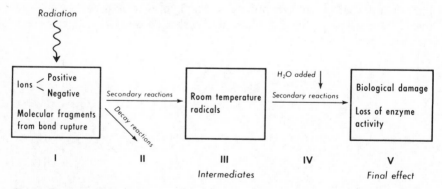

Figure 20. A possible mechanism for radiation damage in an organic solid (from T. Henriksen, 'Electron spin resonance signals in irradiated proteins', in *Electron Spin Resonance and the Effects of Radiation on Biological Systems* (Ed. W. Snipes), Publication 1355, Committee on Nuclear Science, National Academy of Sciences—National Research Council, Washington, D. C., p. 89 (1966))

model, the initial stage is assumed to be ionic. However, Henriksen assumes that the room temperature radicals are formed by interactions of the e.s.r. centres or ions with each other, intact neighbour molecules, or with other groups within the same molecule. One model involves the intramolecular transfer of energy while the other depends on an intermolecular transfer of energy. Experimentally it is difficult to distinguish between these two models. The addition to the protein of molecules which can accept irradiation damage from the protein provides a method for testing the evidence for the intermolecular transfer of energy.

A number of researchers have found evidence for the transfer of energy from one component to another of a mixture.[105] Norman and Ginoza[106] obtained evidence for energy transfer between some amino acid mixtures. A predominantly sulphur type resonance was observed for an irradiated mixture of alanine and cysteine. Less pronounced effects were observed for some other mixtures. Pohlit, Raejewsky, and Redhardt[107] observed a sulphur type resonance in a molecular mixture of the amino acid components of hair after the sample had aged 36 hours following irradiation. Miyagawa and Gordy[103] studied molecular mixtures of cysteine, cystine and certain other chemical protectors with various proteins. Molecular mixtures were prepared by dissolving the components in water and then evaporating the water. A small concentration of these protectors with the protein being studied in many cases drastically altered the e.s.r. spectra of the protein. In some cases a sulphur type resonance was observed and in other cases a singlet resonance different from either the protector or protein was observed. Henriksen, Sanner and Phil[104] have shown that the transfer of the unpaired electron from protein to sulphur protector does not require covalent bonding between the protein and protector. Patten and Gordy[101] studied proteins for which the constitution is known and compared the spectra from the irradiated protein with that obtained from proportional mixtures and solid solutions of the amino acid constituents of the individual proteins. The mechanical mixtures were prepared by simply grinding together proportional amounts of the amino acid residues of the protein being studied. Solid solutions of molecular mixtures were prepared by dissolving similar proportions in water and then evaporating the water. It was assumed that true molecular mixtures are formed through hydrogen bonding between unlike amino acids. The samples were exposed at $77\,^\circ K$ to a γ-ray dosage of 5×10^6 r. Spectra at $77\,^\circ K$ had very little structure, but after the samples were warmed to room temperature, considerable structure appeared. The changes which occurred on warming to room temperature were found to be irreversible since the spectra observed when the samples were returned to $77\,^\circ K$ were not the same as the original $77\,^\circ K$ spectra. The proteins and mixtures each had their own characteristic spectra at room temperature. In general the spectra from the molecular mixtures were not as well resolved or as spread out as the spectra from the mechanical mixtures. It was concluded that there is no meaningful comparison between the spectra obtained for the proteins and the mixtures. However, the experiments on mixtures by different investigators do show that the unpaired electron in many cases may be transferred from one molecule to another. This evidence for the intermolecular transfer of unpaired spins does not rule out the possibility of intramolecular transfer as well. Most likely, both mechanisms play a role in transferring irradiation damage to preferred sites.

G-values, the number of radicals formed per 100 eV of absorbed energy, for proteins and amino acids[70] irradiated in a vacuum and observed at room temperature are usually in the range from 1 to 7. An exception is found for amino acids containing an aromatic ring. Here the G-values are usually less than 1. This result is consistent with experiments in radiation chemistry which have shown that molecules containing aromatic rings do not damage easily. Also G-values for alpha-radiation are lower than those for irradiation with lower LET, such as X- or γ-radiation. Great care must be taken in making absolute measurements of radical concentration.[108–109] A number of measurements should be made at various microwave power levels and dosage in order to avoid errors due to saturation of the e.s.r. signal or saturation in dose. The value may depend on the temperature at which the measurement is made, water content, surrounding gas, and decay during the time interval of irradiation and measurement. Radical concentration in irradiated proteins and their constituents build up exponentially to a saturation level as the irradiation dose is increased. This effect is discussed in section IV B 3.

4. Correlation between Radicals and Biological Damage

In irradiated aqueous solutions secondary chemical damage is produced largely by free radicals. In analogy, it is often assumed that radiation damage on solid biological systems is also caused in some degree by free radicals. Hunt and coworkers[110–111] found that the yield of secondary radicals in ribonuclease was of the same order of magnitude as the inactivation yield, and thus concluded that a large fraction of the biological damage can be accounted for by radicals. They used the oxygen effect for a more quantitative correlation. Exposure to oxygen after irradiation in a vacuum caused a change in the e.s.r. spectrum and inactivation yield. A two step process was proposed in which stable radicals are formed when the ribonuclease is irradiated, and then when the sample is exposed to oxygen, the peroxide radical, $ROO^{\cdot}$, is formed. It is in the second step that a lesion probably occurs.

Correlations which relate the inactivation of a macromolecule to the behaviour of a single e.s.r. centre are somewhat uncertain due to difficulties in determining radical concentrations, obtaining oxygen free samples, eliminating dose effects, etc. Henriksen[3,112] has attempted to avoid some of these difficulties by varying conditions which would alter the yield of secondary radicals. A corresponding change in inactivation yield provides strong evidence for a connection between radicals and inactivation. Simple biochemical systems were irradiated by 6·5 MeV electrons and a number of different ions such as helium, lithium, boron, carbon, oxygen, fluorine, and argon. A ratio representing the variation in the radiation yield from sparsely ionizing radiation to the densely ionizing argon ions was determined for some enzymes and amino acids. The range of this ratio was from 2 to 5 for both the production of secondary radicals and the inactivation of the enzymes. Also, curves having similar shapes were obtained from plots of yield as a function of stopping power for the two measurements. Studies in which the samples were irradiated at different temperatures show a good correlation between radical yield and inactivation yield. For these experiments, reasonable agreement was also found for activation energies calculated from a sum of exponential terms,

$$Y_T = \sum_i y_i\, e^{-E_i/kT} \tag{50}$$

Here Y_T is the yield at the irradiation temperature and E_i is the activation energy.

These experiments establish, particularly in the case of trypsin, a correlation between the production of secondary radicals and the loss of enzymatic activity. Also in mixtures sulphur-containing molecules frequently prevent the formation of the secondary radicals, and may therefore act as protectors against biological damage.

B. Nucleic Acids and Their Constituents

1. Radical Identification in DNA

During the past decade since the initial work by Shields and Gordy[113–114] a number of researchers have studied the effects of ionizing irradiation on the nucleic acids and their components with e.s.r.[115–131] In addition to these original reports several comprehensive reviews have been written.[4,5,9] Only a few of the major achievements in this area will be discussed here due to the lack of space.

The nucleic acids and their components all give detectable e.s.r. signals when they are exposed to ionizing radiation. Spectra have been studied after exposure to high energy electrons, γ-rays, X-rays, u.v. light, and thermal hydrogen atoms. The main features of the observed spectra do not depend on the type of ionizing radiation used to produce the trapped radicals. In the early experiments a single broad asymmetrical resonance was observed for DNA and RNA. Each of the pyrimidine and purine constituents gave their own characteristic spectrum which appeared to be different from the spectrum observed for DNA or RNA. Resonances in DNA and RNA were found to be stable in an atmosphere of O_2, N_2, A, or a vacuum, but unstable in the presence of water vapour.[114,115,126] Researchers doing subsequent studies on DNA and RNA began to notice structure in the e.s.r. resonance which spread beyond the central resonance.[118,132–134] The observation of structure in the resonance seemed to depend on the source of DNA. It was soon discovered that an eight component spectrum similar to that observed in thymine could be most easily observed when the DNA sample was moist (high water content), and u.v. light was used as the irradiation source.[126] A comparison of the spectra observed from calf thymus DNA irradiated with 1 Mev electrons and u.v. light is shown in Figure 21.

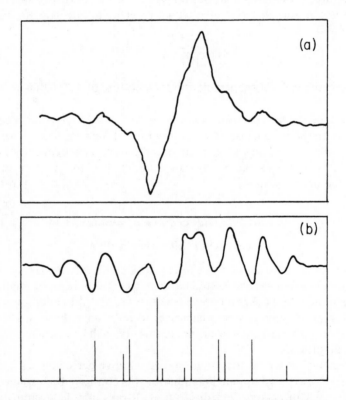

Figure 21. First derivative spectra of calf-thymus DNA irradiated (a) with ionizing irradiation and (b) with u.v. light. (Reprinted with permission from P. S. Pershan, R. G. Shulman, B. J. Wyluda, and J. Eisinger, *Physics*, **1**, 163 (1964), Pergamon Press)

The origin of the eight line spectrum in DNA has been traced to the thymine base. Pershan and coworkers[126] and Pruden, Snipes and Gordy[131] have identified the thymine radical as Formula (18).

$$
\begin{array}{c}
\text{O} \\
\parallel \\
\text{C}
\end{array}
$$

The radical of structure (18) is formed when a hydrogen atom is added to the 6-carbon of the ring. Hydrogen-addition at the 6-carbon breaks the double bond and the unpaired electron is localized mainly on the 5-carbon. The unpaired electron has an equal interaction of 20·5 G with the β-hydrogens of the methyl group. In addition to the interaction with the methyl group hydrogens there is an equal interaction of 37·7 G with the two methylene hydrogens attached to the 6-carbon. The interaction with the methyl hydrogens gives a four component spectrum with intensity ratios of 1:3:3:1. Each component of the quartet is split into a 1:2:1 triplet by the methylene hydrogens. Since the coupling of the methylene hydrogens is approximately equal to twice the methyl hydrogen coupling, a number of the components fall approximately on top of each other. The resulting spectrum is an octet with intensity ratios close to 1:3:5:7:7:5:3:1, Figure 22.

Proof for the existence of the thymine radical in irradiated DNA was obtained by growing a sample of *Escherichia coli* 15 T$^-$, which is incapable of synthesizing its own thymine, on normal thymine and a second sample from thymine containing a deuterated methyl group.[126] The observed spectrum from the deuterated sample was in good agreement with that which was expected from structure (18) with a deuterated methyl group. A poorly resolved triplet from the two methylene hydrogens was observed since the splitting from the deuterated methyl group is $\frac{1}{6}$ that from a normal methyl group (see Table 1).

Further evidence for hydrogen-addition to the 6-carbon atom was obtained by allowing calf-thymus DNA to equilibrate in approximately 100 per cent relative humidity. After equilibration the samples were frozen at 77 °K, irradiated and observed. When H_2O was used the expected octet was observed. However, when the experiment was repeated with D_2O a spectrum expected from deuterium addition at the 6-carbon was observed. This experiment establishes hydrogen-addition as the mechanism by which the thymine radical is formed in DNA. The intensity of the thymine resonance

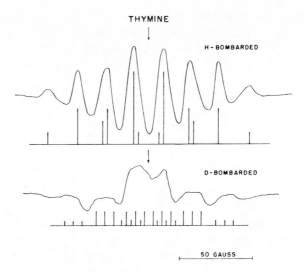

Figure 22. Second derivative spectra from H and D-bombarded thymine. The bars below the spectra indicate the theoretical spectra expected from the proposed thymine radical. An unidentified resonance which is more obvious in the D-bombarded case exists near the centre of the resonance where the arrow indicates the *g*-value of DPPH, 2·0036. (From J. N. Herak and W. Gordy, *Proc. Nat. Acad. Sci. U.S.*, **54**, 1287 (1965))

increases in proportion to the relative humidity in which the DNA is equilibrated before irradiation. The contribution of the thymine radical to the total spectrum at 0 per cent humidity after exposure to ionizing irradiation is 20 per cent[128] while in a moist sample irradiated with u.v. light the contribution increases to about 50 per cent.[126]

An analysis of spectra from a thymidine single crystal by Pruden, Snipes, and Gordy[131] also showed that structure (**18**) is formed by hydrogen-addition, and they concluded that the DNA multiplet structure is from a thymine resonance. Herak and Gordy[135] bombarded thymine with gaseous H atoms at thermal velocities and observed the octet spectrum from radicals formed by the addition of the bombarding hydrogen atoms. All of these experiments seem to prove conclusively that the thymine radical is formed by hydrogen-addition and that a thymine type radical is found in irradiated DNA.

2. Radicals Formed by Hydrogen-addition in the Nucleic Acid Components

Hydrogen-addition radicals have been produced in all of the purines and pyrimidines by bombardment with H atoms at thermal velocities.[136–138]

The radicals formed in this way have been identified by Herak and Gordy[61,135] as structures (19) to (24).

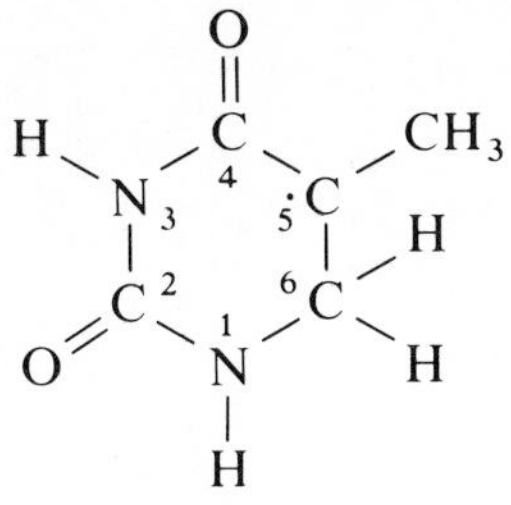

thymine + H

(19)

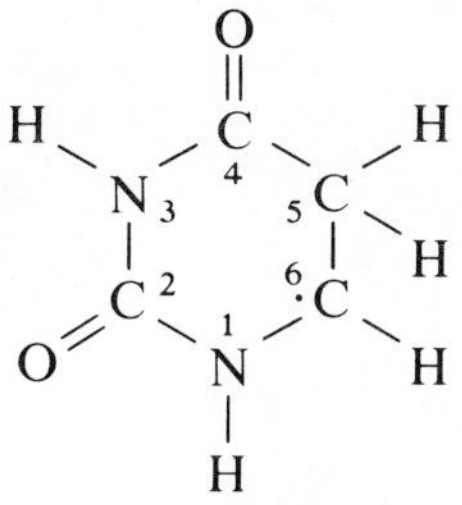

uracil + H

(20)

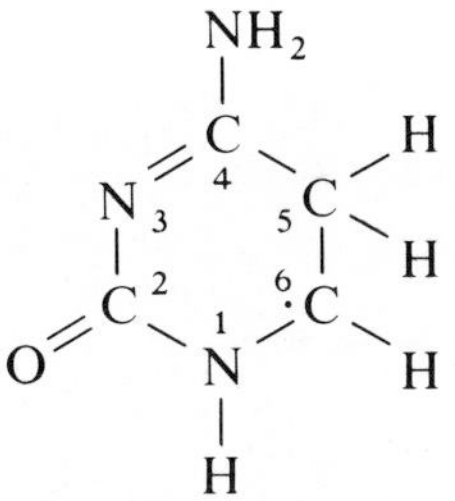

cytosine + H

(21)

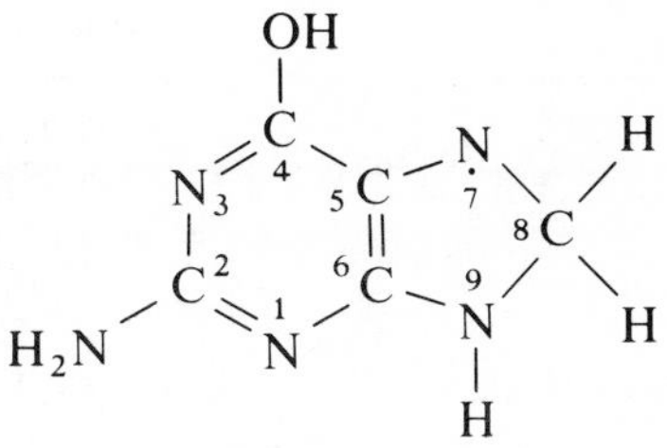

guanine + H

(22)

either   or

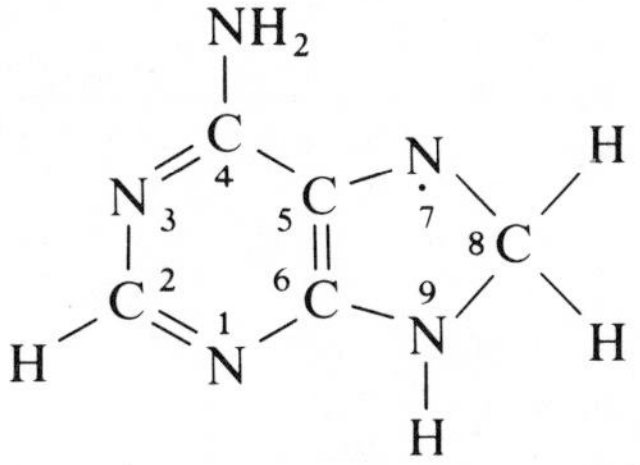

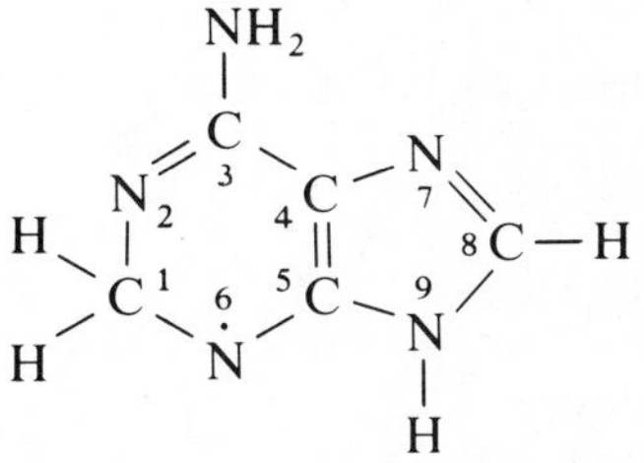

adenine + H

(23)   (24)

Coupling constants for these radicals are given in Table 4. Hydrogen couplings and spin densities are also given for hydrogen-addition radicals in the nucleosides, nucleotides and RNA in Table 5.

Table 4. Hydrogen couplings of H-addition radicals
in purine and pyrimidine bases[a]

Radical source	Coupling group	Coupling in gauss 77 °K		300 °K
Thymine	$>C_{(6)}H_2$	34	42	38
	$-CH_3$		20·5	20·5
Uracil	$>C_{(5)}H_2$	19·5	46	33
	$>C_{(6)}H$		19·5	18·5
Cytosine	$>C_{(5)}H_2$	19·5	48	
	$>C_{(6)}H$		19·5	
Adenine	$>C_{(8)}H_2$			40
Guanine	$>C_{(8)}H_2$	38		38

[a] From J. N. Herak and W. Gordy, *Proc. Nat. Acad. Sci.
U.S.*, **55**, 1373 (1966)

Table 5. Hydrogen couplings and spin densities of H-addition radicals in nucleosides,
nucleotides and RNA*

Compound	Coupling group	Coupling in gauss	Spin densities on C_α
Cytidine	$C_{(6)}H$	$a_1 = 20$	0.76^b
	$C_{(5)}H_2$	$a_2 = 20; a_3 = 50$	
Cytidylic acid	$C_{(6)}H$	$a_1 = 19$	0.73^b
	$C_{(5)}H_2$	$a_2 = 19; a_3 = 46.5$	
Deoxycytidine	$C_{(6)}H$	$a_1 = 20$	0.76^b
	$C_{(5)}H_2$	$a_2 = 14; a_3 = 48.5$	
Deoxycytidylic acid	$C_{(6)}H$	$a_1 = 20; (a_1 = 21)^a$	$0.76^b(0.80)^a$
	$C_{(5)}H_2$	$a_2 = 13; a_3 = 50$	
		$(a_2 = 12.5; a_3 = 50)^a$	
Thymidine	$-CH_3$	$a_1 = a_2 = a_3 = 20.5$	0.70^c
	$C_{(6)}H_2$	$a_4 = a_5 = 39$	
Thymidine 5'-mono-phosphate	$-CH_3$	$a_1 = a_2 = a_3 = 20.5$	0.70^c
	$C_{(6)}H_2$	$a_4 = a_5 = 38$	
Uridine	$C_{(6)}H$	$a_1 = 20$	0.76^b
	$C_{(5)}H_2$	$a_2 = 20; a_3 = 46$	
Deoxyuridine	$C_{(6)}H$	$a_1 = 19.5$	0.74^b
	$C_{(5)}H_2$	$a_2 = 19.5; a_3 = 46$	
Uridylic acid	$C_{(6)}H$	$a_1 = 20$	0.76^b
	$C_{(5)}H_2$	$a_2 = 20; a_3 = 43$	
Guanosine	$C_{(8)}H_2$	$a_1 = a_2 = 37$	
Guanylic acid	$C_{(8)}H_2$	$a_1 = a_2 = 37$	
Deoxyguanosine	$C_{(8)}H_2$	$a_1 = a_2 = 37$	
Deoxyguanosine 5'-monophosphate	$C_{(8)}H_2$	$a_1 = a_2 = 37$	
RNA Pyrimidine	$C_{(6)}H$	$a_1 = 19.5$	0.75
	$C_{(5)}H_2$	$a_2 = 19.5; a_3 = 42$	
Purine	$C_{(8)}H_2$	$a_1 = a_2 = 38$	

* From J. N. Herak and W. Gordy, *Proc. Nat. Acad. Sci. U.S.*, **55**, 1373 (1966); **56**, 7(1966)
[a] At 300 °K
[b] Calculated with the equation $a_\alpha = Q_\alpha \rho_\alpha$ with $Q_\alpha = 26.2$ gauss
[c] Calculated from the equation $a_\beta = \rho_\alpha Q_\beta \cos^2 \theta$ with $Q_\beta = 58.5$ gauss

3. Radical Yields

In general each base and pentose has its own characteristic spectrum. However, the spectra observed from the nucleotides and the nucleosides are characteristic of the particular base they contain. It is, therefore, concluded that the unpaired electron of a radical is localized in a ring of a base group. This conclusion seems to be in contradiction with the measured G-values of the bases and pentoses which are about 1 and 4, respectively. Measured G-values for the nucleotides vary from about 2–5 and those for the nucleosides from 0.5–1.5.[4] The observed G-values for the nucleotides and nucleosides are in approximate agreement with an additive contribution from their components. The predominance of spectra characteristic of the bases in the nucleosides and nucleotides may mean that the G-values of the bases and pentoses are drastically altered in opposite directions. A second and more likely explanation for the characteristic base spectra is a transfer of energy from the pentose to the base as frequently occurs in the formation of a sulphur type radical in the proteins.

Results from initial experiments on nucleic acids and their components indicated that G-values were very low. These experiments were done with high doses of radiation, usually around 10 Mrad. Later experiments showed that in general the radical content in the nucleic acids has a linear dependence on dose up to about 1 Mrad before the onset of saturation (Figure 23).

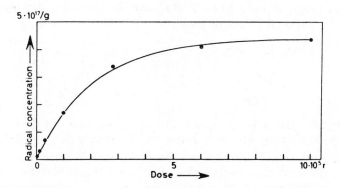

Figure 23. Radical concentration as a function of dose for X-irradiated dry T2-phage DNA in a vacuum at room temperature. The solid line is a best fit of equation (51) to the experimental data. (From A. Müller, *Intern. J. Rad. Biol.*, **6**, 137 (1963))

The G-values determined for the nucleic acids and their components at irradiation doses below 1 Mrad are comparable in values to those found for the proteins and their constituents.

The concentration of radicals in nucleic acids follows the equation[123]

$$C = C_{\infty}(1 - e^{-D/D_0}) \qquad (51)$$

where C is the concentration at dose D, C_∞ is the radical concentration at saturation, and D_0 is the dose at which the concentration is 37 per cent less than at saturation. Differentiation of this equation gives

$$\frac{dC}{dD} = (C_\infty - C)/D_0 \qquad (52)$$

where the initial formation of radicals is C_∞/D_0, and the decrease of radicals is $-C/D_0$. The constant formation of radicals given by equation (52) is in agreement with a linear radical yield as shown in Figure 22. If the loss of radicals occurred by the recombination of species having approximately equal concentrations and mobility the term of equation (52) representing this action would be quadratic. When one of the species in a recombination has a very different mobility and/or concentration, the term is linear. It is believed that atomic hydrogen which is produced by ionizing radiation in organic solids satisfies this condition with respect to the radicals trapped in the nucleic acids and their constituents. Yields in irradiated proteins and their constituents are also described by equation (51). Simmons[139] has used a mass spectrometer to detect atomic hydrogen from irradiated glycine.

Snipes and Horan[140] have used an interesting deuterium–hydrogen exchange effect in irradiated alanine to show that radicals are created and destroyed by irradiation as described by equation (52). When alanine which has the amino hydrogens and hydroxyl hydrogen replaced by deuterium from exchange in heavy water is irradiated the radical in structure (25) is formed by a loss of the amino group.

$$CH_3-\overset{\cdot}{C}-COOD \xrightarrow{\text{heat}} CH_3-\overset{\cdot}{C}-COOD$$
$$\overset{|}{H} \qquad\qquad\qquad \overset{|}{D}$$
$$(25) \qquad\qquad\qquad (26)$$

Heating for several hours around 100 °C causes some of the radicals to exchange the α-hydrogen for a deuterium. The spectrum from (26) is easily distinguished from (25) since the coupling for the α-hydrogen and α-deuterium are very different. If a sample with a mixture of (25) and (26) is given further irradiation the spectrum from (26) disappears and a spectrum entirely from (25) is observed. This experiment clearly demonstrates the formation and destruction of radicals from irradiation.

4. Other Radicals in DNA and RNA

Since the thymine radical does not contribute more than 50 per cent of the radicals formed in DNA, the identity of the remaining radicals is a very important problem. Müller has attributed a doublet structure observed in T2 bacteriophage DNA to an α-hydrogen type radical, $\overset{\displaystyle \diagdown}{\underset{\displaystyle \diagup}{\dot{C}}}-H$.[141] The characteristics of an α-hydrogen radical are discussed in the section on theory. If the α-hydrogen radical is formed in DNA by the loss of a hydrogen atom, the released atom may diffuse to a thymine base and form the thymine radical by addition. Information about other radicals in DNA can be obtained by subtracting the thymine spectrum from the observed spectrum.[126] Examples of this decomposition technique are given in Figure 24. The spectra were obtained by irradiation at 95 °K and then allowing the samples to anneal at 195 °K before the spectra were observed. Since the thymine resonance extends well beyond the centre region of the DNA spectrum, a good fit between the DNA spectrum and a thymine resonance may be made on the high and low field sides. The thymine resonance is then subtracted from the DNA spectrum to give the spectrum from other radicals contributing to the DNA resonance. In addition to the thymine resonance the spectra of Figure 24 decompose into what appear to be various triplets, on the right side of Figure 24. However, Pershan and coworkers[126] observed that the spectra are actually the superposition of a doublet and a central structure which appears to be a singlet. The doublet has a splitting of 39 G and the line width of the components is about 7 G. More than one central structure was identified by measuring the g-factors and widths of the resonances. A narrow centre line of 5 G was observed which anneals at 195 °K. A second centre line with 13 G width was found to be stable at 298 °K. A third centre line was observed whose g-value was slightly different from the first and second centre lines described. This last resonance was similar in appearance to that found in cytidine. It would be expected that the hyperfine patterns of radicals found in solid DNA would depend on the direction of the external magnetic field with respect to the bonds in the radicals. Therefore, it is unlikely that the centre structure in DNA can be attributed to a specific radical until one succeeds in resolving the centre structure by obtaining spectra from oriented samples of DNA. Ehrenberg, Rupprecht, and Ström have made a start in this direction.[142] They observed a definite orientation dependence, but the resolutions of the central portion of the spectra is about the same as that found for polycrystalline samples.

Herak and Gordy[136] have observed a spectrum in hydrogen-bombarded RNA which they attribute to purine and pyrimidine, (Figure 25). The shape and intensity of the observed spectrum agree with the superposition of spectra expected from hydrogen-bombarded purine and pyrimidine groups. Since the spectra from molecules of a group are so similar, it is not possible to distinguish which of the purines or pyrimidines contribute to the RNA spectrum. The spectrum is stable at 77 °K, but decays to a purine type signal at room temperature.

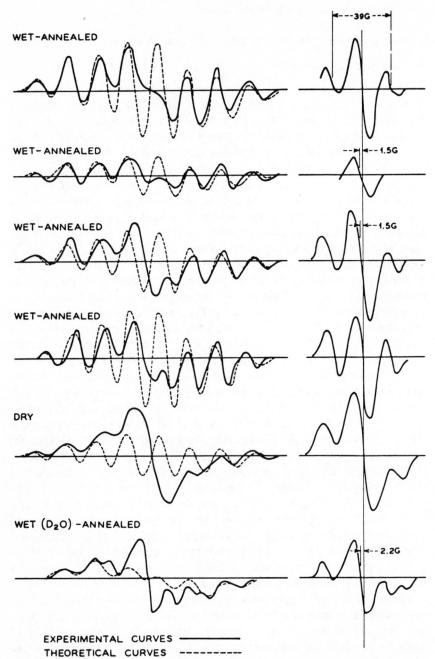

Figure 24. A decomposition of the DNA spectrum into a thymine type spectrum and a spectrum from other radicals. The difference between the thymine spectrum and the DNA spectrum is given on the right side. (Reprinted with permission from P. S. Pershan, R. G. Shulman, B. J. Wyluda, and J. Eisinger, *Physics*, **1**, 163 (1964), Pergamon Press)

H - Bombarded

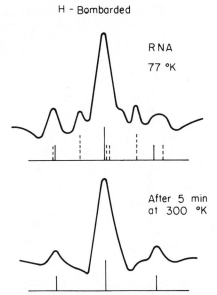

RNA

77 °K

After 5 min
at 300 °K

Figure 25. Second derivative e.s.r. spectra
from H-bombarded RNA. The solid bars
represent the spectrum from guanine or
adenine and the dotted bars the spectrum
from uracil. (From J. N. Herak and W.
Gordy, *Proc. Nat. Acad. Sci. U.S.*, **55**,
1373 (1966))

5. Pyrimidine and Purine Single Crystals

Most of the e.s.r. data on the purine and pyrimidine bases have been
obtained from polycrystalline samples. The spectra from these samples are
often poorly resolved because of the averaging which occurs in both the
g-values and hyperfine coupling constants. In general, spectra from single
crystals are better resolved with much greater possibilities for identifying
the radical structures from which the spectra come. Only in the last few
years have there been reports on single crystal studies of the purines and
pyrimidine bases (Table 6). The reason more of these single crystal studies
have not been done is most likely because suitable single crystals are difficult
to grow. A detailed description of each radical in Table 6 will not be given.
In a radical such as deoxyadenosine, the unpaired electron has a significant
probability of being delocalized over several atoms as indicated by the
coupling constants. The couplings given in the Table are of the α-hydrogen,
β-hydrogen or dipole–dipole type discussed in the section on theory. The
constants determined for thymidine are essentially identical to those observed for thymine. Based on this similarity, Pruden, Snipes, and Gordy[131]

Table 6. Radicals identified in irradiated pyrimidine and purine single crystals

Molecule	Radical	Coupling constants (gauss) and g-values	Reference
Thymidine	(structure of thymidine radical)	CH_3: 20·5 $C_{(6)}H_2$: 40·5 g: 2·0024, 2·0030, 2·0042	[131]
Cytosine	(structure of cytosine radical)	$N_{(3)}$: 2·5 $N_{(1)}$: 5·0	[144]
Cytidine	(structure of cytidine radical)	N_α: 9	[5]
3′ Cytidylic acid	(structure of 3′ cytidylic acid radical)	H_α: 28·7, 16·7, 5·8 H_β: 29·5 g: 2·0063, 2·0039, 2·0026	[145]

[Table continued on p. 477

Table 6—continued

Molecule	Radical	Coupling constants (gauss) and g-values	Reference
5-Nitro-6-methyluracil		CH_3: 10·7, 8·8, 8·8 OH: 9·5, −0·5, −0·5 g: 2·002, 2·005, 2·008	[146]
Alloxan		$OH_{(5)}$: 7·5, 5·5, 3·6 g: 2·0060, 2·0046, 2·0023	[147]
Barbituric acid dihydrate		CH: 30·0, 20·4, 11·2	[148]
Guanine hydrochloride		$N_{(7)}$: 21·0, 11·5, 1·5 $H_{(7)}$: −9·0, 1·0, −15 $H_{(9)}$: −2·9, 0·9, −5 $C_{(8)}H_2$: 36·0 g: 2·0023, 2·0048, 2·0048	[149]

[Table 6 continued on page 478

Table 6—continued

Molecule	Radical	Coupling constants (gauss) and g-values	Reference
Deoxyadenosine		$N_{(1)}$: 19, 0·5, 0·5 $N_{(3)}$: 8·7, 0, 0 $C_{(2)}H_2$: 43·7	[150]
	and		
		Radical postulated No constants measured	

identified the multiplet structure in DNA as a thymine type resonance. Single crystal data for thymidine show a measurable variation in the g-value of the spectra as the orientation of the crystal is changed with respect to the magnetic field. The principal g-values for thymidine are given in Table 6. Based on the single crystal data the identification of the radicals given in Table 6 should be very definite.

6. Radicals in Biologically Active Nucleic Acids

E.s.r. measurements have been made on a viable bacteriophage.[141] Even though 50 per cent of the phage by weight may be protein, spectra are observed which have features of nucleic acid spectra. The shapes of the observed spectra are in good agreement with superimposed nucleic acid and protein spectra. The average number of radicals in a phage particle per inactivating event can be evaluated from biological experiments. Müller[143] found that between 50 and 1,000 radicals are present in a phage particle for each inactivating event. Based on this estimate only 0·1 to 2 per cent of the radicals detected may be involved in biological inactivation.

ACKNOWLEDGMENT

I wish to thank Professor Walter Gordy for introducing me to e.s.r. and for his continuing help and encouragement.

REFERENCES

1. M. G. Androes and M. Calvin, *Biophys. Journ.*, **2**, 217 (1962)
2. J. W. Boag, 'Physical methods in radiation chemistry and in radiobiology', in *Actions Chimiques et Biologiques des Radiations*, Vol. 6, (Ed. M. Haissinsky), Maison, Paris, 1963, pp. 1–70
3. T. Henriksen, 'Electron spin resonance signals in irradiated proteins', in *Electron Spin Resonance and the Effects of Radiation on Biological Systems* (Ed. W. Snipes), National Academy of Science, Washington, D.C., 1966, pp. 81–100
4. A. Müller, 'The formation of radicals in nucleic acids, nucleoproteins, and their constituents by ionizing radiation', in *Progress in Biophysics and Molecular Biology*, Vol. 17 (Eds. J. A. V. Butler and H. E. Huxley), Pergamon, London, 1967, pp. 99–147
5. W. Gordy, *Ann. N.Y. Acad. Sci.*, **158**, 100 (1969)
6. W. Gordy, W. V. Smith and R. F. Trambarulo, *Microwave Spectroscopy*, Wiley, New York, 1953
7. D. J. E. Ingram, *Free Radicals as Studied by Electron Spin Resonance*, Butterworths, London, 1958
8. C. P. Slichter, *Principles of Magnetic Resonance*, Harper and Row, New York, 1963
9. G. Schoffa, *Elektronenspinresonanz in der Biologie*, G. Braun, Karlsruhe, 1964
10. A. Carrington and A. D. McLachlan, *Introduction to Magnetic Resonance*, Harper and Row, New York, 1967
11. P. B. Ayscough, *Electron Spin Resonance in Chemistry*, Methuen, London, 1967
12. C. P. Poole, Jr., *Electron Spin Resonance*, Interscience Publishers, London, 1967
13. S. M. Wyard, *Solid State Biophysics*, McGraw-Hill, New York, 1969
14. G. Fehr, *Electron Paramagnetic Resonance with Application to Selected Problems in Biology*, Gordon and Breach, New York, 1970
15. G. J. Dines and G. H. Vinegard, *Radiation Effects in Solids*, Interscience, New York, 1957, p. 47
16. H. C. Box, H. G. Freund and K. T. Lilga, 'Paramagnetic resonance line width in some irradiated organic crystals', in the *Fifth Annual Symposium on Free Radicals*, Gordon and Breach, New York, 1961, paper No. 9
17. H. N. Rexroad, Y. H. Hahn and W. J. Temple, *J. Chem. Phys.*, **42**, 324 (1965)
18. I. Miyagawa and W. Gordy, *J. Chem. Phys.*, **32**, 255 (1960)
19. C. H. Heller and H. M. McConnell, *J. Chem. Phys.*, **32**, 1535 (1960)
20. N. M. Atherton and D. H. Whiffen, *Mol. Phys.*, **3**, 1 (1960)
21. C. Heller and H. M. McConnell, *J. Chem. Phys.*, **32**, 1535 (1960)
22. D. Pooley and D. H. Whiffen, *Mol. Phys.*, **4**, 81 (1961)
23. H. C. Box, H. G. Freund and K. T. Lilga, 'Radiation-induced paramagnetism in some simple peptides', in *Free Radicals in Biological Systems* (Eds. M. S. Blois and others), Academic Press, New York, 1961, Chap. 18, pp, 239–248
24. M. Katayama and W. Gordy, *J. Chem. Phys.*, **35**, 117 (1961)
25. D. K. Ghosh and D. H. Whiffen, *Mol. Phys.*, **2**, 285 (1959)
26. H. M. McConnell and J. Strathdee, *Mol. Phys.*, **2**, 129 (1959)

27. H. M. McConnell and D. B. Chestnut, *J. Chem. Phys.*, **28**, 107 (1958)
28. S. I. Weissman, *J. Chem. Phys.*, **25**, 890 (1956)
29. J. R. Morton and H. Horsfield, *J. Chem. Phys.*, **35**, 1142 (1961)
30. I. Miyagawa and K. Itoh, *J. Chem. Phys.*, **36**, 2157 (1962)
31. A. D. McLachlan, *Mol. Phys.*, **1**, 233 (1958)
32. D. B. Chestnut, *J. Chem. Phys.*, **29**, 43 (1958)
33. C. A. Coulson and V. A. Crawford, *J. Chem. Soc.*, **1963**, 2052
34. E. W. Stone and A. H. Maki, *J. Chem. Phys.*, **37**, 1326 (1962)
35. G. K. Fraenkel, *Pure Appl. Chem.*, **4**, 143 (1962)
36. H. W. Shields, P. J. Hamrick, Jr. and W. Redwine, *J. Chem. Phys.*, **46**, 2510 (1967)
37. J. R. Morton, *Chem. Rev.*, **64**, 453 (1964)
38. N. F. Ramsey, *Nuclear Moments*, Wiley, New York, 1953, p. 79
39. E. Clementi, J. C. C. Roothaan, and M. Yoshimime, *Phys. Rev.*, **127**, 1618 (1962)
40. G. W. Chantry, A. Horsfield, J. R. Morton, J. R. Rowlands, and D. H. Whiffen, *Mol. Phys.*, **5**, 233 (1962)
41. J. R. Morton, J. R. Rowlands and D. H. Whiffen, *Nat. Phys. Lab. Circular*, No. BPR 13
42. R. E. Watson and A. J. Freeman, *Phys. Rev.*, **123**, 521 (1961)
43. T. Henriksen, *J. Chem. Phys.*, **37**, 2189 (1962)
44. D. H. Whiffen, 'Electron spin resonance of oriented organic radicals', in *Free Radicals in Biological Systems* (Eds. M. S. Blois and others), Academic Press, New York, 1961, Chap. 17, pp. 227–238
45. Y. Kurita and W. Gordy, *J. Chem. Phys.*, **34**, 282 (1961)
46. M. H. L. Pryce, *Proc. Phys. Soc. (London)*, **A63**, 25 (1950)
47. K. D. Bowers, R. A. Kamper and D. R. B. Knight, *J. Sci. Inst.*, **34**, 49 (1957)
48. J. M. Hirshon and G. K. Fraenkel, *Rev. Sci. Inst.*, **26**, 34 (1955)
49. D. T. Teaney, M. P. Klein and A. M. Portis, *Rev. Sci. Inst.*, **32**, 721 (1961)
50. G. Fehr, *Bell System, Tech. J.*, **36**, 449 (1957)
51. G. Fehr, *Phys. Rev.*, **103**, 500 (1956)
52. H. C. Box, E. E. Budzinski and W. Porter, *J. Chem. Phys.*, **55**, 315 (1971)
53. M. A. Collins and D. H. Whiffen, *Mol. Phys.*, **10**, 317 (1966)
54. H. C. Box, H. G. Freund and K. T. Lilga, *J. Chem. Phys.*, **46**, 2130 (1967)
55. J. W. Wells and H. C. Box, *J. Chem. Phys.*, **48**, 2542 (1968)
56. R. W. Fessenden and R. H. Schuler, *J. Chem. Phys.*, **33**, 935 (1960)
57. R. W. Fessenden and R. H. Schuler, *J. Chem. Phys.*, **39**, 2147 (1963)
58. C. K. Jen, S. N. Foner, E. L. Cockran and V. A. Bowers, *Phys. Rev.*, **112**, 1169 (1958)
59. T. Cole and H. C. Heller, *J. Chem. Phys.*, **42**, 1668 (1965)
60. W. Snipes and J. Schmidt, *Rad. Res.*, **29**, 194 (1966)
61. J. Herak and W. Gordy, *Proc. Nat. Acad. Sci. U.S.*, **54**, 1287 (1965)
62. R. Drew and W. Gordy, *Rad. Res.*, **18**, 552 (1963)
63. J. A. Weil and J. H. Anderson, *J. Chem. Phys.*, **28**, 864 (1958)
64. F. K. Kneubühl, *J. Chem. Phys.*, **33**, 1074 (1960)
65. J. Combrisson and J. Uebersfeld, *Compt. rend. Acad. Sci. (Paris)*, **238**, 1397 (1954)
66. W. Gordy, W. B. Ard and H. Shields, *Proc. Nat. Acad. Sci. U.S.*, **41**, 983 (1955)
67. H. Shields and W. Gordy, *J. Phys. Chem.*, **62**, 789 (1958)
68. F. Patten and W. Gordy, *Proc. Nat. Acad. Sci. U.S.*, **46**, 1137 (1960)
69. W. Gordy and H. Shields, *Proceedings of the Royal Academy of Belgium*, **33**, 191 (1961)
70. K. G. Zimmer and A. Müller, in *Current Topics in Radiation Research*, Eds. M. Ebert and A. Howard, Vol. I, North-Holland Publishing Co., Amsterdam, 1961, pp. 1–47

71. R. C. Drew and W. Gordy, *Rad. Res.*, **18**, 552 (1963)
72. J. R. Morton, *J. Am. Chem. Soc.*, **86**, 2325 (1964)
73. Ya. S. Lebedev and G. A. Almanov, *Biofizika*, **12**, 338 (1967)
74. K. Akaska, *J. Chem. Phys.*, **43**, 1182 (1965)
75. J. R. Rowlands, *J. Chem. Soc.*, **1961**, 4262
76. T. S. Jaseja and R. S. Anderson, *J. Chem. Phys.*, **36**, 2727 (1962)
77. H. Shields, P. Hamrick and D. DeLaigle, *J. Chem. Phys.*, **46**, 3649 (1967)
78. H. C. Box, H. G. Freund, and E. E. Budzinski, *J. Chem. Phys.*, **46**, 4470 (1967)
79. J. H. Hadley, *Bull. Am. Phys. Soc.*, **13**, 619 (1968)
80. W. C. Lin, C. A. McDowell and J. R. Rowlands, *J. Chem. Phys.*, **35**, 757 (1961)
81. W. Snipes and J. Schmidt, *Rad. Res.*, **29**, 194 (1966)
82. F. Patten and W. Gordy, *Rad. Res.*, **14**, 573 (1961)
83. Yu. D. Tsvetkov, R. J. Cook, J. R. Rowlands, and D. H. Whiffen, *Trans. Faraday Soc.*, **59**, 2213 (1963)
84. R. E. Drews and J. R. Rowlands, *J. Chem. Soc.*, **1966**, 296
85. K. Akaska, S. Ohnishi, T. Suita and I. Nitta, *J. Chem. Phys.*, **40**, 3110 (1964)
86. H. C. Box and H. G. Freund, *J. Chem. Phys.*, **44**, 2345 (1966)
87. H. C. Box, G. H. Freund, and E. E. Budzinski, *J. Chem. Soc.*, **88**, 658 (1966)
88. J. Sinclair, *J. Chem. Phys.*, **55**, 245 (1971)
89. M. Hanna, R. Schwartz, and B. Bales, *J. Chem. Phys.*, **51**, 1974 (1969)
90. J. Hadley, *Bull. Am. Phys. Soc.*, **13**, 1695 (1968)
91. Yu. A. Kruglyak, M. K. Pulatova, Ye. V. Mozdor, Ye. N. Sud'bina, V. G. Pasoyan, L. P. Kayushin, *Biofizika*, **13**, No. 3, 401 (1968)
92. V. G. Pasoyan, M. K. Pulatova and L. P. Kayushin, *Biofizika*, **15**, No. 1, 12 (1970)
93. V. G. Pasoyan, L. P. Kayushin and M. K. Pulatova, *Biofizika*, **13**, No. 4, 600 (1968)
94. G. McCormick and W. Gordy, *J. Phys. Chem.*, **62**, 783 (1958)
95. W. C. Lin and C. H. McDowell, *Mol. Phys.*, **4**, 333 (1961)
96. I. Miyagawa, Y. Kurita and W. Gordy, *J. Chem. Phys.*, **33**, 1599 (1960)
97. W. Gordy and H. Shields, *Proc. Nat. Acad. Sci. U.S.*, **46**, 1124 (1960)
98. W. Gordy and H. Shields, *Rad. Res.*, **9**, 611 (1958)
99. T. Henriksen, T. Sanner and A. Phil., *Rad. Res.*, **32**, 744 (1967)
100. W. H. Fredrich, Jr., P. Riesz and H. Kon, *Rad. Res.*, **32**, 744 (1967)
101. R. A. Patten and W. Gordy, *Rad. Res.*, **22**, 29 (1964)
102. G. F. Liming and W. Gordy, *Proc. Nat. Acad. Sci. U.S.*, **60**, 794 (1968)
103. W. Gordy and I. Miyagawa, *Rad. Res.*, **12**, 211 (1960)
104. T. Henriksen, T. Sanner and A. Phil, *Rad. Res.*, **18**, 163 (1963)
105. M. G. Ormerod and P. Alexander, *Rad. Res.*, **18**, 495 (1963)
106. A. Norman and W. Ginoza, *Rad. Res.*, **9**, 77 (1958)
107. H. Pohlit, B. Rajewsky and A. Redhardt, 'Electron resonance investigation of X-irradiated fibro-proteins', in *Free Radicals in Biological Systems* (Eds. M. S. Blois and others), Academic Press, New York, 1961, Chap. 29, pp. 367–380
108. W. Köhnlein and A. Müller, *Zsch. Naturforschg.*, **15b**, 138 (1960)
109. K. G. Zimmer, W. Köhnlein, G. Hotz and A. Müller, *Strahlenther*, **120**, 161 (1963)
110. J. W. Hunt and J. F. Williams, *Rad. Res.*, **23** (1964)
111. J. W. Hunt, J. E. Till and J. F. Williams, *Rad. Res.*, **17**, 703 (1962)
112. T. Henriksen, *Rad. Res.*, Suppl. **7**, 87 (1967); *Rad. Res.*, **27**, 694 (1966); *Rad. Res.*, **27**, 676 (1966)
113. H. Shields and W. Gordy, *Bull. Am. Phys. Soc.*, **1**, 267 (1956)
114. H. Shields and W. Gordy, *Proc. Nat. Acad. Sci. U.S.*, **45**, 269 (1959)
115. J. W. Boag and A. Müller, *Nature*, **183**, 831 (1959)
116. P. G. Shen, L. A. Blyumenfeld, A. E. Kalmanson and A. G. Pasynskii, *Biofizika*, **4**, 263 (1959)

117. P. Alexander, J. T. Lett and M. G. Ormerod, *Biochem. Biophys. Acta.*, **51**, 207 (1961)
118. R. Salovey, R. G. Shulman and W. M. Walsh, Jr., *J. Chem. Phys.*, **39**, 839 (1963)
119. W. Köhnlein, *Strahlentherapie*, **122**, 437 (1963)
120. A. Ehrenberg, L. Ehrenberg and G. Lofroth, *Nature*, **200**, 376 (1963)
121. W. Köhnlein and A. Müller, *Int. J. Rad. Biol.*, **8**, 141 (1964)
122. A. Müller and W. Köhnlein, *Int. J. Rad. Biol.*, **8**, 121 (1964)
123. A. Müller, *Int. J. Rad. Biol.*, **8**, 131 (1964)
124. M. Lacroix and C. Williams-Dorlet, *Compl rend. Acad. Sci.* (*Paris*), **259**, 1771 (1964)
125. A. Van De Vorst and F. Villée, *Comp. rend. Acad. Sci.* (*Paris*), **259**, 928 (1964)
126. P. S. Pershan, R. G. Shulman, B. J. Wyluda and J. Eisinger, *Physics*, **1**, 163 (1964)
127. B. B. Singh and A. Charlesby, *Int. J. Rad. Biol.*, **9**, 157 (1965)
128. M. G. Ormerod, *Int. J. Rad. Biol.*, **9**, 291 (1965)
129. K. V. Rajalakshmi, K. Venhateswarlu and A. Van De Vorst, *Comp. rend. Acad. Sci.* (*Paris*), **261**, 4879 (1965)
130. A. Phil and T. Sanner, *Rad. Res.*, **28**, 96 (1966)
131. B. Pruden, W. Snipes and W. Gordy, *Proc. Nat. Acad. Sci. U.S.*, **53**, 917 (1965)
132. C. Dorlet, A. Van De Vorst and A. J. Bertinchamps, *Nature*, **194**, 767 (1962)
133. A. Müller, 'Efficiency of radical production by X-rays in substances of biological importance', in *Biological Effects of Ionizing Radiation at the Molecular Level*, International Atomic Energy Agency, Wien, 1962, pp. 61–71
134. A. Ehrenberg, L. Ehrenberg and G. Lofroth, *Nature*, **200**, 367 (1963)
135. J. N. Herak and W. Gordy, *Proc. Nat. Acad. Sci. U.S.*, **55**, 1373 (1966)
136. J. N. Herak and W. Gordy, *Proc. Nat. Acad. Sci. U.S.*, **56**, 7 (1966)
137. H. C. Heller and T. Cole, *Proc. Nat. Acad. Sci. U.S.*, **54**, 1486 (1965)
138. D. E. Holmes, L. S. Myers, Jr. and R. B. Ingalls, *Nature*, **209**, 1017 (1966)
139. J. A. Simmons, *Nature*, **205**, 697 (1965)
140. W. Snipes and P. K. Horan, *Rad. Res.*, **30**, 307 (1967)
141. A. Müller, *Akad. Wiss. Lit.* (*Mainz*), *Abhandl. Math. Nat. Kl.*, **5**, 139 (1964)
142. A. Ehrenberg, A. Rupprecht and G. Ström, *Science*, **157**, 1317 (1967)
143. A. Müller, 'Radiation-produced electron spin resonance signals in nucleic acids', in *Electron Spin Resonance and the Effects of Radiation on Biological Systems* (Ed. W. Snipes), *National Academy of Science U.S.*, Washington, D.C., 1966, pp. 29–45
144. J. B. Cook, J. P. Elliot and S. J. Wyard, *Mol. Phys.*, **13**, 49 (1967)
145. W. Bernhard and W. Snipes, *Proc. Nat. Acad. Sci. U.S.*, **59**, 1038 (1968)
146. W. Snipes and B. Benson, *J. Chem. Phys.*, **48**, 4666 (1968)
147. B. Benson and W. Snipes, *Rad. Res.*, **31**, 542 (1967)
148. W. Berhard and W. Snipes, *J. Chem. Phys.*, **44**, 2817 (1966)
149. C. Alexander, Jr. and W. Gordy, *Proc. Nat. Acad. Sci. U.S.*, **58**, 1279 (1967)
150. J. J. Lichter and W. Gordy, *Proc. Nat. Acad. Sci. U.S.*, **60**, 450 (1968)

CHAPTER 11

Thermoluminescence in biological materials[*]

L. G. Augenstein† and J. O. Williams‡

Biophysics Department,
Michigan State University,
East Lansing, Michigan, U.S.A.

I. INTRODUCTION

Thermoluminescence (TL) is the term applied to the light emitted when crystalline preparations which were irradiated at low temperatures are subsequently warmed. It is commonly accepted that this is emission arising from the decay of metastable species produced in the sequence of events initiated by radiation—in most cases it is presumed that this reflects the recombination of a trapped electron and a hole. Although u.v. irradiation

* The preparation of this manuscript was supported in part by Contract AT(11-1)1155 between the U.S. Atomic Energy Commission and Michigan State University

† Deceased 1969. Chapter completed *ca.* July 1969

‡ Present Address: Edward Davies Chemical Laboratories, University College of Wales, Aberystwyth, Wales

can lead to TL, most studies have utilized excitation by radiations such as gamma rays, fast electrons or some of the heavier particles. By far the largest number of studies have been conducted on inorganic preparations, and it appears that in those systems a reasonable interpretation of the phenomenon has been achieved (e.g., reference 1). In the organic materials—in particular in biological molecules—the data are more scanty and only tentative interpretations have been advanced.

Studies of TL have both advantages and severe limitations. The most serious problem is that the yields are extremely small: a photon of TL is obtained from powders of amino acids and proteins irradiated at 77 °K for approximately each 10^3 to 10^6 electron volts of energy absorbed, whereas the yields for the immediate luminescence (IL) arising during the irradiation is 10^2–10^3 times as great.[2] It is not known whether this very small yield reflects an inefficient process because of extensive quenching of most of the appropriate metastable states, or whether TL is associated with only a small select group of metastable configurations.

Even so, TL studies can yield valuable information. They can, for example, give quite accurate estimates of the depth of the pertinent metastable states below the conduction band, and thus whether TL arises from a single type or heterogeneous group of metastable states. Further, when such studies are carried on in conjunction with action spectra for photo-bleaching, it is then possible to compare the value for the thermal release of metastable states with the energy required for excitation of the trapped electrons via the lowest lying optically-allowed transition for the metastable configuration. Again, if such studies are done with concomitant electron spin resonance measurements, it should be possible to identify which radicals decay via emission and which ones are thermally quenched. Obviously such information is essential for a complete understanding of the sequence of physical events which is initiated by radiation excitation.

Since so much more work has been done with inorganic materials and particularly with the alkali halides, it seems worthwhile to describe in detail the phenomena and the interpretations which have been reported for those systems. It then will be possible to indicate how TL from biological materials is comparable and how it differs from that in inorganic materials.

II. INORGANIC SYSTEMS

The theoretical model used to describe TL is based on Bloch's collective electron model of a crystalline insulator and has been developed by many workers—notably Randal and Wilkins for the phosphors[3] and Hill and Schwed[4] for alkali halides. In a 'perfect insulator' the allowed energies for electrons are arranged in bands of states separated by forbidden energy gaps according to the scheme shown in Figure 1. In the ideal situation these

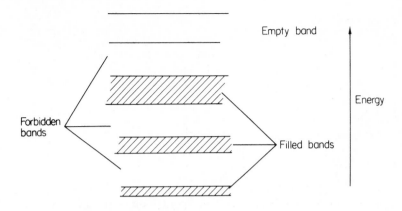

Figure 1. Energy band scheme for a 'perfect' insulator

bands extend throughout the crystal lattice so that electrons which have been excited to one of the allowed states can move through the crystal without further enhancement by an activation energy. In an insulator the allowed bands for electrons are either completely occupied (valence band) by electrons or completely empty (conduction band). Those electrons which occupy the highest filled band in the alkali halides are the valence electrons of the crystal atoms and are associated with the halide ions only. However, for a 'real' crystal the lattice is not in the perfect state that is required by the model of Figure 1. The regularity is disturbed or even destroyed by the presence of both thermodynamic and non-thermodynamic lattice defects—including traces of impurities. These 'trapping centres' introduce additional energy-levels for electrons, which may lie in the forbidden region between the highest filled band and the next unoccupied band. [It is important to note that because of the duality of electrons and holes we could also talk about energy levels for holes; however, for simplicity and convenience we shall consider only electrons.]

These electron-trapping centres can be classified further into (a) shallow traps—those where the captured electrons are essentially in thermal equilibrium with free electrons in the conduction band and (b) deep traps—those from which electrons have only a small probability of being thermally excited into the free state so long as the material is maintained at low temperature.

Figure 2 represents possible rearrangements in a real crystal during the sequence of events accounting for the phenomenon of TL in organic solids. In the normal state shown in (a) the conduction band and the electron trap levels [assumed to be discrete[5]] $T_1 T_2 T_3 \ldots$ are empty as are the luminescence centres $L_1 L_2 L_3 \ldots$ (which can also be thought of as hole-trapping

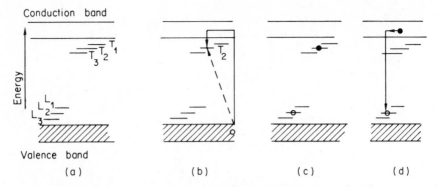

Figure 2. Possible arrangements in a real crystal during the sequence of events leading up to thermoluminescence in inorganic solids. (After reference 13)

levels). By contrast the valence band is filled. To describe the sequence of events we shall consider three general headings:

(1) the initial act of absorption
(2) the recombination process
(3) the kinetics of the thermoluminescence (TL) process.

A. *The Initial Act of Absorption*

Radiation displaces an electron from the valence band of the crystal to the 'metastable states' or traps most readily accessible from the conduction band. This may occur directly or indirectly via the conduction band and leaves a hole in the valence band as shown in Figure 2(b). In the latter case the electrons can move a considerable distance from the initial absorption site. The hole remaining in the valence band also may become trapped at at any one of the levels $L_1 \rightarrow L_3$ (Figure 2(c)).

An important electron trap in the alkali halides consists of a vacancy resulting from the absence of a negative (halide) ion from its normal position (see Figure 3): the term F centre is used to denote a single electron associated with a single negative ion vacancy. Under appropriate conditions a second electron can associate with an F centre producing a so-called F' centre; i.e., two electrons are associated with a single negative ion vacancy. These together with the analogous hole trapping centres and other centres including multiple ion vacancies are shown in Figure 3. (Each centre has associated with it, its own characteristic absorption spectrum.)

In addition to the action of ionizing radiation, F centres may also result from the absorption of light of a wave length where the individual quanta do not have sufficient energy to excite an electron from the valence band directly to the conduction band. However, a bound electron-hole pair or

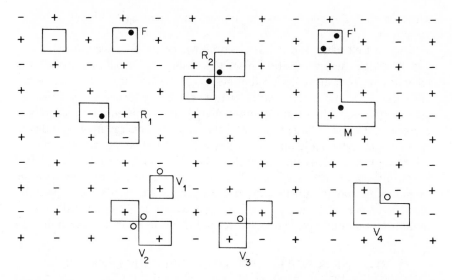

Figure 3. Illustration of various trapping sites in sodium chloride. (After reference 13)

'exciton', having energy levels within the forbidden gap, will be created. This energy packet may move through the crystal as an efficient carrier of excitation but not of electric current. Excitons may generate free electrons by colliding with each other or with a defect, the resulting hole being bound nearby. An important source of such defects are the discontinuities occurring at surfaces or physical breaks in crystals and powders. In those cases where defects occur in pairs or at least in close proximity, an exciton may disassociate if the hole becomes localized at one site and the excited electron tunnels to a nearby trapping centre.

In highly polarizable media the presence of a migrating electron may so distort the surrounding matrix that the electron becomes 'self trapped'.[6] This is particularly facilitated at temperatures so low that once the medium has oriented around the migrating electron relaxation is prohibited and so the distortion is maintained.[1]

B. The Recombination Process

To promote an electron into the conduction band, the metastable configuration of an electron bound at a trapping centre must absorb an amount of energy equal to the difference in energy between the trapped and one of the free states. This may be achieved by either thermal activation or by the absorption of a suitable photon of light. As discussed above the electron may travel through the crystal in the conduction band until it either recombines with a carrier of opposite sign at a luminescence centre (see Figure 2(d)) or is retrapped.

When an electron and a hole recombine, energy is released. Normally not all of the energy resulting from the electronic transitions included in recombination is given off as luminescence. For example, because of the presence of the hole, the nearby lattice will be distorted so that usually recombination will not result immediately in a minimum energy state. Thus, even though a large fraction of the transition energy may be radiated as light, an appreciable amount will appear as lattice vibrational energy: this energy is a likely cause of chemical reactions especially in systems involving weak bonds (see later). As a consequence of the coupling with the surrounding matrix, most of the potential emissive events are apparently quenched since the actual yield of energy released as TL is generally only a small fraction of the total energy absorbed from the incident radiation.[2]

III. THE KINETICS OF THERMOLUMINESCENCE

Most experimental studies of TL have utilized a procedure, due to Urbach.[7] in which the temperature of the sample is raised at a constant rate and the intensity of the emitted light is measured as a function of time: the resulting plot of intensity versus time (or temperature) is designated as a 'glow' curve. In the case of a material having a single trapping species, T_i with an associated energy level E_i below the conduction band, the glow curve has a single peak and can be represented by the luminescence equation (1)

$$I_T = -\frac{dn_T}{dt} = p_T n_T = n_T S \exp(-E_i/kT) \tag{1}$$

where S is a frequency factor, I_T is the intensity of the luminescence emission, n_T is the number of electrons remaining in traps and p_T is the probability of an electron being thermally released from a trap and recombining with a hole after the material has been heated to temperature T. The luminescence from a material such as this will thus decay at an exponential rate at constant temperature and the probability p_T will vary with temperature according to the classical expression

$$p_T = S \exp(-E_i/kT) \tag{2}$$

For a fixed rate of warming β degrees/sec one may write $dT = \beta\, dt$ and on integration, equation (1) gives

$$I_T = n_0 S \exp(-E_i/kT) \exp\left[-\int_0^T \frac{S}{\beta} \exp(E_i/kT)\, dT \right] \tag{3a}$$

where n_0 is the number of electrons in traps at the beginning of the decay process. For such a simple case Garlick and Gibson summarized the essential features of the TL process as follows:[8]

(a) For given fixed values of n_0, S and β the temperature at which maximum emission occurs is approximately proportional to the trap depth, E_i.

(b) For fixed values of n_0 and E_i the emission maximum moves to higher temperature as the rate of warming β increases or as the frequency factor S decreases.

(c) The area under the curve is proportional to n_0, although the shape of the curve for given S, E_i and β values is independent of n_0.

(d) At low temperatures the rising part of the curve follows the relation

$$I_T = n_0 S \exp(-E_i/kT). \tag{3b}$$

Perhaps the most elaborate study of TL in one particular inorganic material is that of Hill and Schwed[4] on NaCl, which was later extended by the work of Bonfiglioli *et al.*[9] Five different 'glow' peaks were measured and it was found that all shared the same activation energy but exhibited frequency factors of considerably different magnitudes. It was suggested that sodium chloride crystals contain only a single trapping energy level and that electrons, thermally activated from there to the conduction band, drop back towards the valence band and are captured by a number of different luminescence centres each having its characteristic depth E_i and recombination probability p_i. Thus with a linear rate of heating β the now second order kinetics of light emission from a given luminescence level L_i has been described as

$$I_i = \frac{-dN_i}{dT} = \frac{p_i N_i n_T}{\beta} \exp\left(\frac{-E_i}{kT}\right) \tag{4}$$

Here I_i is the luminescence emitted by electrons returning only to the L_i centres, N_i is the number of empty L_i centres at temperature T (actually after elapsed heating time t), n_T is the number of electrons in the conduction band at temperature T and p_i is the probability of an electron combining with an empty L_i centre after being released from its trap. According to this model the temperature differences between the various glow peaks are seen to be due to the different recombination probabilities or 'capture cross sections' of the various types of luminescence centres.

IV. ORGANIC SYSTEMS: BIOMOLECULES IN PARTICULAR

While the mechanism of TL in the alkali halides and other inorganic materials seems to be explained satisfactorily,[1,10,11] the study of this process in organic materials has been undertaken only recently. Accordingly, there are as yet no clear cut answers to questions such as: What is the nature of the metastable configurations which persist at 77 °K? What is the molecular species from which emission occurs? And so forth.

Following the report of Arnold and Sherwood on the visible-light-induced TL from dried chloroplasts,[12] Augenstein et al.[2,13,18] made an extensive survey of γ-ray induced TL in proteins and amino acids. In marked contrast to the situation in inorganic systems where TL is essentially a structure-sensitive property, this work showed that TL in the amino acids is determined by molecular structure with little dependence on the degree of crystallinity of the samples even though impurities can change the yields appreciably. However, they observed relatively small changes in TL until deliberately added impurities reached approximately 1 per cent.

On the basis of the analogy between the crystalline and amorphous structures of proteins and polymers, Charlesby's work is in complete contrast.[19] He demonstrated that in the case of polyethylene the glow intensity depends on the crystallinity of the samples and that chemical impurities have no effect. To explain his results Charlesby proposed a model in which the trapped electrons eventually recombine with different types of luminescence centres. Deroulede et al.[20] and Shiomi[21] also attempted to overcome the problem of the intrinsic impurity of organic materials by doping cyclohexane and anthracene with specific impurities for their TL studies. Both groups concluded that the process in these materials is related to the crystalline structure of the materials and they propose that the pertinent traps depend upon lattice imperfections arising in the vicinity of impurities. The work of Deroulede et al. has been extended by Magat[22] and Bullot et al.[23] who utilized TL studies as a tool to investigate the structural properties of molecular solids.

The spectral distribution of TL should be an integral part of any study. However, such measurements have been reported by only a few workers[15,19,21] so that the nature of the actual emission transition following detrapping has only been tentatively identified in trypsin and the amino acids. The interpretation in proteins is complicated by the fact that energy transfer may occur from the recombination site to another part of the molecule from which emission actually occurs. Although TL studies can give valuable information in certain cases its power as a tool in the investigation of radiation effects and energy storage and transfer in biological materials is enhanced by simultaneous consideration of the glow or emission spectra, the e.s.r. spectrum as a function of increasing temperature, the thermoconductivity glow curve and the luminescence decay at constant temperature. Let us briefly consider in turn the additional information which these can provide.

A. Thermoconductivity Glow Curve

Such measurements are the electrical analogue to the thermoluminescence glow curves: it consists of measuring changes in the electrical conductivity as the temperature is increased.[25-27] Since the conductivity is determined by the balance between detrapping and recombination, the peaks in the conductivity curve usually occur at a somewhat lower temperature than the corresponding glow peaks since in TL one of these processes is usually rate limiting.

It is natural to seek an explanation for this phenomenon in terms of a trap-emptying process involving electron migration through the conduction band. This immediately raises the question of the application of such a model to molecular solids, however, because there is appreciable uncertainty about the existence and possible nature of conduction bands in proteins, amino acids, and other organic solids.[28,29] First, Evans and Gergeley[30] using a Hückel LCAO molecular orbital scheme, and later the Pullmans,[31] Mason,[32] and Suard–Sender[33] have proposed that extended energy bands should arise as the result of π-electron delocalization via the hydrogen bond framework in proteins. The discrepancy between the calculated theoretical values of the energy gap and the values determined experimentally from semiconduction activation energies[34,35] has recently been ascribed by Rosenberg and Postow[36] to a polarization stabilization energy. Direct optical transitions to such energy bands, however, are not observed. A similar situation is found in most organic compounds where the threshold for photoconduction (first singlet → singlet transition) is about twice the activation energy for semiconduction (sometimes correlated with the difference in energy between the ground state and the lowest triplet state). Recent work on anthracene[37] where it is still thought that a 'band model' applies,[38] has shown that with equal heating rates maxima in the TL glow curves in the temperature range $4\,°K \rightarrow 77\,°K$ occur at the same temperatures as those in the thermoconductivity 'glow' curve. It is therefore necessary to consider whether the TL process in proteins and amino acids involves either inter- or intramolecular charge migration. This possibility of course implies that trapping and luminescence centres may reside on different parts of a macromolecule.

B. Emission Spectrum of the Glow Curve

Following the decay of the metastable species which persist at the low temperature, one of the crucial steps in the TL process is the transition which produces the emitted photon. In the amino acids the predominance of the TL is at the same wavelength as the u.v.-induced phosphorescence from similar powders. Tryptophan has a very small fraction of its TL at the wavelength typical of fluorescence and the TL from trypsin is peaked at the same wavelength as the TL from tyrosine powders. All of these compounds exhibit appreciable TL at wavelengths longer than that normally observed with u.v. excitation. Further, whereas the glow peaks themselves give only the total light emitted by the sample a study of the emission spectrum as a function of temperature provides information about the homogeneity of the luminescence centres. In the amino acids and trypsin the 'red' components increase with increasing temperature indicating that either the critical luminescence centres are heterogeneous or else an increasing fraction of the recombination energy is partitioned into vibrational modes. This general technique also has been utilized by Augenstein *et al.*, to show that quenching is a function of the macromolecular organization of proteins.[16] By observing how the spectral emission at a given temperature may be changed by impurities or by changes in the molecular environment (e.g., crystallinity) one should be able to study what types of structures can serve as efficient luminescence centres and determine which ones may be crucial in the TL process.

C. *E.s.r. Spectrum and Luminescence Decay*

In addition to electron trapping and recombination, other mechanisms based on molecular excitations and chemical changes warrant attention in molecular solids. A large number of experiments have shown that optical excitation to even the lowest-lying excited states of a crystal can cause fluorescence, phosphorescence, delayed fluorescence, the production of free radicals, as well as TL. With photons or particles of sufficient energy to excite to higher-lying levels, the probability for free radical production, TL and chemical disassociation should be even greater. Thus, appropriate collateral studies should provide insight into which of the various entities and mechanisms are critical in TL. In particular, it should be possible to determine which free radicals may be critical intermediates in TL by (a) joint measurements of e.s.r. spectra, TL and thermoconductivity as a function of temperature; (b) comparing quantitatively the e.s.r. yield with the TL quantum yield; and (c) comparing second-order parameters measured by e.s.r. for radical–radical recombinations or power-law constants reflecting electron–hole recombination at a constant T with the TL decay under the same conditions.

Let us now turn to the experimental methods involved in obtaining answers to questions such as:

What are the conditions required for trapping?

What is the nature and what are the properties of the trapping site?

Are the emitting species different than the trapping species or are they located on the same molecular or crystalline subunit?

To what extent is TL a molecular property?

What effects do such parameters as the nature and amount of excitation, the physical state of the material, the nature of ambient gases and temperature of irradiation have on the entire process?

V. METHODS AND GENERAL CRITERIA FOR STUDYING THERMOLUMINESCENCE

A. Factors involved in the Production of Glow Curves

(a) The source of incident radiation is usually γ-rays, X-rays or u.v. irradiation. For γ-rays extensive use has been made of commercial Co-60 sources having dose rates in the range of 0·5 to 2×10^6 rad/hr, while X-ray beams of effective energy *ca.* 60 keV and delivering dose rates of about 1×10^3 rad/hr are used commonly. Electron beams from a Van de Graff accelerator having energies >1 MeV can also be used. A variety of u.v. sources have been used for excitation.

(b) In most cases the biochemicals are obtained in the purest available form from the supplier (twice crystallized and salt-free) as a polycrystalline powder. Even though the attainable purities are much less than those of the inorganic materials tested, for most biochemical preparations of fairly uniform crystallite size, the reproducibility of TL measurements can be within ± 10 per cent.

The form of the sample must be such that ideally the radiation can penetrate it uniformly (this is a particular problem with u.v.) and that temperature is uniform throughout and any surrounding gas is free to diffuse freely within the material. The experimental evidence so far obtained suggests that TL is a molecular process in biological materials:[13,39] the strongest evidence in this regard comes from the observation that a number of protein powders exhibit no TL (e.g., haemoglobin, ribonuclease). [Unfortunately, most biological molecules cannot be obtained as large single crystals and those few that can exhibit little or no TL.] Accordingly, quantitative studies of the effect of crystal size are difficult. However, extrapolating from a comparison of the results in NaCl and a 'biological-type' molecule such as anthracene indicates that similar glow curves are obtained irrespective of the physical state of biological materials. In NaCl the TL is almost certainly determined by crystal structure imperfections and the TL yields from powders have always been found to be 20 to 500 times greater than that from single crystals. This presumably reflects much greater numbers of imperfections (in terms of dislocations and grain boundary content) in fine powder. By contrast Shiomi[21] found that samples of anthracene which had suffered a high pressure treatment had almost identical glow curves to the single crystal samples.

(c) Precise control and reproducibility of both the temperature during irradiation and the rate of heating later is crucial, of course. The simplest and most reliable way of keeping samples at a constant temperature is by using reservoirs of liquid gases to maintain all of the container at one temperature. In doing this the factors which control the actual temperature of the entire sample are heat losses within the experimental assembly and temperature gradients across the sample. To minimize the former, careful construction and use of low temperature cryostats may suffice and temperature gradients can be minimized by the introduction of a partial pressure of an inert gas that does not condense at the lower temperature (see below). It is usual, therefore, to contain the sample either in or on a block of a heat conducting material (Al or Cu) whose temperature is controlled by immersion in a cryostat and/or electrical heating.

Depending upon what information is desired, heating rates may vary from 1 to 100 °K/minute: although it is not essential, the use of linear heating rates greatly facilitates the resulting calculations. Major difficulties in temperature determinations can arise in two ways. First, the temperature of the sample may be estimated poorly since the temperature usually is measured by a conventional thermocouple either embedded in the underlying metal block or else pressed against the crystallites. Second, under certain conditions the temperature may be quite different in various parts of the samples. For example, Bube's study[40] on the luminescence from thermal insulators in high vacuum demonstrates that the true temperature of the material is mainly determined by radiation of energy to and from the insulator. In this respect it will vary with T^4 and as a result if even a small percentage of the environment surrounding a particle of insulating material is at a relatively high temperature, the cooling effect of the remaining environment at low temperature may be offset. Indeed, Bube's data shows that in certain cases the sample temperature in high vacuum (10^{-6} torr), as recorded by a thermocouple in contact with the material, may be as much as 150 °K higher than that of the container. A similar result was obtained by Carter *et al.*,[17] when studying tyrosine although the large temperature difference could be eliminated in 30 sec by the introduction of a small amount of an inert gas: enough to give only a few mm pressure was sufficient.

Specifically, tyrosine has a glow curve with a maximum intensity at about 115 °K when irradiated and warmed with air present; further, there is little shift in the peak when the pressure of helium is varied. The intensity of TL, however, varied considerably with helium pressure: the greater the pressure the greater the intensity of emitted light (see Figure 4). Furthermore, if the sample was irradiated in a helium atmosphere and then the helium pumped off while the bottom portion of the sample container was held at 77 °K, approximately 95 per cent of the normal TL was emitted during the outgassing. The remaining 5 per cent of the light was emitted when the sample was finally warmed to room temperature. In another experiment (see Figure 5) the sample was maintained in high vacuum during irradiation and the

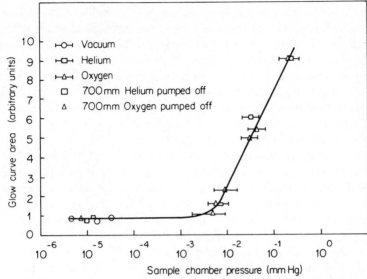

Figure 4. Tyrosine glow curve area *vs.* gas pressure. (After reference 17)

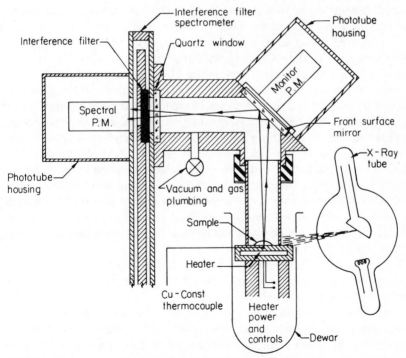

Figure 5. Schematic illustration of the apparatus for luminescence measurements. (After reference 2)

helium gas was introduced prior to warming. The resulting glow curve was identical to that obtained when a sample was irradiated and warmed *in vacuo* showing that the gas must be present during both the irradiation and the warmup to have maximum effect. Thermistors embedded in unusually thick layers of powder showed that when the gas was pumped out the sample temperature increased 25–30 °K with the resulting release of TL—presumably under these conditions convection cooling is abolished so that the top layers of the sample come to black body equilibrium with the top part of the container which was at room temperature.

(d) In addition to maintaining temperature control and homogeneity, within a sample, a gas such as oxygen could have quite a different effect. Specifically, the formation of molecular donor–acceptor type complexes[41] between biological molecules and gases could have marked effects on the resulting glow curves. Since such complexes should form most readily at surfaces, any effect of adsorbed gases should be most pronounced in finely ground powders. Thus, in any study not conducted under inert conditions the ability of gaseous molecules, particularly those of oxygen and water, to act as either trapping and/or recombination centres should be assessed. Unfortunately, many of the earlier studies were conducted in the presence of normal laboratory air—or at least without extensive outgassing at elevated temperatures—so that their absolute validity must be questioned. As might be anticipated, gases adsorbed on NaCl crystals should have a minimal effect compared to the internal imperfection content and He, N_2 and O_2 all produced no demonstrable effects on the TL from NaCl powders.[17]

Any gas which might condense to form appreciable solid aggregates at the low irradiation temperatures could have marked effects on 'glow' curves. In particular, we know that water will be an integral part of any biological system because of the strong and extensive hydrogen bonding potential of amino acids and nucleotides. Even so, it is difficult to specify how much H_2O is merely 'adsorbed' and how much is an integral and essential part of crystalline forms of biological preparations since some water will be included in crystallites simply as a result of the crystal growth process.[42] In fact, the stability of some protein crystals requires the presence of a certain amount of incorporated water. Thus, care must be exercised to minimize TL arising simply from regions of ice.

The position of the TL peak of ice depends greatly on the impurity concentration and the irradiation procedure,[43] but for distilled water of normal purity it occurs around 163 °K.[13]

Although observations of the glow curves for numerous amino acids have demonstrated that most include a maximum in this temperature region the intensity is often too great to interpret the peaks entirely in terms of the water impurity (see Table 1). The study of the TL glow curve and its dependence on dose is not sufficient in these cases because as previously discussed

Table 1. Data on thermoluminescence peaks observed in the 160 °K region

Material	Position (°K)	Intensity	E_A (eV)
H_2O	163	0·003	0·314
L-Alanine	165	0·0004	—
L-Arginine–HCl	167	0·0063	0·224
L-Aspartic acid	166	0·0015	0·213
D-Aspartic	167	0·02	0·448
L-Asparagine–H_2O	167	0·0004	—
L-Cysteine-HCl–H_2O	165	0·003	0·267
L-Cysteine	163	0·0003	—
L-Cysteic acid	168	0·002	0·182
L-Glutamic acid	165	0·003	—
L-Glutamine	167	0·005	0·265
Glycine	167	0·0007	0·225
L-Glycine	165	0·003	0·045
Proline	166	0·005	—
L-Proline	167	0·0006	0·218
L-Valine	164	0·0003	—
DL-Valine	163	0·0007	0·224

it relates only to the trapping process. To date little work—particularly joint TL and e.s.r. studies—has been done on biological materials in an attempt to observe these possible effects, even though it is realized that such entities as peroxide radicals and charged O_2^- molecules—a consequence of the γ-irradiation of O_2 molecules—can act as efficient capturing centres.

(e) The material of which the sample containers are made can affect the TL tremendously. Recent studies with boron carbide, graphite and brass containers have shown that the large differences in TL intensity following X-irradiation must reflect differences in the effective dose received by the thin layers of powders. In these studies the TL intensity—following identical X-ray exposures—in either boron carbide or graphite containers was about 1/7 that measured in the brass containers. This same reduction was observed in the (IL) luminescence emitted during the irradiation and is essentially the same ratio as is found between the number of electrons ejected from graphite and copper respectively.[44]

These results, of course, imply that the electrons ejected from the container walls are producing far greater effects than those electrons ejected directly within the powders by the X-rays. This was readily confirmed in an experiment utilizing thin graphite inserts within a cylindrical brass holder. When the X-rays were incident from the side, the intensity of the IL was reduced to a $\frac{1}{3}$ or $\frac{1}{2}$ by an insert covering the sides but not the bottom on which the powders rested and the intensity was further reduced to about $\frac{1}{7}$ when an insert was added in the bottom. Yet the inserts were so thin that they should not have attenuated the incident X-ray beam by more than 5 per cent.

This general conclusion was buttressed further by studies on a highly-compressed pellet of tyrosine and its fragments as it was broken down stepwise and finally pulverized to a powder. At all stages the IL was directly related to the surface area, but not the volume. Thus, the main effect observed arises not only from the electrons ejected from the brass walls, but those having low penetrance. Presumably, this means that the IL and TL is comprised of photons emitted as a consequence of events occurring at or near the surface of the powder crystallites.

Aluminium containers were found to be poor for studying the TL from amino acids. In addition to the characteristic peaks appearing in the region 100 °K to 140 °K, two additional peaks occur in the regions 190 °K to 210 °K and 95 °K to 120 °K, both having similar intensities (compare Table 1 above and Figure 6 below). However, when blank runs were made on the container itself, a background emission from the aluminium was often observed, as well as poorly defined peaks in the two above-mentioned ranges. The possibility that an oxide coating on the aluminium container was the cause of the effect was discounted on the basis of the markedly different activation energies found for powdered Al_2O_3 and amino acid samples, even though two of the peak positions for each were somewhat similar (98 °K and 208 °K). It was also found that the background emission could be markedly decreased by either coating the interior of the aluminium container with aquadag (a colloidal suspension of graphite in an organic solvent).

Glass sample-containers suffer the disadvantage that appreciable temperature gradients can develop (due to the insulating nature of this material) causing appreciable errors in temperature measurement.

(e) After irradiation at a suitably low temperature the light emission from the sample may be measured as a function of increasing temperature to yield the glow curve. Normally this light is measured using a commercial photomultiplier cooled to liquid N_2 temperature and the amplified output is recorded on a chart recorder. In most cases the light intensity measured is that of the whole TL spectrum as weighed by the spectral sensitivity of the photomultiplier.

When the optical emission spectrum is required a similar experimental arrangement can be used with the inclusion of either a monochromator or a series of band pass filters placed between the sample and the photomultiplier. Unfortunately, few monochromators are optically 'fast' enough to measure the low levels of TL intensity normally observed from the irradiated biochemicals: some monochromators designed for astronomical observations can be adapted for this purpose.[24] The first measurements of the spectral quality of TL from amino acids utilized a 'fast monochromator' made from 50 interference filters in a wheel which could be rotated at a fairly rapid rate so as to traverse the spectral range 200–700 nm in less than 52 minutes, during the warm up. An important ancillary result of this work by Nelson et al.[2] was to measure the importance of self-absorption and scattering of the TL emission in the crystallites.

VI. THE ENERGETICS AND THE KINETICS OF THE LUMINESCENCE PROCESSES

A. Brief Evaluation of the Methods used to Calculate E_i (the Activation Energy)

Even though the validity of the band model to biochemicals such as the proteins and amino acids has not been conclusively proved, the treatment of TL glow curves in terms of electron trapping and subsequent recombination is generally believed to be valid.[29,39] In particular, treatment of the glow curves to yield activation energies for the trapping process has followed the well established schemes for inorganic materials.[4,5,9–11] The early treatment of Randal and Wilkins,[3] although assuming a constant value for a frequency factor, gave $25\,kT^*$ as an approximate value of the activation energy— where T^* is the temperature of the maximum in the glow curve and k is Boltzmann's constant. A later approximate method of determining the E_i related to a given peak was given by Grossweiner[43] in terms of the expression

$$E_i = 1.51kT^* \frac{T_1}{(T^* - T_1)} \tag{6a}$$

where T_1 is the temperature on the low temperature side of a particular glow peak at which the peak intensity attains half its maximum value. Halperin and Braner[45] gave E_i as

$$E_i = \left(\frac{q}{T_2 - T^*}\right)kT^{*2} \tag{6b}$$

where T_2 is the temperature at which the peak attains half intensity on the high temperature side and q is a factor determined from the shape of the glow peak. Usually for a monomolecular process $q \leqslant 1$ and for a bimolecular one $1 < q \leqslant 2$.

Several 'heating-rate' methods have been developed for determining activation energies which are based on the shift of the TL glow maxima with different heating rates. Three ways of plotting the results have been proposed:

(a) $\log I_M$ versus $1/T^*$ by Haering and Adams.[46]
(b) $\log T^{*2}/\beta$ versus $1/T^*$ by Böer et al.[25] and Bube.[27]
(c) $\log \beta^{-1}$ versus $1/T^*$ again by Böer et al.[25] and Booth.[47]

The other general method used widely and considered by many workers as the most reliable involves differential heating rather than running a complete glow curve all at once. This general approach initiated by Garlick and Gibson[8] is based on an analysis of the exponential rise of the glow curve and assumes that at the beginning of the glow peak, far below the maximum temperature, the density of free electrons depends upon an activation energy equal to the thermal trap depth. A variant of this method, generally known as the decayed thermoluminescence method, involves a preliminary heating step to empty partially the trap of interest as well as to empty completely all shallower traps. Recently some workers have utilized a series of sequential heating cycles— where the temperature is raised sufficiently in each cycle to give an intensity of 5–10 per cent of that expected at the peak of a normal glow curve. This usually facilitates the accurate measurement of T^*, and in particular, Boltzmann plots of the data from these successive heating cycles can indicate how homogeneous or heterogeneous the pertinent traps may be: if there is a continual increase in the slope of the successive plots of $\ln I_T$ vs. $1/T$ or if individual plots are not linear, then undoubtedly the TL involves a heterogeneous population of traps, recombination centres or both.

In a recent critical investigation to establish experimentally which of the proposed methods is the most reliable Dussel and Bube[48] conclude that for the case of a single trap in the presence of deeper traps, where a single type of recombination centre accounts for the great variety of recombination and trapping possibilities, the decay method gives the most reliable determination of trap depth.

B. Kinetics of Thermoluminescence

In addition to obtaining values for the activation energy, a knowledge of the kinetic order of the TL process can be helpful in evaluating which mechanism may account for the complete process.[49] This is of prime importance in biochemicals where as stated previously, the band picture has questionable relevance. The kinetics are generally expressable in the form

$$I_T = \alpha r^\theta \exp(-E_i/kT)$$

where I_T is the light intensity measured at any temperature T, α is a proportionality constant, r the concentration of filled traps and θ the order of the reaction. For a first order isothermal reaction, which may be interpreted in terms of a TL process determined only by the number of species being released from traps per unit time, $\log I$ plotted versus time (for a constant heating rate) should yield a straight line. In a similar manner a second order reaction—indicative of either a possible radical–radical recombination process or of a process with a low probability for recombination compared to retrapping—should also give a straight line plot, but with $I^{-1/2}$ plotted versus time. Recently McCubbin[50] pointed out that the usual analysis of the decay process is incapable of distinguishing between the very different physical situations of carrier liberation from a single trapping level followed by recombination with kinetic order > 1, and liberation from a distribution of traps involving first order kinetics. It has often been considered that solids such as anthracene[51] and also partly amorphous materials such as polyethylene will have a heterogeneous group of traps,[52] so that this possibility must be considered with biological materials.

VII. TREATMENT OF EXPERIMENTAL RESULTS

Having considered the fundamentals of the TL process, let us now look briefly at some experimental results. Substantial coverage of the data obtained on luminescence studies in biochemicals has already been given by Lehman *et al.*[39] and Augenstein *et al.*[13,15] Rather than attempting to treat a mass of data, we will try to be illustrative more than exhaustive and concentrate on data available for the enzyme trypsin.

The glow curves for individual amino acids can be separated into two broad categories. Those for tyrosine, phenylalanine and tryptophan—all of which contain ring structures—have predominant peaks in the temperature range $100\,°K$ to $140\,°K$ with maximum intensities ranging between 0.3 and 10 on the relative scale of units used for measurement and activation energies from 0.1 to $0.45\,eV$. Other amino acids, of which cystine and glycine are typical, have a major peak at around $160\,°K$, varying in intensity from 0.0002 to 0.02 units and values of E between 0.2 and $0.4\,eV/molecule$ accompanied by much smaller peaks in the range $105\,°K–135\,°K$ and $200\,°K–240\,°K$, and with smaller values of E. Under the same conditions, the glow curve for trypsin has only one major peak located in the region between $115\,°K$ and $140\,°K$. Typical curves for the 'ring amino acids' plus trypsin are shown in Figure 6 and Table 2 summarizes the glow curve parameters for these materials. For trypsin the peak in the glow curve is broad and the question arises concerning the temperature range in which the initial-rise method (see equation (3b)) should give a meaningful value of E. With the methods used to obtain the results in Table 2, the estimated values of E increase by about 0.0015 to $0.003\,eV$ per $°K$ increase in the mid-temperature point of the initial rise curves. This continual increase in E implies a heterogeneity in either the traps or the recombination centres.

Because of the inherent shape of the trypsin glow peak, analysis of activation energies based on symmetry considerations cannot be applied. However, *if* Halperin and Braner's method[45] is applicable to organic materials, values of $q/S\,(S = T^* - T_1)$ or $q/W\,(W = T_1 - T_2)$ can tell us something about the kinetics of the recombination process. For trypsin $1 \leqslant q \leqslant 2$, and so the

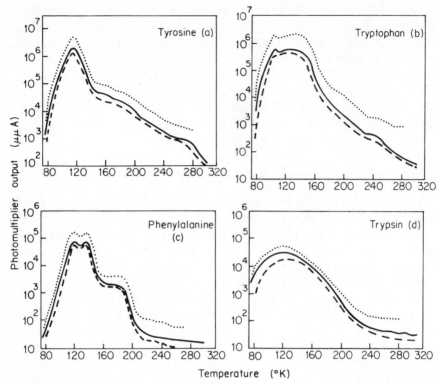

Figure 6. The thermoluminescence of samples of (a) tyrosine, (b) tryptophan, (c) phenyl-alanine, and (d) trypsin which had received about 9×10^5 rad of Co-60 γ-radiation. The solid lines are the averages of at least eight glow curves for each substance; the dashed lines are drawn through the minimum values observed (the maximum observed values deviated from the average by comparable amounts); the dotted lines are equivalent to those which would be measured with an idealized photomultiplier, in which the output per unit radiant flux was the same for all wavelengths as that exhibited by the 1P21 photomultiplier tube at 4100 A. These latter values are calculated by multiplying the total output current at a given temperature by

$$(R_{4100})\left(\sum_{j=1}^{14} S_j\right) \bigg/ \left(\sum_{j=1}^{14} R_j S_j\right),$$

where R_j is the spectral response of the 1P21 in the jth spectral region. (After reference 15)

peak at 122 °K presumably arises from a retrapping dominant, second-order process. On comparison of the curves for the amino acids and trypsin with those available for NaCl one is struck by the relative simplicity of the former. Since macromolecules should contain numerous regions of charge concentration which could serve as potential trapping or recombination centres, one might have expected numerous peaks. Accordingly, it was

Table 2. A tabulation of glow-curve parameters and estimates of activation energies for thermoluminescence. (After reference 15)

	Tyrosine			Tryptophan			Phenylalanine		Trypsin
A. Glow-curve parameters									
$T_g{}^a$	112	151	105	118	177	133	133	174	122
$\sigma_{T_g}{}^b$	1·9	3·3	4·6	2·9	—	6·1	2·3	1·6	4·1
w^c	17·0	47·0	—	13·1	—	—	13·9	27·6	53·6
$\sigma_w{}^b$	0·7	4·0	—	0·5	—	—	0·3	2·8	6·9
δ^d	7·9	33·8	—	5·8	—	—	6·3	12·9	29·2
σ_δ	0·4	4·5	—	0·4	—	—	0·3	1·5	5·9
δ/w	0·47	0·72	—	0·45	—	—	0·45	0·47	0·55
B. Values of E estimated from initial rise curves									
$E_i{}^e$	0·16	0·25	0·19	0·22	0·39	0·25	0·29	0·31	0·088
σ_{E_i}	0·009	0·013	0·012	0·019	0·013	0·018	0·011	0·006	0·024
n^f	25	4	59	59	3	25	46	5	21
C. Values calculated according to the method of Halperin and Braner [see equation (2)] assuming a nonconduction band model									
$E_1{}^g$	0·16	0·21	—	0·24	—	—	0·30	0·27	—
$E_2{}^h$	0·18	0·24	—	0·28	—	—	0·34	0·30	—
D. Values calculated according to equation (2) assuming a conduction band model									
E_1	0·14	0·18	—	0·22	—	—	0·27	0·24	—
E_2	0·16	0·21	—	0·25	—	—	0·32	0·27	—

^{a} T_g—Temperature of peak of glow curve in °K
^{b} σ—Standard deviation
^{c} w—Width of peak in °K at half of peak intensity
^{d} δ—Difference between temperature at half-peak intensity on the high-temperature side and T_g
^{e} E_i—Activation energies estimated from initial-rise curves
^{f} n—Number of initial-rise curves used in determining E_i
^{g} E_1—Activation energies calculated assuming first-order kinetics
^{h} E_2—Activation energies calculated assuming second-order kinetics

concluded that the origin of electron traps in macromolecules is related to the electronic structure rather than to impurities or crystal defects and that radiation effects become preferentially localized at a limited number of sites in such biochemicals. Augenstein[53] points out that this is not unexpected in molecules having biological function.

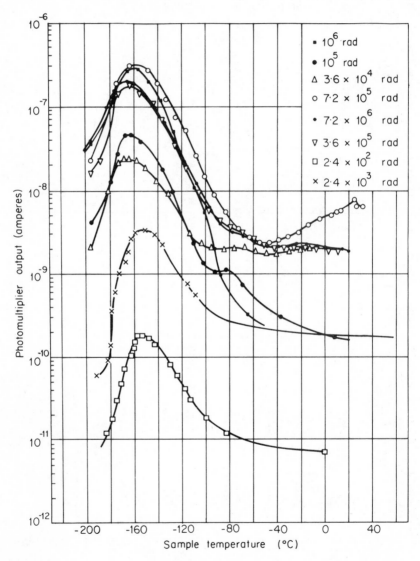

Figure 7. The relation between the thermoluminescence of trypsin and the absorbed dose. (After reference 13)

Figure 7 contains the data obtained for trypsin samples irradiated with doses of γ rays ranging from $2\cdot4 \times 10^2$ to $7\cdot2 \times 10^6$ rad. From this figure one notes the direct dependence of both the maximum intensity and the total luminescence intensity (as given by the area under each curve) on dose and the constancy of shape and position of each peak. Again it is interesting to compare Hill and Schwed's data[4] on NaCl under similar irradiation conditions where the position, shape and intensity of the peaks varied appreciably with increasing dose. The constancy in both the shape and position of the TL peak for trypsin at these dose levels is not entirely unexpected since it requires about 10^7 rad to produce an average of one inactivating event (at room temperature) per trypsin molecule in dried preparations.[54] Hence, since the dose levels commonly used for TL studies were considerably less than this, effects due to saturation and gross disruption of the macromolecular structure would not be expected to be a major factor.

If the critical event in determining the TL properties was simply the initial interaction, then the glow curve for a protein or polypeptide should be the sum of the glow curves for the constituent amino acids, weighted according to their fractional composition. Comparison of the two curves in

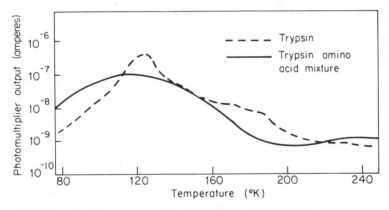

Figure 8. A comparison of glow curves obtained from trypsin and the amino acid mixture of identical composition. (After reference 14)

Figure 8 showed that this is not the case for trypsin. Hence in trypsin radiation-displaced electrons and their associated holes do not become localized at random contrary to what is expected for the initial interaction. Comparable evidence of energy and/or charge migration in proteins is provided by the observation that neither the fluorescence nor phosphorescence spectrum of proteins is the sum of the corresponding spectra for their constituent amino acids.

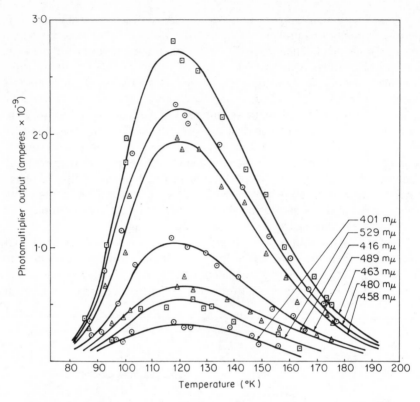

Figure 9. Trypsin thermoluminescence glow through seven interference filters. (After reference 2)

To further test the prediction that the electronic rearrangements initiated by ionizing radiations become localized at a specific group of sites in proteins let us turn to some results obtained using the emission spectra and temperature dependence methods described earlier. By employing an apparatus such as that shown in Figure 5 above, it was possible to obtain the spectral distribution of the TL emission at approximately 5-degree intervals. Such plots of the TL intensity from trypsin as a function of temperature through a number of filters are shown in Figure 9. The maxima in the curves (independent of dose below 10^7 rad) are at 120 °K in agreement with the results in Figure 6 and 7 above. The values at 120 °K in these curves from each of the fifty filters are plotted as the dashed line in Figure 10. The solid line represents the total TL intensity summed over the whole temperature range. Integration of this solid curve leads to the estimate that there are 2.8×10^6 quanta emitted as TL per rad (i.e., 1 quantum per 4.5×10^6 eV absorbed): the quanta emitted for trypsin have an average energy of 2.6 eV.

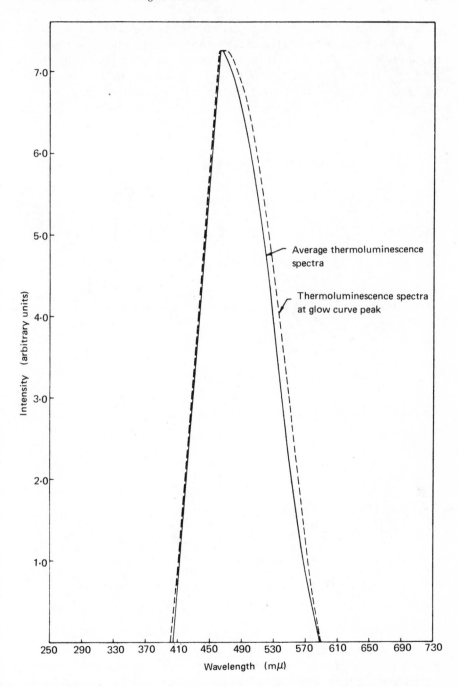

Figure 10. Comparison of trypsin thermoluminescence spectra. (After reference 2)

The intensity of the immediate luminescence emitted during irradiation may also be measured using the above mentioned apparatus. In contrast to the study of TL, the intensity of the immediate luminescence decreases markedly with the accumulation of 10^5 rad or greater so that these measurements utilized doses of 2,500 rad or less. The data for 1 mg powdered samples of trypsin and the three aromatic amino acids in boron carbide containers at

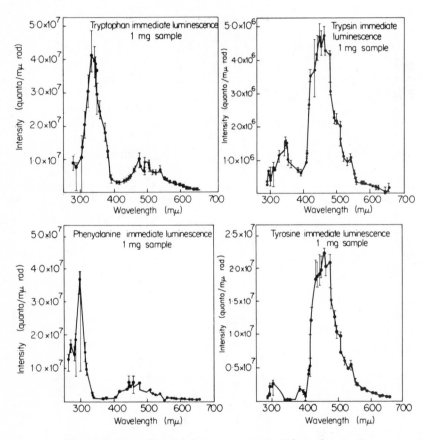

Figure 11. Immediate luminescence spectra. (After reference 2)

77 °K are shown in Figure 11 [these data have been corrected for self absorption and scattering of the emitted fluorescence from the powdered crystallites[2] and for the spectral sensitivity of the detecting system]. Integrating the area under the curve for trypsin leads to the estimate of one quantum of IL emitted per 124 eV absorbed. The average energy of the

Table 3. Yields for the immediate luminescence and thermoluminescence of trypsin and the aromatic amino acids. (After reference 2)

Substance	Immediate luminescence			Thermoluminescence	
	Average energy of emitted quanta (eV/Q)	Average energy absorbed per quantum emitted $\left(\dfrac{1}{Y_{IL}}, \text{in eV/Q}\right)$	Long wavelength/ fluorescence (l.w./F)	Average energy of emitted quanta (eV/Q)	Average energy absorbed per quantum emitted $\left(\dfrac{1}{Y_{TL}}, \text{in eV/Q}\right)$
Trypsin	2·8	$124 \, {}^{+\,11}_{-\,13}$	4·7 ± 0·9	2·6	$4·5 \times 10^6$
Tyrosine	2·7	32 ± 3	25 ± 4·6	2·6	$2·1 \times 10^4$
Phenyl-alanine	3·7	$34 \, {}^{+\,8}_{-\,3}$	0·46 ± 0·14	2·5	$3·7 \times 10^5$
Tryptophan	3·3	18 ± 3	0·45 ± 0·11	2·5	$3·1 \times 10^4$

quanta emitted in this region is 2·8 eV, so that $1·4 \times 10^9$ eV are reemitted per rad absorbed. From the summary of values obtained for trypsin and the three aromatic amino acids (Table 3) it can be seen that both the IL at 77 °K and the TL are peaked at about 460 nm. This wavelength agrees with that previously reported for the phosphorescence from u.v. and X-ray irradiated proteins.[55,56] Furthermore, the trypsin phosphorescence is strongly peaked at the same wavelength as tyrosine phosphorescence and the activation energies for the process in the two materials are almost identical.[16] This close similarity implies that the sequence of rearrangements leading up to TL results in excitation of at least one aromatic amino acid residue. The maximum in the trypsin IL spectrum at 345 nm corresponds closely to that expected for the fluorescence initiated by u.v. or γ-rays in powdered samples of tryptophan-containing proteins or amino acid mixtures.[57] Before discussing possible mechanisms to account for these results in trypsin, it is necessary to consider the available data on the effect of temperature on the induced emission.

The emission from powdered samples of trypsin, phenylalanine, tyrosine and tryptophan at various temperatures is represented in Figure 12. The striking features are (a) the disappearance of the phosphorescence peak of phenylalanine (499 nm) with increasing temperature; (b) the disappearance of the phosphorescence peak at 458 nm for tyrosine with increasing temperature; (c) the relative constancy of the tryptophan peaks at 360 nm and 508 nm; and (d) the disappearance of the phosphorescence peak of trypsin at 458 nm with increasing temperature.

The changes in intensity of these various peaks with increasing temperature are shown in Figure 13, and the values estimated for the activation energies

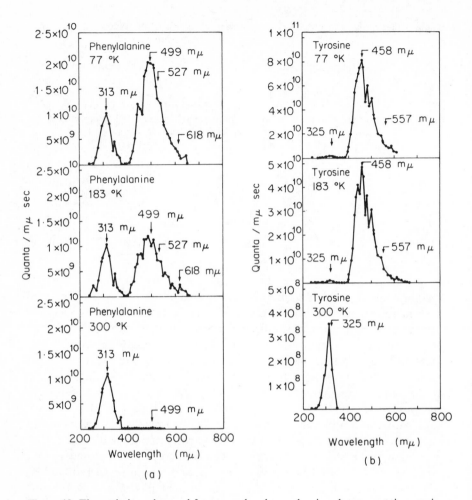

Figure 12. The emission observed from powdered samples, in a brass container main-

on both sides of the transition temperatures are tabulated in Table 4. Clearly all the Boltzmann plots have a marked change in slope in the range 150 to 240 °K—a similar transition in the same general temperature range agrees closely with that obtained from Boltzmann plots of enzyme inactivation.[58] In tryptophan-containing proteins u.v. excitation of any of the aromatic residues normally results in fluorescence from only tryptophan, but in some proteins phosphorescence typical of that from both tyrosine and tryptophan is observed. However, when trypsin is excited by u.v. light in the region 240–280 nm, the phosphorescence appears to arise almost exclusively from the tryptophan residue.[59] With X-rays the trypsin fluorescence occurs

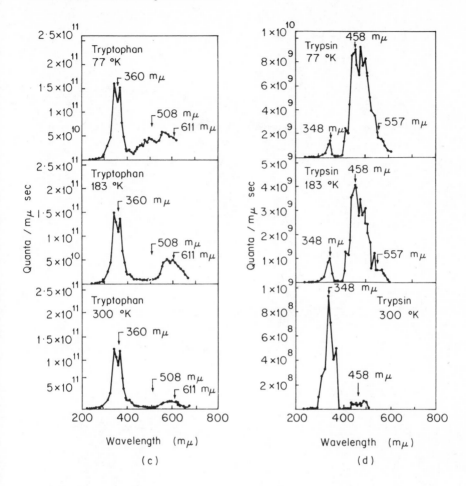

tained during X-irradiation at the temperature indicated. (After reference 16)

at a wavelength corresponding to that from tryptophan; but the thermal coefficients for quenching and the wavelengths of peak intensity indicate that the majority of the X-ray induced emission from trypsin appears as phosphorescence arising from the tyrosine residue.

The observations that TL intensity is a linear function of dose whereas the intensity of IL decreases with absorbed dose is indicative of competitive processes giving rise to these two phenomena in both proteins and amino acids. If both processes were dependent on the same sequence of events then the probability of IL at a given temperature T, would depend on the probability for decay of metastable configurations at that particular temperature.

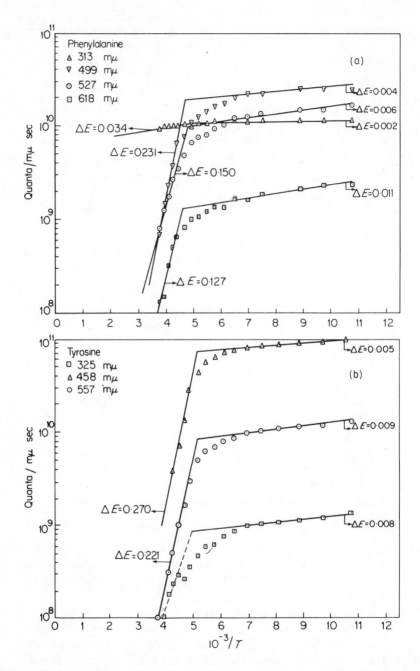

Figure 13. Plots of ln I_T *vs.* $1/T$ for the emission observed through the filters indicated by
phorescent, and 'red'

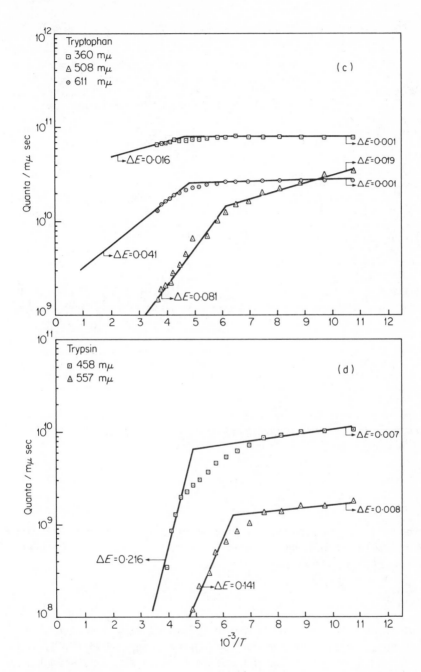

the arrows shown in Figure 12. See the text for identification of fluorescent, phos-
components. (After reference 16)

Table 4. Thermal coefficients and transition temperatures obtained from Arrhenius plots. (After reference 16)

Compound	Fluorescence[a]			Phosphorescence		
	$E_{q,100}$ (eV/molecule)	$E_{q,300}$ (eV/molecule)	T_F (°K)	$E_{q,100}$ (eV/molecule)	$E_{q,300}$ (eV/molecule)	T_p (°K)
Phenylalanine	0·002 ±0·002 3	0·01 ±0·01 3	219 ±30 2	0·006 ±0·003 5	0·15 ±0·05 5	194 ±8 5
Tyrosine	0·005 ±0·002 5	—	—	0·005 ±0·001 5	0·24 ±0·04 5	185 ±9 5
Tryptophan	0·002 ±0·001 6	0·012 ±0·006 6	221 ±17 6	0·016 ±0·004 6	0·06 ±0·02 6	150 ±17 6
Trypsin	—	—	—	0·007 ±0·003 6	0·21 ±0·07 6	200 ±9 6

[a] The E_q's represent the coefficients for the quenching of fluorescence and phosphorescence in the low (near 100 °K) or high (about 300 °K) temperature range. The T's indicate the transition temperature, i.e., the temperature at which the lines for high and low temperature data intersect. The third line in each entry indicates the number of independent experiments for which the average and standard deviation are given

That is, for any given material having an activation energy E_i for TL the IL would be described as follows:

$$\frac{IL}{(IL + TL)} = \frac{I}{1 + \left(\dfrac{TL}{IL}\right)_{i,T}} = S\,e^{-E_i/kT} \qquad (7)$$

where S is the so-called frequency factor (equation 1). The measured values of (TL/IL) and E_i and the calculated values of S for 77 °K are given in Table 5. Since (TL/IL) was found to be very small

$$\ln S \simeq \frac{E_i}{kT} \qquad (8)$$

and values of S would have to vary by 10^6 among these amino acids. This is contrary to the observation that the long-wavelength yields of TL do not differ by more than a factor of 3, and most importantly the observed yields cannot be correlated with the calculated values of S. How then do we reconcile these observations of quite different sequences of events accounting for

Table 5. Measured values of the ratio of thermoluminescence to immediate luminescence. (After reference 2)
(Activation energies (E_i), for thermoluminescence and calculated values of the frequency factors (K_i) for 77 °K are given for the aromatic amino acids)

Substance	TL/IL	E_i (eV)	K_i
Tyrosine	0·007	0·16	$3·0 \times 10^{10}$
Phenylalanine	0·0004	0·25	$2·3 \times 10^{16}$
Tryptophan	0·002	0·19	$2·7 \times 10^{12}$

the IL and TL phenomena in trypsin with the knowledge that the species most predominant in both emission processes is apparently the triplet state of the tyrosine residue? There is little doubt that there is an enhanced production of tyrosine triplets following high energy radiations but we cannot say, with any certainty, whether these are created directly by the absorption of radiation or indirectly following energy or charge migration. Further, we earlier mentioned that a value of $\delta/W = 0·55$ from the trypsin glow curve is characteristic of a recombination obeying a second order process. Triplet involvement would not intuitively be expected to depend upon the concentrations of two trapped species although factors affecting processes of energy or charge transfer could cause the emission of a photon to appear as second-order even though the detrapping and recombination might be first order. Such factors have been considered by Mason[32] to involve the interaction between an excited cystine residue with a neighbouring peptide grouping and by Konev,[60] who postulated a charge transfer mechanism based on the creation of the triplet state from an excited singlet state followed by movement within a triplet-state conduction band to an emitting aromatic amino acid residue.

REFERENCES

1. F. Seitz, *Revs. Modern Phys.*, **26**, 7–94 (1954)
2. D. R. Nelson, J. G. Carter, R. D. Birkhoff, R. N. Hamm and L. G. Augenstein, *Rad. Research*, **32**, 723 (1967)
3. J. Randal and M. Wilkins, *Proc. Roy. Soc.*, **A184**, 366 (1945)
4. J. J. Hill and P. Schwed, *J. Chem. Phys.*, **23**, 652 (1955)
5. A. Rose, *Phys. Rev.*, **98**, 204 (1955)
6. J. Ghormley and H. Levy, *J. Phys. Chem.*, **56**, 548 (1952)
7. F. Urbach, *Wein. Ber.*, **11A**, 139, 363 (1930)
8. G. F. J. Garlick and A. F. Gibson, *Proc. Roy. Soc.*, **A60**, 574 (1948)
9. G. Bonfiglioli, P. Brovetto and C. Cortese, *Phys. Rev.*, **114**, 951 (1959); **114**, 956 (1959)
10. D. Dutton and P. Maurer, *Phys. Rev.*, **90**, 907 (1953)

11. W. L. Medlin, *J. Chem. Phys.*, **38**, 1132 (1963)
12. W. Arnold and H. Sherwood, *Proc. Nat. Acad. Sci (U.S.)*, **43**, 105 (1957)
13. L. G. Augenstein, J. G. Carter, D. R. Nelson and H. P. Yockey, *Rad. Res. Supp.*, **2**, 19 (1960)
14. L. G. Augenstein, J. G. Carter, D. R. Nelson and H. P. Yockey, in *Free Radicals in Biological Systems* (Academic Press, N.Y.), 1961, p. 149
15. C. J. Weinberg, D. R. Nelson, J. G. Carter and L. G. Augenstein, *J. Chem. Phys.*, **36**, 2869 (1962)
16. J. G. Carter, D. R. Nelson and L. G. Augenstein, *Arch. Biochem. Biophys.*, **111**, 270 (1965)
17. J. G. Carter, D. R. Nelson and L. G. Augenstein, *Appl. Phys. Letters*, **2**, 226 (1963)
18. L. G. Augenstein, J. G. Carter, J. Nag-Chaudhuri, D. R. Nelson and E. Yeargers, in *Symposium of Physical Methods in Radiation Biology* (L. G. Augenstein, R. Mason and B. Rosenberg, eds.), Academic Press, N.Y., 1964, p. 73
19. A. Charlesby and R. H. Partridge, *Proc. Roy. Soc.*, **A271**, 170, 188 (1963)
20. A. Deroulede, F. Kieffer and M. Magat, *Israel J. Chem.*, **1**, 509 (1963)
21. N. Shiomi, *J. Phys. Soc. Japan*, **21**, 907 (1966)
22. M. Magat, *J. Chim. Phys.*, **63**, 142 (1966)
23. J. Bullot, A. Deroulede and F. Keiffer, *J. Chim. Phys.*, **63**, 150 (1966)
25. K. W. Boer, S. Oberlander and J. Voigt, *Ann. Phys.*, **2**, 130 (1958)
26. F. J. Bryant, A. Bree, P. E. Fielding and W. G. Schneider, *Disc. Faraday Soc.*, **28**, 48 (1959)
27. R. H. Bube, in *Photoconductivity of Solids* (John Wiley, New York), 1960, p. 292
28. F. Gutmann and L. E. Lyons, in *Organic Semiconductors*, John Wiley, New York, 1964
29. D. D. Eley, in *Horizons in Biochemistry* (M. Kasha and B. Pullman, eds.), Academic Press, New York, 1962, p. 341
30. M. G. Evans and J. Gergely, *Biochim. et Biophys. Acta*, **3**, 188 (1949)
31. B. Pullman, in *Comprehensive Biochemistry* (M. Florkin, E. H. Stotz, eds.) (Elsevier, Amsterdam), 1967, Vol. 22
32. R. Mason, *Radiation Research Suppl.*, **2**, 452 (1960)
33. M. Suard-Sender, *J. Chim. Phys.*, **62**, 79 (1965)
34. M. Cardew and D. D. Eley, *Disc. Faraday Soc.*, **27**, 115 (1959)
35. B. Rosenberg, in *Physical Processes in Radiation Biology* (L. G. Augenstein, R. Mason and B. Rosenberg, eds.), Academic Press, New York, 1964, p. 111
36. B. Rosenberg and E. Postow, Annals New York Acad. of Sci. (1968), in press
37. N. Riehl and P. Thoma, *Phys. Stat. Solidii*, **16**, 159 (1966)
38. M. Pope, *Sci. Amer.*, **2**, 86 (1967)
39. R. L. Lehman and R. Wallace, in *Electronic Aspects of Biochemistry* (B. Pullman, ed.), Academic Press, New York, 1964
40. R. H. Bube and R. Schrader, *Rev. Sci. Inst.*, **25**, 921 (1954)
41. T. N. Misra, B. Rosenberg and R. Switzer, *J. Chem. Phys.*, **48**, 2096 (1968)
42. F. C. Frank, *Advances in Physics*, **1**, 91 (1952)
43. L. I. Grossweiner, *J. Appl. Phys.*, **24**, 1306 (1953)
44. D. R. Nelson, R. D. Birkhoff, R. H. Ritchie and H. H. Hubbell, Jr., *Health Physics*, **5**, 203 (1961)
45. A. Halperin and A. A. Braner, *Phys. Rev.*, **117**, 408 (1960)
46. R. R. Haering and E. N. Adams, *Phys. Rev.*, **117**, 451 (1960)
47. A. H. Booth, *Can. J. Chem.*, **34**, 214 (1954)
48. G. A. Dussel and R. H. Bube, *Phys. Rev.*, **155**, 764 (1967)
49. C. E. May and J. A. Partridge, *J. Chem. Phys.*, **40**, 1401 (1964)

50. W. L. McCubbin, *J. Chem. Phys.*, **43**, 2451 (1965)
51. W. Helfrich and F. R. Lipsett, *J. Chem. Phys.*, **43**, 4368 (1965)
52. J. F. Fowler, *Proc. Roy. Soc.*, **A236**, 464 (1956)
53. L. G. Augenstein, in *Symp. on Information Theory in Biology* (H. Yockey, ed.), Pergamon Press, New York, 1958
54. L. G. Augenstein, *Advances in Enzymology*, **24**, 359 (1962)
55. P. Debye and J. O. Edwards, *Science*, **116**, 143 (1952)
56. I. Sapenzhinskii and N. Emanuel, *Akad. Dok. Nauk (USSR)*, **131**, 1168 (1960)
57. J. Nag-Chaudhuri and L. G. Augenstein, *Biopolymers Symp.*, **1**, 441 (1964)
58. L. G. Augenstein, T. Brustad and R. Mason, *Adv. Radiation Biol.*, **1**, 228 (1964)
59. L. G. Augenstein and J. Nag-Chaudhuri, *Nature*, **203**, 1145 (1964)
60. S. Konev, *Akad. Nauk Isvest. USSR Ser. Fiz.*, **23**, 90 (1959)

VI
Reactions and Interactions of Biomolecules

CHAPTER 12

Nuclear magnetic resonance investigations of the interactions of biomolecules

J. S. Cohen

Physical Sciences Laboratory,
Division of Computer Research and Technology,
National Institutes of Health,
Department of Health, Education, and Welfare,
Bethesda, Maryland, U.S.A.

I. INTRODUCTION

Advances in biophysical chemistry have largely depended on the development of a few physical techniques for the probing of molecular structure. Nuclear magnetic resonance (n.m.r.), which was first detected in bulk matter in 1946,[1,2] has since revolutionized the study of organic chemistry. It has had less impact on molecular biophysics, although this situation is rapidly changing.

Three basic criteria of a useful spectroscopic tool in molecular biophysics are satisfied by high resolution n.m.r. (1) The energy changes involved in the resonance condition are comparatively small, allowing subtle changes to be monitored. (2) There is a high degree of differentiation (resolution) between different chemical groups. (3) Investigations are carried out in solution. Few techniques comply with these requirements, and from the rapidly growing literature it appears that n.m.r. has an almost unique applicability to the study of interactions of biological molecules in solution.

There are a number of excellent general texts on the theory and technique of n.m.r.,[3–8] including some designed to provide a useful background for the biochemically oriented.[9–12] Hence, only a very limited treatment is given here in order to specify the phenomenon and the observable parameters of n.m.r. for those not already familiar with this technique. This is followed by sections of selected application of n.m.r. to the study of molecular interactions.*

II. MEASURED PARAMETERS IN NUCLEAR MAGNETIC RESONANCE

A. Chemical Shift

Certain atomic nuclei absorb radiofrequency radiation when placed in a magnetic field. This resonance condition results from transitions between quantized states of the nuclei induced by the field. If two such states have energies E_1 and E_2 in an applied field then the characteristic resonance frequency v_0 for the system is given by

$$E_1 - E_2 = \Delta E = hv_0 \tag{1}$$

where h is Planck's constant. Nuclei which exhibit this phenomenon possess inherent nuclear spin, characterized by a *nuclear spin quantum number*, I. In the simplest case, for a proton or hydrogen nucleus, $I = \frac{1}{2}$. The two states of the nucleus can then be described by the related quantum number $M_I = \pm\frac{1}{2}$. The value of $M_I = +\frac{1}{2}$ corresponds to the low energy state (E_2), while that of $M_I = -\frac{1}{2}$ corresponds to the higher energy state in the transition.

---

* This chapter was originally written in June 1968 and completely revised in August 1971 for publication.

These represent orientations of the nuclei in the direction and opposed to the direction of the applied field, respectively.

The property of spin results in the presence of a nuclear magnetic moment. The ratio of the magnetic moment to angular momentum is termed the *gyromagnetic ratio*,

$$\gamma = 2\pi\mu/Ih \tag{2}$$

where μ is the nuclear dipole moment. The frequency at resonance is also related to the magnitude of the applied magnetic field H_0 by the important relationship

$$2\pi\nu_0 = \gamma H_0 \tag{3}$$

Here γ represents a proportionality constant, the value of which is constant for a given nucleus, relating the field and frequency at resonance. Some nuclei which exhibit nuclear magnetic resonance, apart from hydrogen, are ^{13}C, ^{15}N, ^{17}O, ^{19}F, ^{31}P. By contrast, ^{12}C and ^{16}O do not exhibit resonance ($I = 0$) and hence do not interfere with the observation of the other species.

Electrons surround the nuclei of atoms and their circulations produce small local magnetic fields which are proportional in magnitude to the applied field. The magnitude of this *shielding* effect will depend on the extent of electronic screening of the nuclei, i.e., on the number and types of the bonds. Thus, different resonance conditions, termed *chemical shifts*, result for nuclei of the same isotope in different electronic (chemical) environments.

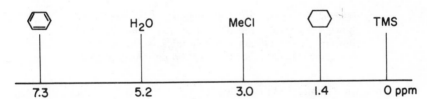

Figure 1. P.m.r. chemical shifts of hydrogens in different chemical environments (δ_{TMS} scale)

The series of H-containing compounds shown in Figure 1 each give a single but distinct resonance line in p.m.r. The spectrum of ethanol, which contains protons in three different electronic environments, methyl (CH_3-), methylene ($-CH_2-$) and hydroxyl ($-OH$), has three separate resonances (Figure 2). Since the methylene hydrogens experience a low electron density due to the proximity of the electronegative oxygen atom they should absorb at lower field than methyl group hydrogens, which are more shielded.

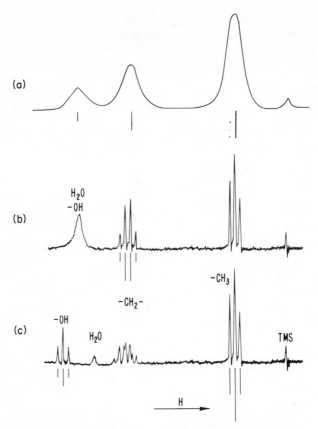

Figure 2. N.m.r. spectrum of ethanol; (a) low resolution,
showing chemical shift differences, (b) high resolution,
showing ethyl group spin–spin splitting, (c) same, but highly
purified ethanol, showing hydroxyl proton coupling

In p.m.r. the chemical shift values for most types of chemically distinct
protons are known (Table 1). The magnitude of the chemical shift is expressed
in terms of the field (in gauss) or the frequency (in Hertz) from the resonance
of a common standard. However, the chemical shift is proportional to the
magnitude of the applied field. Thus, in order to make quoted values independent of the strength of the field at which different spectrometers operate, a
dimensionless parameter is defined,

$$\delta = \frac{\nu_{sample} - \nu_{reference}}{\nu_{reference}} \times 10^6 \tag{4}$$

where δ is the chemical shift expressed in parts per million (p.p.m.). Because
of the smallness of the numerator relative to the denominator the frequency

Table 1. Chemical shift values

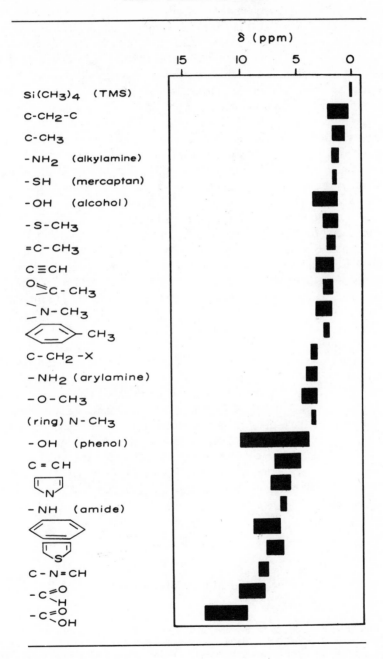

$v_{\text{reference}}$ in the denominator of equation (4) is usually replaced by the frequency at which the spectrometer operates.

Tetramethylsilane (TMS) is the most common reference standard used for reporting p.m.r. data. It is chemically inert and its protons are highly shielded so that most other resonances occur to lower field, usually within a range of 10 p.p.m. (Table 1).* The less volatile hexamethyldisiloxane (HMS), or water soluble 2,2-dimethylsilapentane-5-sulphonate (DSS) are often chosen for biochemical applications. Water is not a useful standard due to the large differences in chemical shift observed in different solutions or at different temperatures. The reference compound may be within the sample as an *internal reference*, or contained within a capillary as an *external reference*. These will usually give slightly different results depending on the difference of bulk diamagnetic shielding between the pure reference compound (external) and the sample solution (internal). Internal referencing is preferably avoided when dealing with biological macromolecules due to their tendency to bind small molecules, and due to micelle formation described for DSS in D_2O.[13] The difficulties inherent in external referencing mainly arise from the glass insert used to contain the reference compound. Spinning of the sample tube to average the magnetic field results in large spinning side-bands[4] associated with the glass insert which may be minimized by the use of precision coaxial tubes.

Any process occurring in solution involving time-dependent changes in the environment of magnetic nuclei will affect the magnetic resonance spectrum (for a review see reference 14). The effects observed will depend on the rate of the process and whether or not it is reversible. For example, the simplest case is that of reversible exchange of magnetic nuclei between two equally populated sites A and B. The shape of the resonance signal(s) observed will depend upon the ratio of τ, the mean lifetime, to the inverse of the relative shift, $(v_A - v_B)^{-1}$. Calculated line shapes are shown in Figure 3, which vary from the limit of slow exchange (a, $\tau \gg (v_A - v_B)^{-1}$) at which separate resonances are seen, to that of fast exchange (h, $\tau \ll (v_A - v_B)^{-1}$) at which a single resonance appears. In the latter case the resonance will be centred at a mean frequency

$$\bar{v} = p_A v_A + p_B v_B \tag{5}$$

where p_A and p_B are the fractional populations in sites A and B and $p_B = (1 - p_A)$.

---

* Various chemical shift scales have been used in the literature. A recommendation of the American Society for Testing and Materials (Manual of Recommended Practices in Spectrophotometry, ASTM, 1966) proposes the adoption of the convention that the chemical shift with respect to TMS to low field be positive. This will be followed as far as possible in this work.

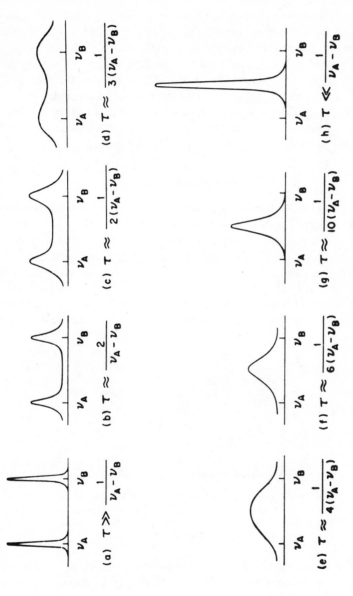

Figure 3. Change of line shape as a result of chemical exchange between two equivalent sites A and B at successively increasing exchange rates. [From reference 9, copyright © 1962 John Wiley & Sons Inc. by permission of John Wiley & Sons Inc.]

B. Relative Areas

The area under a resonance signal will depend on the number of nuclei undergoing transitions at the same frequency. For a molecule containing magnetically nonequivalent nuclei, the ratio of areas in a spectrum will thus reflect the ratio of the number of nuclei in each chemical group. In ethanol this ratio is $3:2:1$ for the methyl, methylene and hydroxyl protons respectively (Figure 2). When the lines in the spectrum are sharp, peak intensities are generally taken to indicate the relative numbers (to the nearest integer) of different nuclei in the compound. For a mixture containing a number of different components the relative concentrations can be determined from the relative areas of their nuclear resonances.

C. Spin–Spin Coupling

Apart from their shielding or deshielding effects on nuclei, electrons also interact by coupling their spins with those of the nuclei. Since the two electron spins are usually antiparallel in chemical bonds, an indirect electron-mediated coupling occurs between adjacent magnetic nuclei.[15,16] The simplest case of coupling is in a diatomic molecule AX in which each nucleus has $I = \frac{1}{2}$. The *spin–spin coupling* gives rise to a splitting in the spectrum of A and X which can be observed at high resolution. Each nucleus experiences two slightly different magnetic fields, consisting of the local field plus and

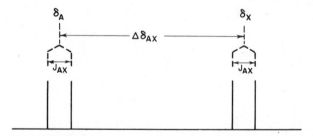

Figure 4. Representation of the spectrum of a molecule AX showing spin–spin splitting. A and X both have $I = \frac{1}{2}$ and the condition $\Delta\delta_{AX} \gg J_{AX}$ is fulfilled, where J_{AX} is the coupling constant

minus the spin components due to the other nucleus. This is equal in both cases and is characterized by a coupling constant J_{AX} as shown in Figure 4. Coupling effects usually extend over no more than a few bond lengths. The magnitude of the coupling constant depends on the electronic nature (degree of hybridization) of the intervening bond(s) and the relative orientations of the nuclear spins (bond angle)[17,18] as well as other factors such as temperature. However the magnitude of the coupling constant is independent of the applied field strength (unlike the chemical shift) and hence is usually expressed in frequency units (Hz), its magnitude being directly measurable from the line splitting in the spectrum. As $\Delta\delta_{AX}$ approaches J_{AX} complex spectra are observed as a result of mixing of energy levels. In such a case the spin system is referred to as AB rather than AX (for a general treatment see reference 19). At higher magnetic fields, such as produced by a super-conducting magnet system, complex spectra become simpler due to the field dependence of the chemical shift but not of the coupling constant. Thus, it is often advantageous to use a n.m.r. spectrometer operating at a higher frequency[20] (see section IIIA2).

D. Line Width and Relaxation Times

The line shape of a resonance peak as derived from the Bloch equations[21] has the form of a Lorentzian curve (Figure 5),

$$f(v) = \frac{2t}{1 + t^2(v_0 - v)^2} \tag{6}$$

For an assemblage of nuclei undergoing resonance the constant t in equation (6) is the *transverse relaxation time*, T_2. This is the time constant governing relaxation of the component of the net magnetization vector of the nuclear assemblage in the direction (defined as the x, y plane) transverse to the applied field (defined as the z direction). Also as a consequence of the Lorentzian form, the width at half-height of the resonance peak (Figure 5) is related to T_2 by the simple expression

$$\Delta v_{1/2} = 1/\pi T_2 \tag{7}$$

Thus, the value of this relaxation time can be determined in the absence of other complicating factors from a measurement of the spectral line width at half-height.

When the exciting radiofrequency is applied with excessive power, equation (6) does not hold, and the relationship in equation (7) is no longer true. This condition is termed *saturation*, since it corresponds to the nuclear magnets being forced into their higher energy states ($M_I = -\frac{1}{2}$), so that fewer transitions can occur, and in the limiting case no signal will be seen.

The characteristic relaxation time in the direction of the field is termed the *longitudinal relaxation time*, T_1. This is also the relaxation time for energy exchange between a nuclear spin and the total assemblage, or lattice, and is termed the spin–lattice relaxation time. This process is related to molecular motion by the relationship

$$1/T_1 \propto \tau_c \propto \eta \tag{8}$$

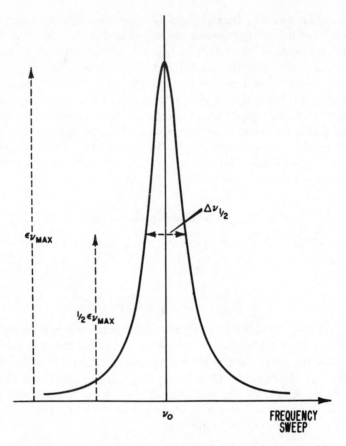

Figure 5. A non-saturated n.m.r. signal in the absorption mode with
Lorentzian shape, such that equation (6) applies

where τ_c is the molecular correlation time (the time taken to move a distance comparable to its dimension) and η is the viscosity.

The transverse relaxation time T_2 is also the relaxation time for energy exchange between two spins, and is termed the spin–spin relaxation time. In liquids local field inhomogeneities are averaged to zero, and in the absence of other factors such as spin coupling or exchange, $T_1 = T_2$. As a consequence of this equality and equations (7) and (8), as the viscosity of a solution increases the line-widths of its resonances increase. Also, as the molecular size increases, so τ_c increases, and the line-widths increase. These are the causes of the broad resonances observed in protein n.m.r. spectra.[22] At very high values of τ_c, $T_1 \gg T_2$, and line widths increase even more rapidly, approaching values as found in the solid state.[23]

As indicated for the chemical shift (equation (5)) the line-width of a resonance for a system under fast exchange conditions between two sites is also a weighted mean, so that from equation (7),

$$1/T_2 = p_A/T_{2A} + p_B/T_{2B} \qquad (9)$$

This property provides an important monitor of the binding of a small molecule to a macromolecule, since the effect of a small amount bound is effectively magnified if the property (chemical shift, line width) is much greater in the bound state or complex.[24,25]

Generally speaking, the relaxation times T_1 and T_2 for a nucleus are determined by pulse or transient n.m.r. methods as opposed to the usual slow sweep or continuous wave (cw) method.[24,26]

III. SELECTED APPLICATIONS

A. Interactions of Small Molecules with Proteins

1. Observation of the Spectrum of the Small Molecule

The theory of nuclear magnetic relaxation for exchange between two sites was developed by Zimmerman and Britten for the exchange of water on silica gel.[27] The first applications to protein-small molecule interactions were those of Jardetzky and coworkers.[28-30]

In general the type of system being studied may be simply represented as,

$$P + S \overset{K}{\rightleftharpoons} PS \qquad (10)$$

where P is a protein, S a small molecule and PS the complex formed between them. The equilibrium constant for the process is given by,

$$K = \frac{[PS]}{[P][S]} \qquad (11)$$

If K is unknown, it may be calculated as follows. When no further change in line width or chemical shift of a resonance of S results in the addition of P to S, the maximum value is that of the bound form. From equation (5) (or (9) for line width)

$$v^i_{obs} = pv^i_{bound} + (1 - p)v^i_{free} \qquad (12)$$

where p is the fraction of S bound and i indicates the individual resonance corresponding to the ith atom of S. This may be used to calculate p, and hence [PS], then from equation (11), to obtain K. An iterative calculation procedure can be used so that high concentrations of protein can be avoided.[31] There are several more extensive mathematical treatments of this particular case.[9,12] The binding process may of course be considerably more complex than indicated by equation (10), particularly if there are multiple binding sites or inhibitor-induced conformational changes in the protein (such as 'induced fit').[32]

This approach is not of course as simple a method for deriving equilibrium constants as some routine laboratory methods, such as equilibrium dialysis or other spectrophotometric methods. However, its value lies in its ability to provide information on individual atoms or groups within the molecule, which these other techniques cannot do.

In general before spectral changes observed in combining two interacting species are judged to reflect only the process of binding, certain criteria must be satisfied.

(1) For equation (9) to apply, fast exchange must be occurring otherwise there will be contributions from the exchange process itself.[14] (2) If line broadening only is examined, values for $1/T_2$ will be obtained (equation (7)). Since there are certain effects on T_2 which are negligible for T_1 it is preferable that both be determined to check that $T_1 = T_2$.[24] (3) Controls for viscosity changes in the medium, affecting τ_c for all nuclei, must be performed. (4) Effects on dipolar relaxation for non-complexing nuclei, must be considered. These could arise from changes in intramolecular distances as a result of a change of molecular conformation on binding. (5) Extraneous paramagnetic substances must be removed because of their great efficiency in promoting relaxation. Since it is often impossible to be sure that these (e.g., oxygen) are totally absent, controls using the same solvents are preferable. (6) Magnetic inhomogeneity must be kept to a minimum. (7) Independent studies of concentration (reflecting molecular association) or titration effects on the spectra of each component should be carried out. Careful control of pH and concentration between the single species and their mixture is essential to obtain meaningful data. Not all of these conditions are satisfied in all the examples presented here. The reader should carefully check the published work for these elementary controls.

While the n.m.r. technique has been used to study the interactions between small molecules, e.g., epinephrine–nucleotide,[33] thiamine–indole[34] and chlorophyll–chloroform interactions,[35] most applications have concerned the binding of small molecules (drugs, haptens, inhibitors) to biological macromolecules (proteins and nucleic acids). The following protein systems have been studied by this technique.

a. Bovine and Human Serum Albumin

Jardetzky and coworkers studied the binding of penicillin[29] and sulphonamides[30] to bovine serum albumin (BSA), measuring the relaxation times T_1 and T_2. The former were obtained from direct saturation experiments and the latter from line broadening measurements.[24] Fast exchange

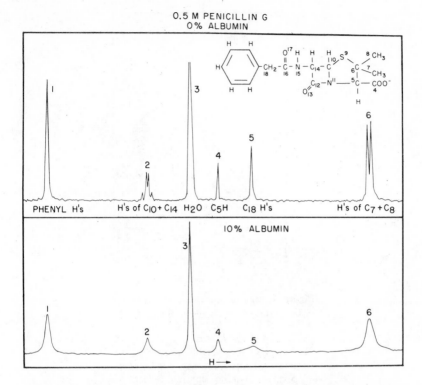

Figure 6. Spectra of potassium penicillin G in D_2O (0·5 M) and in the presence of 10 per cent bovine serum albumin. Line assignments are shown in the figure.[29] [Reprinted from *J. Am. Chem. Soc.*, **87**, 3237 (1965). Copyright 1965 by the American Chemical Society. Reprinted by permission of the copyright owner]

was considered to be operative in all cases so that equations (9) and (12) were applied.

Figure 6 shows the spectra of penicillin in D_2O and in the presence of 10 per cent BSA. Marked *selective* increments in line width were observed (Figure 7), which were greater than the non-selective viscosity effects produced by γ-globulin as a control which does not bind penicillin. The results were consistent with a model of penicillin binding to BSA *via* the phenyl group.

Similarly, addition of BSA to solutions of a number of sulphonamides resulted in greater increments of the relaxation rates for the p-amino-benzenesulphonamide (PABS) parent moiety than for any of the substituents on N-1, with one exception. For sulphaphenazole the phenyl substituent peak was also greatly affected. Phenylpropanol had a differential effect on the absorption of the PABS and phenyl constituents in this case, reducing

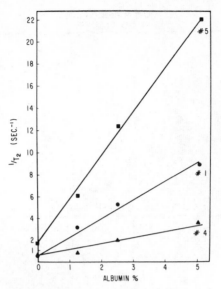

Figure 7. Relaxation rates of penicillin peaks as a function of albumin concentration (see Figure 6 for assignments).[29] [Reprinted from *J. Am. Chem. Soc.*, **87**, 3237 (1965). Copyright 1965 by the American Chemical Society. Reprinted by permission of the copyright owner]

the line width of the former while slightly enhancing that of the latter, which was taken to indicate two independent binding sites.[30]

Sykes *et al.* reported a study of the binding of acetylsalicylic (ASA) acid to human serum albumin (HSA). Line width measurements indicated at least one site on HSA at which rapid exchange occurs. The acetyl protons of bound ASA were found to have a rotational correlation time consistent with that of the HSA molecule itself, indicating a fairly rigid complex.[36]

Fluorine-19 magnetic resonance is useful for monitoring molecular interactions since the ^{19}F nucleus is very sensitive to the nature of the local electronic (dielectric) environment.[38] For example, such properties as the value of the critical concentration of detergent required for micelle formation in a series of detergents of the type $CF_3(CH_2)_nCOONa$ were readily investigated by this technique.[37] The effects on aggregation of increasing chain length in this series was found to be negligible. The application of ^{19}F magnetic resonance is facilitated by the fact that it is almost as sensitive as p.m.r. (83 per cent for equal numbers of nuclei at the same field), and that its resonant frequency occurs close enough to that of protons to be observed on most conventional spectrometers (56·4 MHz compared to 60 MHz for protons at 14,000 gauss). Simple spectra are generally observed since only a few fluorine atoms are usually present in the molecule. The interaction of a fluoro-detergent (8,8,8-tri-fluorooctyl-benzene-*p*-sulphonate) with HSA has been studied using ^{19}F magnetic resonance. As expected relatively non-specific binding was indicated, with up to 17 molecules of detergent being bound per molecule of HSA.[39]

The binding of detergents to HSA has also been studied with [81]Br magnetic resonance, by following the displacement of bromide ion from anionic binding sights on the protein.[40]

b. *Hapten–Antibody Interactions*

The experiments described above demonstrate in some degree the sensitivity of the n.m.r. method for studying weak and multiple binding. Serum albumin is known to have many and varied binding sites, often of a hydrophobic nature. By contrast strong and highly selective binding occurs between haptens and the specific antibodies formed against them. The nature of the binding site of antibody prepared against phenoxycholine has been investigated with n.m.r. for a series of choline derivatives.[41,42] Addition of the specific anti-choline antibody to acetamidophenyl choline ether (Figure 8) and its N-benzyl derivative resulted in selective broadening of lines in the spectra which were greater than the effects of the control γ-globulin (Figure 9). Since the greatest effects were for substituents on nitrogen

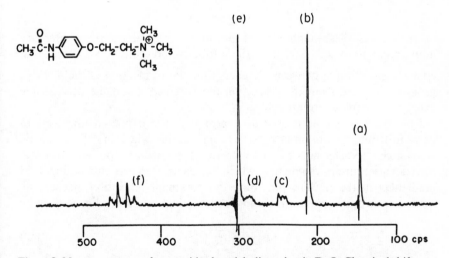

Figure 8. N.m.r. spectrum of acetamidophenylcholine ether in D_2O. Chemical shifts are in cps at 60 MHz downfield from HMS. Line assignments are as follows: (a) singlet 140 cps, acetyl methyl; (b) singlet 208 cps, N-methyl; (c) multiplet 235 cps, N–CH_2 of choline; (d) multiplet 290 cps, O–CH_2 of choline; (e) singlet 300 cps, HDO; (f) multiplet 445 cps, aromatic hydrogens.[41] (Copyright 1968 Academic Press)

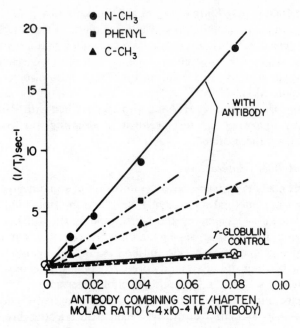

Figure 9. Concentration dependence of the relaxation
rates of the hapten (Figure 8) with anti-choline antibody
and γ-globulin.[41] (Copyright 1968 Academic Press)

in both cases these haptens were presumed to bind *via* the quaternary
ammonium head into a cleft in the antibody molecule.[41] No correlation
could be established between the energy barrier to rotation of the $\overset{+}{R}NMe_3$
group, calculated from the relaxation measurements, and the derived free
energy of complex formation.[41]

Chloride ion has been used as a probe for the hapten binding site of
anti-dinitrophenyl antibody by observing the line width in ^{35}Cl magnetic
resonance.[43] Nuclei such as ^{35}Cl with spin greater than $\frac{1}{2}$ possess a nuclear
electric quadrupole moment q which interacts with the fluctuating field
gradient g at the nucleus providing the dominant relaxation mechanism.
With certain approximations the line width at half height of a line is given by

$$\Delta v_{1/2} = \frac{2\pi}{5}Q^2\tau_c \tag{13}$$

where Q is the quadruple coupling constant which is dependent upon q.
Relaxation *via* quadrupole interaction is sensitive to environment, and for
example large variations in the width of ^{35}Cl resonance lines are found.[44]
Advantage was taken of the fact that chloride ion exchanges at a mercury

atom which is covalently attached to a hapten molecule. Thus, the binding of 2,4-dinitro-4'-(chloromercuri)diphenylamine to anti-dinitrophenyl antibody was monitored by measurement of the ^{35}Cl line width in a solution of 1 M NaCl. The data indicated the degree of accessibility of the mercury atom in the hapten–antibody complex to chloride ion, and as such provides a means of mapping the site by use of suitably designed haptens containing mercury in different relative positions. While this technique has been extended to a number of structural problems[45,46] it is limited for binding studies to those situations where the intermediacy of mercury[47] or another metal ion is possible. Similar line-broadening studies using ^{81}Br magnetic resonance have been reported for interaction of bromide ion with several proteins.[48]

c. *Cholinesterase*

The binding of acetycholine and derivatives to horse serum cholinesterase has been studied by n.m.r.[49] The selective line broadening effects observed indicated that there was more rotational stabilization of the acetyl than of the quaternary ammonium group, in contrast to the binding of acetycholine to the specific antibody discussed above.[42]

As a fortunate result of the relative slowness of the catalytic cleavage of the acetyl group from acetycholine by cholinesterase it was possible to follow the rate of disappearance of the acetyl resonance of acetylcholine and the appearance of free acetate in solution. Plots of the peak intensities as a function of time were then used to calculate K_m for the reaction, and a value of 5×10^{-3} mole^{-1} was obtained which was in good agreement with previous estimates. Experiments were carried out at different temperatures to measure activation energies and in the presence of known inhibitors of esterase activity, such as eserine, so that inhibition constants could be determined.[50] More recently the study has been extended to include such substances as atropine which are believed to bind to a different site than that of eserine and which may be on the regulatory subunit of the cholinesterase.[51] However, the preparations used in this work consisted unavoidably of a mixture of proteins. Time-dependent binding of acetylcholine to possible 'receptor protein' from various sources was also indicated.[52]

d. *Hen Egg-White Lysozyme*

Hen egg-white lysozyme is an enzyme which cleaves glycosidic linkages in cell-wall polysaccharides, and is inhibited by certain mono- and oligosaccharides.[53] The methyl proton absorption of the acetamido group of *N*-acetyl glucosamine (NAG) was found to shift and divide in the presence of the enzyme.[54–57] While the chemical shift changes are small at the enzyme concentrations studied, no such changes resulted in the presence of equivalent concentrations of ribonuclease to which NAG does not bind.[56] To elucidate the origin of the line doubling both the alpha and beta anomers of NAG were freshly prepared, and the enzyme-catalysed equilibration was followed by p.m.r. By comparison of the results with those of the racemate it was shown that the more upfield of the two resonances was assignable to the alpha anomer. Similar magnetic non-equivalence was found for the methyl protons of the non-interconverting α- and β-methyl glycosides in the presence of lysozyme.[54,56] These results could be due either (a) to binding of the two anomers to different sites on the enzyme, or (b) to different affinities or modes of binding at the same site. By observing the loss in signal amplitude on addition of deuterated N-acetamido derivatives it was possible to show that the two methyl glycosides are competitive inhibitors for the same site. Further analysis of the data gave slightly different association constants, but the same overall chemical shift difference for the N-acetyl methyl glycoside between the bound and free forms,[56] indicating (b) to be the case, consistent with the results of other investigations.[58] The considerable upfield shift (0·5 ppm) observed for the bound N-acetyl methyl group is best explained[54,56] in terms of magnetic shielding by tryptophan residues in the cleft of the enzyme.[58] Selective line broadening has been reported for the acetamido methyl proton resonance of the reducing end of di-NAG methyl glycoside in the presence of lysozyme.[55,59] The temperature dependence of the line width and chemical shift of this resonance indicated the effect resulted from relatively slow exchange.[60] A comparative study of the binding properties of several oligosaccharide inhibitors of lysozyme[59,60] gave results consistent with those from the application of other techniques[59,61–63] showing one energetically favoured subsite (C) for NAG, in which the reducing end of oligo-NAG inhibitors always fits.

The pH dependence of the chemical shift of the acetamido methyl group of β-methyl NAG gave a pK of 4·7, which could be identified as that of a carboxyl group in the binding site of NAG. A pK value of 6·1 for a group in the active site was also derived from the pH dependence of the binding constant of β-methyl NAG. From other considerations these pK's were assigned to aspartic acid residue 102 and glutamic acid residue 35 respectively in the amino acid sequence of lysozyme.[64] A transient n.m.r. study of the binding of α and β-methyl NAG to hen egg white lysozyme has provided kinetic data which suggest the binding is not a simple bimolecular process.[65] These results taken together represent the most comprehensive study of a single system by the observation of the spectrum of a small molecule binding to a macromolecule and indicate the kind of detailed quantitative information that n.m.r. can provide.

e. *Alpha-Chymotrypsin*

A number of studies of the binding of inhibitor molecules to α-chymotrypsin have included the application of proton, chlorine-35 and fluorine-19 magnetic resonance techniques. Alpha-chymotrypsin is a proteolytic enzyme with specificity for cleavage adjacent to aromatic amino acids. Tryptophan is an inhibitor of the enzyme, and appreciable line broadening was observed for both the D- and L-forms. Calculation of the respective bound line-widths showed that as for the binding of NAG to lysozyme there are differences between the two isomers, notably that the D-form is bound more tightly.[66] Similar studies have been carried out with *trans*-cinnamate and N-formyl-L-tryptophan and in each case an upfield shift of the aromatic resonances, with little effect on the aliphatic resonances, indicated binding in a hydrophobic pocket by the aromatic moiety,[67,68] consistent with the results of X-ray crystallographic studies.[69,70]

As mentioned above, fluorine-19 chemical shifts are more sensitive monitors of environmental effects than proton chemical shifts, and appropriately substituted inhibitor molecules may be used as selected probes of the binding site. However, ^{19}F is too large (radius = 1·35 Å) to be simply a non-disturbing replacement for [1]H. A substantial downfield chemical shift was reported for the ^{19}F resonance of *N*-acetyl phenylalanine substituted in the phenyl ring,[71] and both downfield[72] and upfield shifts[73] for the acetyl group-substituted inhibitor, on addition of α-chymotrypsin. Upfield shifts are observed for these molecules in organic solvents (due to greater bulk shielding),[72] and the acetamido methyl protons of NAG bound to lysozyme also shift upfield. It is known that the active site region of α-chymotrypsin is strongly hydrophobic.[70,74–76] The quantitative differences reported in the three ^{19}F magnetic resonance studies have been resolved in terms of sensitive pK and concentration effects.[77]

f. *Aspartate Transcarbamylase*

Two reports have appeared from the same laboratory utilizing slow sweep[78] and pulsed[79] n.m.r. methods to study the binding interplay of several inhibitors and substrates to this complex protein. Aspartate transcarbamylase has a molecular weight of 300,000 and consists of regulatory and catalytic subunits. Isolated catalytic subunits have a molecular weight of 100,000 and three binding sites per subunit. In some of the n.m.r. studies only the isolated catalytic subunit rather than the whole protein was employed. This enzyme catalyses the condensation of L-aspartate and carbamyl phosphate to form carbamyl-L-aspartate, an intermediate metabolite in pyrimidine biosynthesis. It is chiefly of interest as a classical case of 'allosterism', that is the transference of an effect of binding at one site through a protein molecule to affect binding at another site.[32] The binding of succinate, a competitive inhibitor of L-aspartate, was studied in the presence of carbamyl phosphate and various analogues. Selective line-width and relaxation time changes were measured to determine the specificity of the interactions. The use of line-widths where small changes occur in the presence of such a large protein are less reliable than the direct measurement of relaxation times by pulse methods. The rate of succinate binding to the carbamyl phosphate-catalytic subunit complex was slow $(2\cdot3 \times 10^5 \text{ M}^{-1} \text{sec}^{-1}$ at $33°)$ and very little difference was observed for the native enzyme. This indicates, rather surprisingly, little influence of the regulatory subunits on the active site under the conditions used.[79]

g. *Alcohol Dehydrogenase*

Binding of the coenzyme nicotinamide adenine dinucleotide (NAD) to the enzyme yeast alcohol dehydrogenase resulted in preferential increase in the line widths of the C2 and C8 hydrogen absorptions of the adenine moiety,[80] in contrast to an earlier report.[28] Some differences in the binding of NAD to equine liver alcohol dehydrogenase were also noted.[80]

h. *Ribonuclease*

A p.m.r. study of the binding of mononucleotide inhibitors to bovine pancreatic ribonuclease has been reported, although this lacked sufficient detail.[81] Several other attempts to obtain quantitative data from the line-widths and chemical shifts of inhibitors[82] have not led to any consistent conclusions. This may result from the line-broadening effect of the tauto-meric exchange observed for cytosine.[83] Line broadening has been observed for the phosphate resonance of cytidine-3′ phosphate in ^{31}P magnetic resonance on the addition of ribonuclease.[84] This effect was significantly greater than the viscosity broadening produced by non-interacting trypsin. The line broadening and chemical shift effects on the ^{31}P resonance of uridine-3′ phosphate have been quantitated to provide a dissociation rate constant and activation energy for the interaction with ribonuclease.[85] Detailed analysis may provide more insight into the nature of the enzyme groups close to the phosphate moiety as seen by its n.m.r. properties.

i. *Membranes*

P.m.r. has been used to study the structure of membranes and their components, although the systems are sufficiently complex to defy straight-forward analysis.[86] In one case benzyl alcohol, an anaesthetic, was used as a probe, and the line width of the aromatic protons were monitored as a function of the alcohol concentration in the presence of 1 per cent membrane ghosts. An initial decrease in line width followed by an increase at a critical concentration was taken to indicate a disordering of the lipid followed by increased binding of the anaesthetic to uncovered protein binding sites.[87] However, complications may arise in such a system due to anisotropic effects on relaxation phenomena as a result of binding to particulate matter.

One of the most significant areas of molecular biology encompasses the correlation of biological activities of small molecules with their chemical structures. Classically this has, of course, been one of the main fields of research in protein biochemistry and pharmacology. From the above review of n.m.r. observation of the small molecule it should be apparent that this technique has a great potential for the study of binding to macromolecules. Its particular advantage is the ability to differentiate between different atoms in the small molecule during a dynamic equilibrium process in solution.

2. Observation of the Spectrum of the Protein

a. *Spectral Assignments in Peptides and Proteins*

The first p.m.r. spectrum of a protein, ribonuclease, was reported in 1957,[88] and was interpreted in terms of the spectra of amino acids as then known.[89] It was believed for some time that little information was obtainable from the broad envelopes observed for protein n.m.r. spectra. Thus, many subsequent studies concentrated on the changes observed in the spectral envelopes of various proteins on heat-denaturation.[90-99] The sharpening of resonances in the spectrum as a result of increased thermal motion and magnetic equivalence were in the earlier work discussed in a general qualitative manner.

Polypeptides and the helix-coil transitions which they undergo have also been extensively investigated by n.m.r.[100-103] although there is some disagreement regarding the interpretation of the phenomena of multiple peaks.[104-106] It is well known that the amino acid sequence of a protein determines its three-dimensional structure.[107] It has also become clear from several protein structures determined by X-ray crystallography that α-helices play only a minor role in the formation of the entire conformation.[58,70] Thus, the phenomena observed in studies of the helix-coil transition are not directly applicable to the interpretation of protein n.m.r. spectra.[92,98] Nevertheless, n.m.r. provides a sensitive probe for conformational effects in oligopeptides, including the determination of the critical size (heptamer) for helix formation.[108]

That magnetic effects on hydrogen nuclei in proteins depend largely on the tertiary structure has been confirmed by two approaches, (a) analysis of chemical shift effects in oligopeptides, (b) comparison of the spectra of proteins with those expected from the known amino acid components (see below). The small magnitude of the nearest neighbour substitution shifts (<0.1 ppm) in oligopeptides shows clearly that primary structure effects are relatively unimportant in determining chemical shifts of residues in peptides and protein.[109-110] These results also indicate the limitation of n.m.r. as a tool for the analysis of primary sequences in proteins, although a certain amount of success have been achieved through the pH-dependencies of resonances (mainly end-effects) in small peptides[111] and by detailed comparisons.[112]

The majority of applications of n.m.r. to small peptides has been concerned with conformational analysis of cyclic peptides. There are several reasons for this; (a) the cyclization greatly reduces the number of degrees of freedom, making successful analysis more likely, (b) the measurement of coupling constants can provide conformational information via the Karplus relationship,[17] (c) cyclic peptides often have biological functions, and are the subject of intensive study by other techniques. Coupling constants for aromatic amino acids,[113] dipeptides[114] and random-coil polypeptides[115] are of course average values dependent on the proportions of each rotamer in solution. In cyclic peptides the NH–C_αH coupling constant can be related to a more restricted value for the rotation angle (ψ) about the NH–C_αH bond.[115,116] When such an analysis is carried out for many of the amino acid residues in the cyclic peptide a limited range of conformational structures or a unique structure may be calculated. The experimental approach includes spin decoupling to determine the relationship between resonances in different parts of the spectrum.[117] In this way N–H resonances can be assigned, and the rate of deuterium exchange reveals information on the extent of hydrogen bonding in the structure.[117,118] The following cyclic peptides have been analysed in this manner, gramicidin-S,[117–119] oxytocin,[120–122] vasopressin,[122,123] actinomycin-D,[124,125] cyclolinopeptide A,[126] antamanide[127] and several synthetic proline peptides.[128] In addition there have been studies of ion-binding to valinomycin,[129] bacitracin[130] and nonactin,[131] and of self-association of tyrocidin-B.[132] In these cases selective changes in chemical shifts and line widths are monitored as a measure of the interaction. While there have been several differences in interpretation in analyses of cyclic peptide spectra, the emphasis on a more quantitative approach[116,127] indicates a greater sophistication in the most recent applications.

Intramolecular interactions in peptides have been investigated utilizing the anisotropic effects of aromatic rings. Thus, the phthalimide group has been termed a 'marker' for the study of conformations of cyclic peptides[133] and polypeptides.[134] Aromatic shielding or deshielding effects have been described for cyclic dipeptides,[135,136] oligopeptides,[137] flavinyl amino acids,[138] vasopressin,[139] and polypeptides.[140] The results in such cases emphasize the secondary structure effects between adjacent amino acids which can be expected in the analysis of more complex systems such as proteins.[137]

The advent of spectrometers operating at higher magnetic fields[20] and the introduction of time-averaging[141] has dispelled the prior belief that protein n.m.r. spectra are unresolvable envelopes. Detailed peak assignments are necessary before structure–function correlations can be made from the effects of small molecule binding on the protein spectrum. Assignments have been made in most published studies, but many have been of limited

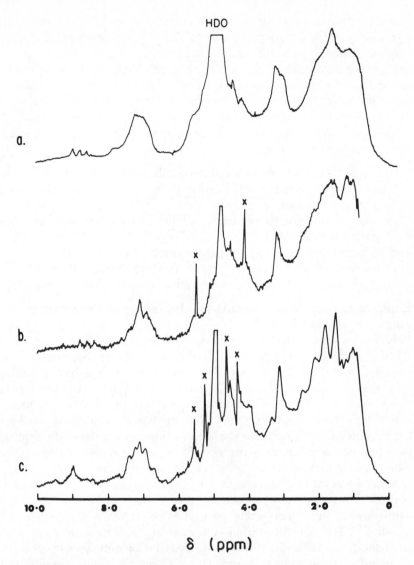

δ (ppm)

Figure 10. Spectra of ribonuclease taken at different operational frequencies:
(a) 60 MHz, 20 per cent solution in D_2O, pH 5·2, 26° (D. Meadows),
(b) 100 MHz, 10 per cent in D_2O, pH 5·8, 5 CAT scans, 32°
(c) 220 MHz, 11 per cent in D_2O, Ph 6·8, 22°.[20] Scale in p.p.m. downfield from external HMS (a) and (b) and internal DSS (c). Histidine C2 protons absorb at *ca.* 8–9 p.p.m., aromatic hydrogens at *ca.* 6·5–7·5 p.p.m. and other resonances are due to aliphatic protons. X indicates spinning side bands of the large HDO peak. ((c) from R. C. Ferguson and W. D. Phillips, *Science*, **157**, 257 (1967). Copyright 1967 by the American Association for the Advancement of Science)

value. This is not surprising in view of the limited resolution in the spectra of native proteins of molecular weight above 10,000, even at the highest magnetic fields (Figure 10). This is due not only to the large number of overlapping resonances, but also to the slow correlation times of macromolecules, resulting in broad resonances.[22] The extent of line broadening depends upon the degree of rigidity with which a given proton is held to the macromolecule. Line widths in general increase with molecular weight.[90,142]

There are several approaches to the analysis of protein p.m.r. spectra: These may be summarized;

(a) comparison with results for amino acids and oligo-peptides,
(b) comparison between very similar proteins which have few amino acid substitutions,
(c) analysis of resolved resonances at the extreme low and high field portions of the spectrum,
(d) selective deuteration to remove unwanted proton resonances,
(e) the use of paramagnetic ions to produce contact shifts allowing resonances to be clearly seen beyond the normal chemical shift range.

Examples of each of these approaches will be given, the last being dealt with in section IIIB2.

There have been several compilations of chemical shift data of amino acids[89,96,143] and correlations with protein spectra (for recent reviews see references 144, 145). The spectra of denatured porcine and bovine insulins (51 amino acids)[146] and glucagon (29 amino acids)[147] have been analysed by comparison with spectra of amino acids and peptides of known sequence. A similar approach has been made to the assignment of resonances in several larger proteins,[95-98] including the use of computer simulation to approximate the spectra of denatured proteins.[154] In these studies no assignments of individual resonances in the native protein were obtained.

Mandel was the first to recognize the potential usefulness of the downfield portion of a protein p.m.r. spectrum arising from the smaller number of hydrogens of the amino acids phenylalanine, tyrosine, tryptophan and histidine.[96] The aromatic region of spectra at 100 MHz of hen egg white lysozyme denatured under various conditions has been analysed in detail into its amino acid components (Figure 11).[142] Comparison with chemical shift values for aromatic amino acids in small peptides under the appropriate conditions (Figure 12) indicated differences in their environments between the heat and urea denatured enzymes, e.g., non-equivalent Phe residues were indicated for the heat denatured form (Figure 11b). This confirms that tertiary structure effects have the greatest influence on protein p.m.r. chemical shifts (see above). Also since the aromatic residues are spread throughout the molecule they act as a monitor of the conformation. Thus, different conformations are indicated for lysozyme under the different denaturing conditions.

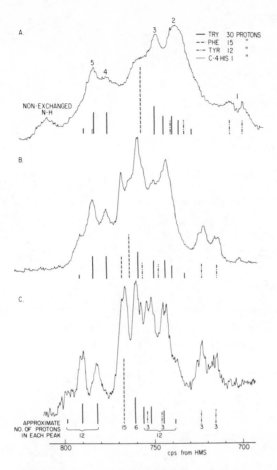

Figure 11. Time averaged 100 MHz n.m.r. spectra of the aromatic region of denatured hen egg white lysozyme and assignments: (a) in 8 M urea, pH 4·5, 32°, (b) 0·1 M NaCl, pH 4·6, 75°, (c) 8 M urea plus excess 2-mercaptoethanol, 65° [142]

The only major change which resulted on increasing the molarity of urea up to 8 M[148] was the downfield shift (*ca.* 0·1 p.p.m.) of a peak which was assigned to strongly shielded Try C2–H in the native enzyme. These may be identified with the Try residues at positions 62 and 63[58] in the peptide chain since the C2-H of the peptide Try–Try absorbs 0·15 p.p.m. to higher field than in the amino acid due to base stacking.[156]

The first unequivocal assignment of an individual proton resonance in a protein spectrum was that of the single imidazole C2 proton of His-15 of hen egg white lysozyme.[149] The histidine ring protons absorb to very low

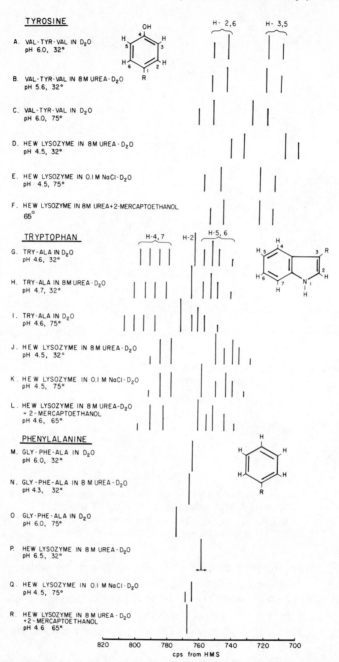

Figure 12. Resonance positions in the aromatic region of spectra of tryptophan, tyrosine and phenylalanine in peptides and denatured hen egg white lysozyme[142]

field due to the deshielding effect of the nitrogen atoms in the imidazole ring. On protonation the aromatization of the ring produces a further deshielding (Figure 13) and a titration curve results for the imidazole C2 and C4 proton resonances.[150] The ionization constant of 5·8 (uncorrected) for the imidazole residue of His-15 of lysozyme[149] was confirmed in another

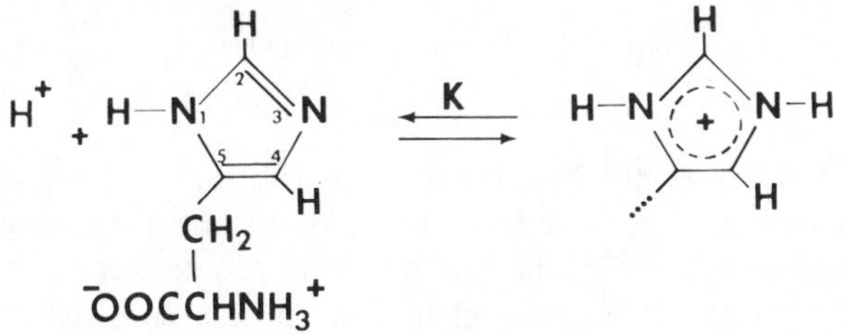

Figure 13. The protonation equilibrium of the imidazole
ring of histidine

study.[151] In comparing pK values derived in this way to other data it is important to remember that they are usually determined in solution in D_2O. Although the deuterium isotope effects on the glass electrode[152] and on the protonation process in solution are almost equal and opposite[153] they are not necessarily identical.[177]

Bradbury and Scheraga[155] partially resolved the C2-H peaks of the four histidine residues of ribonuclease-A (at 60 MHz). The unequivocal resolution (at 100 MHz) of all four peaks and the plotting of their individual titration curves was reported by Meadows, *et al.*,[149] together with titration curves for the four histidine residues of Staphylococcal (Foggi) nuclease. The difference in the microscopic pK values observed reflect differences in the environments of the imidazole residues resulting from the tertiary folding of the polypeptide chain. Similarly, the differences between the pK values of the single histidine residue of hen egg white and human lysozymes of 5·8 and 7·1 indicate positively and negatively charged local environments respectively.[156]

The imidazole ring proton can therefore act as a sensitive non-disturbing monitor of conformational effects in proteins. Ionic strength and temperature effects upon the ionization constant have been observed[153,157] and thermodynamic parameters have been determined as a measure of the nature of the local environment.[153,156]

A more reliable quantitative analysis of these proton n.m.r. titration curves results from the application of a least squares analysis utilizing a theoretical relationship based on a simple proton association (Figure 13),[157,158]

$$K = [HB]/[H][B] \qquad (14)$$

and from equation (5),

$$\delta = ([B]\,\delta_B + [HB]\,\delta_{HB})/([B] + [HB]) \tag{15}$$

where δ is the observed chemical shift and δ_B and δ_{HB} are the values for the unprotonated and protonated species respectively. Substituting for [HB] (or [B]) gives the relationship,

$$\delta = (\delta_B + K[H]\,\delta_{HB})/(1 + K[H])$$
$$= (\delta_B + 10^{(pK - pH)}\,\delta_{HB})/(1 + 10^{(pK - pH)}) \tag{16}$$

This type of analysis has been applied to the n.m.r. titration curves of histidine derivatives[137,158] and the histidine residues in nuclease[157] and carbonic anhydrases.[160,161] The assignment of resonances in the downfield region to imidazole C2 protons when there are several histidine residues in a protein is made more objective by the measurement of relative areas. This can be accomplished by computer curve-fitting utilizing a Lorentzian shape (equation (6)) for the component peaks.[158] There are then two main criteria for an imidazole resonance in a protein, its titratability and its unit area, since each corresponds to a single proton. This approach has clarified the analysis of the imidazole C2 proton resonances in Staphylococcal nuclease[158,162] (Figures 14, 15). There is no doubt that the higher field of a superconducting magnet aids in such an analysis due to the increased sensitivity and resolution. Thus in the case of human carbonic anhydrase C

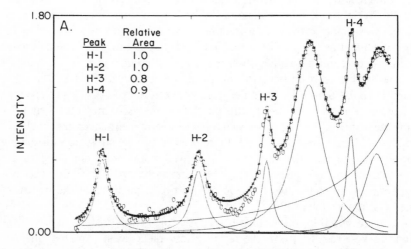

Figure 14. An example of computer curve-fitting of the downfield region of a 220 MHz p.m.r. spectrum of S. nuclease (10 per cent solution at pH 6·4 in 0·1 M NaCl–D₂O using Lorentzian-shaped peaks (equation (6)). The four peaks H-1 to H-4 are the four His C2 proton resonances[158]

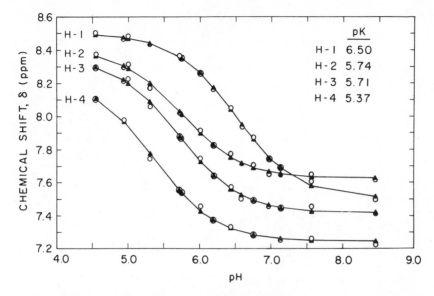

Figure 15. Titration curves of the imidazole C2 proton resonances of S. nuclease in 0·1 M NaCl–D$_2$O as numbered in Figure 14. ○, observed points; ▲, computer calculated points for the best fit to equation (16). The lines connect the calculated points[157]

a total of eight resonances could be resolved at 220 MHz (Figure 16) of which seven gave rise to titration curves (Figure 17).[161] Similarly multiple imidazole resonances have been resolved in the case of sperm whale and horse myoglobins.[163]

Assignments of several imidazole C2 proton resonances have been accomplished. For nuclease one of the four curves (H-3, Figure 15) could be assigned to His-124 by comparison of the results for the enzymes derived from the Foggi and V 8 strains of *Staphylococcus aureus*.[164] There is a single amino acid substitution, a Leu residue in the V 8 enzyme for His-124 in the Foggi enzyme[165] (this approach corresponds to (b) above). The resonance H-2 of nuclease (Figure 15) has been tentatively assigned to His-46 since it is selectively affected by the addition of ions (Ca^{2+}, La^{3+}, Nd^{3+})[166,167] and is the only histidine residue which is near to the ion binding site in the structure of nuclease determined by X-ray crystallography.[168]

The four imidazole C2 proton resonances observed for ribonuclease-A have been assigned to the four histidine residues.[169] They were initially assigned in pairs since carboxymethylation of either His-12 or His-119 affected two of the curves. These two were then assigned individually by exchange of the His-12 C2-H for deuterium in ribonuclease-S peptide

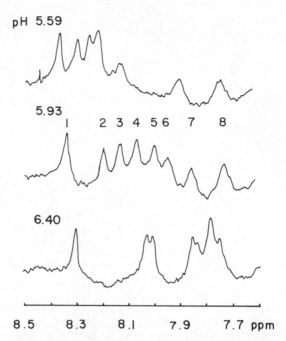

Figure 16. Histidine region of 220 MHz n.m.r. spectra of human carbonic anhydrase C at various pH values.[161] (Reprinted from *Biochem.*, **11**, 327 (1972). Copyright 1972 by the American Chemical Society. Reprinted by permission of the copyright owner)

(residues 1-20) followed by reconstitution of ribonuclease-S'.[170] Comparison with the spectra of ribonuclease-S indicated one titration curve had been lost, and this was assigned to His-12.[169] The extension of these assignments to ribonuclease-A was made on the assumption that both histidine residues are reduced in pK by nearly the same extent in going from ribonuclease-S to ribonuclease-A. It has been pointed out that the alternative assignments are possible if the histidines are affected differently.[171]

The other pair of ribonuclease imidazole C2 proton resonances were assigned[169] to His-105 and His-48 on indirect evidence based on their positions in the structure determined by X-ray crystallography.[172,173] The analysis of a titrating resonance observed in the downfield region of spectra of ribonuclease-A in H_2O, thought to correspond to a slowly exchanging imidazole N-H, may shed further light on these assignments.[174]

Three reports have appeared indicating that the n.m.r. titration curves assigned to histidines-12 and -119 are asymmetric.[158,171,175] This was explained in two of these studies[171,175] by hydrogen-bonding between the

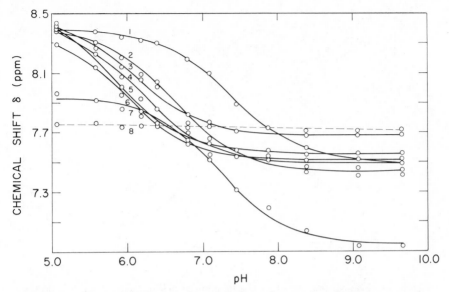

Figure 17. Titration data for the resolved resonances in the histidine region of spectra of human carbonic anhydrase C with curves nos. 1 to 8 corresponding to the numbered resonances in the pH 5·93 spectrum of Figure 16. Solid lines are the best fits to the data for each curve using equation (16). The dashed line represents a non-titrating resonance.[161] (Reprinted from *Biochem.*, **11**, 327 (1972). Copyright 1972 by the American Chemical Society. Reprinted by permission of the copyright owner)

two imidazole groups in the active site. Such a structure would have important consequences for the mechanism of action of the enzyme, and would tend to support that proposed by Witzel involving just such a structure.[176]

However, n.m.r. titration curves showing inflections have been observed for histidine model compounds (Figures 18, 19), and these have been shown to result from the titration of the amino and carboxyl groups[177] (Figure 13). These curves have been analysed by a mathematical model assuming interaction between two adjacent titrating groups[178] (Table 2). Two cases have been delineated, (a) that in which the pK of the imidazole itself and the neighbouring group are well separated, and (b) that in which they are close (<2 pH units). In case (a) a chemical shift plateau exists between the two inflections in the curve, and a sum of single proton association equilibria (equation (16)) may be used to fit the curve (Figure 18). In case (b) an asymmetric curve results and the equation for interaction (Table 2) may be used to obtain the microscopic pK values for the system (Figure 19). In all cases the pK values obtained corresponded to those determined by titrimetric analysis.[177,178]

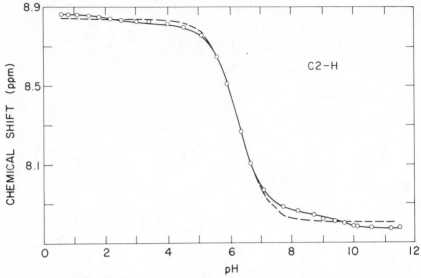

Figure 18. Chemical shift data (○) as a function of pH for the imidazole C2 proton resonance at 60 MHz of L-histidine. The dashed line is the best fit using the simple equation (16). The solid line is the fit of three such equations, giving three pK_a values of 1·98 (carboxyl), 6·21 (imidazole) and 9·24 (amino).[178] (Reprinted from *Biochem.*, **11**, 541 (1972). Copyright 1972 by the American Chemical Society. Reprinted by permission of the copyright owner)

These methods have subsequently been applied to titration data obtained at 220 MHz for ribonuclease-A.[179] The curve which has been assigned to His-105 was found to fit a simple proton association equilibrium (equation

Table 2. Mathematical model for interacting groups[178]

$$K_{A0} = C_{10}/C_{00}H \qquad\qquad K_{B0} = C_{01}/C_{00}H$$

$$K_{A1} = C_{11}/C_{01}H \qquad\qquad K_{B1} = C_{11}/C_{10}H$$

$$K_{A0}K_{B1} = K_{A1}K_{B0}$$

$$\delta_A = \frac{\delta_{A0} + (K_{B0}\delta_{A0} + K_{A0}\delta_{A1})H + (K_{A0}K_{B1})H^2\delta_{A1}}{1 + (K_{B0} + K_{A0})H + (K_{A0}K_{B1})H^2}$$

Where subscripts A and B refer to two sites on C, and 0 and 1 represent the protonated and unprotonated states of a site, respectively

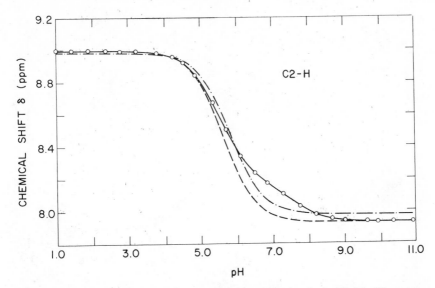

Figure 19. Chemical shift data (○) as a function of pH for the imidazole C2 proton magnetic resonance at 60 MHz of L-histidine methyl ester. The best fit obtained using equation (16) is indicated by a dot-dashed line, that ignoring the data between pH 3 and 9 by a dashed line, which emphasizes the asymmetry in the data. The solid line is the best fit obtained assuming interaction as shown in Table 2. The pK_a values obtained are 5·52 (imidazole) and 7·54 (amino).[178] (Reprinted from *Biochem.*, **11**, 541 (1972). Copyright 1972 by the American Chemical Society. Reprinted by permission of the copyright owner)

(16)), while that of His-119 showed an asymmetry in the low pH region (Figure 20) and that of His-12 showed a smaller asymmetry in the high pH region. Curve fitting procedures indicated these effects to be due to groups having pK's 4·5 and 8·4 respectively. From consideration of the X-ray structure[172,173] as well as earlier studies of ion binding to ribonuclease,[180] it was concluded that these groups are aspartic acid residue-121 and lysine residue-41 respectively (Figure 21).[179] The results of these curve-fitting procedures also were not consistent with a mutual imidazole–imidazole interaction, in the active site,[175] and would tend to support a push-pull type mechanism[181,182] (see page 550).

The single imidazole C4 proton (Figure 13) resonances of hen egg white and human lysozymes could be readily distinguished from the background of other aromatic resonances, since they are the only peaks which titrate in the pH range 4–8.[149,156] Similarly one of the imidazole C4-H resonances in ribonuclease-A was readily identifiable because it was well resolved from the aromatic envelope and had the same pK value as the C2 imidazole resonance assigned to His-105.[156] However, when there are several C4-H peaks shifting differentially with pH through the aromatic region it becomes

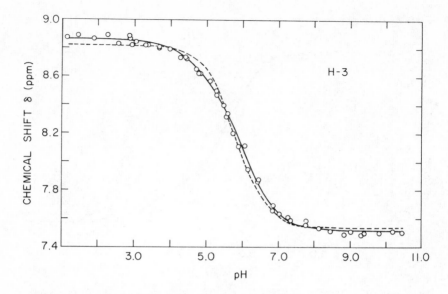

Figure 20. Chemical shift data (○) as a function of pH for the imidazole residue of resonance H-3, identified as His-119. The dashed line is the best fit using equation (16), and the solid line that obtained assuming interaction as shown in Table 2. The pK values obtained are 6·05 (imidazole of His-119) and 4·61 (carboxyl of Asp-121)[179]

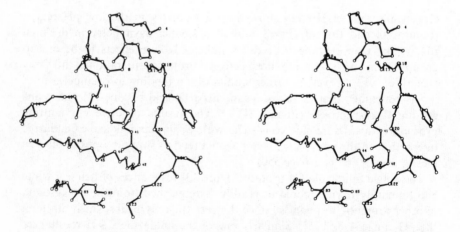

Figure 21. A view of the active site region of ribonuclease-S showing the histidine residues (12 and 119) and the close approach (3·8 Å) of the carboxyl group of Asp-121 to the imidazole group of His-119. The X-ray data of Wyckoff *et al.*[173] were used to generate the picture

very difficult to ascertain the correct continuity of the curves.[162] This problem has been overcome to some extent, for example, in the case of ribonuclease-A, by taking difference spectra utilizing a computer program.[171]

Several peaks have been observed in the extreme high field region of the spectra of native hen egg white lysozyme, but these were absent in the spectra of the heat-denatured form.[97] They have been identified with highly shielded aliphatic hydrogens which are close to the centre of aromatic residues. From possible contacts of this nature in a model based upon the structure determined by X-ray crystallography some tentative assignments were suggested.[97,98] These high-field resonances were further investigated by the use of Co^{2+} binding to hen egg white lysozyme.[183] Since the ion binds stoichiometrically and probably at Asp-35,[58] the magnitude of the shift effects of the paramagnetic ion could be related to the distance from that site to the adjacent shielded hydrogens.

Differences between these shifted resonances in bovine α-lactalbumin and hen egg white lysozyme have been analysed in terms of the sequence homologues between these two proteins. While the number of sequence differences were too great to provide individual assignments ((b) above), it was found possible to relate these differences to the spectral differences between the two proteins.[159]

An alternative approach to the resolution of individual proton resonances in proteins is the use of deuteration ((d) above), a standard method in studies of small molecules.[186,187] Of course, the process of deuterating a protein is a far more difficult task. It has been accomplished by the biosynthetic approach. In one case algae were grown in D_2O in the presence of an excess of leucine. Several protein products were shown to have a simpler p.m.r. spectrum.[188] A more sophisticated approach utilized the fact that *Staphylococcus aureus* requires all amino acids in its growth medium.[189] It was grown therefore on a mixture of deuterated amino acids derived from deuterated algae, with only selected normal amino acids added.[190] As a result of the conditions used for the hydrolysis of deuterated algal protein several of the amino acids were partially re-exchanged with hydrogen.[191] These included the tyrosine α-hydroxyl positions, which were of interest for p.m.r. studies with nuclease in view of the known involvement of tyrosine residues in the activity of this enzyme.[192] As shown in Figure 22 a dramatic simplification in the aromatic region of the spectrum of nuclease resulted for this largely deuterated analogue, with several sharp resonances being resolved.[190] This approach nonetheless suffers from several drawbacks in that it is an extremely expensive process and most of the valuable deuterated amino acid material is lost in the growth medium (which is otherwise normal).

As a means to circumvent some of these problems, and in order to attempt an assignment of the tyrosine resonances, a partially deuterated tyrosine

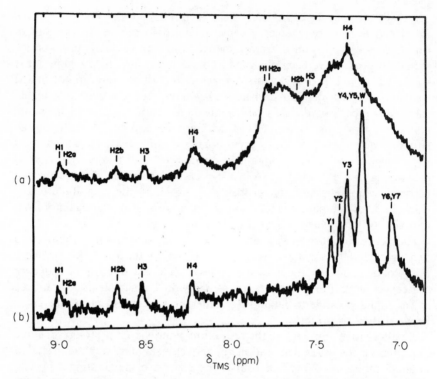

Figure 22. Comparison of the p.m.r. aromatic spectral regions at 100 MHz of (a) S. nuclease, and (b) the deuterated analogue Nase-D4 both at pH 6·0. Assignments are H1 to H4 (δ, 8–9 p.p.m.), His C2-H; H1 to H4 (δ, 7–8 p.p.m.) His C4-H; Trp C2-H, W; Tyr C2, 6-H, Y1-7[190]

analogue of nuclease was prepared.[193] Since this was isotopically normal apart from the positions α to the tyrosine hydroxyl the aromatic region of the spectrum showed intermediate resolution (Figure 23) between those of the normal and largely deuterated analogues (Figure 22). The selective nitration of tyrosine residues[194] allowed only a tentative assignment of two tyrosine resonances, which were both considered to be affected as a result of nitration at a single tyrosine residue.[193]

Several resonances observed in the extreme downfield region of spectra of hen egg white lysozyme in H_2O have been identified as originating from tryptophan NH's. Five of the six such protons present in this enzyme have been assigned on the basis of differential deuterium exchange rates, chemical modification and inhibitor perturbation[184,185] in relation to the known structural roles of tryptophan in the molecule[58] (for a review of this work see reference 145).

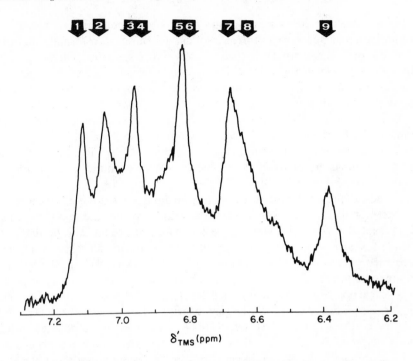

Figure 23. P.m.r. spectrum at 220 MHz of a partially deuterated tyrosine analogue of S. nuclease, [3,5-d]-tyrosyl nuclease, at pH 5·25. Assignments are His C4-H, peaks, 1, 2, 4 and 6; [3,5-d]-Tyr, peaks 3, 5, 7 and 8; Phe, peak 9. The remainder of the 3 Tyr and 2 Phe resonances were unresolved[193]

b. *Carbon-13 N.m.r. Studies of Peptides and Proteins*

As a result of the recent development of the Fourier transform (FT) n.m.r. method,[26,195,196] the application of carbon-13 magnetic resonance (c.m.r.) to peptides and proteins has become feasible. A study of 15 per cent ^{13}C enriched amino acids indicated that the resolution to be expected for the chemically different carbon atoms was *ca.* 15 times greater than for protons.[197] The first c.m.r. spectrum of a peptide, gramicidin, was obtained by slow-sweep methods.[198] However, several studies of peptides and proteins have since been reported utilizing the FT method.[199-202]

In order to carry out c.m.r. studies signal enhancement is necessary since the ^{13}C nucleus has a much lower sensitivity than the proton (1·6 per cent for equal numbers of nuclei at the same magnetic field) and exists in low natural abundance (1·1 per cent). Techniques used to accomplish signal enhancement in c.m.r. include (a) proton noise decoupling, which removes the spin–spin coupling (section B3) of hydrogens bound to carbon atoms,[203] as well as giving a nuclear Overhauser enhancement,[204] (b) the addition of the free induction decay signals resulting from the application of successive radiofrequency pulses to a sample, and (c) the Fourier transformation of the stored signals in a computer to give the slow-sweep spectrum. These methods make it possible to reduce the time taken to obtain a spectrum with the same signal/noise ratio by up to two orders of magnitude compared with conventional methods.

Chemical shift values determined in c.m.r. studies of peptides have indicated little difference from those of the amino acids themselves,[198,200,202] except for the two terminal residues in the chain.[205] In Figure 24 is shown as an example the analysis of the spectra of three amino-terminal oligopeptides of ribonuclease.[206]

In view of previous experience obtained from p.m.r. studies it is important to attempt to resolve resonances of individual carbon atoms in proteins. C.m.r. spectra of proteins which have been published so far indicate that this will not be an easy task. It was originally felt that the longer relaxation times observed for carbon atoms compared to protons would result in sharper ^{13}C resonances in proteins. However, it has been found[207] that $T_2 < T_1$ for ^{13}C, T_{2C} of carbon is largely determined by T_{1H} of the bound hydrogen, and $T_{1H} < T_{1C}$. Since line width is largely determined by T_{2C} (equation (7)), ^{13}C resonance line widths can be expected to be of the order of proton line widths, with the exception of quaternary carbon atoms with no bound hydrogen or with proton noise decoupling at very high power levels (above those normally employed). Some indication of sharp resonances to be expected for macromolecules were derived from c.m.r. studies of amino acids bound non-covalently to ion-exchange resins.[208] However, the ligands are able to exchange sites and rotate and would no doubt have sharper resonances than a covalently linked species. More significantly ^{13}C line widths observed for a large protein such as haemoglobin are of the order of 25–30 Hz.[209]

The use of partially relaxed Fourier transform methods will undoubtedly find useful application in protein studies. This technique enables the determination of relaxation times by the application of pulsed n.m.r. techniques combined with Fourier transform methods.[210] So far it has been applied to the analysis of motional rates in native and denatured ribonuclease.[211]

As in the case of p.m.r. spectra of proteins, the presence of many ^{13}C resonances of the same type defy individual resolution.[199,206] Furthermore, they do not show the same degree of variability of chemical shift for the same chemical type within a protein molecule. This is because ^{13}C chemical shifts are more determined by covalent structural rather than by environmental factors. Put in another way they show more paramagnetic contribution to the chemical shift[212] and less contribution from the external environment than hydrogen. On the other hand absolute changes in ^{13}C chemical shifts as a result of pH or conformational effects may be quite large (several p.p.m.) although smaller percentage-wise than equivalent changes for proton

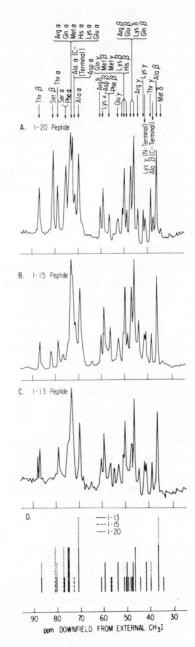

Figure 24. Carbon-13 n.m.r.-Fourier transform spectra at 25·1 MHz with proton noise decoupling of three N-terminal peptides of ribonuclease, showing the aliphatic carbon resonances. The assignments for the 1-20 peptides are shown above, and a stick diagram of the resonances of the amino acids are shown below, with normalized intensities[206]

chemical shifts. This is exemplified by the resonances of the ring carbon atoms of the imidazole group in histidine. It was observed that the quaternary C-5 carbon atom showed the greatest chemical shift difference between the protonated and unprotonated forms (*ca.* 5 p.p.m., 2·5 per cent) while the C2 and C4 atoms, to which protons are attached show smaller titration effects (Figure 25). By contrast these protons show an approximately 1·0 and 0·5 p.p.m. chemical shift change on titration respectively (10 and 5 per cent of the normal proton chemical shift range).[154] This is indicative of the differing factors determining chemical shift effects in p.m.r. and c.m.r.[202,212] It should be noted that titration effects of adjacent groups were also observable in these spectra (e.g. of the amino titration on the carboxyl carbon resonance[209]), and this will no doubt be of value in further studies.

Isotopic labelling of proteins with ^{13}C for c.m.r. studies should undoubtedly have certain advantages. Enrichment increases the sensitivity of the method, reducing the time taken to obtain spectra. Selective ^{13}C enrichment of individual carbon atoms or amino acids in a protein enables studies of individual ^{13}C resonances to be made. In this sense ^{13}C enrichment is 'positive' in that the isotope itself is observed, as opposed to the 'negative' labelling with deuterium in p.m.r. studies to remove the background. A report on the synthesis and c.m.r. study of a selectively enriched peptide $[^{13}C\text{-Phe}_8]$-RNase-(1-15), in which the Phe residue at position 8 in the 1-15 peptide of ribonuclease was 15 per cent enriched in ^{13}C, has recently appeared.[200] The chemical synthesis of peptides is clearly preferable for ^{13}C to the biosynthetic approach as used for deuteration,[190] in that the ^{13}C isotope is much more expensive than ^2H and would be largely diluted in the multitude of metabolic pathways. Nevertheless the recent large-scale production of high yield ^{13}C (>95 per cent) may make such an approach feasible.[213]

Dependent on the success of the approaches outlined above c.m.r. will undoubtedly be used to study molecular interactions. The particular properties of ^{13}C chemical shift effects will probably delineate it as a useful tool for monitoring conformational changes (alterations in the carbon skeleton) on binding.

c. *Denaturation Studies of Proteins*

As mentioned in a previous section (p. 541) several denaturation studies of proteins were carried out in order to attempt the assignment of resonances in denatured and native protein spectra. Meaningful studies of the denaturation process itself could be accomplished only when individual resonances were resolved and identified. These could then be followed during the denaturation process as monitors of *intra*-molecular interactions resulting from tertiary folding (see p. 544).

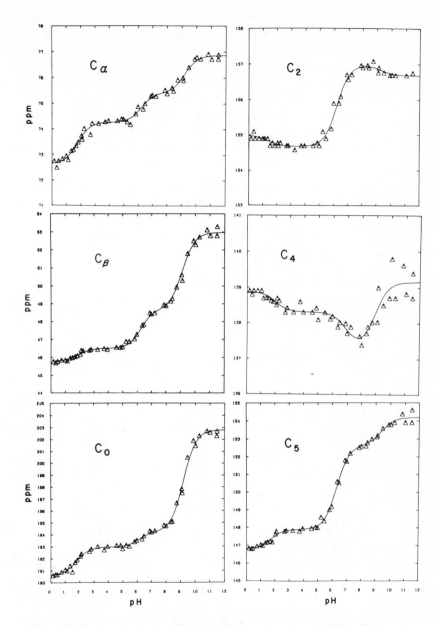

Figure 25. Carbon-13 chemical shift data ($\triangle$) as a function of pH for all the carbon atoms in histidine. It is quite clear that the ionization constants of other groups are reflected in the shifts of the carbon under observation

The conformational transitions of ribonuclease-A and of hen egg white lysozyme produced by several denaturants have been investigated.[214–216] In these cases average peak parameters were used to quantitate the process. The heights of several peaks were measured by Bradbury and King,[214] who reported differential effects in some cases, indicating intermediate states in the denaturation process. However, the measurement of peak height for broad overlapped resonances is not a satisfactorily accurate parameter. Cohen Addad[215] calculated a weighted mean of all the peak positions in the aromatic region of ribonuclease during denaturation, but this specifically

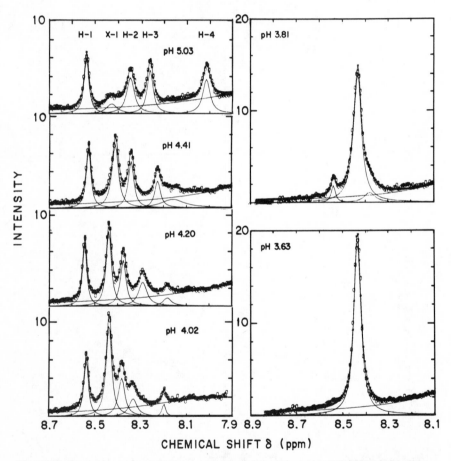

Figure 26. Computer curve fits of time-averaged p.m.r. spectra at 220 MHz of S. nuclease on passing through the acid transition, pH 5·0 fully native form, pH 3·6 fully denatured form. Open circles are experimental points and asterisks calculated points determined by the sum of Lorentzian-shaped components represented by continuous lines. Peaks H-1 to H-4 are the four imidazole C2 proton resonances in the native form

excludes the observation of potential selective differences. McDonald, Phillips and Glickson[216] have studied the transitions of hen egg white lysozyme under several conditions. They measured the intensities of a number of resonances in the spectra, including individual His C2-H and Try NH resonances, but they also averaged the intensities of several peaks. They concluded that a single cooperative transition occurred for hen egg white lysozyme under the conditions used, which is consistent with other studies.[217] Conformational transitions of histones have been studied by Boublik *et al.*[231] using p.m.r.

Two reports have appeared on the denaturation of Staphylococcal nuclease.[218,220] In one case the chemical shift of resolved resonances was used as the parameter to follow the effects of the alkaline transition on tyrosine residues in deuterated analogues.[164,218] It was concluded that distinct conformational equilibria occurred for several residues, and a detailed picture of the denaturation process has been based upon these data.[219] However, this analysis is complicated by the fact that the chemical shift may be a complex function of the concentration of several species, that the tyrosine residues titrate in alkaline pH and the alkaline transition is irreversible.

The acid transition of nuclease is fully reversible and was investigated by Epstein, Schechter and Cohen[220] by the measurement of the areas of the four resolved His C2 protons by computer curve-fitting with Lorentzian peaks (Figure 26). The results indicated that three of the histidine residues follow essentially the same cooperative transition, while a fourth (H4) follows a complex relationship indicating at least two transitions (Figure 27). Kinetic studies of the change in tryptophan fluorescence have also indicated at least two processes for the acid renaturation.[221]

In summary it is quite clear that n.m.r. has the resolving power to enable the observation of thermodynamic intermediates during a conformational transition in solution. However, the only reliable quantitative parameter is the area (intensity) of a resolved resonance.

d. *Enzyme-Inhibitor Binding*

Only a few detailed p.m.r. studies of enzyme-inhibitor binding have been reported in which the spectrum of the enzyme is observed. In most cases chemical shift changes of enzyme resonances resolved in the low field portion of the spectrum (see section IIIA2a) have been monitored. Inhibitors or substrate-analogues were used in these studies, rather than substrates themselves, because of the use of time-averaging, employed to facilitate observation of the enzyme resonances. This usually requires several hours per spectrum, restricting the studies to equilibrium processes in solution. Generally speaking fast exchange processes are studied so that equation (12) applies.

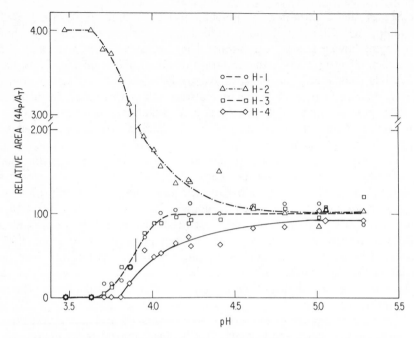

Figure 27. The areas of the four imidazole C2 proton resonances of S. nuclease, determined by computer curve-fitting as shown in Figure 26, plotted as a function of pH. The calculated value for each resonance (A_p) is expressed as a fraction of the total area of the four peaks, normalized to 4·0 at each pH. The curves are hand-drawn approximations to the data. The vertical lines at pH 3·9 represents the pH of half-change of the resonances H-1, H-2 and H-3

Addition of inhibitors to hen egg white and human lysozymes produced no effects upon their single His C2-H p.m.r. titration curves.[142,156] This is consistent with other evidence of the non-involvement of histidine in the active centre of these enzymes.[58] Addition of oligo-saccharide inhibitors, however, did produce extensive changes in the aromatic envelope which could only be attributed to effects upon highly-shielded Try residues.[142,156] The effects at saturation were essentially identical for both di- and tri-*N*-acetylglycosamine, and were strikingly similar for both lysozymes, indicating similar modes of binding in each case.[156,222]

Tryptophan N–H resonances have been monitored upon the binding of NAG and di-NAG to hen egg white lysozyme.[185] Only two of the resonances were significantly affected in each case, one being shifted upfield and one downfield. These effects were attributed to hydrogen bond formation with the saccharide, and the two resonances were assigned to Try-62 and 63 partly from other evidence.[58] The binding constant (equation (11)) was calculated by least squares fitting of the chemical shift *vs.* NAG concentration data, but values were somewhat lower than expected.

In view of the known involvement of two His residues (12 and 119) in the activity of ribonuclease-A several studies of the His C2-H p.m.r. titration curves have been reported (see section IIIA2a). Comparison of the effects of the three cytidine monophosphates upon these curves have been interpreted in terms of detailed structures of ribonuclease-inhibitor complexes. These were based on several lines of evidence, (a) 2'-CMP and 3'-CMP gave rise to deshielding of the protonated form of His-119 (downfield shift at low pH) and shielding for that of His-12; (b) 5'-CMP did not produce deshielding of His-119; (c) the pK value of His-119 was increased differently for each of the three mononucleotide inhibitors, while that of His-12 was increased to the same extent for each. This was taken to indicate a direct interaction between the phosphate moiety in the case of 2' and 3'-CMP and the imidazole residue of His-119, but only an indirect effect upon that of His-12. These results were then qualitatively interpreted in terms of the relative juxtaposition of the groups in the inhibitor–ribonuclease complex.[82,223]

Some re-evaluation of this analysis may be indicated in the light of several more recent findings; (a) the titration curves of His-12 and His-119 are asymmetric[158,171,175] and the asymmetry is greatly accentuated on the addition of phosphate or mononucleotide inhibitors;[175,179] (b) the data on which the above deductions were based was extended down to *ca.* pH 4·3,[223] and this is the pH region where the greatest asymmetry occurs (indicating that relative shielding or deshielding at such a pH is an unreliable guide); and (c) 2' and 3'-uridine monophosphates produce a greater effect upon His-12 than His-119.[224] Further titration data with uridine as well as cytidine mono-nucleotides, and also with dinucleotide substrate-analogues containing a phosphonate bond to prevent cleavage,[225] as well as analysis of the asymmetry in the titration curves,[224] may help to clarify this matter.

The presence of an asymmetry in the curves does not support a mechanism in which the two active-site histidines are mutually interacting[175,176] as indicated above (section IIIA2a). Further the X-ray crystallographic structure indicates that the phosphate group binds between the two active site histidines.[168] Thus a push-pull type (in-line) mechanism[181] is definitely indicated for ribonuclease[182,226] (for a review see reference 227).

A study of the His C2-H resonances of ribonuclease-T_1 has been reported. One histidine residue was considered to interact directly with the phosphate moiety of guanosine-3' phosphate. The three curves were extremely asymmetric, indicating interaction with neighbouring carboxyl groups.[228] The observation of a similar arrangement for ribonuclease-A may indicate a specific function for such an interaction.

A detailed study of the effects of the inhibitor thymidine-3',5' diphosphate (pTp) on the tyrosine resonances in deuterated analogues of staphylococcal nuclease has been reported.[229] Three of the seven tyrosine residues were implicated in the interaction. The marked changes on binding in the absence of Ca^{2+} were surprising in view of an earlier report[230] of a requirement for Ca^{2+} for pTp binding. However, this may be attributed to the much greater concentrations used ($\times 100$) in the n.m.r. studies, and even in this case a low binding constant was calculated (*ca.* 2×10^{-3} M). One of the three tyrosine resonances which was shifted upfield by pTp was shifted further downfield on the addition of Ca^{2+} to form the ternary complex, and a fourth Tyr resonance was also shifted slightly upfield. It was suggested that the former resulted from a residue directly bonded to a phosphate group in pTp while the latter was near the second Ca^{2+} binding site.[229]

Unfortunately, the tyrosine resonances were not well-resolved, and Tyr peak assignments were not made. An attempt to assign the resonance of Tyr-115 by selective nitration indicated the difficulty of such a process.[193] More recent studies indicate that the effect of Ca^{2+} on one of the histidine titration curves (H2) may be explained by a local ionic strength effect as a result of pTp binding.[166]

Two analyses of inhibitor binding to carbonic anhydrases have both indicated conformational effects.[160,161] In both cases the His C2 proton titration curves were monitored, in one case for human carbonic anhydrase-B[160] and in the other for both the B and C isoenzymes.[161] While a titration curve with $pK \simeq 7$ was affected in the latter case, the bulk of evidence tended against implicating this as the group with a similar pK value known to be involved in inhibitor binding and hydrolysis. Most notably in both studies of human carbonic anhydrase-B using three different sulphonamide inhibitors two curves with $pK \simeq 7$ were completely unaffected. It was therefore concluded that the crucial group in the activity is most probably a water molecule.[161]

The direct application of magnetic resonance techniques to studies of proteins, and the effects of substrate analogues upon protein resonances, is one of the most difficult and challenging areas of research. While X-ray crystallography undoubtedly provides more detailed structural information, the results to be anticipated in future years from such n.m.r. studies can be expected to provide increasing insight into the fine structure of molecular interactions.

B. Interactions Involving Paramagnetic Species

1. Observation of the Spectrum of the Solvent or Small Molecule

Paramagnetic species contain unpaired electrons which, by virtue of the large electron magnetic moment, enhance the relaxation of nuclear resonances (section IID). Because of this fact the interactions of paramagnetic metal ions with nucleic acid derivatives and proteins have been extensively studied. One of the commonest techniques employed known as proton relaxation enhancement (p.r.e.) is to measure the spin lattice relaxation rate $(1/T_1)$ of the solvent water protons. When the metal ion is bound to a macro-molecule the effect on the relaxation rate of the bulk water protons is reduced and this provides a sensitive probe of the nature of the metal ion complex.[232,233]

The theory of nuclear relaxation by paramagnetic ions was developed by Bloembergen and Solomon.[234–237] For T_1 in the fast exchange condition (section IIA),

$$\left(\frac{1}{NT_1}\right)_{obs} = kDq\tau_c \tag{17}$$

where,

N is the molar concentration of paramagnetic ions,
k is a numerical constant,
τ_c is the correlation time of the dipolar electron–nuclear spin interaction,
q is the coordination number of water for the ion, and
D is a factor dependent on the inverse sixth power of the ion–proton internuclear distance.

For T_1 to reflect immobilization due to the macromolecular environment τ_c must be approximated by its rotational component τ_{rot}.[238] Since

$$\frac{1}{\tau_c} = \frac{1}{\tau_{rot}} + \frac{1}{\tau_{trans}} + \frac{1}{\tau_e} \tag{18}$$

where τ_e is the electron–spin lattice correlation time, equation (17) is useful only for ions with long τ_e, such as Mn^{2+}, Cu^{2+} and Cr^{3+}.

In a study of ion-binding to DNA, Eisinger, Shulman and coworkers[239,240] defined the *relaxation enhancement factor*,

$$\varepsilon = 1/T_1^* \Big/ 1/T_1 = T_1/T_1^* = \frac{p^*\tau_c^*}{p\tau_c} \tag{19}$$

where,

T_1 is the spin–lattice relaxation time of the water protons of the para-magnetic ion solution,
$p = kDq$ (equation (17)), and
$T_1^*p^*$ and τ_c^* denote the values of those parameters when the ion is bound to the macromolecule.

This allowed a classification of various types of ion binding. Thus, for $\varepsilon > 1$ the main effect was considered to be an increase in τ_c due to the ion being bound externally on the macromolecule, as for Mn^{2+} to DNA ($\varepsilon \simeq 8$). $\varepsilon < 1$ might indicate that the binding site for the metal ion is less accessible to water molecules, indicating $p^* < p$. This could result from a reduction in q or in D, possibly involving increase in the effective ion–proton distance on binding.

For fast exchange of ligand between two environments, one containing a paramagnetic ion, it has been shown that[237]

$$\frac{1}{T_{1P}} = \frac{f}{T_{1M} + \tau_M} \tag{20}$$

where,

> T_{1P} is the paramagnetic contribution to the relaxation time,
> T_{1M} is the relaxation time in the first coordination sphere,
> f is the fraction of ligand bound, and
> τ_M is the average time a ligand is bound before exchanging.

Equations (17) and (20) form the basis of most of the work on molecular interactions involving paramagnetic species. The comparable equations for T_2 are complicated in some cases, such as Mn^{2+}, by an additional factor resulting from an isotropic hyperfine interaction.

There have been a number of studies of Mn^{2+} binding to nucleic acids and derivatives,[239,240] including direct observation of their proton and ^{31}P resonances.[241-244] A reinvestigation which included a complete kinetic analysis of the binding of Mn^{2+} to ATP has indicated the importance of concentration studies when dealing with self-associating species.[245] Studies of Mn^{2+} binding to ribosomal particles[246] and to t-RNA[247] by measurement of the water proton magnetic relaxation have been reported. In the former case no buried binding sites were indicated. In the latter a number of strong cooperative sites were found ($\varepsilon \simeq 19$), which were changed upon aminoacylation or denaturation.[247]

Among the most significant contributions in this area have been the extensive studies of M. Cohn and coworkers on the nature of the substrate-metal–enzyme complexes formed by a number of metallo-enzymes (for recent reviews see reference 248). Large values of ε (*ca.* 10) were used as a monitor to indicate formation of a metal-containing complex due to increase in τ_c of the Mn^{2+} ion. Two categories of enzymes which transfer a phosphoryl group from ATP have been demonstrated,[249] one in which enhancement of the ternary complex is greater than the binary (I), and the other in which the reverse is the case (II). For class II the metal ion binds strongly to the substrate, for class I the metal ion binds directly to the enzyme (Table 3). The actual enhancement value observed for an enzyme in category I is

Table 3. Classification of enzyme–metal-substrate interactions into two groups[248]

Type I	Type II
Enhancement of relaxation rate in ternary complex ≫ binary complex	Enhancement of relaxation rate in binary complex ≫ ternary complex
Creatine kinase (rabbit muscle)	Pyruvate kinase (rabbit muscle)
Creatine kinase (beef brain)	Enolase (rabbit muscle)
Adenylate kinase (rabbit muscle)	Inorganic pyrophosphatase (yeast)
Arginine kinase (lobster muscle)	Phosphoenolypyruvate carboxykinase
Diphosphoglycerate kinase (rabbit muscle)	(pig liver mitochondria)
Hexokinase (yeast)	
Formyltetrahydrofolate synthetase (*Clostridium cylindrosporum*)	

considered to be characteristic of its binding site, and gives quantitative (binding constant) and qualitative (conformational) information on the site. Ionization constants for the groups involved in the binding of the Mn^{2+} ion can be obtained by evaluation of ε at different pH's, and competition studies indicate the nature of the binding of other ions.[248]

From such studies the nature of enzyme–metal-substrate (or inhibitor) complexes for pyruvate kinase[250] and pyruvate carboxylase[251] have been demonstrated. By comparison of the ε values in the presence of various substrates and avidin, a protein which binds strongly to the biotin cofactor, the nature of the coordination around the tightly bound Mn^{2+} in the active site of pyruvate carboxylase has been investigated.[251] Line width measurements in ^{19}F magnetic resonance of fluorophosphate, an inhibitor of pyruvate kinase, have enabled calculation of the $Mn^{2+}-^{19}F$ distance in the ternary complex.[250] The temperature and frequency dependence of proton relaxation rates in water have also indicated the existence of two conformational forms of pyruvate kinase, one of which is inactive and does not bind Mn^{2+}, and has also established the number of ligand water molecules to be three.[252] Of particular interest is the parallel correspondence between the relaxation enhancement factor, the maximum velocity of the enzymatic reaction and the rate constants for the iodo-acetic acid reaction of creatine kinase with various Mn^{2+}-nucleotides. Thus the ε value may be used as a measure of the selective conformational changes at the active site of creatine kinase.[253] It should be noted, however, that the assumption of slow exchange (see below), upon which the quantitative evaluations of rate constants in much of this work was based, has recently been challenged.[254]

Carboxypeptidase A (CPD) is a proteolytic enzyme which cleaves carboxy-terminal peptide bonds and requires a divalent metal ion for activity. Normally this is Zn^{2+}, but the Mn^{2+} complexed enzyme has 40 per cent activity.[255] Shulman and coworkers have studied the relaxation enhancement of the water protons[256] and line broadening of inhibitor molecules[257] and of fluoride ion[258] in the presence of Mn-CPD. They have been able to show that the bound Mn^{2+} has only one coordination site for exchangeable water ($q^* = 1$ in equation (19)). Fluoride ion was also found to bind directly to the Mn^{2+} ion, but not to affect the binding of water. It was therefore concluded that there are five or more coordination sites, three protein ligands (from X-ray crystallography), one water and one fluoride. Both the latter were displaced by inhibitors.[258] These workers also found that the weaker the inhibitor the faster was the exchange and the greater was the broadening of its resonances.[257] Two cases were delineated: (1) *Slow exchange* for strong inhibitors. In this case $\tau_M \gg T_{1M}$ and from equation (20), $1/T_{1P} = f/\tau_M$. Thus τ_M, the reciprocal of the dissociation-rate constant k_{-1} could be calculated, and from it (and the equilibrium constant K obtained routinely by other methods) the association-rate constant k_1 (equation (11), $K = k_1/k_{-1}$). Variations in K, such as the 100-fold increase in going from bromo-acetate to indole-acetate, could then be related to changes in k_1 and k_{-1}, in this case a tenfold increase and decrease respectively. Such values can in turn be interpreted as stronger binding of the aromatic moiety, for which the enzyme has a greater affinity. (2) *Fast exchange* of the ligand gives $\tau_M \ll T_{1M}$. Hence, from equation (20), $1/T_{1P} = f/T_{1M}$, and since T_{1M} is related to D (equation (17)), the distance r between the Mn^{2+} ion and specific inhibitor protons in the complex could be calculated. Values of r of $6\cdot9 \pm 1\cdot4$ Å for the t-butyl protons of butyl-acetic acid, and of $4\cdot3 \pm 0\cdot8$ Å for the methylene protons of methoxyacetic acid were obtained, which were consistent with values from model building,[257] and can be compared to the structure determined by X-ray crystallographic studies.[259]

2. Observation of the Spectrum of the Protein

As well as producing large effects on the resonances of water and small molecule ligands, paramagnetic ions also affect resonances in macro-molecules to which they bind. However, since the resonances in macro-molecules are usually quite broad, one would prefer to work with ions of short electron relaxation time which produce large contact shifts rather than line broadening effects. There is a similar reason for the recent introduction of 'shift reagents', usually based on europium which has a fast relaxation time. These produce simplification of complex spectra of bound sub-stances,[260,261] since they increase the chemical shift separation relative to the coupling constant (see section IIC). However, other lanthanide (rare earth) ions with slower electron relaxation times, such as gadolinium, appear to have wide potential uses in studies of proteins.[262,263]

A number of resolved resonances have been reported shifted to extreme high and low fields for cytochromes, myoglobin and other haem proteins (for a review see reference 145). These may be attributed to three causes; (a) contact shifts (10 to 30 p.p.m.) due to hyperfine interactions with metal ion,[183,264,266] (b) diamagnetic shielding due to the extensively electron delocalized porphyrin ring (2 to 5 p.p.m.)[267-269] and (c) diamagnetic shielding due to aromatic amino acid side-chains (0 to 2 p.p.m.).[97] It is possible to distinguish between these by the temperature dependence of (a) but not of (b) and (c)[97,270] subject to there being no conformational change in the protein.

The shifted and resolved resonances arising from the iron atom, in various states of ligation, have been most extensively studied in the cases of cytochrome *c*, myoglobin and haemoglobin. Some selected examples will be given. In the case of cytochrome *c* the sixth ligand to the iron atom in both the ferric and ferrous states was shown to be a methionyl residue, and the binding of cyanide ion was found to greatly affect the electron distribution in the haem group as evidenced by large changes in the shifted resonances.[271]

Utilizing the resonance assignments in the appropriate prophyrin derivatives,[269] Shulman and coworkers have analysed the shifted resonances in myoglobin by reconstitution with various porphyrin derivatives.[272] Very little difference resulting from these substitutions indicated the importance of the protein environment in determining the distribution of the unpaired electron density in the haem group. A similar conclusion resulted from a comparative study of cyanoferrimyoglobins from various species (seal, whale, porpoise and horse) which showed almost identical patterns of shifted resonances, notwithstanding several amino acid substitutions. This indicates that the important residues in the immediate vicinity of the haem group experience an almost identical electronic environment, and that the haem electronic structure is conservatively maintained.[273] On the other hand small changes in chemical shifts were observed for myoglobin on ligation with different species, indicating ligand-induced conformational changes.[274]

Much of the work on haemoglobin has recently been reviewed and interpreted in terms of subunit–subunit interactions.[275] Several more recent developments warrant mention. Differences in the spectra of haemoglobin variants with one or more amino acid substitutions have been reported.[276-279] This provides a valuable tool for observing the effects on the protein conformation and the haem group as a result of a specific alteration in sequence. Unfortunately, the protein spectral lines considered are quite broad in some cases.

A study of the mixed state haemoglobins has been reported, in which one subunit is kept in the ferric state and complexed with cyanide, while the other is in the normal ferrous state and can be oxygenated or deoxygenated. The spectra show two different quaternary structures for one mixed state species indicating a possible direct observation of a 'modified allosteric transition'.[280] As well as this probe of the conformational characteristics of haemoglobin, low field resonances observed in H_2O and derived from exchangeable tryptophan NH's, appear to provide a means of monitoring structural changes upon ligation.[281]

A further use of paramagnetic ions in studies of proteins involves their substitution for bound diamagnetic ions, followed by observation of the protein spectra. This has been accomplished for staphylococcal nuclease, which requires Ca^{2+} for activity, utilizing Eu^{3+} and Nd^{3+}, in which the effects on the histidine resonances and titration curves have been monitored.[166,167] In such cases it is important to carry out control experiments with rare earth diamagnetic ions. Thus, the line broadening effects observed with Eu^{3+} at some pH values were not due mainly to its paramagnetic contribution (not to be expected in view of its fast electron relaxation time), since they were duplicated by La^{3+}, and must be due to an exchange broadening effect (section IIA). Comparative experiments with several ions,[166] utilizing the histidine and other resonances as monitors, provides a valuable means of exploring the ion binding site of nuclease.[168]

Spin labels are organic radicals, usually containing a nitroxide group, with an unpaired electron. They are ubiquitous as probes, having by now been attached to practically every conceivable type of biomolecule, large and small. The pioneering studies of McConnell with haemoglobin have recently been reviewed.[282] Usually, the electron spin resonance of the spin label is studied. However, there have been a few studies of the effects of the paramagnetic species on n.m.r. spectra.

In view of the intrinsic broadness of protein resonances, an attempt to utilize the distance-dependent line broadening effects of covalently bound spin labels was not notably successful.[283] These studies often suffer from the presence of several spin labels at unknown sites.

A more successful approach has been to utilize the non-covalent binding of inhibitor molecules to introduce the spin label into the active site of the protein. A spin-labelled analogue of nicotinamide adenine dinucleotide was used to probe the binding to alcohol dehydrogenase by measurement of the enhancement of the proton relaxation rate of water (equation (19)). The reduction in the value of ε on the addition of other inhibitors indicates that these do not displace NAD but do displace water molecules.[284,285] The effects of a phosphate free-radical on the line-width of the His C2 proton resonances of ribonuclease have also been monitored, and selective effects reported.[286]

C. Interactions of Nucleic Acid Derivatives

Selective hydrogen-bonded interactions between complementary purine and pyrimidine bases in nucleic acids were originally proposed by Watson and Crick in their double stranded model of DNA.[287] Other forms of hydrogen-bonded base-pairing were observed for purine–pyrimidine co-crystals,[288] and the importance of base stacking interactions were soon indicated.[289–292] N.m.r. has aided in the elucidation of these two basic modes of binding of purines and pyrimidines.

The resonances of purines, their nucleosides and nucleotides shift to higher field (0·25–0·35 p.p.m.) with increasing concentration in aqueous solution.[293–296] This has been interpreted to indicate that the mode of self-association is that of base stacking. This results from the electron shielding effect of the delocalized ring currents, which is maximal at the centre of the aromatic ring.[297] In a free exchange situation this shielding effect would outweigh the deshielding effect experienced at the periphery of the aromatic ring. Very much smaller concentration effects were found for the (non-aromatic) pyrimidines, although addition of purines caused upfield shifts of the pyrimidine absorptions indicating purine–pyrimidine stacking.[298] The presence of these concentration-dependent chemical shift effects require careful control of concentration and temperature in all n.m.r. studies of nucleotide derivatives.

The same criteria of chemical shift changes noted above are applicable to studies of the *intra*molecular interactions of bases. The 5′-nucleotides exist in the *anti* form in aqueous solution as indicated by the specific pH dependent deshielding effect of the phosphate group on the H-8 proton of purine and the H-6 proton of pyrimidine nucleotides (Figure 28),[299–301] an effect not observed with the 2′- and 3′-nucleotides. Di- and trinucleotides have been shown to exist predominantly in the folded conformation in solution, in which the two rings are stacked.[302–308] A detailed analysis of ApA, where there is magnetic non-equivalence of the two bases, shows the bases to be partially stacked in the *anti* conformation. Extensive data on dinucleoside phosphates confirms this to be the predominant conformer in solution.[309–312]

The observation of magnetic non-equivalance of the nicotinamide C4 protons in NADH at 220 MHz was considered to have a potential significance for its mode of stereospecific binding to different dehydrogenases.[313,314] This conclusion has been questioned,[315] although it is clear that there are several possible conformations in solution, and that the reduced and oxidized forms do indeed have different conformations.[313,316]

Addition of purine to pyrimidine dinucleotides resulted not only in upfield shifts, but also a splitting in certain of the pyrimidine absorptions into two sets with equal intensities (Figure 29), indicating purine insertion between the pyrimidine bases.[302] Similarly, purine intercalation in ApA was monitored by the differential shifts of the adenine protons, and provided a very sensitive probe for the conformation of the dinucleotide.[303] Inter-action of purine with poly-U resulted in broadening of the purine and

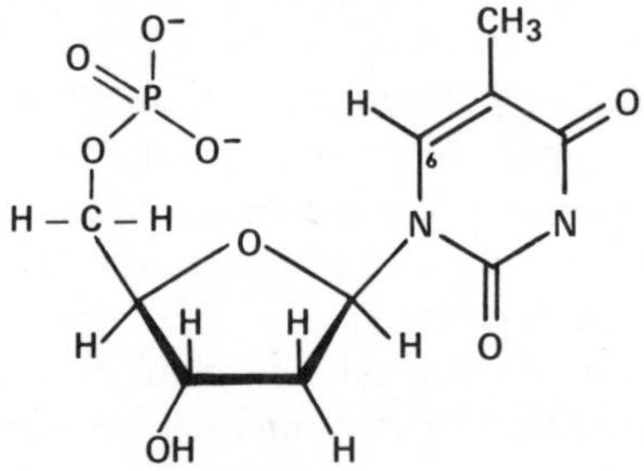

Figure 28. The anti configurations of (a) 5′-AMP and
(b) 5′-TMP indicated by the deshielding of the H-8 and
H-6 hydrogens respectively by the phosphate group

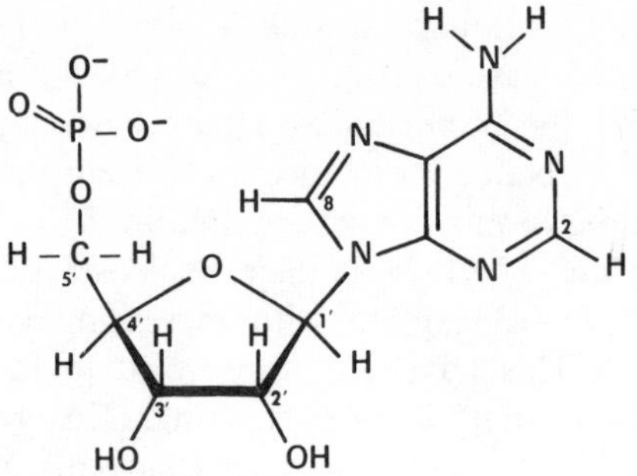

upfield shifts of the uridine resonances, also favouring an intercalation model.[317-319] However, a study of the temperature dependence of the chemical shifts of a mixture of poly-U with adenosine indicated that at temperatures below 26° this mode of association was replaced by another, which from the extreme broadening observed was taken to indicate a rigid triple-stranded 1A:2U complex formed by cooperative base stacking of A together with A-U hydrogen-bonding.[317]

Two separate resonances have been observed for thymine methyl protons in a series of trinucleotides[306] and dinucleotides,[320] and the chemical shift depends upon whether the 5′-neighbour is a purine or a pyrimidine. This would seem to confirm the prior identification of two peaks in the spectrum of single-stranded calf thymus DNA with these absorptions.[321] The relative intensities of these peaks have been used as an indication of the thymidine: purine/pyrimidine nearest neighbour ratios for samples of DNA from different organisms,[320] giving values in good agreement with other calculations.[322]

Temperature denaturation of synthetic homo-polynucleotides results in sharpening of the resonances at the melting temperature.[10,323,324] However,

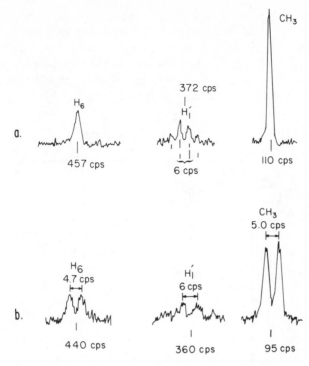

Figure 29. P.m.r. spectrum of (a) 0·13 M TpT, and (b) same in presence of purine at a molar ratio of purine/dinucleotide of 8/1[303]

the results of similar studies carried out with unfractionated t-RNA[321,323,324] have been reinterpreted.[325] The growth of two broad peaks, representing roughly the aromatic and sugar protons, have been attributed to the thermal disaggregation of salt-induced t-RNA aggregates.[325,326] In view of the extreme broadness of the peaks of fractionated t-RNA species, even at 220 MHz at high temperatures, there appears to be little hope that p.m.r. will find useful application in structural studies with the rigid native conformation. However, certain resonances of the rarer bases can be distinguished in the extremes of the spectrum, and their relative broadness and resistance to 'melting' indicates they are in rigid segments of the t-RNA molecule resistant to denaturation.[327,328]

A limitation of n.m.r. to studies of base interactions in aqueous medium is that the large upfield shifts resulting from stacking can mask any smaller downfield shifts due to hydrogen-bonded interactions. However, in non-aqueous media the tendency for hydrophobic interactions is reduced, and that of the weaker hydrogen-bonding[329] is increased. It also becomes possible in non-aqueous solvents to observe the purine and pyrimidine amino protons which are directly involved in the formation of hydrogen bonds, since they are not rapidly exchanging with solvent. Effects upon the chemical shifts of these protons on the addition of other bases in non-aqueous medium indicate the formation of strong hydrogen-bonds only according to the Watson–Crick base-pairing schemes.[330–332] In Figure 30

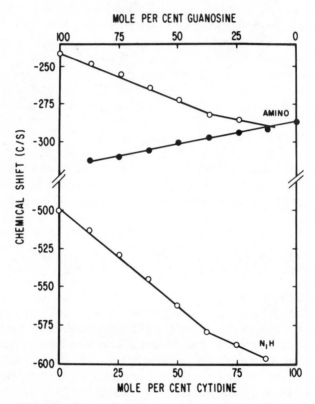

Figure 30. Chemical shifts of the hydrogen-bonding NH groups of guanosine (○) and cytidine (●) in admixtures in dimethyl sulphoxide. The measurements were made at 16° on solutions containing a total nucleoside concentration of 0·5 M[330]

are shown the large downfield shifts of the N-H protons on formation of the G-C pair in d_6-dimethylsulphoxide.[330] Concentration effects gave no upfield shifts as would be expected from the stacking interactions observed in water (see above). Such studies are being extended to association between less common bases.[333]

There have been several n.m.r. studies relevant to the question of the mode of drug binding to DNA. The binding of acridine to DNA has been investigated by observation of the p.m.r. spectrum of acridine.[334] Studies with purines and pyrimidines as controls indicated stacking interactions with acridine.[335] Two types of interaction with DNA have been proposed, tight irrotational binding at low acridine:DNA ratios, thought to be intercalation, and a weaker binding at high acridine concentrations. However, the very fast sweep rate used in those studies precludes the quantitative analysis of line width measurements (section IID).

Linear molecules containing spaced quaternary ammonium groups are being used as reporter molecules to study binding to nucleic acids.[336] The complete assignment of the spectrum[337] of actinomycin D and analysis of the deuterium exchange of the peptide NH's[338] have been reported. An analysis of the interaction with 5′-deoxyguanylic acid indicates that the phenoxasone ring of actinomycin is sandwiched between two guanine rings.[330] Apart from the question of some specific hydrogen bonds involving the peptide residues,[330,331] this would tend to support the proposal that actinomycin D binds to DNA by interaction of the phenoxazone ring.[339]

The binding of many ionic species to nucleotides and derivatives have been studied with some success by n.m.r. methods. The first such study involved metal ion binding to ADP and ATP with observation of p.m.r. and phosphorus-31 magnetic resonance. In this study selective chemical shift effects were observed for the diamagnetic metal ions and selective line broadening effects for the paramagnetic ions (section IIIB) indicating different modes of interaction for different ions.[340] More recently detailed studies of Cu^{2+} binding to nucleotides and polynucleotides,[341] uranyl ion binding to ADP and ATP[342] and Zn^{2+} complexation with nucleotides in dimethylsulphoxide[343,344] have been reported. Also pH-dependent[345] and solvent electrolyte[346] effects on nucleosides and nucleotides have been analysed. The ^{35}Cl magnetic resonance probe technique has been applied to Zn^{2+} binding to ADP[347] and nitrogen-15 enriched pyrimidines have been studied by nitrogen-15 magnetic resonance[348] and p.m.r.[349] to elucidate ion-binding[350] as well as tautomerism effects. While many of these studies throw light on the problem of ion-binding to nucleotides as such, they may also be relevant to structural and interactional functions of native nucleic acids. Of particular interest in this respect is the use of Gd^{3+} to determine distances from the ion binding site to protons of AMP, in order to calculate the most probable conformation of the complex in solution.[351]

IV. CONCLUSION

The number of applications of n.m.r. to studies of molecular interactions have expanded dramatically in the last few years. There appear to be few critical limitations of the n.m.r. method when applied to studies of molecular interactions by observation of the spectrum of the ligand subject to the controls discussed (p. 532). Recent work has indicated the potential of the n.m.r. technique for studies of small proteins (m.w. *ca.* 20,000) (for reviews see references, 144, 145, 254, 361, 362). For example, the plotting of the individual titration curves of the histidine residues of some enzymes, and the observation of the effects of inhibitor binding, represent unique and valuable contributions to molecular biophysics.

As molecular size, and hence correlation times, increase, n.m.r. absorptions become broader. The effective limit of molecular weight for resolution of proton resonances in proteins with presently available spectrometers is *ca.* 35,000. High molecular weight native nucleic acids appear to present an insuperable problem for n.m.r. studies due to their rigidity and low chemical variability. In principle, use of higher magnetic fields should serve to resolve multiple absorptions in proteins, although from existing data it is clear that the line positions of several amino acids will still coincide.

An approach to the observation of individual proton resonances in protein spectra is that of large-scale selective deuteration.[190] This technique will represent a breakthrough if a sufficient number of well-defined small proteins can be prepared by the biosynthetic approach. The chemical synthesis of selectively isotopically labelled peptides or proteins of known sequence is possible, and this approach has been adopted using ^{13}C enriched amino acids.[200] The observation of other magnetic nuclei in selectively substituted proteins and polynucleotides should have certain advantages. Apart from the generally wider range of chemical shifts than for p.m.r. they should provide a more direct insight into the environment and binding at that particular site. Other initial applications of this kind include the use of deuterium[352] and fluorine-19[353] magnetic resonance.

New techniques are having an impact on the applications of n.m.r. to molecular biophysics. The recent burst of publications on carbon-13 magnetic resonance was made possible by the successful application of proton noise decoupling and Fourier transform methods.[26] This greatly decreases the time required to obtain a spectrum and enables carbon-13 magnetic resonance to become a routine tool. There is little doubt that a similar effect will result for nitrogen-15,[354,355] although somewhat delayed due to the extra difficulties. Studies of other magnetic nuclei of potential biological interest which have recently been reported on include oxygen-17,[356] calcium-43,[357] sodium-23,[358] potassium-39[359] and thallium-205.[360] The uses of paramagnetic ions also represents a very fruitful area of research (for reviews see references 145, 248, 275), to which the recent introduction of rare earth elements is a potentially powerful addition.[262,263,351]

The great reduction in time required for data accumulation by the FT n.m.r. technique might also allow a significant extension of the n.m.r. method to the monitoring of enzymatic reactions. Such applications have thus far been restricted to slow processes.[50,363,364] As yet there have been few applications in p.m.r. which have taken useful advantage of the FT method. Both for the rapid accumulation and processing of data and for direct on-line spectrometer control computers have an important role to play in the development of n.m.r. techniques.

Abbreviations used: n.m.r., nuclear magnetic resonance; p.m.r., proton magnetic resonance; c.m.r., carbon-13 magnetic resonance; NAG, N-acetyl glucosamine; BSA, bovine serum albumin; HSA, human serum albumin; TMS, tetramethylsilane; HMS, hexamethyldisiloxane; DSS, 2,2-dimethyl-silapentane-5-sulphonate; ASA, acetylsalicylic acid; DNA, deoxyribo-nucleic acid; AMP, adenosine-5' phosphate; ADP, adenosine-5' diphosphate; ATP, adenosine-5' triphosphate; FT, Fourier transform.

REFERENCES

1. E. M. Purcell, H. C. Torry and R. V. Pound, *Phys. Rev.*, **69**, 37 (1946)
2. F. Bloch, W. W. Hansen and M. E. Packard, *Phys. Rev.*, **69**, 127 (1946)
3. J. A. Pople, W. G. Schneider and J. H. Bernstein, *High Resolution Nuclear Magnetic Resonance*, McGraw-Hill, New York, 1959
4. J. W. Emsley, J. W. Feeney and L. H. Sutcliffe, *High Resolution Nuclear Magnetic Resonance Spectroscopy*, Pergamon, New York, 1965
5. A. Carrington and A. D. McLachlan, *Introduction to Magnetic Resonance*, Harper and Row, New York, 1967
6. L. M. Jackman and S. Sternhall, *Applications of Nuclear Magnetic Resonance Spectroscopy in Organic Chemistry*, Academic Press, New York, 1969
7. E. D. Becker, *High Resolution NMR*, Academic Press, New York, 1969
8. F. A. Bovey, *Nuclear Magnetic Resonance Spectroscopy*, Academic Press, New York, 1969
9. O. Jardetzky and C. D. Jardetzky, 'Introduction to magnetic resonance spectroscopy. Methods and biochemical applications', in *Methods in Biochemical Analysis*, Vol. IX (Ed. D. Glick), Interscience, New York, 1962, p. 235
10. A. Kowalsky and M. Cohn, *Ann. Rev. Biochem.*, **33**, 481 (1964)
11. J. L. Markley, T. C. Holocher, A. S. Brill and O. Jardetzky, 'Magnetic susceptibility and magnetic resonance', in *Physical Methods in Biochemistry*, Academic Press (Ed. D. Moore), 1970
12. J. S. Cohen, *Applications of Nuclear Magnetic Resonance to Molecular Biology*, Academic Press, in preparation
13. B. R. Donaldson and J. C. P. Schwartz, *J. Chem. Soc. (B)*, **1968**, 395
14. C. S. Johnson, Jr., *Adv. Mag. Res.*, **1**, 33 (1965)
15. M. S. Gutowsky, D. W. McCall and C. P. Slichter, *Phys. Rev.*, **84**, 589 (1951)
16. E. L. Hahn and D. E. Maxwell, *Phys. Rev.*, **84**, 1286 (1951)
17. M. Karplus, *J. Am. Chem. Soc.*, **85**, 2870 (1963)
18. A. A. Bothner-By, *Adv. Mag. Res.*, **1**, 195 (1965)
19. J. D. Roberts, *An Introduction to the Analysis of Spin–Spin Splitting in High Resolution Nuclear Magnetic Resonance Spectra*, Benjamin, New York, 1961
20. R. C. Ferguson and W. D. Phillips, *Science*, **157**, 257 (1967)

21. F. Bloch, *Phys. Rev.*, **70**, 460 (1946)
22. J. H. Bradbury, B. E. Chapman and N. L. R. King, *Macromolecular Abstracts*, p. 1132, 23 IUPAC Conf., Boston, July 1971
23. J. S. Waugh, C. H. Wang, L. M. Huber and R. L. Vold, *J. Chem. Phys.*, **48**, 663 (1968)
24. O. Jardetzky, *Adv. Chem. Phys.*, **7**, 699 (1964)
25. J. S. Cohen, *J. Clin. Pharm.*, **9**, 72 (1969)
26. T. C. Farrar and E. D. Becker, *Pulse and Fourier Transform NMR*, Academic Press, New York, 1971
27. J. R. Zimmerman and W. F. Britten, *J. Phys. Chem.*, **61**, 1318 (1957)
28. O. Jardetzky, N. G. Wade and J. J. Fischer, *Nature*, **197**, 183 (1963)
29. J. J. Fischer and O. Jardetzky, *J. Am. Chem. Soc.*, **87**, 3237 (1965)
30. O. Jardetzky and N. G. Wade-Jardetzky, *Mol. Pharm.*, **1**, 214 (1965)
31. M. Nakano, N. I. Nakano and T. Higuchi, *J. Phys. Chem.*, **71**, 2954 (1967)
32. D. E. Koshland, *Adv. Enzymol.*, **22**, 65 (1960)
33. O. Jardetzky and N. Weiner, *Arch. exp. Path. u. Pharmak*, **248**, 308 (1964)
34. H. Z. Sable and J. E. Biaglow, *Proc. Nat. Acad. Sci. (U.S.)*, **54**, 808 (1965)
35. J. J. Katz, H. H. Strain, D. I. Leusing and R. C. Dougherty, *J. Am. Chem. Soc.*, **90**, 784 (1968)
36. D. B. Sykes, *Biochem. Biophys. Res. Comm.*, **39**, 508 (1970)
37. N. Muller and R. M. Birkhahn, *J. Phys. Chem.*, **71**, 957 (1967)
38. K. Jones and E. F. Mooney, *Ann. Rep. NMR Spectra*, **3**, 261 (1970)
39. T. W. Johnson and N. Muller, *Biochem.*, **9**, 1943 (1970)
40. G. G. LaForce and S. Forsen, *Biochem. Biophys. Res. Comm.*, **38**, 137 (1970)
41. J. C. Metcalfe, A. S. V. Burgen and O. Jardetzky, 'On the mechanism of binding of choline derivatives to an anticholine antibody', in *Molecular Associations in Biology* (Ed. B. Pullman), Academic Press, New York, 1968, p. 487
42. A. S. V. Burgen, O. Jardetzky, J. C. Metcalfe and N. G. Wade-Jardetzky, *Proc. Nat. Acad. Sci. (U.S.)*, **58**, 447 (1967). See also S. Joffe, *Mol. Pharm.*, **3**, 399 (1967)
43. R. P. Haugland, L. Stryer, T. R. Stengle and J. D. Baldeschwieler, *Biochem.*, **6**, 498 (1967)
44. T. R. Stengle and J. D. Baldeschwieler, *Proc. Nat. Acad. Sci. (U.S.)*, **55**, 1020 (1966)
45. T. R. Stengle and J. D. Baldeschwieler, *J. Am. Chem. Soc.*, **89**, 3045 (1967)
46. R. G. Bryant, *J. Am. Chem. Soc.*, **89**, 2496 (1967); **91**, 976 (1969)
47. A. G. Marshall, *Biochem.*, **7**, 2450 (1968)
48. M. Zeppezauer, B. Lindman, S. Forsen and I. Lindquist, *Biochem. Biophys. Res. Comm.*, **37**, 137 (1969)
49. G. Kato, *Mol. Pharm.*, **4**, 640 (1968)
50. G. Kato, *Mol. Pharm.*, **5**, 148 (1969)
51. G. Kato, J. Yung and M. Ihnat, *Biochem. Biophys, Res. Comm.*, **40**, 15 (1970)
52. G. Kato, *Canad. Anaes. Soc. J.*, **15**, 545 (1968)
53. M. Wenzel, H. P. Lenk and E. Schutte, *Z. Physiol. Chem.*, **327**, 13 (1962)
54. E. W. Thomas, *Biochem. Biophys. Res. Comm.*, **24**, 611 (1966)
55. E. W. Thomas, *Biochem. Biophys. Res. Comm.*, **29**, 628 (1967)
56. M. A. Raftery, F. W. Dahlquist, S. I. Chan and S. M. Parsons, *J. Biol. Chem.*, in press
57. F. W. Dahlquist and M. A. Raftery, *Biochem.*, **7**, 3269 (1968)
58. D. C. Phillips, *Proc. Nat. Acad. Sci. (U.S.)*, **57**, 484 (1967) and references therein
59. M. A. Raftery, F. W. Dahlquist, S. M. Parsons and R. G. Wolcott, *Proc. Nat. Acad. Sci. (U.S.)*, **62**, 44 (1968)
60. F. W. Dahlquist and M. A. Raftery, *Biochem.*, **8**, 713 (1969)

61. J. A. Rupley, L. Butler, M. Geering, F. T. Hartdegen and R. Pecoraro, *Proc. Nat. Acad. Sci. (U.S.)*, **57**, 1088 (1967)
62. S. S. Lehrer and G. D. Fasman, *J. Biol. Chem.*, **242**, 4644 (1967)
63. D. M. Chipman, V. Grisaro and N. Sharon, *J. Biol. Chem.*, **242**, 4388 (1967)
64. F. W. Dahlquist and M. A. Raftery, *Biochem.*, **9**, 3277 (1968)
65. B. D. Sykes, *Biochem.*, **8**, 1110 (1969)
66. J. T. Gerig, *J. Am. Chem. Soc.*, **90**, 2681 (1968)
67. J. T. Gerig and J. D. Reinheimer, *J. Am. Chem. Soc.*, **92**, 3146 (1970)
68. J. T. Gerig and R. A. Rimerman, *Biochem. Biophys. Res. Comm.*, **40**, 1149 (1970)
69. B. W. Matthews, P. B. Sigler, R. Henderson and D. M. Blow, *Nature*, **214**, 652 (1967)
70. P. B. Steitz, R. Henderson and D. M. Blow, *J. Mol. Biol.*, **46**, 337 (1969)
71. T. M. Spotswood, J. M. Evans and J. H. Richards, *J. Am. Chem. Soc.*, **89**, 5052 (1967)
72. E. Zeffren and R. E. Reavill, *Biochem. Biophys. Res. Comm.*, **32**, 73 (1968)
73. B. D. Sykes, *J. Am. Chem. Soc.*, **91**, 949 (1969)
74. J. Kallos and K. Avatis, *Biochem.*, **5**, 1979 (1966)
75. B. F. Erlanger, *Proc. Nat. Acad. Sci. (U.S.)*, **58**, 703 (1967)
76. W. B. Lawson, *J. Biol. Chem.*, **242**, 3397 (1967)
77. B. D. Sykes, *Biochem. Biophys. Res. Comm.*, **33**, 727 (1968)
78. H. G. Schmidt, G. R. Stark and J. D. Baldeschwieler, *J. Biol. Chem.*, **244**, 1869 (1969)
79. B. D. Sykes, P. G. Schmidt and G. R. Stark, *J. Biol. Chem.*, **245**, 1180 (1970)
80. D. P. Hollis, *Biochem.*, **6**, 2081 (1967)
81. Y. Inoue and S. Inoue, *Biochem. Biophys. Acta.*, **128**, 100 (1966)
82. D. H. Meadows and O. Jardetzky, *Proc. Nat. Acad. Sci. (U.S.)*, **61**, 406 (1968)
83. G. C. Y. Lee, J. H. Prestegard and S. Chan, *Biochem. Biophys. Res. Comm.*, **43**, 435 (1971)
84. J. S. Cohen, G. C. K. Roberts and D. Z. Denny, *Proc. Fourth Int. NMR in Biology Conf.*, Warrenton, Va., 1968
85. G. C. Y. Lee and S. Chan, *Biochem. Biophys. Res. Comm.*, **43**, 141 (1971)
86. D. Chapman, V. B. Kamat, J. De Gier and S. A. Penkett, *J. Mol. Biol.*, **31**, 101 (1968)
87. J. C. Metcalfe, P. Seaman and A. S. V. Burgen, *Mol. Pharm.*, **4**, 87 (1968)
88. M. Saunders, A. Wishnia and J. Kirkwood, *J. Am. Chem. Soc.*, **79**, 3289 (1957)
89. O. Jardetzky and C. D. Jardetzky, *J. Am. Chem. Soc.*, **79**, 5322 (1957)
90. C. C. McDonald and W. D. Phillips, 'NMR studies of biological macromolecules', in *Magnetic Resonance in Biological Systems* (Ed. A. Ehrenberg, B. G. Malmstrom and T. Vanngard), Pergamon, London, 1967, p. 3
91. M. Saunders and A. Wishnia, *Ann. N.Y. Acad. Sci.*, **70**, 870 (1958)
92. F. A. Bovey, G. V. D. Tiers and G. Filipovitch, *J. Polymer Sci.*, **38**, 73 (1959)
93. A. Kowalsky, *J. Biol. Chem.*, **237**, 1807 (1961); 239 (1964)
94. A. Wishnia and M. Saunders, *J. Am. Chem. Soc.*, **84**, 4235 (1962)
95. M. Mandel, *Proc. Nat. Acad. Sci. (U.S.)*, **52**, 736 (1964)
96. M. Mandel, *J. Biol. Chem.*, **240**, 1586 (1965)
97. C. C. McDonald and W. D. Phillips, *J. Am. Chem. Soc.*, **89**, 6332 (1967)
98. H. Sternlicht and D. Wilson, *Biochem.*, **6**, 2881 (1967)
99. D. P. Hollis, G. McDonald and R. L. Biltonen, *Proc. Nat. Acad. Sci. (U.S.)*, **58**, 758 (1967)
100. J. L. Markley, D. H. Meadows and O. Jardetzky, *J. Mol. Biol.*, **27**, 25 (1967)
101. W. E. Steward, L. Mandelkern and R. E. Glick, *Biochem.*, **6**, 143 (1967)

102. J. A. Ferretti, *Chem. Comm.*, **1967**, 2030
103. E. M. Bradbury, C. Crane-Robinson, H. Goldman and H. W. E. Rattle, *Nature*, **217**, 812 (1968)
104. R. Ullman, *Biopolymers*, **9PP**, 471 (1970)
105. J. Tan and I. M. Klotz, *J. Am. Chem. Soc.*, **93**, 1313 (1971)
106. J. A. Ferretti, *Accounts Chem. Res.*, in press
107. C. B. Anfinsen, *Harvey Lectures*, **61**, 95 (1967)
108. M. Goodman, A. S. Verdini, C. Toniolo, W. D. Phillips and F. A. Bovey, *Proc. Nat. Acad. Sci. (U.S.)*, **64**, 444 (1969)
109. A. Nakamura and O. Jardetzky, *Proc. Nat. Acad. Sci. (U.S.)*, **58**, 2212 (1967)
110. A. Nakamura and O. Jardetzky, *Biochem.*, **7**, 1226 (1968)
111. M. Sheinblatt, *J. Am. Chem. Soc.*, **88**, 2845 (1966)
112. A. Nakamura and O. Jardetzky, 'NMR study of an octapeptide', in *Magnetic Resonance in Biological Systems* (Eds. S. Fujiwara and P. H. Piette), Hirokawa, Tokyo, 1968, p. 10
113. L. R. Cavanaugh, *J. Am. Chem. Soc.*, **92**, 1488 (1970)
114. A. E. Tonelli, A. I. Brewster and F. A. Bovey, *Macromolecules*, **3**, 412 (1970)
115. A. E. Tonelli and F. A. Bovey, *Macromolecules*, **3**, 410 (1970)
116. W. A. Gibbons, G. Nemethy, A. Stern and L. C. Craig, *Proc. Nat. Acad. Sci. (U.S.)*, **67**, 239 (1970)
117. A. Stern, W. A. Gibbons and L. C. Craig, *Proc. Nat. Acad. Sci. (U.S.)*, **61**, 734 (1968)
118. A. M. Liquori and F. Conti, *Nature*, **217**, 635 (1968)
119. Y. A. Ovchinnikov, V. T. Ivanov, V. F. Bystrov, A. I. Miroshnikov, L. N. Shepel, N. D. Abdullaev, E. S. Efremov and L. B. Benjamina, *Biochem. Biophys. Res. Comm.*, **39**, 217 (1970)
120. L. F. Johnson, I. L. Schwartz and R. Walter, *Proc. Nat. Acad. Sci. (U.S.)*, **64**, 1268 (1969)
121. D. W. Urry, M. Ohnishi and R. Walter, *Proc. Nat. Acad. Sci. (U.S.)*, **66**, 111 (1970)
122. J. Feeney, G. C. K. Roberts, J. M. Rockey and A. S. V. Bergen, *Nature*, **232**, 108 (1971)
123. P. N. von Dreele, A. I. Bernstein, H. A. Scheraga, M. F. Ferger and V. du Vigneaud, *Proc. Nat. Acad. Sci. (U.S.)*, **68**, 1028 (1971)
124. T. A. Victor, F. E. Hruska, K. Hituchi, S. S. Danyluk and C. L. Bell, *Nature*, **223**, 320 (1969)
125. F. Conti and P. de Santis, *Nature*, **227**, 1240 (1970)
126. A. I. Brewster and F. A. Bovey, *Proc. Nat. Acad. Sci. (U.S.)*, **68**, 1199 (1971)
127. A. E. Tonelli, D. J. Patel, M. Goodman, F. Naider, H. Foulstich and Th. Wieland, *Biochem.*, **10**, 3211 (1971)
128. C. M. Deber, F. A. Bovey, J. P. Carver and E. R. Blout, *J. Am. Chem. Soc.*, **92**, 6191 (1970); C. M. Deber, D. A. Torchia and E. R. Blout, *J. Am. Chem. Soc.*, **93**, 4893 (1971)
129. D. M. Haynes, A. Kowalsky and B. C. Pressman, *J. Biol. Chem.*, **244**, 502 (1969)
130. N. W. Cornell and D. G. Guiney, Jr., *Biochem. Biophys. Res. Comm.*, **40**, 530 (1970)
131. J. M. Prestegard and S. I. Chan, *Biochem.*, **8**, 3921 (1969); *J. Am. Chem. Soc.*, **92**, 4440 (1970)
132. A. Stern, W. A. Gibbons and L. C. Craig, *J. Am. Chem. Soc.*, **91**, 2794 (1969)
133. R. Schwyzer and U. Ludescher, *Biochem.*, **7**, 2519 (1968)
134. R. Schwyzer and U. Ludescher, *Biochem.*, **7**, 2514 (1968)
135. K. D. Kopple and K. H. Mann, *J. Am. Chem. Soc.*, **89**, 6193 (1967)
136. K. D. Kopple and M. Ohnishi, *J. Am. Chem. Soc.*, **91**, 962 (1969)

137. J. S. Cohen, *Biochem. Biophys. Acta.*, **229**, 603 (1971)
138. W. Fory, R. E. McKenzie, F. Y-H. Wu and D. B. McCormick, *Biochem.*, **9**, 515 (1970)
139. R. Deslauriers and I. C. P. Smith, *Biochem. Biophys. Res. Comm.*, **40**, 179 (1970)
140. D. N. Silverman and H. A. Scheraga, *Biochem.*, **10**, 1340 (1971)
141. M. P. Klein and G. W. Barton, Jr., *Rev. Sci. Instr.*, **34**, 754 (1963)
142. J. S. Cohen and O. Jardetzky, *Proc. Nat. Acad. Sci. (U.S.)*, **60**, 92 (1968)
143. B. Bak, C. Dambmann, F. Nicolaisen, E. J. Pedersen and N. S. Bhacca, *J. Mol. Spect.*, **26**, 78 (1968)
144. G. C. K. Roberts and O. Jardetzky, *Adv. Protein Chem.*, **24**, 447 (1970)
145. C. C. McDonald and W. D. Phillips, 'Proton magnetic resonance spectroscopy of proteins', in *Fine Structure of Proteins and Nucleic Acids*, Vol. 4 (Ed. G. D. Fasman and S. N. Timasheff), Dekker, New York, 1970, p. 1
146. B. Bak, E. J. Pedersen and F. Sundby, *J. Biol. Chem.*, **242**, 2637 (1967)
147. B. Bak, J. J. Led and E. J. Pedersen, *J. Mol. Spect.*, **32**, 151 (1969)
148. K. Hamaguchi and K. Rokkaku, *J. Biochem. (Tokyo)*, **48**, 358 (1960)
149. D. H. Meadows, J. L. Markley, J. S. Cohen and O. Jardetzky, *Proc. Nat. Acad. Sci. (U.S.)*, **58**, 1307 (1967)
150. C. C. McDonald and W. D. Phillips, *J. Am. Chem. Soc.*, **85**, 3736 (1963)
151. J. H. Bradbury and P. Wilairat, *Biochem. Biophys. Res. Comm.*, **29**, 84 (1967)
152. P. K. Glasoe and F. A. Long, *J. Phys. Chem.*, **64**, 188 (1960)
153. D. M. Meadows, G. C. K. Roberts and O. Jardetzky, *Biochem.*, **8**, 2053 (1968)
154. C. C. McDonald and W. D. Phillips, *J. Am. Chem. Soc.*, **91**, 1513 (1969)
155. J. H. Bradbury and H. A. Scheraga, *J. Am. Chem. Soc.*, **88**, 4240 (1966)
156. J. S. Cohen, *Nature*, **223**, 43 (1969)
157. J. S. Cohen, A. N. Schechter, R. I. Shrager and M. McNeel, *Nature*, **228**, 642 (1970)
158. J. S. Cohen, A. N. Schechter, R. I. Shrager and M. McNeel, *Biochem. Biophys. Res. Comm.*, **40**, 144 (1970)
159. D. A. Cowburn, E. M. Bradbury, C. Crane-Robinson and W. B. Gratzer, *Europ. J. Biochem.*, **14**, 83 (1970)
160. R. W. King and G. C. K. Roberts, *Biochem.*, **10**, 558 (1971)
161. J. S. Cohen, C. T. Yim, M. Kandel, A. G. Gornall, S. I. Kandel and M. Freedman, *Biochem.*, **11**, 327 (1972)
162. J. L. Markley, N. M. Williams and O. Jardetzky, *Proc. Nat. Acad. Sci. (U.S.)*, **65**, 645 (1970)
163. J. S. Cohen, H. P. Hagenmaier, H. Pollard and A. N. Schechter, *J. Mol. Biol.*, in press
164. J. L. Markley, Thesis, Harvard University, 1968
165. C. L. Cusumano, H. Taniuchi and C. B. Anfinsen, *J. Biol. Chem.*, **243**, 4769 (1968)
166. E. Nieboer, D. East, A. N. Schechter and J. S. Cohen, unpublished results
167. M. N. Williams, *Fed. Proc.*, **30**, 1398 (1971)
168. A. Arnone, C. J. Bier, F. A. Cotton, V. W. Day, E. A. Hazen, Jr., D. C. Richardson, J. S. Richardson and A. Yonath, *J. Biol. Chem.*, **229**, 404 (1971)
169. D. H. Meadows, O. Jardetzky, R. M. Epand, H. H. Ruterjans and H. A. Scheraga, *Proc. Nat. Acad. Sci. (U.S.)*, **60**, 766 (1968)
170. F. M. Richards and P. J. Vithayathil, *J. Biol. Chem.*, **234**, 1459 (1959)
171. N. L. R. King and J. H. Bradbury, *Nature*, **229**, 404 (1971)
172. G. Kartha, J. Bello and D. Harker, *Nature*, **213**, 862 (1967)
173. H. W. Wyckoff, D. Tsernoglu, A. W. Hansen, J. R. Knox, B. Lee and F. W. Richards, *J. Biol. Chem.*, **245**, 305 (1970)
174. D. J. Patel, C. K. Woodward and F. A. Bovey, *Proc. Nat. Acad. Sci. (U.S.)*, **69**, 599 (1972)

175. H. Ruterjans and H. Witzel, *Europ. J. Biochem.*, **9**, 118 (1969)
176. H. G. Gassen and H. Witzel, *Europ. J. Biochem.*, **1**, 36 (1967)
177. D. H. Sachs, A. N. Schechter, J. S. Cohen, *J. Biol. Chem.*, **426**, 6576 (1971)
178. R. Shrager, J. S. Cohen, S. R. Heller, D. H. Sachs and A. N. Schechter, *Biochem.*, **11**, 541 (1972)
179. J. S. Cohen, R. Shrager, S. R. Heller, D. H. Sachs and A. N. Schechter, *Fed. Proc.*, **30**, 1404 (1971)
180. H. A. Saroff, *J. Theoret. Biol.*, **9**, 229 (1965)
181. D. Findlay, D. G. Herries, A. P. Mathias, B. R. Rabin and C. A. Ross, *Biochem. J.*, **85**, 152 (1961)
182. G. C. K. Roberts, E. A. Dennis, D. H. Meadows, J. S. Cohen and O. Jardetzky, *Proc. Nat. Acad. Sci. (U.S.)*, **62**, 1151 (1969)
183. C. C. McDonald and W. D. Phillips, *Biochem. Biophys. Res. Comm.*, **35**, 43 (1969)
184. J. D. Glickson, C. C. McDonald and W. D. Phillips, *Biochem. Biophys. Res. Comm.*, **35**, 492 (1969)
185. J. D. Glickson, W. D. Phillips and J. A. Rupley, *J. Am. Chem. Soc.*, **92**, 4031 (1971)
186. E. W. Garbisch, Jr. and M. G. Griffith, *J. Am. Chem. Soc.*, **90**, 6543 (1968)
187. A. T. Blomquist, D. M. Rich, V. J. Hruby, L. L. Nageroni, P. Glose and V. dur Vigneaud, *Proc. Nat. Acad. Sci. (U.S.)*, **61**, 688 (1968)
188. H. L. Crespi, R. M. Rosenberg and J. J. Katz, *Science*, **161**, 795 (1968); see also J. J. Katz and M. L. Crespi, *Science*, **151**, 1187 (1966)
189. J. L. Markley, I. Putter and O. Jardetzky, *Science*, **161**, 1249 (1969); *Z. Anal. Chem.*, **243**, 367 (1968)
190. I. Putter, A. Barretto, J. L. Markley and O. Jardetzky, *Proc. Nat. Acad. Sci. (U.S.)*, **64**, 1396 (1969)
191. J. S. Cohen and I. Putter, *Biochim. Biophys. Acta.*, **222**, 515 (1970)
192. P. Cuatrecasas, M. Edelhoch and C. B. Anfinsen, *Proc. Nat. Acad. Sci. (U.S.)*, **58**, 2043 (1967)
193. J. S. Cohen, M. Feil and I. Chaiken, *Biochim. Biophys. Acta.*, **236**, 468 (1971)
194. P. Cuatrecasas, S. Fuchs and C. B. Anfinsen, *J. Biol. Chem.*, **243**, 4787 (1968)
195. R. R. Ernst and W. A. Anderson, *Rev. Sci. Insts.*, **37**, 93 (1966)
196. F. J. Weigert, M. Jautelat and J. D. Roberts, *Proc. Nat. Acad. Sci. (U.S.)*, **60**, 1152 (1968)
197. W. Horsley, H. Sternlicht and J. S. Cohen, *J. Am. Chem. Soc.*, **92**, 680 (1970)
198. W. A. Gibbons, J. A. Sogn, A. Stern, L. C. Craig and L. F. Johnson, *Nature*, **227**, 840 (1970)
199. A. Allerhand, D. W. Cochran and D. Doddrell, *Proc. Nat. Acad. Sci. (U.S.)*, **67**, 1093 (1970)
200. M. Freedman, J. S. Cohen and I. Chaiken, *Biochem. Biophys. Res. Comm.*, **42**, 1148 (1971)
201. J. C. W. Chien and J. F. Brandts, *Nature*, **230**, 209 (1971)
202. F. R. N. Gurd, P. J. Lawson, D. W. Cochran and E. Wenkert, *J. Biol. Chem.*, **246**, 3725 (1971)
203. R. R. Ernst, *J. Chem. Phys.*, **45**, 3845 (1966)
204. K. F. Kuhlmann and D. M. Grant, *J. Am. Chem. Soc.*, **90**, 7344 (1968)
205. M. Christl and J. D. Roberts, *J. Am. Chem. Soc.*, **94**, 4565 (1972)
206. M. H. Freedman, J. R. Lyerla, Jr., I. M. Chaiken and J. S. Cohen, in press
207. R. R. Shoup and D. L. VanderHart, *J. Am. Chem. Soc.*, **93**, 2052 (1971)
208. H. Sternlicht, G. L. Kenyon, E. L. Packer and J. Sinclair, *J. Am. Chem. Soc.*, **93**, 199 (1971)
209. L. Johnson, *Proc. First. Nat. Conf. on Carbon-13*, Los Alamos, N.M., June 1971

210. D. Doddrell and A. Allerhand, *Proc. Nat. Acad. Sci. (U.S.)*, **68**, 1083 (1971)
211. A. Allerhand, D. Doddrell, V. Glushko, D. W. Cochran, E. Wenkert, P. J. Lawson, F. R. N. Gurd, *J. Am. Chem. Soc.*, **93**, 544 (1971)
212. E. F. Mooney and P. H. Winson, *Ann. Rev. NMR Spect.*, **2**, 153 (1969)
213. C. T. Gregg, *Proc. First. Nat. Conf. on Carbon-13*, Los Alamos, N.M., June 1971
214. J. H. Bradbury and N. R. L. King, *Nature*, **223**, 1154 (1969); *Aust. J. Chem.*, **22**, 1083 (1969)
215. J. P. Cohen Addad, *J. Mol. Biol.*, **50**, 595 (1970)
216. C. C. McDonald, W. D. Phillips and J. D. Glickson, *J. Am. Chem. Soc.*, **93**, 235 (1971)
217. A. Ikai and C. Tanford, *Nature*, **230**, 100 (1971)
218. J. Putter, J. L. Markley and O. Jardetzky, *Proc. Nat. Acad. Sci. (U.S.)*, **65**, 395 (1970)
219. O. Jardetzky, *Fed. Proc.*, **30**, 1041 (1971)
220. H. Epstein, A. N. Schechter and J. S. Cohen, *Proc. Nat. Acad. Sci. (U.S.)*, **68**, 2042 (1971)
221. A. N. Schechter, R. F. Chen and C. B. Anfinsen, *Science*, **167**, 886 (1970)
222. C. C. F. Blake and I. D. A. Swan, *Nature*, **232**, 12 (1971)
223. D. H. Meadows, G. C. K. Roberts and O. Jardetzky, *J. Mol. Biol.*, **45**, 491 (1969)
224. J. S. Cohen and A. Schechter, unpublished results
225. G. H. Jones, H. P. Albrecht, N. P. Damodaran and J. G. Moffatt, *J. Am. Chem. Soc.*, **92**, 5500 (1970)
226. D. A. Usher, D. I. Richardson, Jr. and F. Eckstein, *Nature*, **228**, 663 (1970)
227. F. W. Richards and H. W. Wyckoff, 'Ribonuclease', in *The Enzymes* (Third Edition), Vol. IV (Ed. P. D. Boyer), p. 647, 1970
228. H. Ruterjans and O. Pongs, *Europ. J. Biochem.*, **18**, 313 (1971)
229. J. L. Markley and O. Jardetzky, *J. Mol. Biol.*, **50**, 223 (1970)
230. P. Cuatrecasas, S. Fuchs and C. B. Anfinsen, *J. Biol. Chem.*, **242**, 3063, 4759 (1967)
231. M. Boublik, E. M. Bradbury and C. Crane-Robinson, *Europ. J. Biochem.*, **14**, 486 (1970)
232. J. King and N. Davidson, *J. Chem. Phys.*, **29**, 787 (1958); N. Davidson and R. Gold, *Biochem. Biophys. Acta*, **26**, 370 (1957)
233. A. Wishnia, *J. Chem. Phys.*, **32**, 871 (1960)
234. N. Bloembergen, E. M. Purcell and R. L. Pound, *Phys. Rev.*, **73**, 679 (1948)
235. I. Solomon, *Phys. Rev.*, **99**, 559 (1955)
236. N. Bloembergen, *J. Chem. Phys.*, **27**, 572 (1957)
237. Z. Luz and S. Meiboom, *J. Chem. Phys.*, **60**, 2686 (1964)
238. S. Broersma, *J. Chem. Phys.*, **24**, 659 (1956)
239. J. Eisinger, R. G. Shulman and W. E. Blumberg, *Nature*, **192**, 963 (1961)
240. J. Eisinger, R. G. Shulman and B. M. Szymanski, *J. Chem. Phys.*, **36**, 1721 (1962)
241. J. Eisinger, F. Fuwaz-Estrup and R. G. Shulman, *J. Chem. Phys.*, **42**, 43 (1965)
242. H. Sternlicht, R. G. Shulman and E. W. Anderson, *J. Chem. Phys.*, **43**, 3123 (1965)
243. M. Cohn and T. R. Hughes, *J. Biol. Chem.*, **237**, 176 (1962)
244. R. G. Shulman, H. Sternlicht and B. J. Wyluda, *J. Chem. Phys.*, **43**, 3116 (1965)
245. H. Sternlicht, D. E. Jones and K. Kustin, *J. Am. Chem. Soc.*, **90**, 7110 (1968)
246. B. Sheard, S. H. Miall, A. R. Peacocke, I. O. Walker and R. E. Richards, *J. Mol. Biol.*, **28**, 389 (1967)
247. M. Cohn, M. Grunberg-Manago and A. Danchin, *J. Mol. Biol.*, **39**, 199 (1969)
248. A. S. Mildvan and M. Cohn, *Adv. Enzymol.*, **33**, 1 (1970); M. Cohn, *Quart. Rev. Biophys.*, **3**, 61 (1970)
249. M. Cohn, *Biochem.*, **2**, 623 (1963)

250. A. S. Mildvan, J. S. Leigh and M. Cohn, *Biochem.*, **6**, 1805 (1967)
251. A. S. Mildvan and M. C. Scrutton, *Biochem.*, **6**, 2978 (1967); **7**, 1490 (1968)
252. J. Reuben and M. Cohn, *J. Biol. Chem.*, **245**, 6539 (1970)
253. W. J. O'Sullivan and M. Cohn, *J. Biol. Chem.*, **241**, 3104, 3116 (1966)
254. B. Sheard and E. M. Bradbury, *Prog. Biophys.*, **20**, 187 (1970)
255. J. E. Coleman and B. L. Vallee, *Biochem.*, **3**, 1874 (1964)
256. R. G. Shulman, G. Navon, B. J. Wyluda, D. C. Douglass and T. Yamane, *Proc. Nat. Acad. Sci. (U.S.)*, **56**, 39 (1966)
257. G. Navon, R. G. Shulman, B. J. Wyluda and T. Yamane, *Proc. Nat. Acad. Sci. (U.S.)*, **60**, 86 (1968)
258. G. Navon, R. G. Shulman, B. J. Wyluda and T. Yamane, *J. Mol. Biol.*, **51**, 15 (1970)
259. G. N. Reeke, J. A. Hartsuck, M. L. Ludwig, F. A. Quiocho, T. A. Steitz and W. N. Lipscomb, *Proc. Nat. Acad. Sci. (U.S.)*, **58**, 2220 (1967)
260. C. C. Hinckley, *J. Am. Chem. Soc.*, **91**, 5160 (1969)
261. J. K. M. Sanders and D. H. Williams, *Chem. Comm.*, **1970**, 422
262. K. G. Morallee, E. Nieboer, F. J. C. Rossotti, R. J. P. Williams and A. V. Xavier, *Chem. Comm.*, **1970**, 1132; R. A. Dwek, R. E. Richards, K. G. Morallee, E. Nieboer, R. J. P. Williams and A. V. Xavier, *Europ. J. Biochem.*, **21**, 204 (1971)
263. J. Reuben, *Biochem.*, **10**, 2834 (1971)
264. D. R. Eaton and W. D. Phillips, *Adv. Mag. Res.*, **1**, 119 (1965)
265. A. Kowalsky, *Biochem.*, **4**, 2382 (1965)
266. R. J. Kurland, D. G. Davis and C. Ho, *J. Am. Chem. Soc.*, **90**, 2700 (1968)
267. E. Becker and R. B. Bradley, *J. Chem. Phys.*, **31**, 1413 (1959)
268. W. S. Caughey and W. Koski, *Biochem.*, **1**, 923 (1962)
269. K. Wuthrich, R. G. Shulman, B. J. Wyluda and W. S. Caughey, *Proc. Nat. Acad. Sci. (U.S.)*, **62**, 636 (1969)
270. K. Wuthrich, R. G. Shulman and J. Peisach, *Proc. Nat. Acad. Sci. (U.S.)*, **60**, 373 (1968)
271. K. Wuthrich, *Proc. Nat. Acad. Sci. (U.S.)*, **63**, 1071 (1969)
272. R. G. Shulman, K. Wuthrich, T. Yamane, E. Antonini and M. Brunori, *Proc. Nat. Acad. Sci. (U.S.)*, **63**, 623 (1969)
273. K. Wuthrich, R. G. Shulman, T. Yamane, B. J. Wyluda, T. E. Hugli and F. R. N. Gurd, *J. Biol. Chem.*, **245**, 1947 (1970)
274. R. G. Shulman, K. Wuthrich, T. Yamane, D. J. Patel and W. E. Blumberg, *J. Mol. Biol.*, **53**, 143 (1970)
275. R. G. Shulman, S. Ogawa, K. Wuthrich, T. Yamane, J. Peisach and W. E. Blumberg, *Science*, **165**, 251 (1969); E. Antonini and M. Brunori, *Ann. Rev. Biochem.*, **39**, 977 (1970)
276. D. G. Davis, N. L. Mock, V. L. Laman and C. Ho, *J. Mol. Biol.*, **40**, 311 (1969)
277. S. Ogawa, R. G. Shulman, P. A. M. Kynoch and H. Lehmann, *Nature*, **225**, 1042 (1970)
278. D. G. Davis, N. H. Mock, T. R. Lindstrom, S. Charache and C. Ho, *Biochem. Biophys. Res. Comm.*, **40**, 343 (1970)
279. T. Yamane, K. Wuthrich, R. G. Shulman and S. Ogawa, *J. Mol. Biol.*, **49**, 197 (1970)
280. S. Ogawa and R. G. Shulman, *Biochem. Biophys. Res. Comm.*, **42**, 9 (1971)
281. D. J. Patel, L. Kampa, R. G. Shulman, T. Yamane and M. Fujiwara, *Biochem. Biophys. Res. Comm.*, **40**, 1224 (1970)
282. H. McConnell, *Ann. Rev. Biochem.*, **40**, 227 (1970)

283. H. Sternlicht and E. Wheeler, 'Preliminary magnetic resonance studies of spin-labeled macromolecules', in *Magnetic Resonance in Biological Systems* (Eds. A. Ehrenberg, B. G. Malmstrom and T. Vanngard), Pergamon, New York, 1967, p. 325

284. A. S. Mildvan and H. Weiner, *Biochem.*, **8**, 552 (1969)

285. A. S. Mildvan and H. Weiner, *J. Biol. Chem.*, **244**, 2465 (1969)

286. G. C. K. Roberts, J. Hannah and O. Jardetzky, *Science*, **165**, 504 (1969)

287. J. D. Watson and F. H. C. Crick, *Nature*, **171**, 737 (1953)

288. K. Hoogsteen, 'Hydrogen bonding between purines and pyrimidines', in *Molecular Associations in Biology* (Ed. B. Pullman), Academic Press, New York, 1968, p. 21 and references therein

289. S. A. Rice, A. Wada and E. P. Geiduschek, *Disc. Farad. Soc.*, **25**, 130 (1958)

290. A. M. Michelson, *Nature*, **182**, 1502 (1958)

291. B. H. Zimm, P. Doty and K. Iso, *Proc. Nat. Acad. Sci. (U.S.)*, **45**, 1601 (1959)

292. W. Kauzmann, *Adv. Protein Chem.*, **14**, 1 (1959)

293. O. Jardetzky, *Biopolymers*, **1PP**, 501 (1964)

294. P. O. P. Ts'o, G. K. Helmkamp and C. Sander, *Proc. Nat. Acad. Sci. (U.S.)*, **48**, 686 (1962)

295. S. I. Chan, M. P. Schweizer, P. O. P. Ts'o and G. K. Helmkamp, *J. Am. Chem. Soc.*, **86**, 4182 (1964)

296. A. D. Broom, M. P. Schweizer and P. O. P. Ts'o, *J. Am. Chem. Soc.*, **89**, 3612 (1967)

297. C. E. Johnson, Jr. and F. A. Bovey, *J. Chem. Phys.*, **29**, 1012 (1958)

298. M. P. Schweizer, S. I. Chan and P. O. P. Ts'o, *J. Am. Chem. Soc.*, **87**, 5241 (1965)

299. M. P. Schweizer, A. D. Broom, P. O. P. Ts'o and D. P. Hollis, *J. Am. Chem. Soc.*, **90**, 1042 (1968)

300. S. S. Danyluk and F. E. Hruska, *Biochem.*, **7**, 1038 (1968)

301. A. Dugaiczyk, *Biochem.*, **9**, 1557 (1970)

302. S. I. Chan and J. H. Nelson, *J. Am. Chem. Soc.*, **91**, 168 (1969)

303. S. I. Chan and B. W. Bangerter and H. H. Peter, *Proc. Nat. Acad. Sci. (U.S.)*, **55**, 720 (1966)

304. W. L. Meyer, H. R. Mahler and R. H. Baker, *Biochim. Biophys. Acta*, **64**, 353 (1962)

305. O. Jardetzky and N. G. Wade-Jardetzky, *J. Biol. Chem.*, **241**, 85 (1966)

306. K. H. Scheit, F. Cramer and A. Franke, *Biochim. Biophys. Acta*, **145**, 21 (1967)

307. Y. Inoue and S. Aoyagi, *Biochem. Biophys. Res. Comm.*, **28**, 973 (1967)

308. N. S. Kondo, H. M. Holmes, L. M. Stempel and P. O. P. Ts'o, *Biochem.*, **9**, 3479 (1970)

309. B. W. Bangerter and S. I. Chan, *J. Am. Chem. Soc.*, **91**, 3910 (1969)

310. P. O. P. Ts'o, N. S. Kondo, M. P. Schweizer and D. P. Hollis, *Biochem.*, **8**, 979 (1969)

311. I. C. P. Smith, B. J. Blackburn and T. Yamane, *Canad. J. Biochem.*, **47**, 513 (1969)

312. F. R. Hruska and S. S. Danyluk, *Biochim. Biophys. Acta*, **157**, 238 (1968)

313. R. H. Sarma and N. O. Kaplan, *J. Biol. Chem.*, **244**, 771 (1969); *Biochem.*, **9**, 539 (1970)

314. D. P. Patel, *Nature*, **221**, 1241 (1969)

315. D. P. Hollis, *Org. Mag. Res.*, **1**, 305 (1969)

316. R. H. Sarma and N. O. Kaplan, *Biochem.*, **9**, 551 (1970)

317. B. W. Bangerter and S. I. Chan, *Biopolymers*, **6**, 983 (1968); *Proc. Nat. Acad. Sci. (U.S.)*, **60**, 1144 (1968)

318. S. I. Chan and G. P. Kreishman, *J. Am. Chem. Soc.*, **92**, 1102 (1970)

319. P. O. P. Ts'o and M. P. Schweizer, *Biochem.*, **7**, 2963 (1968)

320. C. C. McDonald, W. D. Phillips and J. Lazer, *J. Am. Chem. Soc.*, **89**, 4166 (1967)
321. C. C. McDonald, W. D. Phillips and S. Penman, *Science*, **144**, 1234 (1964)
322. J. Josse, A. D. Kaiser and A. Kornberg, *J. Biol. Chem.*, **236**, 864 (1961)
323. J. P. McTague, V. Ross and J. H. Gibbs, *Biopolymers*, **2**, 163 (1964)
324. C. C. McDonald, W. D. Phillips and J. Penswick, *Biopolymers*, **3**, 609 (1966)
325. I. C. P. Smith, T. Yamane and R. G. Shulman, *Science*, **159**, 1360 (1968)
326. T. Schleich and J. Goldstein, *Proc. Nat. Acad. Sci. (U.S.)*, **52**, 744 (1964)
327. I. C. P. Smith, T. Yamane and R. G. Shulman, *Canad. J. Biochem.*, **47**, 480 (1969)
328. J. E. Crawford, S. I. Chan, M. P. Schweizer, *Biochem. Biophys. Res. Comm.*, **44**, 1 (1971)
329. H. De Voe and I. Tinoco, *J. Mol. Biol.*, **4**, 500 (1962)
330. L. Katz and S. Penman, *J. Mol. Biol.*, **15**, 220 (1966)
331. R. R. Shoup, H. T. Miles and E. Becker, *Biochem. Biophys. Res. Comm.*, **23**, 194 (1966)
332. R. A. Newmark and C. R. Cantor, *J. Am. Chem. Soc.*, **90**, 5010 (1968)
333. D. J. Blears and S. S. Danyluk, *Biopolymers*, **5PP**, 535 (1967)
334. K. H. Scheit, *Angew. Chem.*, **6**, 179 (1967)
335. F. E. Hruska and S. S. Danyluk, *Biochim. Biophys. Acta*, **161**, 250 (1968)
336. E. J. Gabbay, B. L. Gaffney and L. A. Wilson, *Biochem. Biophys. Res. Comm.*, **35**, 854 (1969)
337. B. H. Arison and K. Hoogsteen, *Biochem.*, **9**, 3976 (1970)
338. T. A. Victor, F. E. Hruska, K. Hikichi, S. S. Danyluk and C. L. Bell, *Nature*, **223**, 302 (1969)
339. W. Muller and D. M. Crothers, *J. Mol. Biol.*, **35**, 251 (1968)
340. M. Cohn and T. R. Hughes, *J. Biol. Chem.*, **235**, 3250 (1960)
341. N. A. Berger and G. L. Eichhorn, *Biochem.*, **10**, 1857 (1971)
342. K. E. Rich, R. T. Agarwal and I. Feldmann, *J. Am. Chem. Soc.*, **92**, 6818 (1970)
343. S. M. Wang and N. C. Li, *J. Am. Chem. Soc.*, **90**, 5069 (1968)
344. L. S. Kon and N. C. Li, *J. Am. Chem. Soc.*, **92**, 281 (1970)
345. S. S. Danyluk and F. E. Hruska, *Biochem.*, **7**, 1038 (1969)
346. J. M. Prestegard and S. I. Chan, *J. Am. Chem. Soc.*, **91** 284 (1969)
347. R. L. Ward and J. A. Happe, *Biochem. Biophys. Res. Comm.*, **28**, 785 (1967)
348. B. W. Roberts, J. B. Lambert and J. D. Roberts, *J. Am. Chem. Soc.*, **87**, 5439 (1965)
349. E. D. Becker, H. T. Miles and R. B. Bradley, *J. Am. Chem. Soc.*, **87**, 5575 (1965)
350. J. A. Happe and M. Morales, *J. Am. Chem. Soc.*, **88**, 2077 (1966)
351. C. D. Barry, A. C. T. North, J. A. Glasel, R. J. P. Williams and A. V. Xavier, *Nature*, **232**, 236 (1971)
352. P. R. Srinivasan, S. Hendler and J. A. Glasel, *Proc. Nat. Acad. Sci. (U.S.)*, **60**, 1038 (1968)
353. W. H. Huestis and M. A. Raftery, *Biochem.*, **10**, 1181 (1971)
354. M. Alei, Jr., A. E. Florin and W. M. Lichtman, *J. Am. Chem. Soc.*, **92**, 4248 (1970)
355. R. L. Lichter and J. D. Roberts, *J. Am. Chem. Soc.*, in press
356. J. Reuben, *J. Am. Chem. Soc.*, **91**, 5725 (1969)
357. R. G. Bryant, *J. Am. Chem. Soc.*, **91**, 1870 (1969)
358. T. L. James and J. H. Noggle, *Proc. Nat. Acad. Sci. (U.S.)*, **62**, 645 (1968)
359. R. G. Bryant, *Biochem. Biophys. Res. Comm.*, **40**, 1162 (1970)
360. F. J. Kayne and J. Reuben, *J. Am. Chem. Soc.*, **92**, 220 (1970)
361. O. Jardetzky and N. Wade-Jardetzky, *Ann. Rev. Biochem.*, **40**, 605 (1971)
362. J. J. M. Rowe, J. Hinton and K. L. Lowe, *Chem. Rev.*, **70**, 1 (1970)
363. P. A. Srere, *Biochem. Biophys. Res. Comm.*, **26**, 609 (1967)
364. M. S. Feather and M. J. Lybyer, *Biochem. Biophys. Res. Comm.*, **35**, 558 (1969)

CHAPTER 13

Spin labels in the e.p.r. investigation of biological systems

A. E. Kalmanson and G. L. Grigoryan

D. I. Ivanovsky Institute of Virology,
The U.S.S.R. Academy of Medical Science, Moscow, U.S.S.R.

I. INTRODUCTION

During the last 10–15 years the studies made by means of electron paramagnetic resonance spectroscopy (e.p.r.) have yielded much information
which showed the important role of free radicals and other paramagnetic
centres in the energetics of biological systems as well as in the radiobiological
and photobiological processes.[1–6]

However the modern phase of development of the molecular biology is
characterized by the ever increasing interest of researches into questions of
conformational changes of biopolymers determining their functional activity.
It is precisely in this direction that theoretical and experimental methods
of modern physics and chemistry are most widely used. The study of

biopolymers labelled by free radicals is one of the new and extremely promising methods of approach to solve these problems.

Iminoxyl relatively stable organic free radicals used in these studies were named 'spin labels'. This term was put forward and employed for the first time in the work by H. M. McConnell and associates[7] who used the iminoxyl free radicals to examine biopolymers.

This method may be extended to study an even wider range of paramagnetic substances. Semiquinone free radicals, for instance, were used[8,8a] to investigate thermal and chemical denaturation of proteins and nucleic acids. Ion-radicals of chlorpromazine were employed to study the orientation of molecules of dye adsorbed on DNA;[9] paramagnetic cupric ions were used[10] to examine protein structure. It should at last be mentioned that both naturally occurring free-radicals involved in energetic metabolism of biological systems and metal ions of variable valency participating in the process of electron transfer may be referred to as spin labels of 'natural' origin.

It is necessary to point out that Soviet authors frequently[11-13] use the term 'paramagnetic zond' instead of 'spin label'. As a rule, the term 'paramagnetic zond' is used in case of noncovalently bound free radicals.

Thus one can see that various paramagnetic substances can be regarded as spin labels.

However, it is beyond any doubt that iminoxyl free radicals (called 'nitroxides' by American authors) are the species most convenient to study biopolymers.

The discovery and systematic investigation of the physical and chemical properties of iminoxyl free radicals were the main reason for the development of the spin label method.

The work by M. B. Neiman, E. G. Rozantsev and their coworkers in the USSR[14-16] and that by A Rassat and his coworkers in France[17-18] have made the greatest contribution to the development of the chemistry of iminoxyl free radicals. A great number of studies devoted to the study of iminoxyl free radicals by the e.p.r. method were performed in the Soviet Union by A. L. Buchachenko and his coworkers.[19,20]

It should be particularly stressed that the employment of the spin label method in biology is very closely connected with the application of modern achievements in the theory of the e.p.r. spectra to organic free radicals in solution. H. M. McConnell[21-23] has made the greatest contribution to the development of this theory.

The spin label method is based on the fact that hyperfine structure of the e.p.r. spectra of iminoxyl free radicals as well as of other organic free radicals depends in a regular way on the hindering of rotation of these paramagnetic particles. This change in tumbling of radicals may take place under the influence of different factors which will be analysed later on.

The paramagnetism of stable iminoxyl radicals is due to the presence of an odd electron localized on sterically *screened (ecranized)* NO group which may be a component of both linear and/or cyclic molecules. (See formulae (1), (2) and (3).)

(1)   (2)   (3)

At present the research work with application of iminoxyl radicals is carried on the three main directions:

1. *The employment of spin labels bound by means of intermolecular forces to the molecules of biopolymers (the paramagnetic zonds).* With the help of spin labels which do not form strong covalent bonds with the polymers it is rather convenient to study the initial stages of hydration of biopolymers and other biological material and analyse the processes of penetration of spin labels through various biological membranes. Thus one can for instance follow the spin label in different organs of the animals injected into their peritonial cavity; or follow the spin labels in different parts of plants when the former is added to a nutrient solution and enters the plants through the roots. The initial phases of hydration of various nucleic acids were studied with the help of a loosely bound spin label.[11] The results of our experiments in studying initial stages of hydration of serum albumin show that the increase in tumbling of iminoxyl radicals causes changes in the e.p.r. spectrum. These changes are observed only at a certain water content of the sample (about 10–12 per cent). With some intermediate humidities (10–24 per cent) we observe superposition of one signal on another. One signal is due to a strongly immobilized spin label, the other to a loosely bound one. With increasing water content of the sample the first signal becomes weaker and the second stronger. Once the sample possesses sufficient humidity all the radicals become relatively loose. In these experiments dependence of spectra of iminoxyl radicals on the level of humidity in the sample may be explained if we take into account the competition between the molecules of water and radicals that occupy polar parts of the protein surface.

2. *The use of iminoxyl free radicals as paramagnetic redox indicators for reduction–oxidation changes of active centres of electron transfer chains during biological oxidation and photosynthetic reactions.* In these experiments the ability of iminoxyl radicals to change reversibly into hydroxylamines by interaction with sufficiently strong donor–acceptor systems which take part

in the transfer of electrons during oxidative metabolism was used.[24] V. M. Chumakov and coworkers[25] have recently shown that the reversible and simultaneous changes in the concentration of the 'natural' free radicals and iminoxyl spin labels in certain lyophilized tissues are due to a low concentration of adsorbed water. This result supports the idea of the special role of adsorbed water in the electron transport systems of the biooxidation.

3. The third and undoubtedly most important direction of application of spin labels in biology concerns *the employment of spin labels strongly fixed on the molecules of polymers by means of covalent bonds.*

This survey will be mainly devoted to the description of work carried out in this respect because of its great interest for molecular biology.

II. SOME THEORETICAL BASES FOR STUDYING BIOPOLYMERS BY MEANS OF IMINOXYL SPIN LABELS

The researchers in modern biology pay great attention to the study of details of structure of biopolymers, to the study of specific conformation and conformation changes of biological macromolecules.

Iminoxyl free radicals are extremely convenient for application under different conditions of biological experiments due to the following reasons:

1. They are sufficiently stable at high temperatures and extreme values of pH.

2. They are soluble both in water and non-polar solvents.

3. Hyperfine structure of the e.p.r. spectra of these radicals is characterized by the following features:

(a) it is simple (it consists of three narrow lines of equal intensity);

(b) it is anisotropic (it depends on the orientation of the orbit of a free electron in the constant magnetic field and on the rate of the irregular movements of free radicals in solution which average the anisotropic interaction;

(c) it depends on exchange and dipole–dipole interactions of free radicals with each other.

4. Iminoxyl free radicals give a possibility to carry on reactions without involving free valencies. It is possible, to a great extent, to synthesize iminoxyl free radicals with given properties and, what is particularly important, even such ones which specifically interact with certain functional groups of biopolymers.

As we have already mentioned the hyperfine structure of the e.p.r. spectra of iminoxyl free radicals in solution shows three equal components and their h.f. splitting is about 15·5 Oe. (Figure 1(a)). Due to anisotropy of the tensor of the dipole–dipole interaction and the g-tensor the restriction of tumbling of these radicals, for instance, as a result of increase of the solvent viscosity

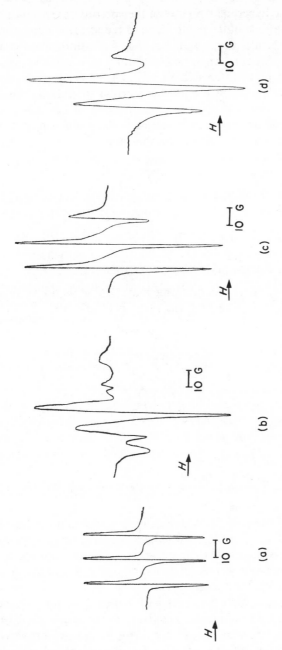

Figure 1. (a) E.p.r. spectrum of iminoxyl free radical in a water solution. (b) E.p.r. spectrum of iminoxyl free radical attached to molecules of bovine serum albumin, 0·1 M phosphate buffer, pH 6·8. (c) E.p.r. spectrum of iminoxyl free radical attached to molecules of bovine serum albumin in a 10 M solution of a urea. (d) 50 per cent dioxane. (After reference 31)

brings about certain changes in the e.p.r. spectra. As a consequence the components of the hyperfine structure of the spectra are irregularly broadened and the amplitude of the components is reduced while the area below the e.p.r. signal which is proportional to the concentration of free radicals remains constant. The tumbling rate of free radicals is characterized by the correlation time—a value depending on the effectiveness of averaging the anisotropic dipole–dipole interaction by irregular thermal motions of free radicals.

Bloembergen and his coworkers[26] using a rotating sphere model obtained the following expression for the correlation time

$$\tau = \frac{4\pi\eta a^3}{3kT} \qquad (1)$$

where

τ = correlation time;
η = viscosity;
a = radius of the sphere;
T = absolute temperature;
k = Boltzmann's constant.

On the other hand, the correlation time may also be found by an analysis of the width of the components of the e.p.r. spectra.[27]

$$\tau = 8 \cdot 402 \Delta H_{max}\alpha \times 10^{-10} \qquad (2)^{[11]}$$

Here ΔH_{max} is the half width of the component of the hyperfine structure of the low magnetic field spectra
and

$$\alpha = 1 - \sqrt{\frac{I_{(+1)}}{I_{(-1)}}}$$

where

$I_{(+1)}$ is the intensity of the component of the hyperfine structure of the spectra at low magnetic field and
$I_{(-1)}$ is the intensity of the component of the hyperfine structure of the spectra at high magnetic field.

Here, a relative width of the components (the α parameter) and not the absolute one is analysed. ΔH_{max} is usually considered to be a constant.

The given expression is true when non-secular contributions to the width of the e.p.r. line may be neglected. This assumption is justified with $\omega_0\tau \geqslant 1$ where ω_0 is the Larmor frequency.

Thus the analysis of the e.p.r. spectra of iminoxyl free radicals makes it possible to obtain quantitative characteristics of the rate of their tumbling which as we shall show later may considerably depend on weak intermolecular interactions playing an extremely important role in biological processes at the molecular level. (Figure 1(c), (d).)

In the case of firm fixation of iminoxyl radicals there appears a new and different type of e.p.r. spectrum which is characterized by a large widening of the components of the hyperfine structure and by a considerable increase in the value of hyperfine splitting (Figure 1(b)). We shall call this spectrum a 'wide' e.p.r. signal.

In studies of monocrystalline samples of free radicals the value of the anisotropic splitting displays its dependence upon the relative orientation of the crystallographic axes of the system in the external magnetic field. The values of the constants of anisotropic hyperfine interaction in parallel and perpendicular orientation of the π-orbital of a free electron of the iminoxyl free radicals in the constant magnetic field as well as the components of the g-tensor were determined by O. N. Griffith, D. N. Cornel and H. M. McConnell by studying single crystals of iminoxyl free radicals.[28]

III. IMMOBILIZATION OF A SPIN LABEL. STUDIES OF CONFORMATION AT TRANSITIONS OF ALBUMIN IN SOLUTION

In their first work[7] on application of the iminoxyl free radicals for studying biopolymers Stone, Buckman, Nordio and McConnell used a radical (**4**) which may be bound to the $-NH_2$ albumin groups as a paramagnetic label.

(**4**)

When the paramagnetic label joined the amino-poly-L-lysine groups the components of the hyperfine structure of the e.p.r. spectra were broadened a little, testifying of the fact that the mobility of the free radicals had decreased to a certain extent. The authors analysed the e.p.r. spectra of the labelled serum albumin and found out that the major part of free radicals was in the same state as it was in the case with poly-L-lysine. However in the analysed spectra one could observe a component of the e.p.r. signal with other parameters as compared with the analogous e.p.r. signal of the iminoxyl free radicals in rigid glasses.

The intensity of the irregular movement of the albumin molecules in solution is no longer sufficient for an effective averaging of anisotropic interaction when a small part of the free radicals is firmly fixed to the rigid tertiary structure of serum albumin. This phenomenon was called a 'strong' immobilization of the spin label as compared to a 'weak' one when the components of the hyperfine structure of the e.p.r. spectra were only slightly broadened.

Proceeding from the fact that each molecule of albumin bound about 1·5–2 free radicals the authors suggested that the serum albumin possesses two different amino groups reacting with the iminoxyl free radical. When the paramagnetic label joins the NH_2 group situated on the surface of the albumin molecule there appears the e.p.r. spectrum of the weakly immobilized label. The other amino group is in the depth of the albumin globule and the label at this position yields the e.p.r. spectrum of the strongly immobilized free radicals. The authors presume that in the second case the free radical bound to the albumin molecule by a covalent bond interacts with the adjacent fragments of the polypeptide chain by means of hydrophobic forces.

On acid and alkaline denaturation of the serum albumin as well as after digesting the labelled albumin by pepsin the wide components of the e.p.r. spectrum of the strongly immobilized free radicals disappeared and the e.p.r. signal of the analysed system approached the e.p.r. spectrum of the spin labelled poly-L-lysine.

In the work which followed[29] it was pointed out that the described label is of little effectivity and easily hydrolysed. That is why Griffith and McConnell[29] employed the maleimide derivative 2,2,5,5-tetramethyl-3-aminopyrolydine-I-oxyl as a paramagnetic label (structure (5)).

(5)

After carrying out the reaction which resulted in the binding of the spin label by serum albumin one could observe a mixture of the two e.p.r. signals belonging to the strongly and weakly immobilized free radicals. If the label was applied after blocking of the SH-albumin group by a specific reagent the e.p.r. spectrum did not possess wide components characteristic of the strongly immobilized free radicals. At pH 2·1 the components of the wide signal were not observed and the amplitude of the e.p.r. narrow signal was enhanced by a factor of ~ 2. Proceeding from these facts, the authors came to a conclusion that about 60 per cent of the spin label is specifically bound to a sulphhydril group of serum albumin producing the e.p.r. spectrum of the strongly immobilized free radicals and the remaining part of the free radicals reacts with the $-NH_2$ groups. Since the time of inversion of the anisotropic electron and nuclear interaction has a value of about 10^{-8} sec the wide e.p.r. signal must correspond to the free radicals strongly limited in tumbling in relation to the albumin molecule. Griffith and McConnell[29] point out that the components of the e.p.r. wide signal can be observed while studying a large number of SH-containing proteins.

From the fact that in case of strong immobilization the paramagnetic label must be located somewhere in a hole on the albumin surface corresponding roughly to the outline of the label molecule the authors draw two important conclusions:

First, the albumin structure must be still sufficiently open for the sulphhydril group to react with the label.
Second, as soon as the probability of initial steric correspondence between the label and the albumin is extremely small the albumin structure should be elastic enough.

A maleimide derivative 2,2,6,6-tetramethyl-4-aminopyperidine-I-oxyl of the free radical[30–31] is a more convenient and specific label (configuration (6) and Figure 1).

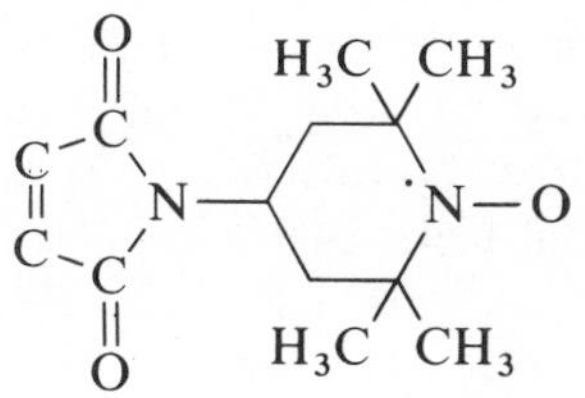

(6)

It was shown that more than 90 per cent of this label during its binding to serum albumin is transformed into a strongly immobilized species.[31] The reaction of the paramagnetic label with the sulphhydril groups of the albumins (7) may be represented in the following way:

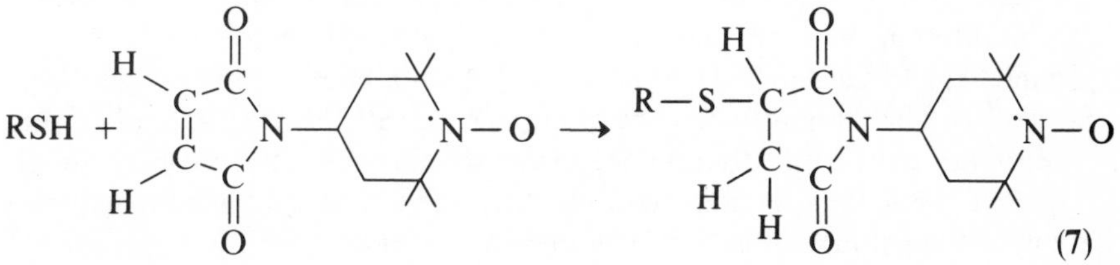

(7)

If we may neglect the influence of the thermal movement of the albumin molecules in solution on the e.p.r. spectra of the label we should take into account only the oscillations of the free radicals around the covalent bond and the configuration of a fragment of the macromolecule adjacent to the free radical. Owing to this the application of the spin labels to study the conformation changes of biopolymers turned out to be very convenient. Taking into account the above-mentioned suppositions about the mechanism of strong immobilization of the paramagnetic label one can reach the conclusion that even small changes in the protein tertiary structure may bring about the observed alteration of mobility of the free radicals. It is known that the tertiary structure of the proteins can be changed, within certain limits, independent of the secondary one.[32,33] In this respect the effect of non-polar solvents of the dioxane type is particularly characteristic. In the proteins which possess a great number of intermolecular S–S bonds these solvents modify the tertiary structure without considerable influence on the melting point of the α-helix.

On the other hand, proteins completely lose their secondary structure in concentrated solutions of urea and assume the state of a random coil. That is why it was interesting to compare the influence of the urea and dioxane on the mobility of the spin label in the serum albumin. When sufficient quantities of urea and dioxane are added to the solutions of serum albumin labelled with iminoxyl free radicals the broadened spectrum disappears and changes into the e.p.r. spectrum of the free radicals possessing considerably greater mobility. However urea which destroys the secondary and tertiary albumin structure brings about a more considerable increase in the mobility of the free radicals. The correlation time calculated for these two cases turned out to be 1.09×10^{-10} s for serum albumin in the presence of 10 M of urea and 2.04×10^{-9} s for serum albumin in a 50 per cent water–dioxane mixture.

It is interesting to point out that during the process of denaturation only the narrow part of the e.p.r. signal was increasing whereas the ratio of its components remained constant. These results testify that under the influence of both urea and dioxane serum albumin undergoes conformational changes only between two states—native and modified—without intermediate phases.

Since the described label cannot be fixed on to proteins which possess no sulphhydril groups capable to react it can be assumed that in the case of serum albumin the paramagnetic label is specifically bound to sulphhydril groups. At the same time we know that serum albumin has only one free SH-group. That is why, when the paramagnetic label is used to analyse the conformation of serum albumin one can speak about the state of only a small fragment of the polypeptide chain to which the free radical is attached and about the portions which are in immediate contact with the free radical.

According to the information obtained from studies of protein secondary and tertiary structures by optical and hydrodynamic methods[32,33] it is quite natural to presume that the states of the polypeptide chain fragment to which the spin label is attached when albumin is treated with urea and dioxane are characterized by the absence and presence of the α-helix structure respectively. In the latter case (when dioxane is added) the presence of α-helix structure considerably increases the firmness of part of the macromolecule and due to this fact the mobility of the paramagnetic label in the presence of dioxane is considerably less than in the presence of urea. The comparison of the results obtained by means of paramagnetic labels with the results of experiments carried out by other methods which allow one to judge the state of the macromolecule secondary structure, on the whole, is of considerable interest.

We have already spoken about the conformation changes in the proteins in different model systems. But the macromolecule conformation changes carried out as a result of various functional interactions are of special interest.

S. Ogawa and H. M. McConnell have studied, by means of the spin label method, the conformational changes undergone by haemoglobin when oxygenated.[34] In this work they used I-oxyl-2,2,5,5,5-tetramethyl-3-pirrolidinyl iodoacetamide as a spin label (Formula (8)).

$$\text{(pyrrolidine ring)}-\overset{\overset{\displaystyle H}{|}}{N}-\overset{\overset{\displaystyle O}{\|}}{C}-CH_2I \tag{8}$$

The transformation of haemoglobin into oxyhaeomoglobin was accompanied by a transition of one state of the paramagnetic label into another. The absence of intermediate states was proved by the presence of isobestic points and the superposition of the e.p.r. spectra. The authors drew the conclusion that the conformational change as a result of the oxygenation takes place simultaneously in all the four subunits of haemoglobin and is a cooperative process.

S. Ohnishi and coworkers,[35] H. M. McConnell and C. L. Hamilton,[36] G. Lichtenstein and coworkers,[37] studied various derivatives of haemoglobin and some subunits of this protein labelled by SH groups. It turned out that the addition of CO to haemoglobin causes the reduction of mobility of the spin label. The e.p.r. spectra of the labelled subunits considerably differed from the tetramer spectra. The association of the subunits in the tetramer was accompanied by a sharp decrease in the mobility of the spin label. The conformation change during the formation of the tetramer affected mainly the haemoglobin β-chains.

Thus, we can obtain extensive information about the conformation changes in the proteins with the help of the spin label method. In our opinion the most valuable thing is to use spin labels to study small changes in the tertiary structure of the proteins, because the method of measuring optical activity, very popular in biology, cannot be used here extensively. However, in many cases, the study of similar effects may be carried out by measuring the polarization of luminescence. Study of the rate of depolarization of fluorescence in solution can yield information about the mobility of the luminescence centre, for instance, of the dye molecules bound to the protein. Still, the e.p.r. method is more suitable since it supplies information on many other important parameters of the analysed systems. Besides, as we shall show later, one of the advantages of the e.p.r. method is the fact that it allows us to study such effects even in the non-transparent condensed systems which is particularly important in biological studies.

IV. SPIN-LABELLED NUCLEIC ACIDS

I. C. P. Smith and T. Yamane[38] carried out the studies of the nucleic acids labelled by iminoxyl free radicals. They used three different paramagnetic labels (9), (10), (11).

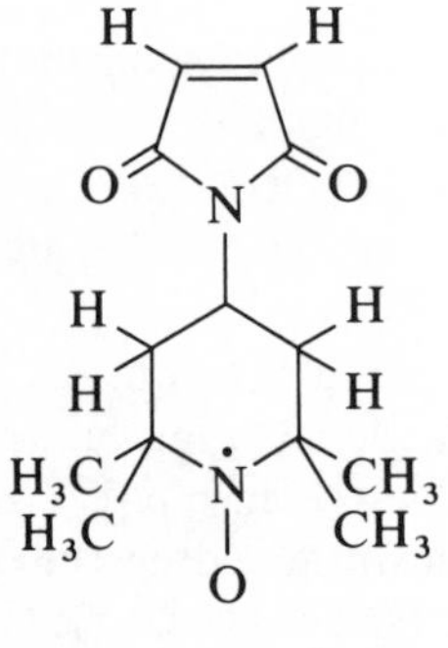

(9)

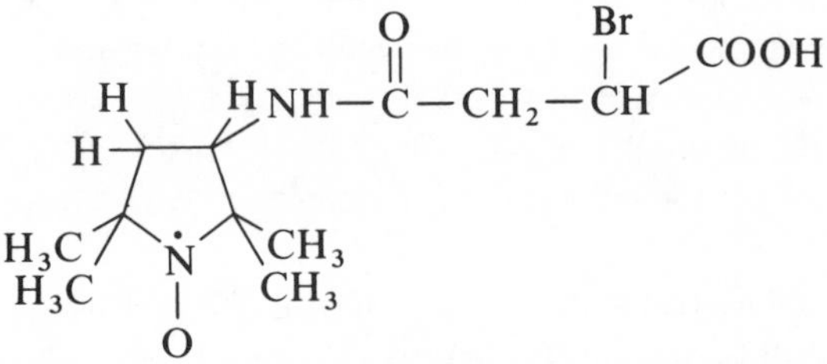

(10)

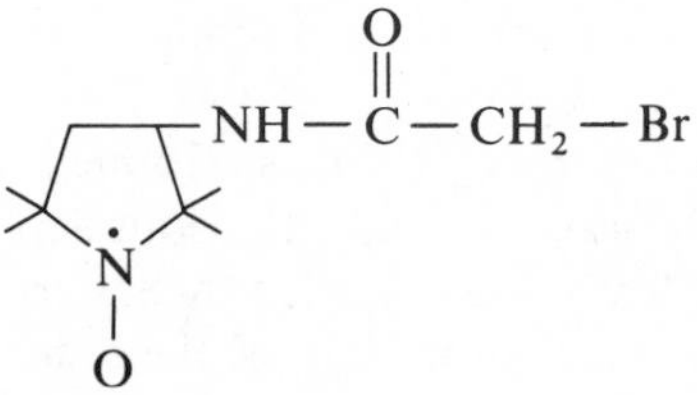

(11)

In the case of interaction of all three labels with molecules of the ribonucleic acids the mobility of the free radicals became slightly less. The largest amount of binding was obtained with *N*-(I-oxyl-2,2,5,5-tetramethyl-3-pyrrolinyl)bromoacetamide (**11**). Besides, when this label was used the rate of immobilization of the iminoxyl free radicals was greater than with other labels. The treatment of the preparations of labelled ribonucleic acid by nuclease brought a sharp increase of mobility of the paramagnetic label. At the same time the e.p.r. spectrum of the studied system was getting closer to the spectrum of the free paramagnetic label in solution.

Addition of the paramagnetic label to desoxyribonucleic acid induced the appearance of a mixture of two signals corresponding to different types of immobilization of the free radicals. The results obtained by these authors in their study of the binding of the spin label (**11**) to polynucleotides are extremely instructive. With polyguanine and polyadenine at pH 5·5 they observed the e.p.r. spectrum of the strongly immobilized free radicals. At pH 11·5 the mobility of the paramagnetic label sharply increased as a result of the destruction of the regular secondary structure of the polynucleotides.

The occurrence of isobestic points shows that this transition was co-operative.

When the authors studied polycytidylic acid they observed a mixture of the two e.p.r. signals. It followed from this that only a portion of the molecules had the regular secondary structure. The study of polyuridine showed that this molecule was in the statistic coil state even at pH values close to neutrality.

Proceeding from the results of studying of acid–alkaline denaturation of polyguanine and polyadenine by means of the e.p.r. method the authors determined the pK values of this transition for polynucleotides. The pK values turned out to be in agreement with those obtained by optical methods.

V. USE OF A SUBSTRATE LABELLED BY STABLE RADICALS IN THE STUDY OF ENZYME–SUBSTRATE COMPLEXES

Y. E. Berliner and H. M. McConnell[39] used as a substrate a compound (**13**) of an ester with iminoxyl free radical (**12**)

radical
(**12**)

substrate
(**13**)

The hydrolysis of esters by α-chymotrypsin takes place according to the following reaction:

$$E + S \overset{K_1}{\rightleftharpoons} ES \overset{K_2}{\rightarrow} ES + P_1 \overset{K_3}{\rightarrow} E + P_2$$

After the equilibrium between the enzyme E and the substrate S has been reached the enzyme–substrate complex is decomposed producing P_1, p-nitrophenol and acylcoenzyme, which in turn is decomposed giving the substrate and P_2 (acid).

At acid pH (pH 4·5) when K_3 is negligible a small quantity of the acylcoenzyme was separated. The e.p.r. spectrum of the enzyme–substrate complex showed that the acyl group together with the iminoxyl free radical was rigidly fixed on the enzyme molecule. At pH 6·8 the complex was getting decomposed and the e.p.r. signal characteristic for the free iminoxyl label in solution was observed.

In 10 M urea the e.p.r. wide signal disappeared changing into the e.p.r. spectrum of the free radicals which possess considerably greater mobility. However the e.p.r. signal components were broadened to a certain extent which proved that the transition of the enzyme into the random coil state does not bring about dissociation of the substrate, despite the fact that during this process the frequency of its tumbling was considerably increased. The paramagnetic label was not removed during the dialysis in acid medium. Consequently the acyl group is bound to the enzyme by covalent bonds, evidently by means of simple ester linkage and in addition it is immobilized by van der Waals forces. It is very likely that the most important result of this work is the rigid orientation of the enzyme and substrate molecules relative to each other during the formation of the enzyme–substrate complex. According to the present view[40] a high catalytic ability of the enzymes can be explained by the fact that the specific configuration of the enzyme molecule ensures a certain mutual orientation of the substrate and catalytic groups of the enzyme during the formation of the complex. Since the probability of an accidentally correct orientation of the molecules of the substrate and of the catalytic groups in solution is extremely small and decreases sharply with increasing number of substrate and catalytic groups taking part in the reaction, the orientational function of the enzymes evidently greatly contributes to the ensuring of their high catalytic activity.

VI. UTILIZATION OF THE SPIN LABELS IN THE IMMUNOLOGIC REACTIONS

As Pauling suggested the interaction between antigen and antibody is due to short-range intermolecular forces and to an exact steric correspondence of the determinant groups of these molecules.

During recent years antibodies have been obtained for many simple haptens. Dinitrophenyl haptens proved to be an especially convenient model substance. Their interaction with antibodies could be studied by fluorescent methods.

L. Stryer and O. N. Griffith[41] used a dinitrophenyl antigen which serves as a dual label, fluorescent and paramagnetic. The composition of this hapten comprises an iminoxyl free radical as shown in structure (14).

Dinitrophenyliminoxyl hapten (2-4-dinitrophenyl hydrazone-2,2,6,6-tetramethyl-4-aminopiperidine-I-oxyl).

(14)

When hapten solution was added to the immune serum they observed a strong anisotropic broadening of the e.p.r. spectrum showing that the hapten was bound to the antibodies.

The authors[41] studied the binding of hapten by antibodies at different ratios and showed that hapten labelled by the iminoxyl free radicals could be used to determine the titre of antibodies. Because of the strong immobilization of the paramagnetic label they failed to determine the correlation time on the basis of the analysis of the width of the e.p.r. spectral components.

It was possible to determine the analogous value by a method of luminescence polarization. For this purpose an aqueous–glycerol solution was selected in which the e.p.r. spectra of the dinitrophenyliminoxyl hapten were analogous to the e.p.r. spectrum of the hapten–antibody complex (90 per cent glycerol, 5 per cent H_2O, 5 per cent alcohol). In this solution by a method of the polarization of luminescence Stryer and Griffith[41] determined the relaxation time for the Brownian movement, the value of which was 3.6×10^{-8} sec.

In the literature[42] the antibodies labelled by the iminoxyl free radicals have been studied. As a paramagnetic label the maleimide derivative of 2,2,6,6-tetramethyl-4-aminopiperidine-I-oxyl were used. The γ-globulin molecule contains four polypeptide chains (2 L-chains and 2 H-chains) attached remain intact.

The spin label was attached to the sulphhydril groups of interchain disulphide bonds of man and rabbit γ-globulins (the bonds were reduced with mercaptoethanol). In both cases the e.p.r. spectra correspond to a relatively large mobility of the free radicals (the correlation time τ was 1.1×10^{-9} s for the human globulin and 7.4×10^{-10} s for rabbit globulin). Comparing the correlation time of the iminoxyl free radicals bound to the sulphhydril groups of the cleaved disulphide bonds of the γ-globulins and to that of the serum albumin SH-group treated with urea ($\tau = 1.09 \times 10^{-9}$) and dioxane ($\tau = 2.04 \times 10^{-9}$) it is quite natural to come to the conclusion that the fragments of the polypeptide chains of the γ-globulins to which the free radicals are attached do not possess an ordered secondary structure. The mobility of the iminoxyl labels is especially great in the rabbit γ-globulin where it even exceeds the one for the serum albumin denaturated with urea. These results are in agreement with the data about a rather low content of helical structures in γ-globulins.

The nature of the e.p.r. spectrum of the rabbit immune γ-globulin when it conserved its specific activity allowed the study of conformational and phase transitions during the specific antigen–antibody reaction. During the studies of precipitation of the rabbit antibodies by the specific antigen (egg albumin) and non-specific precipitation of the γ-globulin as a result of the salting out with ammonia sulphate a sharp difference in the mobility of the spin label attached to the sulphhydril groups of the cleaved interchain disulphide bonds of the γ-globulin was found in these two cases. The precipitation of the antibodies by the specific antigen brought only a small decrease in the mobility of the paramagnetic label, whereas the precipitation of the γ-globulin with ammonia sulphate brought about strong immobilization of the free radicals. These results are of a certain interest for an understanding of the precipitation reaction mechanism.

The precipitation theory by J. Bordet[43] based on the resemblance of serological and colloid phenomena failed to explain the surprising specificity of this reaction. J. R. Marrack[44] suggested that the forces responsible for interaction of the antigen and antibody are of a specific nature and not connected as it was suggested by J. Bordet with charge neutralization and correspondingly with the decrease in solubility of the interacting molecules. In their works M. Heidelberger and F. E. Kendall[45] advanced a theory of specific precipitation on according to which the precipitate formation was related to the immunological polyvalency of the antigen and antibody regarding each other. Usually this theory is called a theory of grate or an alteration theory.

The results described above can be regarded as a direct corroboration of the grate theory. In fact if we do not place the spin labels in the active centre of the antibody we can conclude that a relatively mobile state of the spin labels in the antigen–antibody precipitate may be preserved only on condition of absence of strong dehydration of the antibodies at the expense of the intermolecular interactions. Consequently, in contrast to the γ-globulin precipitated by ammonia sulphate the specific antigen–antibody precipitate according to the grate theory has a microcellular structure.

If the antigen–antobody precipitate was kept for a long time in the absence of stabilizers, the rate of mobility of the imminoxyl free radicals decreased sharply. This fact can be connected with the increasing dehydration of the antibodies due to the close interaction of the protein molecules in the precipitate. This dehydration is a secondary process.

The reversible nature of the analysed changes was shown in the case of non-specific precipitation by solution of proteins precipitated with salt as a result of the decrease of the ion concentration in solution, and in case of specific precipitation by a limited hydrolysis of the antibodies in the precipitate, by means of R. R. Porter's method. The latter showed that the rabbit antibodies hydrolysed by papain are cleaved into three fragments; two of them preserve their ability to interact with the antigen but they do not form a precipitate in complex with it. After the treatment of the suspension of the antigen–antibody precipitate by papain about 80 per cent of the protein passed into solution. Then by means of gel-filtration through KM cellulose the purified complexes of the monovalent fragments of the antibodies with the antigen were separated. The e.p.r. spectrum of the complexes was getting closer to the spectrum of the labelled γ-globulin in solution.

VII. STUDY OF BIOMEMBRANES AND MODEL SYSTEMS BY MEANS OF SPIN LABELS

A. S. Waggoner, O. N. Griffith and C. R. Christensen[46] studied the e.p.r. spectra of the paramagnetic labels added to micelles of dodecylsulphate (lauryl sulphate). The study of these systems is important to clarify the organization principles of the biomembranes. The authors employed two kinds of iminoxyl free radicals (15) and (16)

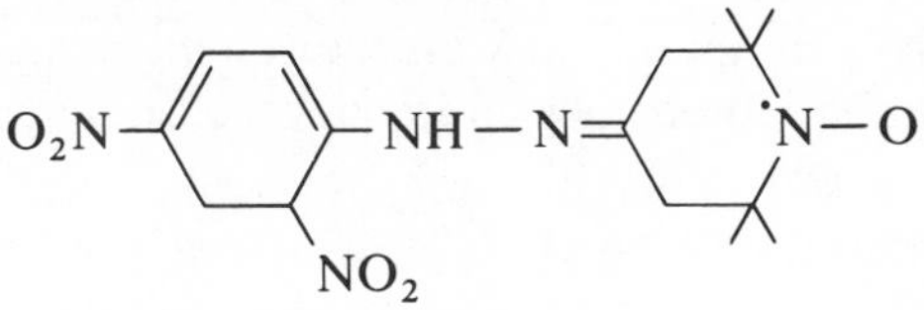

2-4-dinitrophenyl hydrazone-2,2,6,6-tetramethyl-4-piperidine-nitric oxide

(15)

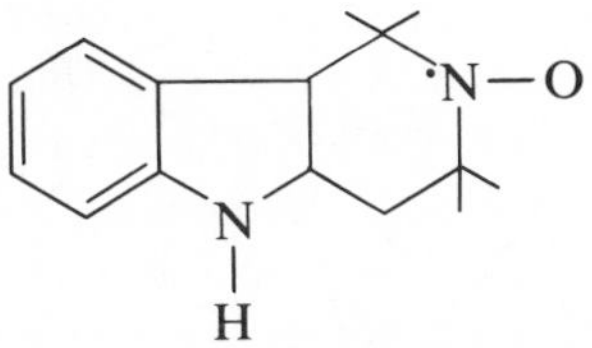

2,2,4,4-tetramethyl-1,2,3,4-tetrahydro-carboline-3-oxyl

(16)

With increase of sodium lauryl sulphate concentration (NaDS) from 0 to 5 per cent (in solution of 5 per cent concentration almost all molecules NaDS enter the micelles), the solubility of both iminoxyl radicals increased more than 100-fold. Correspondingly the mobility of the free radicals in this system increased from $1 \times 10^{-10}\,s^{-1}$ (a value characteristic of the free radical in solution) to $7 \times 10^{-10}\,s^{-1}$. The study of the dependence of τ on concentration allowed one to determine the critical point of formation of the micelles.

At present three models for inclusion of the solubilized molecules in the micelles of the detergents are suggested:

   I. Adsorption on the micelle surface.
  II. Radical emplacement of the solubilized molecules in the micelle in the way that the polar groups are placed near the micelle surface.
 III. Solution in the hydrocarbon cavity of the micelle.

Model I implies rigid fixation of the radical with the assistance of the van der Waals forces. In this case the radical can tumble in solution only with the micelle. Evidently, model I is not valid, because the calculation of the correlation time yields a value 18-times higher than the experimental value obtained on the basis of the analysis of the width of the hyperfine structure components. Relatively high mobility of the iminoxyl radicals can be explained in models II and III but according to these hypotheses microscopic environment of the radical should not be polar. Polarity of the microscopic environment of the iminoxyl group and chromophore was studied by analysis of change in the hyperfine splitting of the e.p.r. spectra and displacement of the optical absorption. The comparison of the spectra of the iminoxyl free radicals in water and decanol permitted one to conclude that both the iminoxyl group and the chromophore of the radical in the micelle have a rather polar environment (about 80 per cent of the H_2O polarity).

Thus none of the given models is in agreement with the results of the experiments. The authors suggest a new model which implies the dynamic nature of solubilization.

M. J. Goldfield, V. K. Coltover and coworkers[13] in 1969 used radicals with the general formula (17)

$$O-N^{\cdot}\!\!\diagup\!\!\!\bigcirc\!\!\!\diagdown\!\!-O-\overset{\overset{\textstyle O}{\|}}{C}-(CH_2)_nH \tag{17}$$

to study the micelles of the sodium lauryl sulphate.

Because of the non-polar tail these radicals are insoluble in water but they join the micelles easily. With a certain ratio of water and sodium lauryl sulphate in the system the authors observed a structural transition.

The work of V. K. Coltover, M. J. Goldfield and coworkers[12] is of particular interest. The authors found a conformational transition during a change of the physiological state of the respiration chain of the mitochondria. The radical (**18**)

$$O-N^{\cdot} \diagdown \diagup O-\overset{\overset{\textstyle O}{\|}}{C}-C_7H_{15} \qquad (18)$$

was solubilized in a suspension of electron transport particles evolved from the mitochondria. In this system they observed an anisotropic e.p.r. spectrum which in their opinion was a mixture of two signals.

After substrate (succinate) was added the e.p.r. spectrum parameters considerably changed. The correlation time decreased fivefold. At the same time they observed a decrease in the e.p.r. signal amplitude. When potassium ferricyanide was added the initial e.p.r. signal was restored.

Thus the authors showed that the reduction of the respiration chain by means of substrate was accompanied by a change of the paramagnetic label which was localized probably in the lipid layer of the biomembranes of the mitochondria.

VIII. STUDY OF PROTEIN SINGLE-CRYSTALS LABELLED BY IMINOXYL FREE RADICALS

Onishi, Boeyens and McConnell[47] studied protein single-crystals labelled by iminoxyl free radicals and showed that unique information on the structure of the analysed molecules could be obtained. Basic results were obtained by the authors after the study of horse oxyhaemoglobin.

In the work by Y. S. Y. Boeyens and H. M. McConnell[48] it was found out that the maleimide spin label reacted both with haemoglobin and oxyhaemoglobin in solution giving in the first case the e.p.r. spectrum of weak immobilized free radicals and in the second the e.p.r. spectrum of a corresponding strong immobilization. The authors showed that the spin label in both cases reacts with the sulphhydril group at the amino acid residue No. 93 of the chain. Later S. Onishi, J. C. Boeyens and H. M. McConnell studied single crystals of oxyhaemoglobin labelled this way.[47]

As far as the π-orbit of a free electron is perpendicular to the plane of the iminoxyl ring and the heterocyclic plane itself is orientated in a certain way with respect to the three-dimensional structure of the protein, the study of the e.p.r. spectra at different positions of the single crystal in the external magnetic field allows us to determine a position of the π-orbit of the free electron in relation to the crystallographic axes. It is prerequisite to determine the constants of the anisotropic splitting and the values of the g-tensor for parallel and perpendicular orientation of the π-orbit of the free electron of the paramagnetic label in the magnetic field. As it was stated above for the iminoxyl free radical this information was obtained during the studies of single crystals of the free radicals. The isotropic splitting constant was determined by an analysis of the hyperfine structure of the e.p.r. spectra of the iminoxyl free radicals in solution.

The study by means of the e.p.r. method of the protein single crystals labelled with free radicals permits us to find certain elements of symmetry which exist when the protein molecule consists of two or more subunits. One can speak of rotational twofold axes of such elements of symmetry. Since the proteins contain only left direction of the helix, twisting the centre of symmetry of mirror planes does not occur in proteins. The determination of rotating twofold axes by means of paramagnetic resonance is based on the following principle: one can proceed from the fact that the protein molecule contains two equal subunits which can be related to a twofold axis of rotation. In this case one can judge about the presence of a twofold axis of rotation by the fact that extreme values of the constant of the anisotropic hyperfine splitting and of the g-factor corresponding to the perpendicular and parallel orientation of the π-orbital of the free electron in the external magnetic field will not be shown simultaneously by the two equivalently placed spin labels at any position of the single crystal with respect to the external magnetic field. Thus the study of the dependence of the e.p.r. spectra on the single crystal orientation in the magnetic field allows us to determine the presence of these elements of symmetry.

Twofold axes of rotation are present in the molecules of haemoglobin, of α-chymotrypsin, and evidently of lactic dehydrogenase. They can also appear, for instance, during agglomeration of viral coats consisting of protein subunits. If these elements of symmetry coincide with the crystallographic axes they can be easily found by means of X-ray diffraction. If it is not the case considerable difficulties are encountered to find them. In both situations by means of the e.p.r. method the twofold axes of rotation are found considerably easier than by a method of X-ray structural analysis. H. M. McConnell and coworkers found the twofold axes of rotation in haemoglobin (where they coincide with the crystallographic ones) and in the dimer of the α-chymotrypsin. The possible biological importance of these elements of symmetry is discussed as related to a theory of interalles complimentation and the mechanism of allosteric transition.[47]

Applying this method under favourable conditions one can determine even small changes in the protein conformation, for instance, as H. M. McConnell presumes the replacement of an amino acid residue or allosteric changes in structure taking place as a result of the interaction of the protein molecule with the substrates, activators, inhibitors or coenzymes.

Finally, we would like to emphasize that the application of the rather fine and specific spin label method in molecular biology appears to be very promising.

IX. CONCLUSION

Our confidence was strengthened by the appearance of a lot of new important works on this subject in the course of preparing this article.

The general questions of the application of e.p.r. in biology were considered carefully in the books *Magnetic Resonance in Biological Systems*, 1968[49] and *Solid State Biophysics*, 1969.[50]

The comprehensive reviews on physics and chemistry of spin labels were given by Rosantsev,[51,52] Rosantsev and Scholle,[53] McConnell, McFarland.[54]

The great potential in the field of structure of proteins, including conformational transitions caused by such interactions as substrate or inhibitor–enzyme and hapten–antibody ones was clearly demonstrated in reviews by Hamilton, McConnell,[55] Lichtenstein[56] and Smith.[57]

Theoretical analysis of parameters of spin-labelled biological systems attracted special attention.

Itzkowitz[58] gave the analysis of anisotropy of spin-label spectra having slow rotational diffusion.

This subject was developed further in the works of Lazarev and Korst,[59] Alexandrov and coworkers[60] and Lazarev and Strjukov.[61]

The theoretical investigations of anisotropic rotation of spin labels conducted by Kuznetzov[62] were of great interest. Conclusions of his work were confirmed experimentally by the results of the work of Rosantsev and coworkers.[63] The latter showed that the character of anisotropy of spectra in viscous solution depended on shape of the spin-label molecule.

The following works[64,65] are devoted to measuring distances between spin labels attached to biopolymers using dipole–dipole broadening of spectral lines.

It paid special attention to investigation of structure-function relationships of biomembranes. Spin labels presented new approaches for understanding of these important systems. The comprehensive review of early works in this field was published by Griffith and Waggoner.[66] The more recent data on this subject were reviewed in other works.[54,57]

Syntheses of different spin-labelled lipids from analogues of fatty acids to steroid hormones permitted exciting experiments to be carried out on the investigation of structural transition in membranes upon changing of pH, temperature and binding of ions, antibiotics and other physiological substances.

The unique information about orientation of spin-labelled lipids in membranes and the extreme sensitivity of spectral parameters to the changing of the vicinity of spin labels allowed fine details to be elucidated about structure and function of biomembranes.

The use of spin labels as redox-indicators turned down the convenient tool for studying electron transport systems in the course of respiration and photosynthesis. By using inhibitors of electron transport system of mitochondria, information was obtained about the part of the respiratory chain which was responsible for reducing the spin label to hydroxylamine.[67] These works contain new data on the changing of membrane structure during oxidative phosphorylation.

Finally, the perspective of application of n.m.r. and double n.m.r. for investigations of spin-labelled biological systems must be noted.[68,69,70,71]

We are sure that new exciting examples of using spin labels for solution of problems in molecular biology will have appeared by the moment of publication of this book.

REFERENCES

1. L. A. Blumenfeld, V. V. Voevodsky and A. G. Semenov, *Primeneniye EPR v khimii* (in Russian), Izd. 'Nauka', Novosibirsk, 1962
2. *Svobodno-radikalniye protsessi v biologicheskikh sistemakh* (in Russian), Izd. 'Nauka', Moscow, 1966
3. A. E. Kalmanson, 'Primeneniye metoda EPR v biokhimii' (in Russian). In *Uspekhi Sovremennoyi Biokhimii*, Izd. 'Mir', Moscow, 1962
4. A. E. Kalmanson, 'Radiospektroskopiyia EPR v biologii i Meditsine'. In *Novosti Meditsinskoyi Radioelektroniki* (in Russian), Izd. VNIIMTI, Moscow, 1968
5. M. S. Blois (ed.), *Free Radicals in Biological Systems*, Academic Press, New York and London, 1961
6. C. Nicolau and Z. Simon, *Biofizica Moleculara*, Editura Ştiinţifică, Bucureşti, 1968
7. D. J. Stone, T. Buckman, P. L. Nordio and H. M. McConnell, *Proc. Nat. Acad. Sci. (U.S.)*, **54**, 1010 (1965)
8. G. L. Grigoryan, J. G. Charitonenkov, T. J. Tichonenko and A. E. Kalmanson (in Russian), *Doklady Akad. Nauk SSSR*, **165**, 224 (1966)
8a. C. G. Charitonenkov, G. L. Grigoryan and A. E. Kalmanson (in Russian), *Biofizika*, **10**, 1085 (1965)
9. S. Ohnishi and H. M. McConnell, *J. Am. Chem. Soc.*, **87**, 2293 (1965)
10. G. G. Lichtenstein (in Russian), *Molekuliarnaya Biologiya*, **2**, 234 (1968)
11. B. I. Sukhorukov, A. M. Vasserman, L. A. Kozlova and A. L. Buchachenko, (in Russian), *Dokladi Akad. Nauk SSSR*, **177**, 454 (1967)
12. V. K. Coltover, M. G. Goldfield, L. Hendel and E. G. Rosantsev, *Biophys. Biochem. Res. Comm.*, **32**, 421 (1968)
13. M. G. Goldfield, V. K. Coltover, E. G. Rosantsev and V. I. Siskina, *Kolloid Z.*, **178**, No. 2 (1969)
14. E. G. Rosantsev (in Russian), *Uspekhi Khimii*, **35**, 1549 (1966)
15. M. B. Neiman, Yu. G. Mamedova and E. G. Rosantsev (in Russian), *Azerb. Khim. Zurnal*, **6**, 37 (1962)
16. E. G. Rosantsev and M. B. Neiman, *Tetrahedron*, **20**, 131 (1964)
17. B. M. Dupeyre, H. Lemaire and A. Rassat, *Tetrahedron Letters*, **1964**, 1781

18. R. Briere, H. Lemaire and A. Rassat, *Tetrahedron Letters*, **1964**, 1775
19. A. L. Buchachenko, *Stabilnie Radikaly* (in Russian), Izd. 'Nauka', Moscow–Leningrad, 1963
20. A. L. Buchachenko and A. M. Vasserman (in Russian), *Zhur. Strukt. Khim.*, **7**, 27 (1967)
21. H. M. McConnell, *J. Chem. Phys.*, **24**, 764 (1956)
22. H. M. McConnell, *J. Chem. Phys.*, **25**, 709 (1956)
23. H. M. McConnell, *Ann. Rev. Phys. Chem.*, **8**, 105 (1957)
24. J. A. Corcer, M. P. Klein and M. Calvin, *Proc. Nat. Acad. Sci. (U.S.)*, **56**, 1365 (1966)
25. V. M. Chumakov, G. L. Grigorian, V. I. Suskina, E. Y. Rosantsev and A. E. Kalmanson (in Russian), *Biofizika*, **XVI** (3), 564 (1971)
26. N. Bloembergen, E. M. Purcell and R. V. Pound, *Phys. Rev.*, **73**, 679 (1948)
27. J. H. Freed and G. K. Fraenkel, *J. Chem. Phys.*, **39**, 1326 (1963)
28. O. H. Griffith, D. W. Cornell and H. M. McConnell, *J. Chem. Phys.*, **43**, 2909 (1965)
29. O. H. Griffith and H. M. McConnell, *Proc. Nat. Acad. Sci. (U.S.)*, **55**, 8 (1966)
30. G. L. Grigoryan, V. I. Suskina, E. G. Rosantsev and A. E. Kalmanson (in Russian), *Molekuliarnaya Biologia*, **2**, 148 (1968)
31. G. L. Grigoryan, A. E. Kalmanson, E. G. Rosantsev and V. I. Suskina, *Nature*, **216**, 5118, 927 (1967)
32. S. E. Bresler, V. P. Kushner and S. Ya. Frenkel (in Russian), *Biokhimia*, **24**, 685, (1959)
33. S. Ya. Frenkel and P. Orn (in Russian), *Vysokomolekuliarniya Soedineniya*, **3**, 541 (1961)
34. S. Ogawa and H. M. McConnell, *Proc. Nat. Acad. Sci. (U.S.)*, **58**, 19 (1967)
35. S. Ohnishi, T. Maeda, T. Ito, K. Hxant and G. Guima, *Biochem.*, **7**, 2662 (1968)
36. H. M. McConnell and C. L. Hamilton, *Proc. Nat. Acad. Sci. (U.S.)*, **60**, 776 (1968)
37. G. J. Lichtenstein, P. C. Bobodjanov, E. G. Rozantsev and V. J. Suskina (in Russian), *Molekuliarnaya Biologiya*, **2**, 334 (1968)
38. I. C. P. Smith and T. Yamane, *Proc. Nat. Acad. Sci. (U.S.)*, **58**, 884 (1967)
39. J. E. Berliner and H. M. McConnell, *Proc. Nat. Acad. Sci. (U.S.)*, **55**, 708 (1966)
40. D. Koshland. In *Horizons in Biochemistry* (ed. M. Kasha and B. Pullman), Academic Press, New York, 1962, p. 196
41. L. Stryer and O. H. Griffith, *Proc. Nat. Acad. Sci. (U.S.)*, **54**, 1785 (1965)
42. G. L. Grigoryan, S. G. Tatarinova, A. Ya. Kulberg, A. E. Kalmanson, E. G. Rosantsev and V. A. Suskina (in Russian, *Doklady Akad. Nauk SSSR*, **178**, 930 (1968)
43. J. Bordet, *Traité de l'Immunité*, Masson, Paris, 1939
44. J. R. Marrack, *The Chemistry of Antigens and Antibodies*, Med. Res. Council, Special Report series, No. 230, London, 1958
45. M. Heidelberger and F. E. Kendall, *J. Exptl. Med.*, **61**, 563 (1935)
46. A. S. Waggoner, O. H. Griffith and C. R. Christensen, *Proc. Nat. Acad. Sci. (U.S.)*, **57**, 1198 (1967)
47. S. Onishi, J. C. Boeyens and H. M. McConnell, *Proc. Nat. Acad. Sci. (U.S.)*, **56**, 809 (1966)
48. J. S. A. Boeyens and H. M. McConnell, *Proc. Nat. Acad. Sci. (U.S.)*, **56**, 22 (1966)
49. A. Ehrenberg *et al.* (ed.), *Magnetic Resonance in Biological Systems*, Pergamon Press, New York–London, 1967
50. S. Wyard (ed.), *Solid State Biophysics*, McGraw-Hill, New York, 1969
51. E. G. Rosantsev, *Free Nitroxyl Radicals*, Plenum Press, New York–London, 1970
52. E. G. Rosantsev, *Stabilniye Svobodniye Radicali* (in Russian), Izd. 'Chimiya', Moscow, 1971

53. E. G. Rosantsev and V. D. Scholle (in Russian), *Uspekhi Khimii*, **40**, 417 (1971)
54. H. M. McConnell and B. C. McFarland, *Quart. Rev. Biophys.*, **3**, No. 1, 91 (1970)
55. C. L. Hamilton and H. M. McConnell, 'Spin labels', in *Structural Chemistry and Molecular Biology*, ed. G. Rich and R. Davidson, W. H. Freeman and Co., San Francisco, 1968, p. 115
56. G. I. Lichtenstein, 'Investigation of structure and function of enzymes by spin-label methods', in *Uspekhi Biologischteskikh Nauk* (in Russian), Vol. 12, Izd. 'Nauka', Moscow, 1971
57. I. C. P. Smith, 'The spin label method', in *Biological Application of Electron Spin Resonance Spectroscopy*, ed. J. R. Bolton, D. Borg and H. Swartz. John Wiley–Interscience, 1972
58. M. S. Itzkowitz, *J. Chem. Phys.*, **46**, 3048 (1967)
59. N. N. Korst and A. V. Lazarev, *Mol. Phys.*, **17**, 481 (1969)
60. I. V. Alexandrov and A. N. Ivanova *et al.*, *Mol. Phys.*, **18**, 681 (1970)
61. A. V. Lazarev and V. B. Strjukov (in Russian), *Doklady Akad. Nauk SSSR*, **197**, 627 (1971)
62. A. N. Kuznetsov (in Russian), *Zhurnal Strukturnoi Khimii*, **11**, 535 (1970)
63. E. G. Rosantsev, V. P. Ivanov and G. L. Grigoryan (in Russian), *Izvestiya Akademii Nauk SSSR*, **9**, 2036 (1969)
64. V. I. Kokorin, K. I. Zamaraev, G. L. Grigoryan, V. P. Ivanov and E. G. Rosantsev (in Russian), *Biofizika*, **17**, No. 1, 1972
65. A. V. Kulikov, G. I. Lichtenstein, E. G. Rozantsev and A. B. Shapiro, *Biofizika*, **17**, No. 1 (1972)
66. O. H. Griffith and A. S. Waggoner, *Acc. Chem. Res.*, **2**, 17 (1969)
67. L. S. Jaguzschinskiy, W. M. Chumakov, W. P. Ivanov, W. W. Chistyakov, E. G. Rosantsev and A. E. Kalmanson (in Russian), *Doklady Akad. Nauk SSSR*, **197** (4), 969 (1971)
68. G. C. K. Roberts, J. Hannah and O. Jardetsky, *Science*, **165**, 504 (1969)
69. H. Sternlicht and E. Wheeler in *Magnetic Resonance in Biological Systems*, ed. A. Ehrenberg *et al.*, Pergamon Press, New York–London, 1967
70. J. S. Hyde in *Magnetic Resonance in Biological Systems*, ed. A. Ehrenberg *et al.*, Pergamon Press, New York–London, 1967
71. R. E. Richards in *Magnetic Resonance in Biological Systems*, ed. A. Ehrenberg *et al.*, Pergamon Press, New York–London, 1967

CHAPTER 14

Thermodynamic study of antigen–antibody reactions

R. Wurmser

Institut de Biologie Physico-Chimique,
Paris, France

I. INTRODUCTION

The combination of antigen with antibody is a paradigm of specific inter-action between a protein and a ligand. For this reason the thermodynamic study of immunological reactions should provide information relevant to many important problems of molecular biology.

Two distinct steps may be recognized in these reactions: First, the association of antigen with antibody followed by a secondary process such as precipitation or agglutination. From the thermodynamic point of view relatively little is known about the secondary reaction by itself while many determinations of thermodynamic quantities concerning the primary or the total process have been made.

These parameters are of importance because of their relationship to the structure of the reagents. The measurements of enthalpy change (ΔH), free energy change (ΔG) and entropy change (ΔS) are employed in two ways: for identification of specific substances or information on their structure and for checking theoretical calculations of molecular interactions.

A. Free Energy Change

The standard free energy change, ΔG^0, in a reaction, $X + Y \rightleftharpoons XY$ is obtained from the relation:

$$\Delta G^0 = -RT \ln K \tag{1}$$

where R is the gas constant, T the absolute temperature and K the experi-mental equilibrium constant. The value of ΔG^0 corresponds to the formation of one mole of XY when the initial activities of the substances X, Y and the final activity of the product XY are unity. Note that in the case of antigens and antibodies, the determinations of K are always made in very dilute solutions. Thus no distinction between activities and molar concentrations is necessary. In the field we have to consider these are generally expressed in molarities (numbers of gram formula weights per litre).

A reaction for which the standard free energy change ($-\Delta G^0$) has a high value is usually said to take place between bodies having a high affinity. However it should be noted that affinity is defined in a strict sense as the amount of free energy decrease corresponding to an infinitesimal advance-ment of the reaction.

In accordance with the fundamental relation

$$\Delta G = \Delta H - T\Delta S \tag{2}$$

and the condition $\Delta G < 0$ for a spontaneous reaction at constant temperature and pressure, two classes of spontaneous reactions are distinguished. In the first case, the reaction is enthalpy-driven; i.e. the term ΔH is predominant.

Alternatively the reaction, non-exothermic or even endothermic, is entropy-driven.

The discrimination is of particular interest when considering the nature of the forces implicated in the bonding $X-Y$.

B. Enthalpy Change

The enthalpic quantity, ΔH, is obtained directly by calorimetric measurement of the heat of reaction at constant pressure or indirectly from the displacement of equilibrium with temperature. For substances in solution ΔH is a measure of the change in internal energy brought about by the reaction. This change is the sum of two terms:

(a) the change of enthalpy ΔH_0 which would occur if the reaction proceeded at absolute zero;

(b) the difference in thermal energy (heat content) of the product XY and the reactants $X + Y$.

For example simply from the destruction of translational degrees of freedom in the association of X and Y a decrease in heat capacity is produced, contributing to the difference between ΔH_0 and ΔH. This difference creates some difficulties when attempting to compare the experimental heat of reaction ΔH with the relatively weak bonding energies calculated as arising from coulombic attractions, hydrogen bonds or dispersion forces.

C. Entropy Change

From the experimental determinations of ΔG^0 and ΔH^0, and by application of the equation (2) the value of ΔS^0 is obtained. This quantity is an essential parameter for the interpretation of the antigen–antibody reaction in terms of structure since these reactions are very often entropy-driven.

For the purpose of biochemical applications the entropy is advantageously expressed in terms of statistical mechanics. Consider N molecules in a volume V. In the equilibrium distribution, the probability that any molecule will have an energy of a particular kind i is an exponential function of the temperature. The number of molecules having this energy ε_i is:

$$N_i = \text{Constant} \times e^{-\varepsilon_i/\kappa T}$$

In this relation, κ is the Boltzmann constant, and the other constant is equal to N/f where f is the sum

$$\sum g_i e^{-\varepsilon_i/\kappa T}$$

a quantity known as the molecular partition function for the given volume. The factor g_i is the number of states corresponding to a same energy ε_i and, in spite of that, distinguishable.

The thermal energy of a molecule is distributed among the various degrees of freedom, for example each of the three components of translational motion, each of the particular internal vibrations. It is convenient to collect into a particular partition function all the terms referring to the same type of energy for the various degrees of freedom. One has therefore to specify a translational partition function f_{tr}, a rotational partition function f_{rot}, and a vibrational partition function f_{vib}. It is easy to show that f is equal to the product f_{tr}, f_{rot}, f_{vib}.

In principle its value can be calculated provided knowledge of the dimensions of the molecule and its vibration frequencies is available. An example useful for the application of statistical procedure to problems concerning the antibody–antigen combination is the calculation of the product f_{tr}, f_{rot}. The partition function for the three degrees of translational energy per unit volume is:

$$f_{tr} = \frac{(2\pi m \kappa T)^{3/2}}{h^3}$$

where m is the mass of the molecule and h is the Planck constant.

The rotational partition function for a polyatomic molecule with three moments of inertia A, B, C is:

$$f_{rot} = \frac{8\pi^2 (8\pi^3 ABC)^{1/2}}{h^3} (\kappa T)^{3/2}$$

If the molecule is considered as a rigid sphere ($A = B = C$) with the radius r and the density ρ, the term $(ABC)^{1/2}$ becomes $8(\pi \rho/15)^{3/2} r^{15/2}$, and the product $f_{tr} f_{rot}$ is proportional to m^4 (see application in section IV A).

Now the entropy of a gram molecule in 1 litre can be calculated by adding to $R \ln f$ the content in thermal energy divided by T. Let S_X^0, S_Y^0, S_{XY}^0 be respectively the molar entropy of X, Y, XY, the standard entropy change for the reaction $X + Y \rightarrow XY$ is

$$\Delta S^0 = S_{XY}^0 - S_X^0 - S_Y^0$$

The greater are the number of states over which the total energy is distributed, the greater is f. Thus the value of ΔS^0 experimentally determined permits derivation of information about the number of degrees of freedom appearing or disappearing during the reaction. For instance the rupture of hydrophobic bonds in aqueous phase brings about a decrease in entropy caused by the fixation of water molecules in the vicinity of non-polar groups, in quasi-crystalline structures. Conversely, inside a complicated molecule the rupture of a bond may result in liberating hindered rotation of some group. Entropy changes may also reflect transconformations of regions of a molecule distant of the reacting site.

For such interpretations it is convenient to eliminate the part of ΔS^0 which does not depend on structural modifications and has the same value whatever may be the nature of the molecules which appear or disappear. This part of the change of entropy strictly related to the mixing of the molecules with the solvent is easy to calculate. One takes in account the number of the states in which a molecule may belong owing to the fact that it is inside a solvent. This number is obtained by considering that, at a given instant, the molecule could have been in the place occupied by any molecule of the solvent.

The result is that the entropy of mixing for one gram molecule per litre of water (55·6 moles of water) is $R \ln(1/55\cdot6) = 7\cdot98$ cal deg^{-1}. The molar entropy of a substance, corrected by 7·98 cal deg^{-1}, is called the unitary entropy and is characteristic of the structure. Thus for the reaction $X + Y \rightarrow XY$ the unitary entropy change ΔS_u will be $\Delta S^0 + 7\cdot98$ cal deg^{-1}. The unitary free energy, dependent only on the structure, is given, with respect to the relation (2):

$$\Delta G_u = \Delta G^0 - 7\cdot98T$$

Although a large body of information has been acquired from such interpretations of the thermodynamic studies of antigen–antibody reactions, an important limitation exists. All the quantities measured $\Delta G, \Delta H, \Delta S$, are always global quantities. Each of them depends on many events occurring simultaneously. For example, the association of an antigen site with an antibody site implies not only a formation of bonds, but also the possible extrusion of water bound to the sites and intramolecular changes at more or less large distances from the reactant sites.

II. METHODS

A. Measurement of the Heat of Reaction

The quantity ΔH can be determined directly with a calorimeter at constant pressure. The calorimetry has been used in immunochemistry only in a few cases. Most often the value of heat of reaction itself is derived from measurements of equilibrium constant K at two temperatures. According to the van't Hoff relation:

$$\frac{d \ln K}{dT} = \frac{\Delta H^0}{RT^2} \qquad (3)$$

and over small ranges

$$\ln \frac{K_2}{K_1} = \frac{\Delta H^0}{R}\left(\frac{1}{T_1} - \frac{1}{T_2}\right) \qquad (3a)$$

where ΔH^0 is the sum of the enthalpies of the products minus the enthalpies of the reactants when each is at its reference state.

B. Equilibrium Studies

1. Reversibility

The first point to establish for the thermodynamic study of a process is its reversibility. In the case of immunological reactions, arguments for the reversibility are provided by many experiments concerning their first step, i.e. the association of antigen with antibody, according to the law of mass action. In most cases the complexes produced in these experiments remain in the dissolved state. However the reversibility of the total process, including the aggregation, has been quantitatively proved for the phenomenon of agglutination.

This demonstration has been made by S. Filitti-Wurmser and Y. Jacquot[1] for the human isohaemagglutination of B red cells.

(1) In a series of experiments it was shown that the same yield for agglutination is obtained when the reaction is carried out directly at 37° or in two steps, the first at a low temperature 4 °C and the second at 37 °C.

(2) Instead of dissociating the agglutinates by raising them to a higher temperature, they may be dissociated by dilution. One then finds that

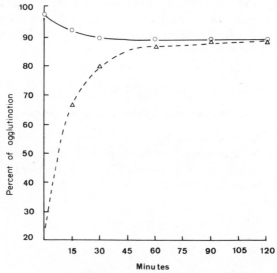

Figure 1. Evolution towards equilibrium in the forward and backward directions: ——— agglutination; – – – dissociation of the agglutinate (from M. H. Wilkie and E. L. Becker[3])

the same yield of agglutination is attained if a mixture of red cells and serum is made directly in a definite volume V or if the operation is carried out in two steps: mixing in a small volume and then adding buffer to the volume V.

The reversibility of the isohaemagglutination of A red cells by sera of group B has been demonstrated in the same manner by S. Mavrides.[2] An excellent experiment is due to M. H. Wilkie and E. L. Becker.[3] A mixture of serum and red cell suspension was placed on a shaker. At appropriate intervals of time, aliquots were withdrawn, and the number of free cells per mm^3 determined. The lower curve of Figure 1 represents the results. Duplicates of the same mixture were centrifuged for one minute, resuspended, and placed in a shaker; at appropriate intervals, the free cell count was determined. The yield of agglutination was shown independent of the direction from which it was approached.

Another demonstration of reversibility is reported by R. S. Evans, W. K. Mebust and M. T. Hickey[4] for the reaction between Rh antibody and its antigen.

2. Expression of the Law of Mass Action

The major difficulty encountered in thermodynamic studies of immunological reactions is that the antiserums are generally heterogeneous with respect to the affinity of its constituent antibodies. Methods of purification have been reported efficient for selecting antibodies which react with a particular chemical structure, but not for separating antibodies differing in affinity for the same structure, excepting the extreme cases.

A second complication arises from the plurality of binding sites present on the same molecule (antigen and antibody). This multifunctionality, generally called multivalence, may be accompanied by quantitative differences among these groups.

Let $G_nA_{n'}$ be a complex constituted by n molecules of antigen and n' molecules of antibody. It is eventually possible to measure the concentrations of three complexes with different but relevant values of n or n' in the mixture and determine an equilibrium constant. However, this constant is not characteristic of the nature of reacting sites. If the sites are not identical, the observed constant is an average. Even if the sites are identical, the constant depends on the number of sites, due to statistical factors. Nevertheless one is sometimes able to obtain a constant characterizing the sites by studying the proportion of bound antigen or antibody as a function of their free concentration. In such a case the theoretical model which gives the best description of the experimental results must be looked for.

The most simple case occurs when one of the two reactants X, antigen or antibody, possesses a single reactive site, i.e. is unifunctional (univalent), the other, Y, being multifunctional. Consider the case of identical and independent sites Y. Let m be their number per molecule. At equilibrium there is a more or less large number of molecules Y with $1, 2 \ldots m$ sites occupied. These molecules are $YX_1, YX_2 \ldots YX_m$. Although all the sites on Y are assumed to be identical, they may be distinguished by their position on the molecule. Then there will be a large number of equilibria;

$$YX_{n-1,i} + X \rightleftharpoons YX_{n,j}$$

and an intrinsic equilibrium constant,

$$K = \frac{(YX_{n,j})}{(YX_{n-1,i})(X)} \tag{4}$$

The terms in parentheses represent molar concentrations which are here assumed to be equal to activities, as already mentioned.

The constant K characterizes the X and Y sites of the molecules X and Y.

Now it can be shown (see J. Wyman[5]) that the ratio of the concentration (X_b) of bound X to the sum of the concentrations of Y and all its complexes is:

$$r = \frac{(X_b)}{(Y)_0} = \frac{mK(X)}{1 + K(X)} = mp \tag{5}$$

As (X_b) is equal to the concentration (Y_b) of the sites Y bound, p represents the probability for one site Y to be occupied. This probability is the ratio $(Y_b)/(Y)_0$ denoting by $(Y)_0$ the molarity of free and bound sites Y.

The equation (5) could be obtained directly (under the conditions assumed) by assimilating the sites X, Y and the product YX of the bonding to independent molecules, and the intrinsic constant K to a classical equilibrium constant expressed in terms of molar concentrations of sites.

$$K = \frac{(YX)}{(X)(Y)} \tag{4a}$$

Several straight line representations have been employed to determine the constant K and the number of sites m from analytical data on the composition of the mixture at equilibrium.

(a)
$$\frac{1}{r} = \frac{1}{m} + \frac{1}{mK(X)} \tag{6}$$

The plot of $(1/r)$ against $1/(X)$ gives a straight line. The extrapolated intercept on the ordinate is $1/m$ and the value of the slope is $(1/mK)$.

(b)
$$\frac{r}{(X)} = mK - rK \tag{7}$$

The straight line (Scatchard representation) obtained by plotting $r/(X)$ against r intercepts the ordinate at the value of mK and the abscissa for $r = m$.

Other linear relations can be used but do not lead to the determination of m

$$\frac{(X_b)}{(X)} = Km(Y)_0 - K(X_b) \tag{8}$$

A plot of $(X_b)/(X)$ versus (X_b) yields a straight line with a slope of $-K$ and ordinate intercept of $Km(Y)_0$.

Similarly we have:

$$(X) = \frac{(X)}{(X_b)} m(Y)_0 - \frac{1}{K} \tag{9}$$

The straight line obtained by plotting (X) against $(X)/(X_b)$ intersects the ordinate at the value $(-1/K)$. The slope gives the molarity of the sites Y.

The equations (6) to (9) permit us to determine K without knowledge of the concentration of any particular complex YX_n. Now the classical association constant K_n for the equilibrium between the unifunctional component X and two complexes such YX_{n-1} and YX_n can be obtained from the relation:

$$K_n = \frac{1}{n}(m - n + 1)K$$

evidently only valuable when the intrinsic affinity of all the sites is the same.

Important information about the active region of the antibody can be obtained from inhibition by structurally related compounds. The quantitative study of the association of an inhibitor to an antibody is quite comparable with that of the association of an inhibitor to an enzyme.

Let K and K_I be the intrinsic constants for the combination of an antibody Y with two unifunctional ligands, the hapten X and the inhibitor I. The effect of the added inhibitor can be formulated as follows: If r and r' are the fractions of Y sites combined with X, respectively, in the presence and in the absence of I, for the same concentration (X), we have:

$$r = \frac{(YX)}{(Y) + (YX)} \qquad r' = \frac{(YX)'}{(Y)' + (YX)' + (YI)'}$$

$$K_I = \frac{(r/r' - 1)(1 + K(X))}{(I)} \tag{10}$$

and the average number of inhibitor molecules bound per antibody molecule is

$$r_I = \frac{mK_I(I)}{1 + K(X) + K_I(I)} \tag{11}$$

where $m = 2$ for a bifunctional (bivalent) antibody.

3. Heterogeneity

The linear relations expressing the mass action law imply not only the absence of interaction between the binding sites present on a molecule of the multifunctional component Y but also the identity of all the molecules of the other component X. This last condition is rarely realized.

Heterogeneity of the antibodies present in an antiserum is currently admitted. However many attempts have been made to obtain antibodies with molecular uniformity and recently some cases of restricted heterogeneity have been quoted. Note that homogeneity as tested by thermodynamical method does not exclude the presence of antibodies differing by structural details not influencing the affinity.

An example of the departure from the linearity for the relation (6) is given in Figure 2 which represents the data of H. N. Eisen and F. Karush[6]

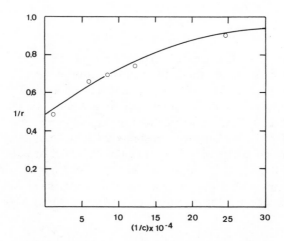

Figure 2. Binding of homologous haptenic dye *p*-(*p*-hydroxyphenylazo)-phenylarsonic acid by purified antibody (equation (6)). *r* mole bound hapten per mole antibody, *c* free hapten concentration (from H. N. Eisen and F. Karush[6])

for the interaction of purified antibody with its homologous hapten, the *p*-azophenylarsonic acid.

Among the various antibodies which are present in an antiserum can exist populations within each of which a distribution of affinities is mathematically describable. In this case parameters with thermodynamical signification can be obtained for characterizing a given population.

L. Pauling, D. Pressman and A. L. Grossberg[7] have considered the case where the distribution of antibody sites with respect to their free energy of combination with hapten can be described by a Gauss error function.

For K, the intrinsic association constant for a particular antibody, there is an average intrinsic constant K_0, such that the fraction of antibody molecules for which $\ln(K/K_0)$ lies between $\ln(K/K_0)$ and $\ln(K/K_0) + d \ln(K/K_0)$ is given by

$$\frac{1}{\sigma\sqrt{\pi}} \exp\{-[\ln(K/K_0)]^2/\sigma^2\} \cdot d \ln(K/K_0)$$

in which σ is the heterogeneity index.

The fraction of antibody sites occupied as a function of the concentration (X) of free hapten is

$$\frac{r}{m} = 1 - \frac{1}{\sqrt{\pi}} \int_{-\infty}^{+\infty} \frac{e^{-\alpha^2}}{1 + K_0(X) e^{\alpha\sigma}} d\alpha \tag{12}$$

in which α is equal to $\ln(K/K_0)/\sigma$, and m is the number of combining sites per antibody molecule (F. Karush).[8]

For the half-saturation of the antibody $r/m = \frac{1}{2}$, $K_0 = 1/(X)$ for all values of σ. The integral of the equation (12) is evaluated numerically, and corresponding curves are constructed for several values of σ; the best value is obtained by matching the experimental results with the theoretical curves.

Results of the application to the binding of an azohapten by a purified antibody are shown on Figure 3 (F. Karush).[9] The points are experimental and the curves are calculated for $\sigma = 2\cdot3$.

The determination of the value of σ by this method is a laborious procedure. A. Nisonoff and D. Pressman[10,11] have found more convenient to use the distribution function of Sips. This function represents the distribution of the adsorption energies of the sites on a solid surface exposed to a gas. In fact, this distribution is extremely similar to a gaussian distribution. Its advantage is to permit an analytic integration of the relation between the concentration of the free hapten and the average number of sites occupied. This gives the binding equation,

$$\frac{r}{m} = \frac{[K_0(X)]^a}{1 + [K_0(X)]^a} \tag{13}$$

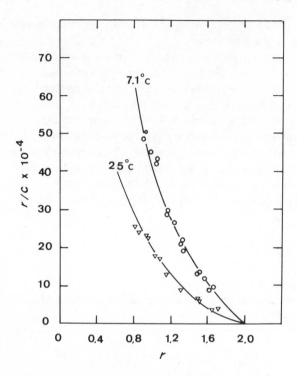

Figure 3. Binding results at 25 °C and 7·1 °C for the
reaction between the azo dye D-phenyl- (*p*-(*p*-dimethyl-
aminobenzeneazo)-benzoylamino) acetate and the
homologous antibody. The points are experimental
and the curves are theoretical (equation (12) *r* and *c*
as for Figure 2. (F. Karush[9])

where *a* is the Sip's heterogeneity index. The value of *a* may be obtained by
plotting experimental values of $(1/r)$ against $[1/(X)]^a$ for various values of *a*
and finding the best straight line.

A more convenient plot is $\log(r/(m - r))$ against $\log(X)$ (Karush).[12] Note
that for a homogeneous system of sites, $a = 1$ as in equation (5). On the
other hand for the small values of *c*, equation (13) becomes the Freundlich
equation.

An example where *a* is found unity within experimental error is the binding
of a tritium-labelled octasaccharide (fragment derived from *Pneumococci*
Type VIII) to antitype VIII antibody (E. Haber).[13]

An important question is whether there is only a normal population or
many normal populations of sites, each characterized by representative
parameters. When two populations have a small index of heterogeneity and

possess sites of sufficiently different mean affinity it is in principle possible to separate them by fractionation. A small amount of antigen, added to an heterogeneous antiserum, will combine preferentially with antibodies of higher affinity.

We shall mention here the role of this competition in the difference of aspect of the titration curves according to whether serum or antigen is used as the constant reactant (K. Hummel).[14] It must be discussed how far linearity obtained in the graphical representation of relations (6) to (9) is a test of homogeneity of the studied antiserum. It is easy to see that a given measured value of K may represent an average of many constants. This is the case if for experimental difficulties the linear representation is limited to a range of concentrations where the components of an eventual mixture do not enter notably in competition.

As an example, consider the case of the isohaemagglutinins of the normal human sera. A great deal of work has been done with this material, precisely because a very large proportion of the agglutinins present in a normal serum —in contrast to antisera generated by immunization—combine with red cells like a homogeneous substance. The thermodynamic properties depend only on the genotype, A_1A_1, OO, A_1O and so on, of the individual who synthetized the agglutinin (S. Filitti–Wurmser, Y. Jacquot-Armand and R. Wurmser;[15] S. Filitti-Wurmser and Y. Jacquot-Armand;[16] R. Wurmser and S. Filitti-Wurmser).[17] Figure 4 shows the Scatchard representation (equation (7a)) of the experimental data obtained with isohaemagglutinins anti-B designated as $\beta(A_1O)$ and $\beta(OO)$.

Here r_A is the ratio of the combined agglutinin to the total agglutinin and (G) the concentration of the free agglutinogen sites as explained in a further section IIIB2. It is evident that the plots representing the measurements present no deviating from linearity, and can be used for calculating an equilibrium constant. Nevertheless in this test of equation (7) there exist a large gap. Thus, possibility is not excluded that measurements made with an extremely small number of red cells, corresponding to a small value of r_A would deviate from linearity. This would be in accordance with M. B. Gibbs and H. A. Toro[18] who have found heterogeneity of the binding affinity of anti-B isohaemagglutinins by experiments with low concentrations of erythrocytes. However many facts permit us to exclude the presence of any large proportion of agglutinins of different affinities. No change of affinity is observed in sera after successive absorptions and elutions under conditions where drastic changes occur in immun anti-B sera. Moreover the demonstration of a relatively high homogeneity of natural isoagglutinins is strengthened by the fundamental fact that all the human normal sera of the same genotype give values of (r_A/c) which when plotted against r_A fall on the same line. It is unlikely that all these sera possess a constant proportion of different aggluti-nins.

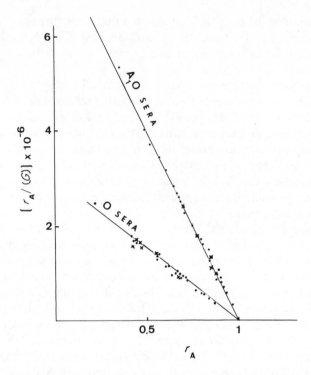

Figure 4. Scatchard plots of binding data for anti-B isohaemagglutinins from A_1O and O sera at 37 °C. ● untreated sera, × eluates or partially absorbed sera. (From R. Wurmser and S. Filitti-Wurmser.[17]) Two more recent data have been added (equation (7a)). Abscissa: r_A is the number of moles of sites G combined per mole of agglutinin. Ordinate: ratio of r_A to the molar concentration of free G sites. Intercepts with ordinate: $K_O = 3 \times 10^6 \, 1 \, \text{mole}^{-1}$, $K_{A_1O} = 8 \times 10^6 \, 1 \, \text{mole}^{-1}$

When no linear representation of the mass action law can be obtained from the experimental data, the interpretation of these is difficult. We have seen above the effect of a gaussian distribution of affinities in a population of antibodies. It is generally necessary to derive theoretical relations from a number of possible models and to try the best fitting of the data.

An example* is the study of the formation of complex insulin–anti-insulin (Berson and Yalow).[19] The most satisfying model has been found to be an unifunctional insulin reacting with two different orders of antibody sites A

---

* Different notations have been used by various authors in their original publications. They had to be modified in some cases in order to follow a given convention throughout the course of the present review.

and A'. If the sites are independent the situation is the same as it would be if two different species of antibody molecules were present. The ratio (b/f) of bound insulin to free insulin is:

$$\frac{b}{f} = K[(A)_0 - (A_b)] + K'[(A')_0 - (A'_b)]$$

where K and K' are the respective equilibrium constants, and b is the sum of the molar concentrations of combined sites A and A', let $(A_b) + (A'_b)$. Solutions for these constants and the total concentrations of the sites $(A)_0$ and $(A')_0$ can be obtained (equation (8)) from the plot of (b/f) against b. The most convenient method has been found to be a curve-fitting with assumed trial values.

4. Unifunctional Behaviour of Some Systems

Many experimental data on the association of unifunctional haptens to antibodies have been obtained and treated by means of the various relations indicated above. However these relations are not exclusively applicable to systems where one reactant at least carries one unique site. Steric effects can lead other systems to the same situation. The type of immunological processes where these effects can be expected is the agglutination. Indeed considering the spatial arrangement of the aggregated cells, it is evident that, for geometrical reasons, even if the agglutinin is bifunctional, a very small number of its molecules would have more than one specific site combined with an antigen group. Thus one may expect that the agglutinin will behave as a unifunctional reactant, as has been demonstrated for the isohaemagglutinins. (Figure 4).

On the other hand a natural antigen behaving as a unifunctional reagent is, as mentioned already, the insulin. It is possible that, as for the haemagglutination, steric restrictions make two or more sites unable to bind antibody.

5. Multifunctional Reactants

When both antibody A and antigen G are multifunctional many complexes of various composition may form and coexist, even when limiting the equilibrium studies to the zones of solubility. It is generally believed that in precipitin reaction antibody and antigen are bound together in the form of a lattice as a result of series of bimolecular reversible reactions $G + A$, $GA + G$, and so on. A thermodynamic treatment of the process is possible provided that one admits the equivalence and thus independence between all the sites A of the antibody and between all the sites G of the antigen. In other words all the bonds $G-A$ have the same energy (Goldberg).[20,21]

Consequently there is an intrinsic association constant $K = (GA)/(G)(A)$ and its determination is relatively easy in certain conditions. Indeed a very simple probability calculation is possible (Watson,[22] Talmage and Cann).[23]

Consider the case where the concentration (G) of the free antigen has been measured. Moreover the total concentrations $(G)_0$, $(A)_0$ respectively of antigen and antibody and the number m of sites present on a molecule G, are known. Finally the antibody is taken as bifunctional. The probability that any given G site in the system has reacted with a site A is $p = (GA)/m(G)_0$, which is called the extent or the degree of advancement of the reaction. The probability that all of the sites G are unreacted is $(1 - p)^m$. Thus the value of p is known from the value of $(G) = (G)_0(1 - p)^m$ and the intrinsic equilibrium constant is given by the expression:

$$K = \frac{pm(G)_0}{[m(G)_0 - pm(G)_0][2(A)_0 - pm(G)_0]} = \frac{p}{(1 - p)2(A)_0(1 - pr)} \tag{14}$$

where r is the ratio of total antigen sites to antibody sites.

By following arguments, as above, it is possible to determine K from the concentration of free antibody or very small complex of antigen and antibody. For even moderately large complexes calculation of the relation between the concentration and the extent of the reaction is excessively unwieldly. But Goldberg has given a theory which permits us to derive a general relation between p and the details of composition of the system.

The theory is based on the statistical methods of Flory and Stockmayer which had been conceived for the study of branched-chain polymers. It attempts to take into account all phases of the reaction, including the precipitation. The antigen carries a number m of sites. The antibody may be bifunctional or unifunctional or a mixture of both. The precipitates are formed by the aggregation of these units in three dimensions. Reactions which result in cyclic structures and limit the number of bonds in an aggregate of given size are not considered. Finally it is recalled that all the bonds between a site G on the antigen and a site A of any antibody are of equal energy and the reaction proceeds by the formation of an increasing number of G—A bonds. The most probable distribution of sizes and compositions of the aggregates may be calculated for all degrees of advancement p of the reaction as long as the overall composition of the system is known.

Applications of the theory for determination of equilibrium constants have been made by S. G. Singer and D. H. Campbell[24,25,26] and by S. I. Epstein, P. Doty and W. C. Boyd.[27] A more general theory (F. Aladjem and M. T. Palmiter)[28] takes in account the possibility of differences between the various sites of antigen and even the heterogeneity of the antibodies. This heterogeneity is described by a multivariate probability density function which could be, in principle, obtained from the experimental data by an iterative procedure involving the method of least squares. But the range and the precision of the experimental measurements does not yet permit an application of the method.*

By a method which gives only ratios of equilibrium constants Pauling and his collaborators obtained results, which will be mentioned in the section IV concerning the nature of the bonding forces. (L. Pauling, D. Pressman, D. H. Campbell and C. Ideka;[29] L. Pauling and D. Pressman;[30] D. Pressman, A. L. Grossberg, L. H. Pence and L. Pauling).[31] The addition of an unifunctional hapten, inapt to make a lattice to a mixture of antigen and bifunctional antibody, inhibits competitively the formation of the specific precipitate. By measuring the quantities of the two haptens which reduce the concentration of antibody to a point corresponding to reduction of the precipitate by one-half, a ratio of inhibiting powers is obtained which can be interpreted as representing the relative values of the average association constants of these haptens with antibody.

Kabat used a similar interpretation in his studies on the antibody-combining region for a natural antigen, namely dextran (E. A. Kabat;[32] E. A. Kabat and M. M. Mayer).[33]

III. TECHNIQUES

A. Calorimetry

Two types of apparatus are currently in use:

(a) Adiabatic calorimeters in which the heat produced or absorbed raises or lowers the temperature of the container in which the reaction takes place. Heat exchange between the outer wall of the container and its environment is reduced to a minimum. One measures either the change of temperature of the container or the amount of electric energy necessary to produce an equivalent temperature change.

Differential methods are especially useful: A twin calorimeter consists of two containers as identical as possible in their construction and their environment and linked by thermocouples. If heat is evolved in one of them

---

* The aim of the Goldberg theory was to describe and predict the precipitation. It is shown that precipitation can occur only if p has attained a critical value

$$p_c = \left[\frac{1}{\rho} \frac{1}{m-1} \frac{2(A)_0}{m(G)_0} \right]^{1/2}$$

and if the ratio $(A)_0/(G)_0$ lies inside limits depending on m and p. Here ρ is the fraction of antibody sites in the system which belong to bifunctional antibody molecules. Though unable to take into account many particular properties of the complexes, for example differing solubilities, the theory is consistent with the general features of the precipitation reaction.

by a chemical reaction, the other (the tare container) is heated sufficiently to eliminate the difference of temperature revealed by the thermocouples. The apparatus can be used as a single calorimeter: in this case the tare container supplies a reference temperature.

W. C. Boyd, J. B. Conn, D. C. Gregg, G. B. Kistiakowsky and R. M. Roberts[34] have measured with a calorimeter of this type the heat evolved when anti-hemocyanin reacts with its antigen.

(b) Conducting calorimeters with which one measures the heat flux which passes from the container to the environment or vice versa. Modern microcalorimeters based on this principle are sensitive to heat production of the order of several micro-calories sec^{-1} and permit sufficiently precise measurements on heats on the order of 10^{-3} cal with volumes of about one centilitre.

In the microcalorimeter of E. Calvet,[35] the flux of heat between one container and the environment, a cavity in a block of metal, is maintained equal to the flux of heat between a second container and a second cavity in the same block. The transfer of heat crosses two opposed thermocouples, each enclosing the walls of one or the other of the containers. A galvanometer placed in the circuit is maintained at zero by Peltier effect in the same container where heat is produced by the reaction.

In the calorimeter of C. Kitzinger and T. Benzinger[36] the reaction takes place in a thin layer; thus the heat transfer to thermocouple is rapid. The reactant solutions are accommodated in a bicompartimented cell containing 7 to 8 ml of each with a separating air space. The mixing is obtained by a rotating motion. The surface of the vessel is completely encompassed by a thermoelectric pile (of several thousand junctions) which is itself placed in a cavity within a copper block. Another identical vessel, located in the same block, is filled with a blank solution. The thermopile which surrounds it, wired oppositely in series with the first one, suppresses the effect of disturbances common to both systems. The terminals of the combined thermopiles are connected with a d.c. amplifier. The output (90,000 microvolts to heat flow of 1 calorie per sec) is recorded on a potentiometer. The area under the curve over time represents the total calories evolved by the reaction.

R. F. Steiner and C. Kitzinger[37] have measured with this apparatus the heat of reaction for the associations of serum albumin and ovalbumin with their respective antibodies.

B. Analytic Techniques

It is not possible to enter into a detailed description of the analytical techniques used in the studies of equilibrium between antigens and antibodies. This will be found in the original papers cited.

1. Analysis by Direct Study of the Mixture at Equilibrium

a. *Equilibrium Dialysis*

This method is a classical process for the study of the interaction of proteins with low molecular weight ligands. A membrane, permeable only to small molecules, separates the solution of protein from the solution of ligand. The experiment can be realized with sacks or with cells containing two compartments separated by a disc. A convenient technique of titration of the ligand must be available, e.g. spectrophotometry, radioisotope procedure. In calculating the amount of bound-ligand correction must be made for its adsorption to the membrane.

The first application of equilibrium dialysis to antibody binding was achieved by J. R. Marrack and F. C. Smith.[38] Later numerous works were conducted by this method—for example, that already cited of H. N. Eisen and F. Karush.[6] In this case, the diffusible substance is a haptenic dye. A cellophane sack contains a rabbit antibody, specific for *p*-(*p*-hydroxy-phenylazo)-benzene arsonic acid and purified to 90 per cent. The haptenic dye is measured spectrophotometrically in the exterior solution and the amount bound per mole of antibody r, is obtained by subtracting the concentration of hapten at equilibrium (H) from the initial concentration.

When one plots, in the normal way, $1/r$ as a function of $1/(H)$, one obtains the curve of the Figure 2, not a straight line. Nevertheless the intersection with the ordinate corresponds well to a value of $1/r = 1/m = 0.5$, that is $m = 2$, which indicates a bifunctional antibody.

In more recent work, already mentioned, (F. Karush)[9] the experiments were carried out with the optically isomeric forms of the azohapten phenyl-(*p*-(*p*-dimethylamino benzene azo)-benzoylamino)-acetate. A two compartment cell was used. The concentration of the unbound dye at equilibrium was determined by spectrophotometry. The results for the D-dye are shown in Figure 3 where the data are given in the form of a Scatchard plot, r/c vs. r. In addition, inhibition experiments have been made and association constants of the inhibitors K_I derived from the equation (10). The equation (11) is used for calculating the average number of inhibitor molecules bound per antibody molecule. It is convenient indeed for the amounts of dye and bound inhibitor to be in the same range. Then the value of K_I is close to the value which would be obtained by direct measurement of the inhibitor binding.

In the experiments of J. M. Dubert,[39] the concentration of free hapten (*m*- and *p*-sulphonylazotyrosine) is measured by labelling with radioactive elements ^{14}C and ^{35}S. This permits experiments at concentrations (0.3 to 30×10^{-7} M), much lower than possible with spectrophotometry. Results are represented by plots of these concentrations against the ratio of free to bound hapten (equation (9)).

We have mentioned above (section IIB2) the measurements of Haber[13] with a tritium-labelled octasaccharide.

b. *Utilization of Distinctive Physical Properties of the Complex*

i. *Light Scattering.* The change in molecular size resulting from the formation of antigen–antibody complex has been measured by light scattering (S. I. Epstein, P. Doty and W. C. Boyd).[27] The system employed was the association of rabbit antiarsanilic antibody with the acids terephthanilide-*p-p'*-diarsonic and adipanilide-*p-p'*-diarsonic.

Antibody and haptens are both bivalent; the authors have derived from the Goldberg equilibrium an approximate relation between the advancement of reaction *p*, the ratio of haptenic groups to antibody sites r and the ratio of molecular weight of the antibody M_A to the average molecular weight of the aggregates M_W:

$$\frac{M_W}{M_A} = \frac{1 + p^2 r}{1 - p^2 r}$$

M_W was measured from the intensity of scattered light and the intrinsic equilibrium constant obtained from equation (14).

ii. *Fluorescence.* Other techniques are founded on changes in intensity or polarization of fluorescence as a consequence of the association of antigen with antibody. These methods permit work with a few micrograms of antibody per ml. F. Perrin[40,41] established the relationship between molecular size and the polarization of light emitted by fluorescent molecules within a solution. The velocity of rotary Brownian motion and thus the amount of rotation during the excited life time, decreases as molecular size increases. Biochemical applications of this effect have been made by G. Weber,[42] and, for the study of antigen–antibody reaction, by W. B. Dandliker, H. C. Schapiro, J. W. Meduski, R. Alonso, C. A. Feigen and J. R. Hamrick.[43] Either the antigen or antibody is isolated, purified and then labelled with a fluorescent dye, for example fluorescein isothiocyanate (J. D. Marshall, W. C. Eveland and C. W. Smith).[44] Let ΔV and ΔH the excess of fluorescence relatively to a blank for the vertically and horizontally polarized light, the polarization is:

$$p = (\Delta V - \Delta H)/(\Delta V + \Delta H)$$

Let (F_f), (F_b) respectively represent the molar concentration of the fluorescent labelled material free or bound, and M the total concentration $[(F_f) + (F_b)]$. The maximum value of (F_b), namely $(F_b)_{max}$ is equivalent to the concentration of combining sites in the unlabelled component. Let Q be the molar fluorescence $[(\Delta V + \Delta H)/M]$ and Q_f, Q_b, p_f, p_b respectively the values of Q and p for completely free or completely bound material.

The ratio of bound to free fluorescent-labelled material concentrations is:

$$\frac{(F_b)}{(F_f)} = \frac{Q_f}{Q_b} \frac{p - p_f}{p_b - p}$$

Measurements on the labelled material alone give directly Q_f and p_f. The determination of Q_b and p_b is obtained by a double extrapolation procedure. The ratio $[(F_b)/(F_f)]$ can be used for determination of the value of the association constant K, and (F_b) maximum from the Scatchard equation (8):

$$K = \frac{(F_b)/(F_f)}{(F_b)_{max} - (F_b)}$$

In the experiments of Dandliker *et al.* with fluorescein labelled ovalbumin a mean intrinsic association constant K_0 and an index of heterogeneity a were determined from the relation (13). These values were used only as first approximations. A method was established for the ready determination of the best numerical values permitting us to represent the measured polarization p as a function of M.

A number of compounds show increased fluorescence when bound to proteins. It has been suggested that this effect is due to decreased dispersion of excitation energy by internal rotation.

A. A. Amkraut[45] has studied by this fluorescence enhancement method the binding of an anti-N-2,4, dinitrophenyl (DNP) insulin with the fluorescent hapten consisting of a conjugate of DNP-lysine and rhodamine isothiocyanate. Let q, q' respectively be the net fluorescence intensity (after subtracting the blanks) in presence and in absence of antibody; the concentration of bound hapten is:

$$(F_b) = \frac{q - q'}{((Q_b/Q_f) - 1)Q_f}$$

Q_f is determined by a standard plot of fluorescence against the concentration of free hapten. (Q_b/Q_f) is obtained from the maximum enhancement found during the course of the measurements in large antibody excess. Finally (F_f) can be calculated from (F_b) and $q = Q_f[(F_f) + (F_b)]$.

The quenching of fluorescence of the tryptophan residues is observed as a result of association of protein with certain kinds of organic molecules. The excitation energy is transferred to the ligand and here dissipated as heat or reemitted at a characteristic frequency of the ligand. The tryptophan fluorescence shows an intensity maximum at 350 mμ. It is quenched in complexes or conjugates of the proteins with molecules that have absorption bands in the 300 to 400 mμ region.

In the experiments of S. F. Velick, C. W. Parker and H. N. Eisen[46] on the antibody specific against the 2,4-dinitrophenyl group (DNP) fluorescence was excited usually at 290 mμ where protein absorption is strong and absorption by the hapten minimal. The fluorescence quantum yield of the antibody is diminished about 70 per cent by the binding of ε-N- DNP lysine. This permits the determination of a dissociation constant as small as $2 \cdot 4 \times 10^{-9}$ M. The antibody is titrated and the intensity of fluorescence plotted against the amount of added hapten. The equivalent combining weight of the purified antibody (nearly $\frac{1}{2}$ mole γ-globulin per mole of hapten) is obtained by extrapolation of the linear initial portion of the curve to the base line. The intrinsic dissociation constant is calculated for individual points.

The quenching of fluorescence has been used in another thermodynamic study of antigen–antibody equilibrium by P. Tengerdy.[47] Conalbumin was labelled with fluorescein. The ratio $[(F_b)/(F_f)]$ is calculated from the observed molar fluorescence of the reaction mixture and that of the free Q_f and bound Q_b forms by the relation,

$$(F_b)/(F_f) = (Q_f - Q)/(Q - Q_b).$$

As indicated by Dandliker, the value of Q_b is obtained by a double extrapolation. The constant K is determined by plotting $[(F_b)/(F_f)]$ against (F_b).

2. Analysis by Separation of Components of the System

a. *Paper Chromatographic Method*

One example is found in the work of S. A. Berson and S. Yalow.[19]

b. *Ultracentrifugation and Electrophoresis*

These techniques have been used by S. A. Singer and D. H. Campbell[24,25] in their study of the systems serum–albumin–antiserumalbumin and ovalbumin–antiovalbumin. Here the separation itself, owing to the change in refraction or absorption produced, furnish a possibility of estimation of the components of the system. Moreover, ultracentrifugation gives some information of their respective sizes, and hence a possibility of identifying these components.

The difficulty arising from the fact that removing one of the components of the mixture may lead to a disturbance of the equilibrium has been considered. This depends on the rates of the dissociation and association as compared to the rate of separation. As Singer and Campbell remark, the various components cannot be in a very rapidly readjusting equilibrium with one another if separate peaks appear on the schlieren diagrams of ultracentrifugation and no pronounced differences are observable in electrophoresis between the ascending and descending limbs. The authors have measured the area encompassed under the peak corresponding to free antigen. In this manner the concentrations of the other components were not available but have been calculated as previously described according to the theory of Goldberg.

c. Centrifugation of Cells

The separation of free antibody by centrifugation of the cells carrying bound antibody has been used in the thermodynamic study of haemagglutination (S. Filliti-Wurmser, Y. Jacquot-Armand and R. Wurmser,[15] R. Wurmser and S. Filitti-Wurmser).[17] Here the concentrations of the antibody in the supernatant varying between 10^{-8} to 10^{-10} moles per litre were not directly measurable. Fortunately it is possible to take advantage of the amplificatory power of the agglutination (the binding of about a hundred of agglutinin molecules per cell produces a visible effect). The method consists of measuring under reproducible conditions the power of agglutination, expressed as the maximum number of agglutinated red cells. In order to know the concentration in absolute value it is necessary to establish the coefficient φ describing the relation between the agglutination power and the molarity of the agglutinin. This determination must be done for each type of isoagglutinin in a zone of concentrations sufficiently high to permit good micronitrogen determinations. Since the agglutination is an exothermic process there is an advantage in choosing for the relative measure of the isoagglutinin concentration, the maximum number of agglutinated red cells at a temperature as low as possible, but sufficiently distant from the freezing point. As experimental conditions the temperature of 4° C and a mixture of 0.6 mm^3 of serum with 0.3 mm^3 of a suspension of red cells in a buffer were chosen. The shaking of the mixture was continued until the equilibrium was reached (90 minutes at 4 °C). The maximum number of agglutinated red cells N_4 is proportional to the concentration of the isoagglutinin in a good range of concentration. The number of agglutinated red cells is obtained by the difference between the total number N_t of cells present and the number of cells remaining free.

The visual counting of free red cells in an haemocytometer is evidently very tedious but can be now replaced by electronic counting. In the Coulter apparatus the suspension runs through a micro-hole so that the crossing of a particle changes the conductivity and produces an impulsion. The model B Coulter lets us enumerate single cells in a population of agglutinates. This model has indeed two thresholds, the upper of which can be used to eliminate the counts due to aggregates of two or more cells. (See H. S. Godman,[48] M. Gibbs and E. Becker).[49]

Let us consider a mixture at equilibrium containing N_t red cells per mm^3 in the presence of a more or less diluted serum at a given temperature.

The relative agglutinin concentration of the serum was, before the addition of red cells, measured by the number N_4. One centrifuges and then one looks for the maximum number N'_4 of agglutinated red cells in the supernatant. N'_4 is proportional to the molar concentration of free agglutinin. The difference $(N_4 - N'_4)$ or N_f is proportional to the number (GA) of agglutinogen sites G combined with the agglutinin A in one mm³. Moreover, N_t is proportional to the total number of sites G, combined or not, present in this same volume.

If then one considers a red cell as a large molecule containing m_G agglutinogen sites, one should be able to apply the mass action law, following the theory of multiple equilibria. For steric reasons, as explained above (IIB4) the agglutinin is to be considered as an unifunctional ligand. Thus the relation (6) can be applied. The number of agglutinogen sites bound per mole of red cells is $r_H = (6 \times 10^{17} N_f / \varphi N_t)$, where φ is the ratio N_4/(molarity of the agglutinin).

The relation (6) may be written:

$$\frac{1}{r_H} = \frac{N_t}{N_f} = \frac{6 \times 10^{17}}{m_G \varphi} + \frac{6 \times 10^{17}}{m_G K N'_4} \qquad (6a)$$

If one mixes different proportions of red cells to serum, one has, once the equilibrium is attained, different values of N_t/N_f and of $1/N'_4$. Accordingly if N_t/N_f is plotted against $1/N'_4$ one ought to obtain a straight line whose slope is $6 \times 10^{17}/m_G K$.

Figure 5 gives an example of straight lines obtained at two temperatures 25 °C and 37 °C, by plotting $1/r_H$ against $1/N'_4$ for normal anti-B human sera of genotype $A_1 O$.

The ratio of the two slopes lets us determine the value of ΔH. But as regards the equilibrium constant it is necessary to know the value of m_G. In the experimental conditions of the work of Wurmser and Filitti–Wurmser [17] the determination of m_G by extrapolation was not possible. The number 5×10^5 has been obtained from experiments of saturation of red cells with antiserum of known agglutinin concentration. This leads us to find the value of the coefficient φ from the nitrogen fixed on stromas of red cells when they are mixed with a specific serum and reduce the relative concentration of this serum from N_4 to N'_4. Finally φ is expressed in mole/litre, taking into account the evaluation of molecular weight, hitherto effected only by ultracentrifugation in partition cell since normal isohaemagglutinin has not yet been isolated in sufficient amount for optical method to be used.

If the fraction of antigen groups combined is small, it is possible to take the total concentration of the groups agglutinogen present as the concentration (G) of the free groups. Hence this is $m_G N_t / 6 \times 10^{17}$. The ratio r_A of the combined isoagglutinin to the total isoagglutinin is N_f/N_4. Thus, since the

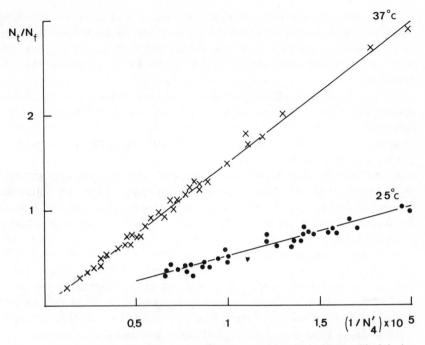

Figure 5. Binding at two temperatures of the anti-B (A₁O) isohaemagglutinin by B red cells (equation (6a)). Abscissa: reciprocal of the free agglutinin concentration (in relative value). Ordinate: reciprocal of the amount of agglutinin fixed (in relative value) per red cell. (From R. Wurmser and S. Filitti-Wurmser[17])

isoagglutinin reacts like it is unifunctional, $m_A = 1$, one has, in the Scatchard representation (equation (7a)):

$$\frac{r_A}{(G)} = \frac{N_f}{N_4} \times \frac{6 \times 10^{17}}{m_G N_t} = K - K\frac{N_f}{N_4} \tag{7a}$$

When the values of $(N_f/N_4)(6 \times 10^{17}/N_t)$ are plotted against the values (N_f/N_4), the results at 37 °C appear as Figure 4.

Ch. Salmon[50] in the applications of the thermodynamic method to the study of genetic machinery employs a variant form of the relation (6a) advantageous for the application of the least squares method and valid when the concentration of free agglutinin is much less than $1/K$:

$$N_4/N_4' = 1 + (1/a).N_t \times 10^{-5}$$

where $a = (1/mK)$. For plots of N_4/N_4' against N_t the slope of the straight line is evidently more steep as the affinity is higher.

In the experiments outlined above all the measurements of concentration of agglutinin were made by determination of the maximum number of agglutinated red cells. Another procedure for determining in relative value the haemagglutinin titre of serum has been described by M. M. Wilkie and E. L. Becker.[51]

The per cent agglutination given at two fold increasing concentrations of anti serum is converted into probits and the probits plotted against the logarithm of the concentration of serum. The 50 per cent effective concentration interpolated from the obtained straight line, is taken as a measure of the strength of the serum.

Some researches have been based on the study of the yield of agglutination as a function of the initial, and not equilibrium, agglutinin concentration. Wilkie and Becker have introduced the use of probits not only as a procedure for estimating the agglutinin concentration but also as an identification method based on the slope of the regression line probit–log concentration. Interesting qualitative conclusions have been obtained. Numbers of these are summarized in review of S. Filitti–Wurmser and R. Wurmser.[52] But a difficulty already emphasized by M. B. Gibbs and J. H. Akeroyd[53] who were the first to apply this method, resides in the interpretation of the slope of the regression line. Indeed the experimental data depend on both the primary and secondary reactions of the agglutination and thus the interpretation in terms of thermodynamical quantities is complicated.

C. Kinetic Measurements

1. Equilibrium Constant

Following the customary treatment of chemical kinetics, the equilibrium constant can be obtained from the rates in the forward and backward directions. For an equilibrium $X + Y \rightleftharpoons XY$ the value of K is equal to the ratio k_a/k_d of the association and dissociation rate constants.

The values of the second order constant k_a can be of the order 10^7 to 10^8 M^{-1} sec^{-1}. Thus they must be measured at very feeble concentrations of the reactants or with the special techniques used for the study of rapid reactions.

S. A. Berson and S. Yalow[19] have shown, as already mentioned (section II.3) by equilibrium studies that unifunctional radioactive insulin reacts with two different orders of antibody sites. These authors have also measured the dissociation rate constants k_d, k_d' of the two sorts of complexes. For this purpose the association was inhibited by adding high concentrations of non radioactive insulin. The concentrations (A_b), (A_b') of the bound antibody sites decreases exponentially:

$$(A_b) = (A_b)_i\, e^{-k_d t}, \qquad (A_b') = (A_b')_i\, e^{-k_d' t}$$

where $(A_b)_i$ and $(A'_b)_i$ are the concentrations of bound antibody sites at the inhibition time. By resolving the two-component curve representing the evolution of b, that is to say $[(A_b) + (A'_b)]$, the values of k_d, k'_d, $(A_b)_i$ and $(A'_b)_i$ are obtained—if the reaction between insulin and antibody is allowed to reach equilibrium prior to inhibition, and when insulin concentration is very low relative to that of antibody:

$$\lim_{b\to 0} \frac{(A_b)_i}{f} = K(A)_0, \qquad \lim_{b\to 0} \frac{(A'_b)_i}{f} = K'(A')_0$$

where f is the concentration of free insulin and $(A)_0$, $(A')_0$ are respectively the total concentration of the sites A and A'.

Let k_a, k'_a respectively be the association rate constants. It is possible to compare $k_d K(A)_0 + k'_d K'(A')_0$ with the sum $k_a(A)_0 + k'_a(A')_0$ as determined from the zero time slopes of the forward reaction. The two sums are found to agree reasonably well.

The difficulty in the determination of the equilibrium constant by kinetic measurements arises principally from the heterogeneity of the reactants. For example, D. W. Talmage[54] studied the kinetics of the reaction between antibody and bovine serum albumin, using the Farr method i.e. using half-saturated ammonium sulphate for separating the free albumin antigen (labelled with ^{131}I) from antigen–antibody complex. The rate constants k_a and k_d were determined from measurements of the precipitate fraction at the beginning of the experiment, at equilibrium and at timed intervals. When the ratio (mk_a/k_d), where m is the number of sites on an antigen molecule, has a value 30×10^8, the association constant obtained from equilibrium experiments is 0.3 to 6×10^8 l mole^{-1}. This observation is explained as follows: 'the equilibrium data reflect the average of all the antibody binding at the concentrations used, whereas the rates of dissociation were obtained from the slowly dissociating component only'.

As regards the special methods for rapid reactions the stopped flow technique with fluorometry recording has been used by L. A. Day, J. M. Sturtevant and S. J. Singer.[55]

Another method permits measurements within a time as short as one microsecond. A. Froese, A. H. Sehon and M. Eigen[56] and then A. Froese and A. H. Sehon[57] have applied the temperature-jump method which is available for the study of reactions involving a change in enthalpy. When the equilibrium system $X + Y \rightleftharpoons XY$ is disturbed by a sudden change of temperature the reequilibration can be characterized by the relaxation time. The relation of this constant to the rate constants can be expressed as:

$$1/\tau = k_d + k_a[(X)_e + (Y)_e]$$

where $(X)_e$ and $(Y)_e$ represent equilibrium concentrations of X and Y.

The disturbance is obtained by a fast input of electric energy (discharge of a high-voltage capacitor). The relaxation effect can be measured by spectrophotometric recording.

The values of k_a and k_d are calculated from the slope and the intercept of the straight line representing the concentration dependence of $(1/\tau)$.

As example for the combination of rabbit anti-p-nitrophenyl antibodies with the dye hapten 4,5-dihydroxy-3-(p-nitrophenylazo)-2,7-naphthalene disulphonic acid, the values of k_a and k_d were found respectively $1\cdot8 \times 10^8$ $M^{-1} \sec^{-1}$ and $760 \sec^{-1}$. Thus, an association constant equal to $2\cdot4 \times 10^5 M^{-1}$ is to compare with the value of $5\cdot8 \times 10^5 M^{-1}$ derived from equilibrium experiments.

2. Activation Energy

There remains to mention another approach which gives information leading to considerations of structure. This concerns the effect of temperature on the reaction rates. One obtains experimentally the quantities A and E of the Arrhenius equations:

$$k = A\,e^{-E/RT} \quad \text{and} \quad \frac{d \ln k}{d\,T} = \frac{E}{RT^2}$$

In principle these quantities A and E can be interpreted according to either the absolute reaction rates theory or the collision theory. The first of these would let us deal with thermodynamics of the transition state. As regards the antigen–antibody reactions the data up to now reported have been discussed only from the point of view of the collision theory.

For more complete information the reader is referred to the works of N. C. Hughes-Jones, B. Gardner and R. Telford,[58,59] J. Economidou, N. C. Hughes-Jones and B. Gardner,[60] concerning the reaction on the one part between erythrocytes and ^{131}I labelled anti-D, and on the other hand between erythrocytes and ^{125}I labelled anti-A and anti-B.

IV. APPLICATION TO MOLECULAR BIOLOGY

In this last section some illustrative examples of studies where the immunochemical thermodynamics is applied to problems of molecular biology will be presented.

Thermodynamic data may be considered in two ways as a source of information for a detailed analysis of the specific structures and as a means of identification in research concerning the biosynthesis of these structures.

A. Problems of Structure

Besides their use for the determination of the number of sites implicated in the antigen–antibody association the binding measurements are used for the comparison of energy data with calculations based on well defined model compounds. Difficulties already emphasized are generally encountered in these studies.

(1) antibody heterogeneity: for many experiments it is necessary to proceed by comparing the association of various ligands with a same sample of antibody;

(2) the overall character of the energy data, already emphasized in the first section.

An example will illustrate the non-univocal nature of the information derived from these data.

The interaction of apolar molecules and substituents with their aqueous environment is considered of major importance (F. Karush)[13] among the bonds which play a role in the association antigen–antibody. The transfer to water of an apolar molecule from an apolar solvent is associated with a decrease of the unitary entropy, and a small change of enthalpy. For example for benzene $\Delta S_u = -14$ e.u. and $\Delta H = 0$. Thus the change of free energy is positive ($\Delta G_u = 4.1$ kcal mole^{-1}) and there is a tendency for apolar molecules to acquire a non-aqueous environment and establish hydrophobic bonds. Hence, an entropy-driving and nearly athermal reaction between antigen and antibody should be interpreted as implying a strong contribution from apolar interactions. A remarkable confirmation of this idea has been presented by Karush[9] in his study of the inhibition of the binding of D-phenyl-[p-(p-dimethylaminobenzene azo) benzoylamino] acetate. The difference between the free energies for the binding of D-phenyl-(benzoylamino) acetate and benzoylaminoacetate corresponds to the contribution of the phenyl group to the antibody–hapten interaction. The value of this difference 4.1 kcal mole^{-1} is equal to the value of ΔG_u for the benzene. Now, from fluorescence quenching experiments on the specific binding of 2,4-dinitro benzene (DNP) by purified anti-DNP, H. N. Eisen and G. W. Siskind[61] obtained for one of their preparations a value of ΔH^0 as high as -19.6 kcal mole^{-1}. ΔS is negative and attains -30.4 e.u. mole^{-1}. The occurrence of a conformational change could explain the inconsistency with the view generally accepted. But the rapidity of the reaction seems to the authors difficult to reconcile with this hypothesis. They suggest that extrusion of interstitial water between ligand and the side chains lining the proteins binding site facilitates the formation of hydrogen bonds. In any case the sole energetic data does not permit us to decide between the two possibilities.

The reader is referred to the reviews of W. C. Boyd,[62] F. Karush[12] and S. I. Singer[63] where thermodynamic values along a scale ranging from 10^5 to 10^9 litre mole^{-1} are reported for various serological reactions. But for the reasons just outlined conclusions can usually be drawn only from thermodynamic quantities ascribable to strictly defined structures. It is a great advantage that artificial antigens can be produced. By suitable choice of haptens introduced into proteins the nature of the forces acting in the combination of antibody has been investigated. The method used by the school of Pauling consists of comparing data obtained with various haptens as nearly identical as possible in regard to their structure except for a determined substituent group.

For example D. Pressman, A. L. Grossberg, L. H. Pence and L. Pauling[31] used an antiserum directed against the *p*-azophenyltrimethylammonium group. They studied the inhibition by two haptenic groups similar in size and shape. One contained the ionic group phenyltrimethyl ammonium which carries a positive charge. The other hapten contained the group tertiary butylbenzene which is uncharged. The difference between the free energies of association, corrected for the unequal binding of the two haptens to the normal serum proteins (D. Pressman and M. Siegel),[64] is a measure of the Coulomb interaction energy between the charged haptenic group and the antibody site. It corresponds to a distance of 2 to 3 Å between the group and the site.

By the same comparative method L. Pauling and D. Pressman[30] have studied quantitatively the role of van der Waals attractions. The theory of London permits a calculation of the electronic dispersion forces between two neutral molecules. The interaction energy between two atoms or groups, A and B, may be evaluated on the basis of molar refractions and ionization potentials. In this way the effect of replacing in a hapten a hydrogen by a chlorine or another substituent can be calculated and the result compared with the observed effect. Some experiments concern the inhibition of precipitation of anti-*p*-azobenzoate by various *p*-substituted benzoates. A table may be found in the review by F. Karush[12] which shows the correlation between the relative free energies of association (taken from Pressman) and the calculated interaction. On the whole the calculations are in reasonable agreement with experimental values. The exceptions may be explained by the occurrence of hydrogen bonds.

The comparative method has furnished not only a quantitative basis for the concept of complementarity but also information on the size of the antibody combining region.

The large size of the haptenic groups used by F. Karush,[8,65] is important for minimizing the possibility that the antibody is directed not only against the hapten but also against some amino-acids of the protein carrying the hapten. This possibility was shown in experiments of S. B. Hooker and W. C. Boyd.[66] The finding that dextran is antigenic has permitted this complication to be disregarded (E. A. Kabat)[32,33] for the determination of the combining area of the antibody produced in this case.

The dextran is composed of molecules of glucose. One needs only to determine how many such units are in contact with the site. The inhibition of the precipitation of antidextran by oligosaccharides of increasing length has been studied and the contribution of each glucose residue to the binding energy has been computed as a $\Delta(\Delta G^0)$ by comparing each of the oligosaccharides at 50 per cent inhibition. It was found that the inhibition increases with the number of units up to a maximum at six, usually a 6 carbon sugar hexose. This represents the limit of the size of the antigenic determinant against which an antibody specific site can be formed.

Other relevant studies with the comparative methodology include comparison of structures where differences have been produced by an ionization or change in ionic strength. Such experiments have been reported by S. G. Singer and D. H. Campbell.[67] From the variation of the equilibrium constant with pH it appears very probable that a single ionized carboxyl group is involved in the bond between the serum albumin and its rabbit antibody.

In another field of works evidence of delocalized structural change in the reaction antigen–antibody is furnished by the comparative study of the human natural isohaemagglutinins (R. Wurmser and S. Filitti–Wurmser).[17] There are characteristic differences between the variations of entropy corresponding to the fixation of various kinds of isoagglutinins on the same B agglutinogen. Owing to the high specificity of the combination these differences must be ascribed not to the sites but to the carriers of these sites. The effect of the inequality of the molecular weights has been calculated for two agglutinins, $\beta(A_1O)$ and $\beta(OO)$ as indicated in section I.3. The result is a difference of free energy of approximately 3·4 kcal, which corresponds to a variation of entropy of 11 units, whereas the observed variation approaches 44 units. Therefore the respective molecular dimensions of these isoagglutinins do not explain entirely the differences of ΔS if one only considers the contribution of the terms of translation and rotation of the molecule as a whole in the expression of affinity: changes of intramolecular structure influencing vibrational degrees of freedom may occur at a distance from the bonding site itself.

B. Problems of Biosynthesis

Immunological reactions are highly specific, their range of specificity being evaluated from the number of structures tested and the proportion that are non-reacting. But among the reactive structures, some of them react more or less strongly. The measurement of the affinity and the two quantities which determine its value, enthalpy and entropy changes, constitute a very important means of identification of antigens and antibodies. There is accumulation of such data concerning the heterogeneity of the antibodies against a single antigen, present in one serum or in the serum of a same individual at various times or arising from different immunization procedures. For example when relatively small amounts of antigen are present during the induction of the immune response H. N. Eisen and G. W. Siskind[61] find higher average affinity for the anti-2,4-dinitrophenyl-L-lysine. Another example is the comparison by J. M. Dubert[39] of the affinity of antibodies before and after a secondary stimulation, especially when this latter is made with an antigen different from the primitively injected one.

A field where the thermodynamical method has led to novel views is the genetic functions concerning the human blood groups.

We have outlined above how the equilibrium between isohaemagglutinins and agglutinogen group of erythrocytes can be studied. The uniqueness of the results is that natural human isohaemagglutinins in contrast to provoked antibodies present for a given class of sera, as characterized by the genotype of the donors, enough homogeneity to permit identification of this class (R. Wurmser and S. Filitti–Wurmser).[17]

As pointed out in section IIIB data on the equilibrium between agglutinin and agglutinogen can be obtained by the simple technique of numeration of cells and the application of the relations (6a) and (7a).

The first experiments concerning natural anti-B isoagglutinin have been made with different sera of A_1O individuals, different sera of A_1A_1 individuals, different sera of A_2 individuals and different sera of group O individuals. According to the relation (6a) all the normal human sera of the same genotype when mixed with B red cells give values of (N_t/N_f) which when plotted against $(1/N_4')$ fall on the same line, as shown Figure 5. The difference in the behaviour of two sorts of β-agglutinins is seen from Figure 4 where the results are represented by the relation (7a).

The isohaemagglutinins anti-B contained in each of these sera are called respectively $\beta(A_1A_1)$, $\beta(O)$... In the case of $\beta(A_1A_1)$, $\beta(O)$, $\beta(A_1O)$ it has been possible to evaluate ΔG^0 and ΔS^0 from these experiments owing to an independent determination of the number of sites per cell m_G. But for many purposes it is sufficient to know the relative values of ΔG measured—if relation (6a) is applied—by the slopes $(1/m_G K)$ and the values of ΔH^0 derived from measurements of these slopes at two temperatures.

It is clear that the value of ΔH^0 which does not depend on the number of sites is specially convenient for the identification of the substances. At present a number of values of ΔH^0 for the binding of agglutinins from various genotypes have been determined. They are summarized in a table (S. Filitti–Wurmser and R. Wurmser)[52] from which the following data are extracted as examples.

	$-\Delta H^0$ (kcal mole^{-1})
$\beta(A_1O)$ + B red cells	16 ± 2
$\beta(A_1A_1)$ + B red cells	$6{\cdot}5 \pm 1$
$\beta(OO)$ + B red cells	$1{\cdot}7 \pm 0{\cdot}4$
$\alpha(B)$ + A_1 red cells	19 ± 3
$\alpha_1(B)$ + A_1 red cells	$31 \pm 0{\cdot}5$

It is of interest to note that the agglutinin produced by A_1O persons differs from those produced by A_1A_1 and OO. The molecule $\beta(A_1O)$ is characteristic of the heterozygote. This finding has led to hypotheses as to the agglutinin synthesis process. In an individual A_1O, there is a cooperation of the two genes in the formation of the anti-B, or as suggested by J. B. S. Haldane[68] the synthesis is determined by the nature of the antigen present. In fact, there is some evidence that this takes place. The most striking argument is founded on experiments by Ch. Salmon, D. Salmon and B. Saint-Paul[69] of which only the design can be given here. Red cells A_1O and O are alternatively transfused to an erythroblastopenic child of group A. The plots corresponding to the anti-B agglutinin reveal affinity alternatively normal and abnormal.

To sum up, important information has been brought by the thermodynamical study of the isohaemagglutination. As it is pointed out by R. R. Race and R. Sanger:[70] 'This suggests that the finished antibody is not under complete and direct genetic control but rather that the antibody making apparatus is reflecting the state, which may be transient, of the antigen with which it is faced.' The question arises, at what step and how does the antigen act in order to switch the synthesis towards one protein or another of different affinity. Presently we know only that the two species possess respectively high and low molecular weights. Some light probably would be thrown by an immunological study of the chains constitutive of the agglutinins.

The thermodynamic method can be used for the study of antigens as well as for the study of the agglutinins. A similar antibody is opposed to various antigens A_1, A_2 (S. Mavrides)[71] (Y. Jacquot-Armand).[72] A considerable work has been accomplished by Salmon in the study of various types of B antigen (Ch. Salmon;[73] A. Bouguerra-Jacquet and coworkers).[74]

Finally in order to illustrate the scope of the method, its application to the detection of changes caused by leucemic diseases in the operations of the A B O locus should be mentioned. (Ch. Salmon).[75]

These examples show the value of thermodynamic study in a wide variety of immunological problems. More generally it can be concluded that methods described in this chapter have contributed notably to the advancement of biology along quantitative lines.

REFERENCES

1. S. Filitti-Wurmser and Y. Jacquot-Armand, *Arch. Sci. Physiol.*, **1**, 151, (1947)
2. S. Mavrides, *J. Chim. Phys.*, **51**, 600 (1954)
3. M. H. Wilkie and E. L. Becker, *J. Immunol.*, **74**, 192 (1955)
4. R. S. Evans, W. K. Mebust and M. T. Hickey, *Blood*, **16**, 1469 (1960)
5. J. Wyman in *Proteins, Aminoacids and Peptides*, ed. Cohn and Edsall, Reinhold Publ. Corp., New York, p. 451 (1943)
6. H. N. Eisen and F. Karush, *J. Am. Chem. Soc.*, **71**, 363 (1949)
7. L. Pauling, D. Pressman and A. L. Grossberg, *J. Am. Chem. Soc.*, **66**, 784 (1944)
8. F. Karush, *J. Am. Chem. Soc.*, **71**, 1369 (1949)
9. F. Karush, *J. Am. Chem. Soc.*, **78**, 5519 (1956)
10. A. Nisonoff and D. Pressman, *J. Immunol.*, **80**, 417 (1958)
11. A. Nisonoff and D. Pressman, *J. Immunol.*, **81**, 126 (1958)
12. F. Karush, *Adv. in Immunol.*, **2**, 1 (1962)
13. E. Haber, *Federation Proceed.*, **29**, 66 (1970)
14. K. Hummel, *Zeitschr. Immunitätforsch. exp. Therapie*, **122**, 180 (1961)
15. S. Filitti-Wurmser, Y. Jacquot-Armand and R. Wurmser, *J. Chim. Phys.*, **47**, 419 (1950)
16. S. Filitti-Wurmser and Y. Jacquot-Armand, *Rev. Hémat.*, **15**, 25 (1960)
17. R. Wurmser and S. Filitti-Wurmser, in *Progress in Biophysics*, ed. Butler and Noble, Vol. 7, Pergamon, Oxford, 1957, p. 90
18. M. B. Gibbs and H. A. Toro, *J. of Immunol.*, **98**, 461 (1967)
19. S. A. Berson and R. S. Yalow, *J. Clin. Inv.*, **38**, 1996 (1959)
20. R. J. Goldberg, *J. Am. Chem. Soc.*, **74**, 5715 (1952)
21. R. J. Goldberg, *J. Am. Chem. Soc.*, **75**, 3127 (1953)
22. G. S. Watson, *J. Immunol.*, **80**, 182 (1958)
23. D. W. Talmage and J. R. Caim, *The Chemistry of Immunity in Health and Disease*, ed. Charles C. Thomas, Springfield, 1961
24. S. G. Singer and D. H. Campbell, *J. Am. Chem. Soc.*, **75**, 5577 (1953)
25. S. G. Singer and D. H. Campbell, *J. Am. Chem. Soc.*, **77**, 3499 (1955)
26. S. G. Singer and D. H. Campbell, *J. Am. Chem. Soc.*, **77**, 4851 (1955)
27. S. I. Epstein, P. Doty and W. C. Boyd, *J. Am. Chem. Soc.*, **78**, 3306 (1956)
28. F. Aladjem and M. T. Palmiter, *J. Theoret. Biol.*, **8**, 8 (1965)
29. L. Pauling, D. Pressman, D. H. Campbell and C. Ideka, *J. Am. Chem. Soc.*, **64**, 3003 (1942)
30. L. Pauling and D. Pressman, *J. Am. Chem. Soc.*, **67** 1003 (1945)
31. D. Pressman, A. L. Grossberg, L. H. Pence and L. Pauling, *J. Am. Chem. Soc.*, **68**, 250 (1946)
32. E. A. Kabat, *J. Am. Chem. Soc.*, **76**, 3709 (1954)
33. E. A. Kabat and M. M. Mayer in *Experimental Immunochemistry*, ed. Charles C. Thomas, Springfield (1961)
34. W. C. Boyd, J. B. Conn, D. C. Gregg, G. B. Kistiakowsky and R. M. Roberts, *J. Biol. Chem.*, **139**, 787 (1941)
35. E. Calvet and H. Prat, *Microcalorimétrie*, ed. Masson et Cie, Paris (1956)
36. C. Kitzinger and T. Benzinger, *Z. Naturforsch.*, **10b**, 365 (1955)
37. R. F. Steiner and C. Kitzinger, *J. Biol. Chem.*, **222**, 271 (1956)
38. J. R. Marrack and F. C. Smith, *Brit. J. Exper. Path.*, **13**, 394 (1932)
39. J. M. Dubert, *Thèse Fac. Sc. Paris*, (1959)
40. F. Perrin, *Ann. Phys. Paris*, **12**, 169 (1929)
41. F. Perrin, *J. Phys. Radium Ser.*, **7**, 7 (1936)

42. G. Weber, *Biochem. J.*, **51**, 145 (1952)
43. W. B. Dandliker, H. C. Schapiro, J. W. Meduski, R. Alonso, G. A. Feigen and J. R. Hamrick, *Immunochemistry*, **1**, 165 (1964)
44. J. D. Marshall, W. C. Eveland and C. W. Smith, *Proc. Soc. Exp. Biol.*, **98**, 898 (1958)
45. A. A. Amkraut, *Immunochemistry*, **1**, 231 (1964)
46. S. F. Velick, C. W. Parker and H. N. Eisen, *Proc. Natl. Acad. Sci. US.*, **46**, 1470 (1960)
47. P. Tengerdy, *Immunochemistry*, **3**, 462 (1966)
48. H. S. Godman, *Nature*, **193**, 385 (1962)
49. M. Gibbs and E. Becker, *Nature*, **198**, 190 (1963)
50. Ch. Salmon and M. Hautenauve, *Nouv. Rev. fr. Hemat*, **1**, 847 (1960)
51. M. N. Wilkie and E. L. Becker, *J. Immunol.*, **74**, 199 (1955)
52. S. Filitti-Wurmser and R. Wurmser, *Acta Haemat.*, **36**, 239 (1966)
53. M. B. Gibbs and J. H. Akeroyd, *J. Immunol.*, **82**, 577 (1959)
54. D. W. Talmage, *J. Infect. Dis.*, **107**, 115 (1960)
55. L. A. Day, J. M. Sturtevant, S. J. Singer, *Ann. N.Y. Acad. Sci.*, **103**, 611 (1963)
56. A. Froese, A. H. Sehon and M. Eigen, *Canad. J. Chem.*, **40**, 1786 (1962)
57. A. Froese and A. H. Sehon, *Immunochemistry*, **2**, 135 (1965)
58. N. C. Hughes-Jones, B. Gardner and R. Telford, *Biochem. J.*, **88**, 435 (1960)
59. N. C. Hughes-Jones, B. Gardner and R. Telford, *Immunology G.B.*, **7**, 72 (1964)
60. J. Economidou, N. C. Hughes-Jones and B. Gardner, *Immunology*, **13**, 227 (1967)
61. H. N. Eisen and G. W. Siskind, *Biochemistry*, **3**, 996 (1964)
62. W. C. Boyd, *Introduction to Immunochemical Specificity*, Interscience, New York, 1962
63. S. J. Singer in H. Neurath (ed.), *The Proteins*, vol. 3, Academic Press, New York (1965)
64. D. Pressman and M. Siegel, *J. Am. Chem. Soc.*, **75**, 686 (1953)
65. F. Karush, *J. Am. Chem. Soc.*, **79**, 3380 (1957)
66. S. B. Hooker and W. C. Boyd, *J. Immunol.*, **25**, 61 (1933)
67. S. G. Singer and D. H. Campbell, *J. Am. Chem. Soc.*, **77**, 3504 (1956)
68. J. B. S. Haldane, *The Biochemistry of Genetics*, Allen, London (1952)
69. Ch. Salmon, D. Salmon and B. Saint-Paul, *Nouv. Rev. Franc. Hémat.*, **5**, 95 (1965)
70. R. R. Race and R. Sanger, *Blood Groups in Man*, Blackwell Sc. Publ., Oxford and Edinburgh (1968)
71. S. Mavrides, *J. Chim. Phys.*, **52**, 1 (1955)
72. Y. Jacquot-Armand, *Rev. Hémat*, **13**, 305 (1958)
73. Ch. Salmon, *Nouv. Rev. Franc. Hémat.*, **5**, 191 (1965)
74. A. Bouguerra-Jacquet, M. Hrubisko, D. Salmon and Ch. Salmon, *Nouv. Rev. Franc. Hémat.*, **10**, 173 (1970)
75. Ch. Salmon and D. Salmon, *Nouv. Rev. Franc. Hémat.*, **3**, 653 (1963)

Author Index

Subject Index

673